F. Hasan

Practical Dispersion

Practical Dispersion
A Guide to Understanding and Formulating Slurries

Robert F. Conley

VCH

Robert F. Conley
Mineral and Resource Technology, Inc.
3131 Adams NE, Bldg. F-48
Albuquerque, NM 87110

This book is printed on acid-free paper.

Library of Congress Cataloging-in-Publication Data
Conley, Robert F., 1936-
Practical Dispersion : a guide to understanding and formulating slurries / Robert F. Conley.
p. cm.
Includes bibliographical references and index.
ISBN 1-56081-931-6 (alk. paper)
1. Emulsions. 2. Suspensions (Chemistry) I. Title.
TP156.E6C66 1996
660′.2945—dc20 95-38037
CIP

Printed in the United States of America

ISBN 1-56081-931-6 VCH Publishers, Inc.

Printing History:
10 9 8 7 6 5 4 3 2 1

Published jointly by

VCH Publishers, Inc.
220 East 23rd Street
New York, New York 10010

VCH Verlagsgesellschaft mbH
P.O. Box 10 11 61
69451 Weinheim, Germany

VCH Publishers (UK) Ltd.
8 Wellington Court
Cambridge CB1 1HZ
United Kingdom

Preface

Over the past three decades a number of texts have been written about surface active agents, dispersion dynamics, and general dispersed-state effects. The greater portion of these are academic in origin, or are the results of technical conferences on the topics. Most of these are excellent efforts at elucidating research into complex surface chemistry and phenomena and their associated mathematical descriptions. While these equations are very much operative in the real world of dispersed systems, they represent a degree of sophistication well beyond applications of interest to industrial technicians—those individuals whose day-to-day activities lie in manufacturing broad categories of dispersed products: paints, pigment premixes, treated metallic, inorganic, and organic powders, food products, cosmetics, pharmaceuticals, dyes, inks, and the like.

These technical personnel have a greater need for a handbook or working text that will familiarize them with the manner in which dispersants function and enable them to modify their products as commercial, customer, and field needs frequently dictate. It will also allow them to continue to produce effective and stable materials within economic constraints and governmental regulations.

This does not suggest that such a handbook is too simplistic for academic use. The author has for over 16 years given a series of guest lectures on dispersion, pigment preparation, surface properties and other aspects of dispersion at Kent State University. In addition, since 1976 he has held annual seminars throughout the United States, Canada, and Europe on various aspects of dispersion, especially those pertaining to high-technology processes, hazardous materials properties, and their effects. These are part of a material handling and processing series conducted by The Center for Professional Advancement (East

Brunswick, NJ) and Mineral & Resource Technology (Albuquerque, NM). It is because of the response to those lectures and the recommendations from other lecturers and the publishers that this formalization of lecture material and field investigations was commenced. Significant portions of the materials cited appear in this text in graphs, tables, and descriptive materials.

Much of the technical data and information that field processing personnel require are available, but lie buried in obscure data compilations, sophisticated journals, and manufacturers' brochures, that is, in broad scattering of technical publications covering a period of over 50 years. Even with today's computer information bank access services, these are not often, unfortunately, adequately utilized by industry.

The university, with its phenomenal data bank access and scholarly research on surfaces and surfactants, has much of this information available to its faculty and students. Several of these institutions offer excellent academic programs, periodic seminars, and short courses dealing with new developments and basic concepts. However, much of the information on practical dispersion of systems does not appear to disseminate adequately into the industrial workplace.

The purpose in writing this text, or guidebook, is to provide hands-on knowledge of sufficient technical depth to allow those personnel in a variety of occupations where actual dispersion practices are ongoing to feel more proficient in making system modifications and to enable them to meet specific mechanical, chemical, and other requirements of their customers. To this end a broad description of dispersants, their function, and field applications have been cited to make these selections less guesswork and more scientific for the plant chemist.

Dispersants herein are defined in a very broad categorization—those agents that increase the wetting energy of specific liquids on specific solid surfaces. Some readers may contend that this breadth encompasses surface modification, which indeed it does. The author does not discriminate dispersants on the basis of the energy with which they adsorb upon solids. This enables a broader and deeper understanding of the principles involved in producing slurries of all varieties of solids in all varieties of liquids.

Contained throughout this book are structural representations of representative dispersants and an appendix section that lists manufacturers of dispersants and dispersing equipment, as well as an example from McCutcheon's Guide. The combination of these should enable formulators to design stable, functional, dispersed systems more intelligently and more confidently than by time-consuming trial-and-error techniques.

Not every known dispersant is listed in this book, nor even all applications of common ones, for a variety of reasons, not the least of which is book size and manufacturers' proclivity to change trade names periodically. Also, the excellent trade volume of McCutcheon, as well as the U.S.F.D.A. registries and the Cosmetics, Toiletries and Fragrance Association Dictionary, are dedicated to keeping up with tradenames and structures of such materials. Even as this book is being prepared, new materials and variations on old ones are forthcoming. However, these products fall into certain functional classes, which are described in detail throughout this book. An understanding of functional classes and functional surfaces holds the key to successful utilization of dispersants, which will enable the formulator, the materials manufacturer, and the production manager to develop and to modify products more efficently and economically than through massive laboratory testing procedures.

The author must be pardoned for using what the reader may regard as a disproportionate reference to clays throughout this book. This derives from the wealth of information provided him by workers in that industry, as well as 17 years of his own industrial employment in that field. Clays possess many unique properties as well as serving as excellent general models for surface activity, both of which have been employed herein to illustrate dispersion principles.

Mathematical equations have been kept to a minimum and used only to illustrate dispersant phenomena. For those sufficiently interested in equation derivation, adequate references are provided to the original literature source. The chemistry described is kept at as low an academic level as possible within the ability to explain surface and dispersant interactions. Explanations, definitions, examples, and illustrations are given as much as practical to ensure the reader has an adequate concept of the technology being described.

The units recommended by the International Union of Pure and Applied Chemistry have been employed throughout the text, such as micrometers and nanometers (and occasionally inches!), rather than the more vernacular microns and Angstroms, to keep the material current with technology and in step with other sciences.

Because a broad variety of technical terms are employed pertaining to rheology, surface chemistry and dispersion, with which some readers may be unfamiliar, definitions are included where these first appear. However, to assist the reader and minimize the need to search through the text for specific definitions when such terms repeat, an extensive glossary is provided in the Appendix section of the book.

The description and explanation of emulsions, those dispersions of liquids within liquids, have purposely been omitted in this work because those characteristics whereby wetting solids can be measured and predicted are of a greatly different nature than liquids, both chemically and mechanically. Also, a number of superb texts already have been published on the practicalities of emulsion formation and stabilization.

Dedication and Acknowledgments

Dispersion is a complex topic, commonly a trial-and-error technique where errors significantly dominate the trials. Its practitioners are akin to Shakespeare's "greats," paraphrased by, "Some are born to dispersion, some achieve it through hard work, and some have dispersion thrust upon them!" For most it is a field of endeavor for which they have been but marginally trained, in college or in industry.

The author's first memorable exposure to dispersion occurred a year or so following graduate school while he was employed as a process chemist for a pigments company. Needing assistance at one particular juncture in a project, he was sent to the laboratory of a consulting chemist, a gentleman who will be called, simply, Harold.

Harold's facility was situated in perhaps the grimiest, most chemically contaminated spot on the face of the Western world—along the Arthur Kill riverfront in pre-EPA northern New Jersey. The ramshackle building, which served as both laboratory and office, was surrounded by old tires, parts of hopelessly dismembered industrial reactors and encrusted plumbing, scores of rusting, dented, unlabeled barrels, and a ground stench indescribable with the English language arising from carelessly disposed chemicals whose composition encompassed perhaps 90% of the periodic table.

Inside the small, corrugated iron building lay a large darkened chamber serving as a laboratory from which a small office had been partitioned with plywood.

This room had but a single window through which rays of morning sunshine struggled vainly to penetrate decades of condensate. With a bit of cleaning, the laboratory could have doubled as a Medieval painting of the alchemist's lair. One threaded one's way past old leaking drums, benches with uncountable unwashed beakers, and mixers corroded beyond functioning to the office. There, awaiting guests, sat two rusting, plastic upholstered chairs whose orange hue graded from center to edge, courtesy of exposure to the general atmosphere, both interior and exterior, of a myriad of corrosives and decades of neglect.

Along one laboratory wall ran two large sections of seriously rusting metal shelving, eight tiers high, on which were piled hundreds of dust-laden bottles bearing such esoteric names as Minisperse J, Alcophobe 10T, Merlene B, truly foreign words, and an occasionally pencilled, nonscientific description. The liquids ranged from the viscous to the watery, from dark brown to eerie yellow, and each bore a discolored label disintegrating slowly from laboratory fumes.

Upon being shown the material of interest and hearing a general description of what was needed, Harold gazed at the powder and pensively caressed his jaw and cheek with his left hand, both covered with a two-day grizzled stubble. After a few moments he reached to a shelf and pulled down a bottle. Rinsing out what appeared to be an old mayonnaise jar, he added perhaps an inch of the amberish elixir, filled it with tap water, stirred it to a cloudy consistency and poured the contents into a clean, used medicine bottle.

After capping the bottle and wiping away the greater portion of the dust, he handed me the container and said with some authority, "Use about a half-percent and the slurry should be stable for a good five days." Upon being asked how he knew what would work, which might add to a recent college graduate's store of textbook knowledge, Harold replied simply, "Dispersion is an art, not a science. It's not a black art, mind you, but it's awfully gray!"

Harold had been born at the turn of the century and had worked, either as employee or consultant, for almost 100 companies in his nearly 50 technical years. Furthermore, he seemed never to have forgotten a fact he had learned or a material he had acquired. They were there, every last experiment, every trial and every error. And those bottle names, now fading as fast as their contents were congealing, were as scientific to Harold as transuranium elements to an aspiring nuclear physicist.

Years later, at Harold's funeral, the author remarked, "There went the last of the twentieth century's alchemists!" Harold left those of his association with a highly wary sense of dispersion and is as responsible as anyone for the present effort to render scientific sense to a massive collection or trials, errors, and tradenames.

This work is dedicated to the spirit of Harold and the many of his walk who laid the foundation, however poorly mortared, for a better approach to dispersion.

The writing of this text represents the culmination of some three decades of association by the author with diverse, fine particle systems—pigments, metal powders, ceramics, cosmetics, magnetic recording materials, and a host of esoteric products and applications. Such experience and association is never the product of but one research individual. The path has been illuminated by a host of scientists and engineers whose expertise, assistance, encouragement, and specialities have been drawn upon liberally to organize the material set forth in this text into meaningful order.

More than anyone, Dr. Haydn H. Murray, first CEO of the Georgia Kaolin Company, then Chairman of the Geology Department at Indiana University and now retired, has my sincere appreciation and respect for his insight and unstinting and unselfish effort to advance dispersion technology in the clay industry and move it from a "gray art" to a fine science.

I thank Mr. Carl Knauss of Kent State University (Ohio) for his encouragement and suggestions on my lectures given at that institution over 15 years on dispersion practices. Mr. Francis Albus of the Aljet Company and a course director at the Center for Professional Advancement in New Jersey has contributed to the effort by introducing the science of surface chemistry and dispersion to the fields of size reduction and classification and by permitting me to participate in seminars and short courses for nearly 20 years on these topics. Many of the text figures derive from these two lecture sources.

Two scientists at the Georgia Kaolin Company R & D Laboratories who contributed heavily with laboratory data and technical effort on which numerous studies quoted in this book rely are Dr. Mary K. (Lloyd) Gest and Mr. Harry G. Golding, both of whom have gone on to advance their careers in other endeavors.

Many people at many institutions and industries also have contributed their time and effort to thread science through the maze of mills, materials, and manufacturing by supplying dispersants, pigments, test work, and reports of investigations, information from which has been invaluable in the organization of this book. It is not possible to name all of those or even all of the companies throughout the many years of cooperation. However, those who have contributed heavily during that time include Brookfield Corporation (instrumentation), Byk-Chemie USA (organic dispersants), Chicago Boiler Co. (mills), FMC Corporation (polyphosphates), Georgia Kaolin Co. (now Dry Branch Kaolin), IMC Commercial Solvents (alkanolamines), Kerr-McGee Co. (titanium dioxide pigments), Pfizer MPM Division (calcium carbonate surface chemistry), Premier Mill Co. (dispersion equipment), Union Process Co. (dispersion equipment), Utah Clay Technology (silicate pigments), and Witco Corporation (specialty dispersants). They have given of their time and technical assistance to provide information for the many projects that have developed over the years in the broad arena of dispersion application and product development.

Finally, I sincerely appreciate the assistance and forbearance of my wife, Jean, whose journalistic talents in editing the manuscript and suggestions for its literary clarity and technical structure have proved invaluable in this writing.

Robert F. Conley
Albuquerque, NM

Photographic Credits

The author sincerely appreciates the superb talents and efforts of the many individuals who have assisted with the numerous photographs, micrographs, and electron micrographs used throughout this book to illustrate dispersion phenomena.

The illustrations are too extensive to cite individually, and more than one individual has contributed to many of these. However, especially recognized are the efforts of Mr. John Brown and the electron micrography staff at Georgia Tech University (Atlanta, GA) in providing superb transmission, thin section, and replication micrographs. Similarly, the staff at the University of Utah (Salt Lake City, UT) are commended for their scanning electron micrography. Special thanks are extended to Mr. Brian L. Conley (Scotch Plains, NJ) and Mrs. Barbara Hill of Hill Associates (Bloomington, IN) for providing excellent photomicrography, photographic prints, and special electronic imaging techniques to make the illustrations clear and definitive.

Where specific equipment or products have been illustrated, the manufacturing organizations have been credited in the individual figures and chapter references. Because of the wealth of equipment available for performing dispersant, surface structure, mechanical, and rheological measurements, unfortunately only a few representative examples could be incorporated into this book. Thus the illustration of individual pieces of equipment is not to be construed as a specific endorsement of that equipment, nor is the omission of any equipment to be interpreted as a criticism thereof.

Contents

CHAPTER

1

The Dispersed State, Its Form and Formation

Like most other chemical fields, dispersion technology has expanded dramatically during the past quarter-century. In the years around 1970, fewer than one patent annually on average was granted by the U.S. Patent Office and Trademark Office on dispersants. By 1990 this number had risen to over 25 annually. In 1970 most articles and patents were written by American technical staff from U.S. companies. By 1990 over 95% were written by Japanese and European personnel from companies in those countries.

This should not suggest the United States has fallen woefully behind in this important technological field. American industry has led the way for 50 years in the area of dispersancy and has arrived at a position where it can rely on products and materials which have stood the test of time and production economies. Laurels aside, however, there is always room for improvement!

A short review of patents, especially foreign patents, shows most current dispersant innovations involve complex chemistry to build complex molecules. Many of these substances will never be commercialized. First, heavy complex organic molecules require greater weight percentages in many instances to perform a given dispersancy task than common materials in the marketplace. Second, as methods of formation become more complex, production costs increase. Any new dispersant entering the industrial arena must compete with well-established products whose engineering and raw materials have been thoroughly scrutinized for commercial practicality. Major changes in the United States involving dispersant development have been wrought, not through improved efficiency or reduced production costs, but because of secondary aspects, such as environmental concerns and handling toxicity.

1.1 Coming to Terms with Dispersion

The term *dispersion* is employed most often in a relative or semiquantitative sense to describe the state of a two-phase system. The verb *to disperse* derives from the Latin, meaning to remove and redistribute and relates to a close collection of entities moved to a more remote and separated position. It was often employed by the Roman army to define troop encounters and is still used by law enforcement agencies to describe crowd control. Clouds, or haze, are said to disperse when clarity resumes, although, technically, this usage of the verb is incorrect.

Several words tend to be employed interchangeably in industry, both U.S. and foreign, with regard to slurry stabilization practices. These include surfactants, dispersants, wetting agents, grinding aids, and pigmentation aids. Subtle technical differences exist between each of these expressions.

Surfactants (corrupted from "surface active agents") are agents that assist the distribution of one phase in another. They are not limited to solid–liquid slurries but may include liquid–liquid emulsions, for example.

Dispersants are agents employed solely to stabilize particulate distributions in liquid systems (a subclass of surfactants).

Wetting agents are substances employed to reduce the surface tension of a solid for the liquid into which it is suspended. They may or may not assist in dispersion of that solid.

Grinding aids are chemicals that reduce the energy required to achieve size reduction (comminution) by mechanical means. Such aids may function as dispersants, as lubricants, or as preventatives for particle fusion and adhesion within the mill, especially ball mills.

Pigmentation aids are agents that maintain pigment particles sufficiently separated in a liquid or polymer phase that individual photons of light can be absorbed or diffracted efficiently, thereby increasing color intensity or opacity of the particle–liquid film.

Other terms associated with dispersion are defined in the glossary section of this book.

The most thorough definition of dispersion is:

> "The distribution and concentration of particulates suspended in a liquid in which these remain homogeneous with progressively decreasing volumes insofar as the number of solid particles per unit liquid volume is statistically representative."

It has been defined somewhat more succinctly[1] as:

> "The ultimate state of homogeneous suspension of one phase within another."

Dispersion, even though frequently a qualitative term, is capable of being defined rigorously, that is, mathematically, by a number. The dispersed state can be quantified thermodynamically by the Gibbs–Helmholtz equation[2] (derived from the Second Law of Thermodynamics). While this equation is commonly associated with gas activities, it applies to a broad variety[3,4] of energy interactions with matter. The general form of the Gibbs–Helmholtz equation is given by:

$$\Delta G = \Delta H - T\Delta S \tag{1.1}$$

where ΔG is the system free energy, ΔH the enthalpy, T the temperature, and ΔS the system entropy or disorder. Rearranging, this yields:

$$\Delta S = \frac{\Delta H - \Delta G}{T} \tag{1.2}$$

Applied to dispersion, ΔG becomes the system energy provided by a mechanical agitator, ΔH represents the binding energy of the aggregates in suspension, and ΔS is the entropy of the particle–liquid system (the state of particle disorder). When the mechanical free energy introduced into the slurry exceeds the binding energy of the floccules, individual particles are removed and redistributed. This increases disorder (positive ΔS). However, part of this energy introduction is wasted, not going to disaggregation but appearing as heat, that is, an increase in T. Thus a change in entropy ΔS, quantitatively defines dispersion.

The Gibbs equation has been employed to calculate directly the surface tension of ionic species in solution,[5] although this must be performed for dilute conditions. The data are verified by experimental measurements. The method represents, however, an excellent statistical-mechanical verification of the application of the laws of thermodynamics to surface chemical phenomena.

The situation is represented in the energy diagram of Figure 1.1. Following mechanical separation beyond the distance where van der Waals forces (induced dipolar fields and interparticle gravitation) are operative, a value approximately 4 times the kinetic energy of the liquid, the system remains in a dispersed state until particles return close enough that the kinetic energy of the liquid phase (kT) causes the particles to recoalesce with time and return to the trough in the energy diagram.

The total energy at any point on the curve, E_Σ, is simply:

$$E_\Sigma = E(\text{attraction}) + E(\text{repulsion})$$

If the maximum in the E_Σ curve is $15kT$ or larger, dispersion of the system is stable.[6] Location of the maximum along the distance axis represents the absolute closest approach of particles, including their entourage of adsorbed liquid-phase molecules.

However, by introducing a barrier that prevents a return to the aggregated state (e.g., electrostatic repulsion), scattered particles will then remain in random suspension throughout the liquid system. The mechanism is shown schematically in Figure 1.2.

Several difficulties arise in these or any other dispersion quantifications and descriptions because of the nature and definition of what represents the particulate in suspension. Aggregates, for example, occur with varying degrees of interparticle bonding. As a consequence, dispersions obtained through differing

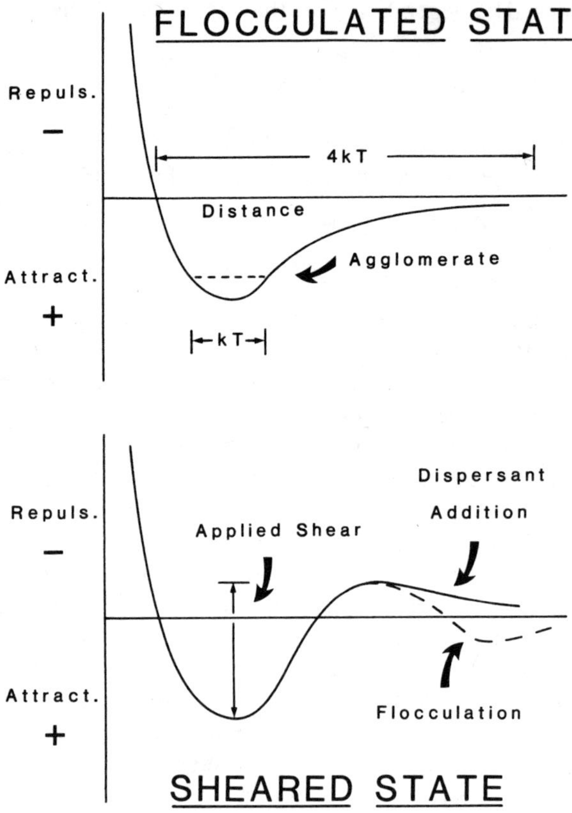

Figure 1.1. Attraction–repulsion dispersion energy diagram.

intensities of shearing, or input energies, may produce different distributions (ΔS values), not necessarily discrete particles, especially where the slurry is diluted to low solids levels.

While dispersion may also apply to liquids in liquids, gases in liquids, and solids in solids, in this book the term will be restricted to solids in liquids, that is, to slurry formation.

1.2 Characteristics of Dispersed States

Any definition of dispersion carries with it an assumed structure, or particle assemblage stability, of the solid phase being distributed within the liquid phase.

Ludicrous situations obviously develop at either end of particle size distributions. A simple dispersed state, for example, is represented by a single golf ball in a mug of beer. The ball settles immediately to the bottom, yet is fully

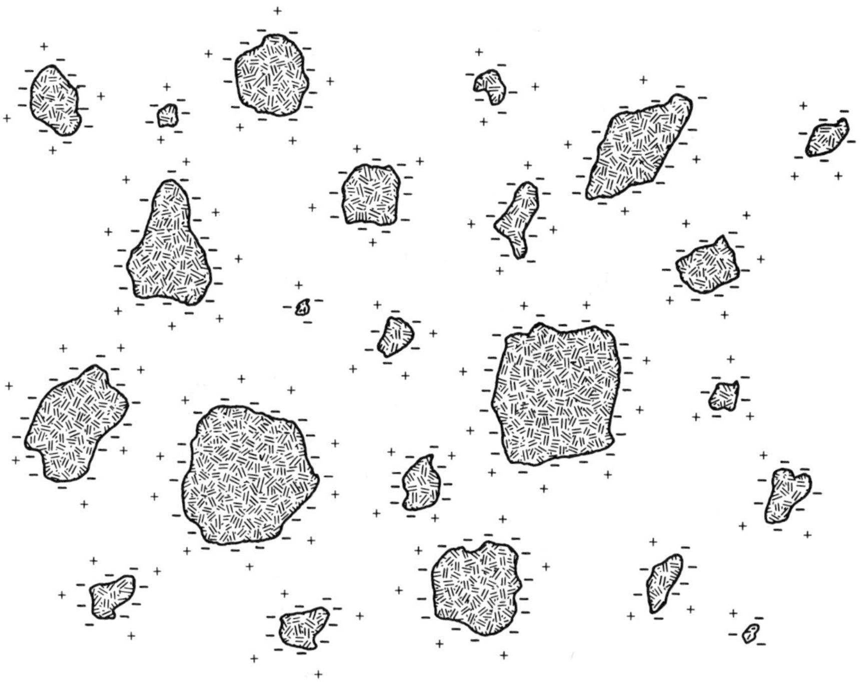

Figure 1.2. Electrostatic repulsion from charged dispersants.

dispersed. By contrast, on the extremely fine end of the particle size scale, at sufficient solids level, the slurry system becomes immobilized through surface solvation. This, too, is a dispersed state—it goes by the more common name of "mud." Thus, implied in all such working definitions of dispersion is the condition that particle count and liquidity are both of sufficient magnitude to enable the system to be recognized as a fluid, that is, a slurry.

All dispersed slurry states have definable flow properties, a phenomenon termed *viscosity*. However, liquid–solid interactions generating viscosity are more complex than simple arithmetic representations of resistance to flow and may change when external parameters applied to the system are altered.

Shear is the term applied to differential velocity, v, between two mechanical surfaces separated by a distance, s, in a liquid (or slurry) system, as illustrated by Figure 1.3.

Where the induced stress on the particle assemblage increases linearly with applied shear (viscosity is constant with shear), the system is said to be *Newtonian*. Water and most organic liquids (kerosene, acetone, toluene, etc.) exhibit Newtonian behavior. If the resistance to shear increases faster than the applied shear (system thickens with applied shear), the system is termed *dilatant*. Dilatancy is often associated with narrow particle size distributions. Finally, where the resistance to flow actually decreases with shear, a situation termed *pseudo-*

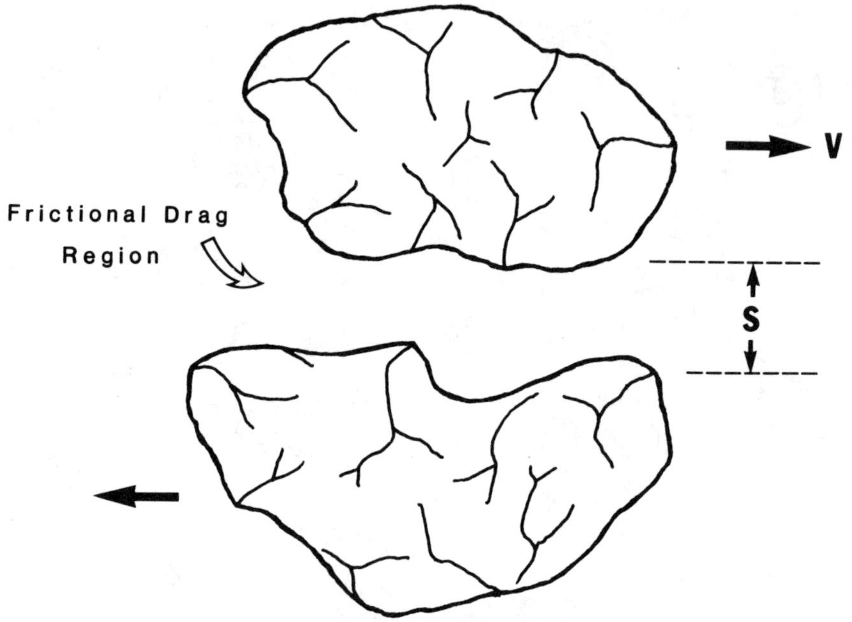

Figure 1.3. Shear schematic and viscosity relationship.

plasticity (system thins with shear) develops. In this state the steepness of the shear thinning curve is great, and the state may be partially retained during increasing–decreasing shear cycles; i.e., hysteresis may occur.

If viscosity change upon removal of shear (shear thinning) is time dependent, the system is said to be *thixotropic*. Upon reducing applied shear, thixotropic systems retain a "memory" of their previous state and display a higher viscosity at the same value of declining shear than was displayed as during shear increase. Such systems may actually "set," resulting in a quasisolid termed a *gel*. Pseu-

Table 1.1 Aqueous Sedimentation Velocity for Highly Dispersed Spheres

	Spherical Particle Diameter						
Sp. Gr.	100 μm	50 μm	20 μm	10 μm	5 μm	2 μm	1 μm
6.0	160.1	40.0	6.40	1.60	0.40	0.064	0.016
5.0	128.1	32.0	5.12	1.28	0.32	0.051	0.013
4.0	96.1	24.0	3.84	0.96	0.24	0.038	0.010
3.0	64.0	16.0	2.56	0.64	0.16	0.026	0.006
2.5	48.0	12.0	1.92	0.48	0.12	0.019	0.005
2.0	32.0	8.0	1.28	0.32	0.08	0.013	0.003
1.5	16.0	4.0	0.64	0.16	0.04	0.006	0.002

1. Velocities (in cm/min) are calculated from the Stokes Law for water at 25°C and at solids $<0.5\%$.
2. Table velocities above about 10 cm/min will in reality be larger because nonlaminar flow (eddy currents) occurs.
3. Table velocities below about 0.01 cm/min will be retarded by Brownian motion.

doplasticity is commonly associated with particle systems having large anisometries (greatly disparate dimensions). Thixotropy is more often associated with gels, gelatinous hydroxides and high cation exchange capacity ultrafine particles (bentonites, for example). True gels, unlike thixotropic systems, have a yield point, that is, a ΔH of organized structure (somewhat analogous to the latent heat of melting of ice at 0°C), beyond which shear reduces them to a random system of thixotropic slurry. These systems are often classified as "Bingham plastic," and the transition point between set structure and fluid is termed the Bingham yield point.* Yield point is a direct measure of gel strength.

Shear thickening occurs in situations where shear forces override the interparticle forces. Under these conditions critical shear rate is proportional to interparticle distance and inversely related to the dispersion medium viscosity. Also, it is inversely related to the mean particle radius.[7] Dispersions of polydisperse particles (broad size range) will exhibit a pronounced shear thickening under such applied stress. However, the increase in viscosity with shear will be far less dramatic than with monodisperse particles (single sized). Shear thickening, in spite of its physical complexity, has been modeled by sophisticated mathematics and verified by work with collections of spheres.

Other viscodynamic states include *rheopectic*, which is the time-independent form of dilatancy.

Numerous factors influence which of these viscodynamic systems will develop from fully dispersed states. These include particle size distribution (sigma or breadth), surface area (particle size fineness), particle morphology (shape), and the magnitude of the induced field created by the dispersant on the particle surface.

The dispersed state may be fully stabilized, metastable, or even unstable, depending upon interactions of surface forces, molecular movements, and gravity. For example, clay particles below about 0.2 micrometers in diameter may remain in aqueous suspension for days or even weeks. This condition is a consequence of Brownian motion, where water molecule kinetic energy acts nonuniformly upon extremely small particles to override settling forces resulting from gravity. This situation is often referred to as *Brownian* or *colloidal stabilization* of ultrafine, low-solids slurries.

As particles increase in size, some sedimentation occurs upon quiescent standing. Not only does sedimentation diminish the quantity of solids in suspension, it changes the particle size distribution throughout the system because coarser particles gravitate out of suspension more rapidly than finer ones. Settling velocities for particles of different size and density are shown in Table 1.1.

---

* The Bingham yield point was named in honor of E. C. Bingham, an English chemical engineer at Cambridge University (UK), who first proposed this concept of particle–liquid structure in the early 1900s to explain gel activity. Bingham later joined Lafayette University (PA), where his prolific engineering studies and publications on plasticity, elasticity, fluidity, and viscosity earned him worldwide recognition. His rheological efforts and publications spanned nearly a half-century (1900–1946). He is said to have developed his early interest in these phenomena, especially those associated with natural products, from observing (and consuming) his mother's home-cooked jams and jellies.

In situ microscope studies[8] have shown that large and small particles do not mix in the same manner with laminar flow mixers, especially at different heights within the mixing vessel, independent of the sedimentation effect. These are momentum effects. They allow larger, more dense particles to increase in concentration toward the vessel's walls while finer particles concentrate in the annulus near the mixer shaft. Cylindrical tanks experience, as a consequence, a horizontal stratification of sizes in addition to the more common vertical one. This effect becomes of special importance where dispersions are drained from a bottom wall port.

In turbulent mixing sedimenting particles (free fall) will exhibit a component of lateral motion.[9] The streamline velocity fluctuation, however, is much greater than the cross-stream fluctuation, by a factor of approximately 5.

For these reasons, particle size analysis, with the sample removed intact from a homogeneous suspension and not further treated or sheared, is an excellent method of assessing state of dispersion. This technique will be described in more detail in Chapter 13.

Particles possessing anisometry settle slower than an "equivalent" spherical diameter would suggest. Also, at higher solids (i.e., above 20%), where the particulate component within the system exerts rheological influence, fluidity of isometric particles improves with dispersion simply because individual small particles move more readily in an assemblage than as a larger agglomerate. As a consequence, viscosity is often employed as a gauge for assessing degree of dispersion. While this is a reasonable assumption at moderate solids, it is less accurate at high solids (i.e. above 60%). The reason behind this is complex and will be dealt with here only in generalities.

Particulate mobility is a function of system free space. Mathematically, fluid volume around particles maximizes when the size distribution is such that ratios of equal amounts are equally likely. If size is plotted on a logarithmic scale and weight percent on a probability, or gaussian, scale (log–probability graph), this mobility function will generate a nearly straight line with a 45 degree slope, depending on the scale expansion. This relationship is termed a log-normal distribution[10]. Diameter ratios can be shown to be exponentially related to their cube (volume function) by the relationship:

$$D_{50}/D_{25} = D_{75}/D_{50} = 1.77 \tag{1.3}$$

where 25, 50, and 75 are the weight finer percentages at those specific diameters. The value, D_{50}, is commonly defined as the "median" size. These relationships are shown graphically in Figure 1.4.

A corollary follows from this, that the packing density of dry powders also maximizes with log-normal distributions, that is, those following a 45° slope on a log particle size, probability function weight percent graph. True 45° slopes are not realized on all graph papers because of line and size of construction, however.

The consequence of a size distribution effect upon viscosity (the ability to

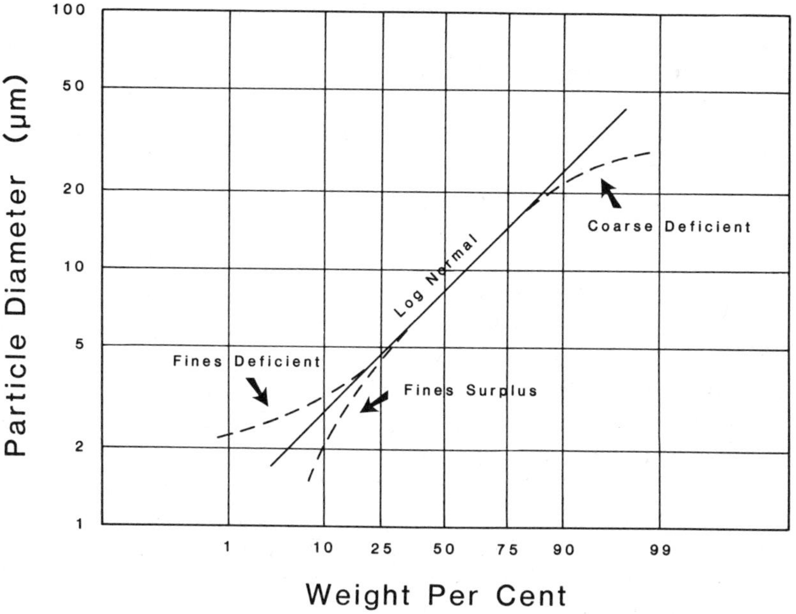

Figure 1.4. Plot of logarithmic size versus exponential weight.

flow) is that distributions in slightly flocculated states (singlet particles and minifloccules), may more closely approach log normality than will fully dispersed systems. This holds true for both fines-rich and fines-poor systems at full dispersion.

Kaolinite slurries during shear thinning have been shown[11] to experience a decrease in effective volume. As shear progresses, immobilization of the particle assemblage passes through a maximum then decreases; that is, shear causes the dispersion to change from one system or state to another. The transition between these two states occurs at the viscosity maximum.

At zero shear a hydrostatic network forms in a particle matrix, essentially a zero-strength gel. This network has a shifting viscoelastic spectrum of energies associated with it. During flow the network will not persist, but moves in progressively smaller units of the network, the size of which diminishes with shear rate. It is not possible to attribute a Newtonian viscosity to each level of this structure. Although attempts have been made to model this behavior,[12] the rheological transients which occur involve too complex kinetics.

The "tightness" or stability of concentrated flocs is directly related to their permeability, a relationship that can influence the effectiveness of chemical reactions conducted on slurries in a flocculated state. When a flocculated system begins to break down, first to subfloccules, then to progressively smaller floccules, to individual particles, the overall packing factor of solids in the liquid phase is approximately equal to that of the loosest packing structure of a bed of equivalent spheres.

The ''ensemble averaging theory'' has been applied to assess these various aspects of rheological behavior. One of the more detailed and realistic of approaches is by Zubarev et al.,[13] who employ as primary variables the system modulus of elasticity and Young's modulus acting on such a packed bed of spheres in a fluid medium. Their approach, though complex, can explain the rheological changes that occur in these transitions from a flocculated to a dispersed state.

To understand this disparity between dispersion and rheology, the dispersion phenomenon must be examined a bit more closely. As particles reduce in size, individual surface area decreases as the square of the diameter, while mass decreases as the cube. Thus, on an individual particle basis the charge-to-mass ratio increases. As particles become finer, van der Waals forces (mass dependence) become progressively dominated by charge (surface dependence). Thus small particles require a lower charge density (number of anions per unit area) than coarse ones to counteract van der Waals forces. At limited dispersant levels all particles receive proportionally less agent. Flocculation then occurs preferentially with fine particles, both between fine particles and between fine particles and coarse particles. This process tends to proceed in the direction of log normality, based on statistical probabilities of interparticle interactions.[14] Thus, for a system having a surplus of fine particles (compared to a log-normal distribution), slight flocculation diminishes their fine population in the distribution. Conversely, coarse-rich distributions appear yet more dilatant with underdispersion.

A similar mathematical study by Urev,[15] based on the packed bed of spheres concept, predicts that aggregation and disaggregation (flocculation and dispersion) kinetics progress such that the population of particle assemblages under shear changes by a log-normal rate process, not a linear one.

If the particles interact only hydrodynamically, a condition that is approximated with a well-dispersed assemblage, the mathematical relationship between the viscosity of a dispersed system and the concentration of discrete particles dispersed in a homogeneous continuum will remain the same for the whole concentration region, assuming the concentration is expressed as the volume fraction, not weight fraction. This has been shown to be true[16] both for monodispersed systems, where particles may vary in size from one system to another but have the same morphology, as well as polydisperse systems, where particles have the same qualitative distribution of sizes and shapes, but where every particle can be scaled to a corresponding particle in another system by a constant factor.

Rate equations for the rheology of concentrated, non-Newtonian dispersed systems under shear also demonstrate[17] a progressively changing internal structure, part interparticle in character and part ability of individual, anisometric particles to orient under applied shear to minimize resistance to flow. This equation, Eq. (1.4), fits pseudoplastic behavior reasonably well.

$$n_r = (1 - \tfrac{1}{2} k\, \theta)^{-2} \tag{1.4}$$

where θ is the volume concentration of particles; $k = (k_o + k_\infty P)/(1 + P)$ is the structural intrinsic viscosity; k_o is the first structural parameter; k_∞ is the second structural parameter; and $P = 1.0$ for spheres and 0.5 for platelets and needles.

During high applied shear, the ability of any given particle to move in a matrix is also influenced by its shape because of frictional forces associated with surface eddies. This is distinct from the particle's ability to orient under stress. Smooth particles, for example, display lower viscosities than asperic ones, spheres lower than ellipsoids, which are lower in turn than cylinders or needles. The greater the eccentricity or anisometric character, the greater the dispersant loading required to attain the minimum in viscosity.[18]

As with all matter, thixotropy, dilatancy, and Newtonian behavior may have limited size ranges in which this activity dominates. Newtonian systems at low-to-moderate shear may well exhibit dilatancy at much higher shear. Thixotropy is a metastable state involving quiescence. As such it does not exist at high shear unless the system is returned to rest. Often a specific shear range is specified in quantifying thixotropy. For example, viscosity ratio at 10 RPM Brookfield shear and 100 RPM is often referred to as a dimensionless *thixotropic index* (T.I.). These are arbitrary[19] but industry-common assignments for quantifying flow behavior.

As an example, a 20% solids slurry of hectorite or halloysite, (fine, acicular aluminum silicate minerals) when shaken in a bottle has the consistency of milk. After standing a few seconds this slurry becomes so viscous it cannot be poured from the bottle, resembling more a paste than a liquid. Such a slurry will possess a thixotropic index from 5 to 15 for 1 μm, 10:1 aspect ratio particles.

Having defined dispersion, albeit empirically, other terms dealing with dispersant activity must also be defined more precisely:

Agglomerate: A weak association of particles, often fines attached to coarse, held together by charge, polar attraction, or van der Waals forces.

Aggregate: A slightly stronger association of particles held together by charge forces or cemented by chemical or physical agents. Aggregates may be ordered, that is, particles having parallel crystalline axes, or random.

Deflocculent: Another term for a dispersant.

Dilatancy: A state in which viscosity increases with applied shear (''shear thickening'') on a time-dependent basis.

Dispersant: Any agent that produces an improvement in particulate dispersion in a liquid compared with no agent at all.

False body: The development of apparent structure in a solid–liquid suspension and disappearance of mobility.

Flocculent: Any agent that reduces the dispersed state of particulates in suspension through agglomeration.

Flocculation: State of low-energy agglomeration between particles as a consequence of charge attraction, van der Waals force, or dipole attraction.

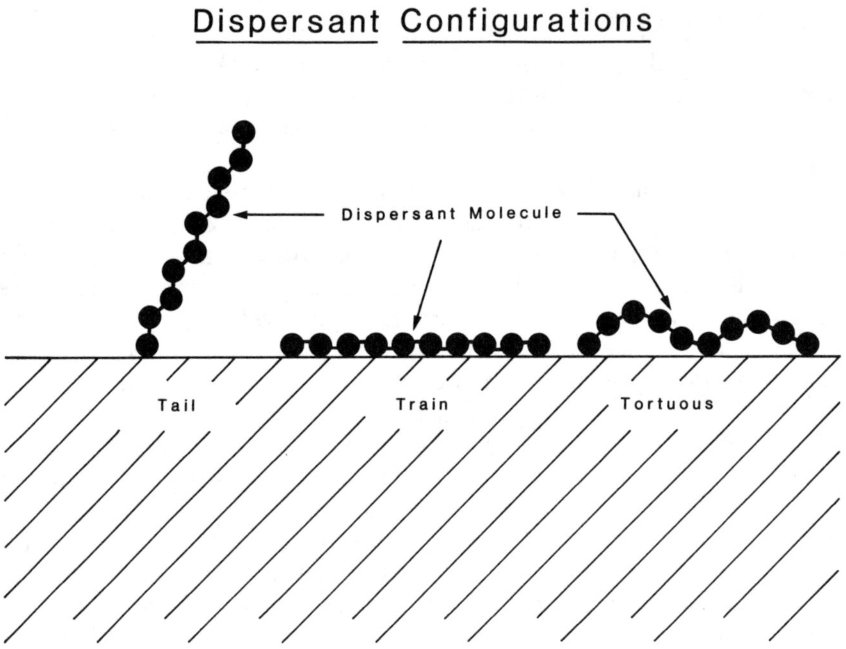

Figure 1.5. Dispersant orientations on solid surfaces.

Floccule: An agglomeration unit of discrete particles bound together by electrostatic, polar, and/or van der Waals forces.

Newtonian: A system characteristic in which the induced stress is proportional to the applied shear (viscosity is constant).

Peptizing agent: Another term for a dispersant.

Rheology: The state of flow characteristic of a liquid (unquantified).

Rheogram: A plot of viscosity versus applied shear, concentration of dispersant or solids level in the system.

Rheometer: An instrument which measures flow under variable conditions, that is, a sequential series of viscosities at various shear forces.

Sinter: An association of particles held together by fused contact regions. Sinters require elevated temperatures to form and high mechanical energies to break down.

Sol: A dispersed state of very fine particulates, less than 100 nm in diamter.

Thixotropy: A system in which viscosity decreases with applied shear (''shear thinning'') on a time-dependent basis.

Viscometer: An instrument that measures the drag force incurred with a set quantity of applied shear force. Often one measurement at a time is provided.

Viscosity: The ratio of the applied shear in a fluid system to the induced stress into that system (measured in poise units).

1.3 Dispersant Orientation

The manner in which dispersant molecules affix themselves upon solid surfaces influences several factors relating to dispersion.

Because dispersant molecules are nonisometric, they possess a variety of modes of attachment to a solid surface, and their orientation can be varied. Attachment is classed into three specific geometric categories, ''tail,'' ''train,'' and ''tortuous.''* These are illustrated in Figure 1.5.

Molecules possessing a ''head'' whose charge or polarity is located solely at the molecular termination attach in the tail mode. Dispersants with periodically repeating attachment groups configure in the train mode. Agents with variable and nonuniform structures, often very large molecules, attach in a tortuous mode.

Mode of attachment may change as a consequence of solution pressure, competition for other adsorbing species, or because of differential site affinities. This is illustrated in Figure 1.6 for a low-molecular-weight polyacrylate adsorbed as dispersant on a silicate pigment surface.** Note the Langmuir isotherm is initially steep (quantitative), peaks, diminishes slightly, then resumes in a revised Langmuir pattern, one having reduced slope (less adsorption energy). This phenomenon is typical of tail-to-train transitions upon surfaces with multiattachment group dispersants. By contrast, train-to-tail shifts, while theoretically possible, are much rarer.

Because enthalpy of adsorption of solid surfaces by many dispersants is low, competition may arise between the aqueous phase and the solid surface for these molecules. The lower the liquid-phase solvation energy compared with the solid surface adsorption energy, the higher the proportion of dispersant species will be localized upon the surface. Surprisingly, even quality dispersants are shifted eventually to the solution phase. Compare the polyacrylate dispersant equilibrium from Figure 1.6 (about 28% adsorbed at optimum dispersion) with that for butylamine adsorption on sequential grinding stages ($A \rightarrow D$) of silica in Figure 1.7. Butylamine adsorption, though less efficient in dispersivity (discussed in detail in Chapter 5) than polyacrylates, is essentially quantitatively adsorbed.

Adsorption of the dispersant molecule on the solid surface causes structural changes in the hydration layers. When such systems are diluted, equilibrium shifts occur that may reduce the necessary charge or protective envelope the dispersant provided. Gel-like condensates[20] will form that build to visible-size

* The word is derived from the Latin ''tortus,'' meaning rambling or meandering. However, use of the literal Latin word is confusing because of its similarity to ''tortoise,'' a word also derived from this base, but which would give quite a misleading analogy!

** A graph of quantity or concentration of a species added (abscissa) against that quantity adsorbed (ordinate) for a specific amount of solid provides a measure of adsorption affinity of the species on the solid surface. Where this is high, the adsorption curve will rise at about 45 degrees (100% adsorption), then quickly flatten. Extrapolation of the flat top back to the ordinate will yield the monolayer adsorption value. Such plots are named after the brilliant physical chemist, and Nobel Laureate, Irving Langmuir, who developed their use in 1910–1916 to explain adsorption of gases on solids. His work greatly improved the design of gas masks during and following World War I.

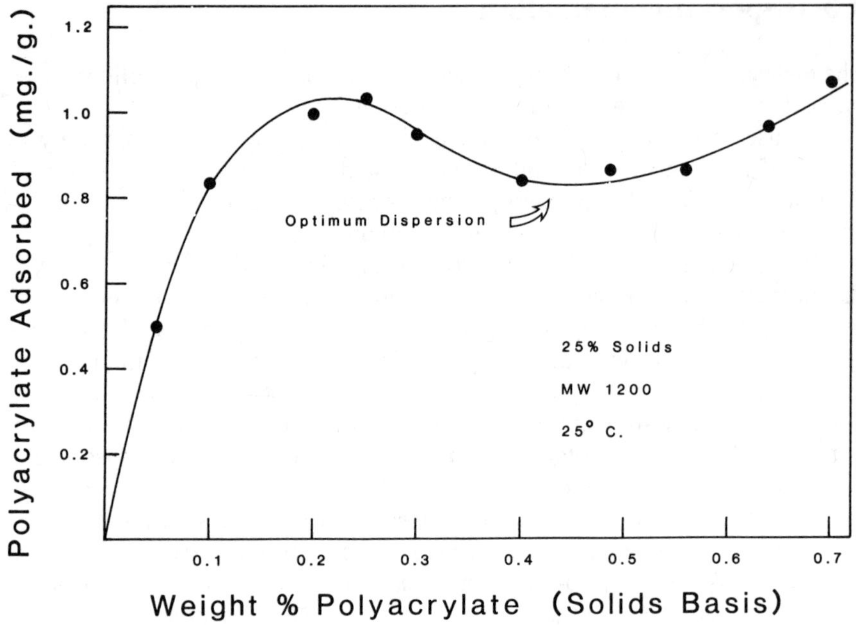

Figure 1.6. Langmuir isotherm for polyacrylate on silicate surface.

floccules. A fully flocculated network rarely develops. This phenomenon often becomes a problem in sampling a well-dispersed slurry and diluting by 1000:1 or greater for particle size analysis. The process in toto may require several minutes, and the sample be in the instrument before size enlargement commences.

A similar phenomenon occurs in reverse during the drying of dispersed cellulosic pulps. Natural dispersoids impair the ability of fibers to form aggregates during water loss, with the result that highly dusting products emerge from the drier.

Mechanical wear of slurry handling equipment, especially pump impeller blades, pipe elbows, and slurry flow impingement surfaces, is generally more severe with a flocculated state slurry than with a dispersed state because the mass-to-surface area of the impacting entities is greater.

In a plant employing mineral pigment processing,[21] the impeller pump employed for system makedown (agitation mixing of dispersant with feed material) required regular impeller blade replacement on a weekly basis. Farther downstream a booster pump, used for feeding this slurry to size classification equipment, was yet in service after a year of operation. Finally, a third pump, employed to convey the slurry following postprocessing flocculation for filtration, also wore out every few weeks. The common denominator in the wear equation was state of dispersion.

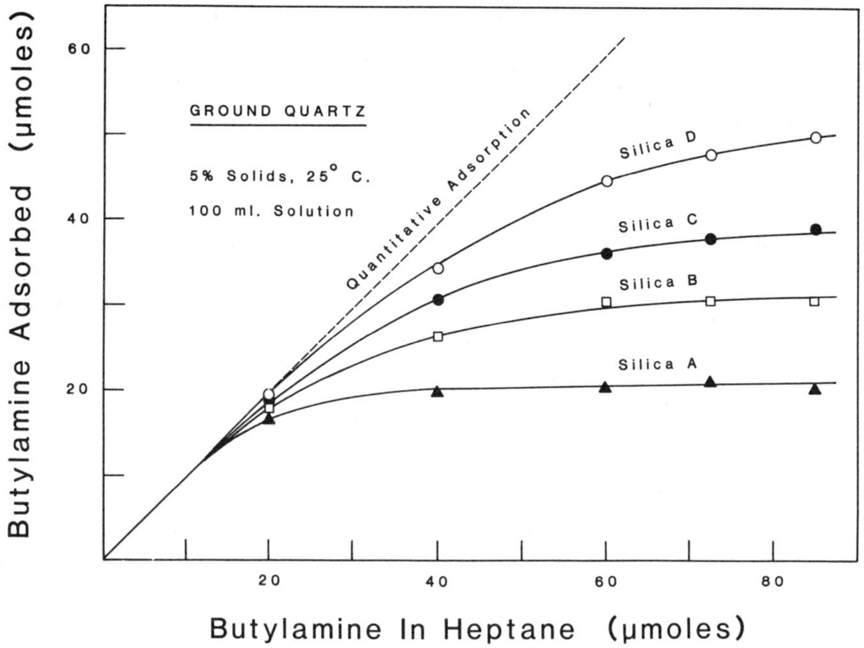

Figure 1.7. Langmuir isotherms for butylamine on stage ground quartz.

Yet another phenomenon associated with flow in dispersed slurries relates to the application of electrostatic fields. Charged particles behave differentially in such environments. As noted, with smaller particles surface charge diminishes as the square of the diameter, but mass decreases as the cube. Electrostatic attraction is proportional to the charge in a uniform electric field, but momentum, derived from pumping, is mass dependent. The consequence of this is that particles become differentiated in size during the flow, becoming fines-poor in the core of the fluid. This is turn shifts the rheological character from Newtonian to dilatant. Advantage can be taken of this phenomenon, for example, in coupling-variable clutches, where the coupling efficiency is controlled through an applied variable electric field on a viscous, charge-dispersed slurry.

This phenomenon and its engineering ramifications fall under the general term *electrorheology*.[22] Viscosity is affected not only by applied field strength, but by AC frequency, shear rate, and dielectric constant of the fluid.

1.4 Uses of Dispersion

The topic of discussion usually conjures up immediate images of paint formulations and liquified food products. However, the range of applications is far broader and often more subtle than is commonly realized. Those characteristics which derive from dispersion are those that optimize when all particles are discretely separate in suspension. Many of these are optical in nature, some are electromagnetic, and yet others are solely mechanical.

Disproportionately more applied shear energy is required to separate floccules composed of 0.5 μm size units for dispersion than with 5 or 10 μm units (see Chapter 8). As an example, finger-thumb pressure is sufficient to break up aggregates in the 44 μm range, but shear with a 6 in. diameter blade rotating at 2000 RPM may be required to break up 0.5 μm aggregates. Dispersion requiring high-energy disaggregation may not result in full particle separation. Thus, not only will viscosity reduce with increased time of makedown, but so will the high/low shear ratio rheology. Because fine particles are those first to agglomerate during flocculation, they are those last to release upon redispersion, especially following a drying stage. Rheological measurements of these systems will commonly display a dilatant character until all components in the system are fully dispersed. Thereafter, they approach Newtonian character.

Both the scattering function of opacity and color reflection/absorption degrade with dispersion because fewer randomly oriented surfaces are available to interact with impinging light rays. A partially flocculated paint pigment in suspension will require a reduced applied film thickness, or fewer coats, to effect acceptable hiding power than will a fully dispersed one. Heavier film thicknesses in turn tend to run during application requiring additional brushing and longer drying periods, both of which are disadvantageous to customers.

Good dispersions of very fine, primary pigments result in optical mass inefficiency at higher pigment concentrations when these are dried into films. As a consequence, partial replacement of primary pigments with "extender" pigments can be employed without sacrificing opacity or photon absorption. Because primary white pigments, which include titanium dioxide and zinc oxide, are significantly more expensive than white extender pigments (calcium carbonate, barite, kaolinite), even partial substitution of the former by the latter can result in significant cost savings in a formulated product.

Paint pigments are almost universally small, often submicrometer in size and usually equidimensional in shape. Fineness enables their long-term suspension in the spray tanks employed on assembly lines. Minimal agitation is required. However, when metallic paints were introduced, the aluminum flake component creating the "glitter" was not only not submicrometer, particles were frequently in the 100 to 1000 μm size range. Now settling became a major problem. Any flocculation would result in regions of paint having less sparkle, which showed up dramatically in the sunlight. Also, where coarse particles do not align during application, surface texture is altered, creating regions with surfaces varying between ripple and glossiness. To maintain aluminum flakes in suspension, in a fluid state, and capable of selective orientation, some pigment manufacturers have altered the alumina surface with chemically "anchored" dispersants to prevent agglomeration and settling while in the spray can. Fatty and reactive organic acids (acrylic, phthalic, etc.) may be employed in treating aluminum flakes to create this dispersive stability. Other methods of maintaining uniformity include additions of thixotropic additives.

In any particle size distribution, system mass is controlled by the coarse size end, while surface area is controlled by the fine size end. Thus the greater portion of light scattering derives from the lowest 20–30% of a pigment's mass. Size control and thoroughness in dispersion can bring about cost efficiencies in pigment utilization. However, dispersion difficulties increase with surface area and, hence, with finer-size particles.

Some applications of dispersion border upon emulsification and emulsion stabilization. For example, waxes and other noncrystalline organic solids are often made into dispersions with high shear equipment and special dispersants. In the process much of the input energy (ΔG) goes into a temperature rise (ΔT) with but a small portion into particle breakup and separation (ΔS). As a result, the organic component melts or softens and becomes spherical from surface tension and hydrodynamic forces. At this stage it qualifies as an emulsion. Upon cooling, however, the organic phase solidifies within the carrier liquid phase and the system then reverts to a dispersed state.

With cosmetic pigments and similar applications, these dispersions can break under shear as the liquid phase evaporates. The individual waxlike species in suspension again melt with rubbing to form very thin films of the organic component upon the skin.

Magnetic tape and disk industries rely heavily upon optimum dispersion of pigments for their product quality. Fluidity of anisometric needles of gamma ferric oxide is an absolute necessity. During formulation iron oxide particles must be dried, deagglomerated from clusters resulting from production technology, without breaking the needles, and dispersed with appropriate agents until a discrete system appears. The particles are then oriented parallel and solvent is evaporated sufficiently to "freeze" the particles in parallel orientations. The result is a receptive field film capable of being magnetized with high remanent strengths in small locales. Typically, a magnetization region is no more than 4 micrometers square (40 million fields per square inch).

Magnetic pigment dispersion can be accomplished through a number of techniques, either by the pigment producer or the formulator. Dispersants effective for obtaining efficient disaggregation during milling include salicylic acid, acrylic acid, and benzoic acid. Figure 1.8(a), a casting thin section from a 6 hour ball-milled oxide without agent, shows aggregates 40–50 μm in size. Figure 1.8(b), an equivalent milling, but with 0.5% polyacrylic acid added, is considerably finer (D_{max} − 10 μm).

During the 1970s magnetic pigments were developed that were self-dispersive[23,24] because of mode of manufacture. Particles were extremely fine, 0.1 μm or less, with surfaces having organophilic coatings to enable high-density magnetic data recording. However, the particles had little *XY* anisometry (within the tape plane), being platelike in morphology. While particles were capable of being dispersed readily, maintaining them in position with parallel magnetic axes (also *XY* oriented) became very difficult. Ambient kinetic energy, especially during

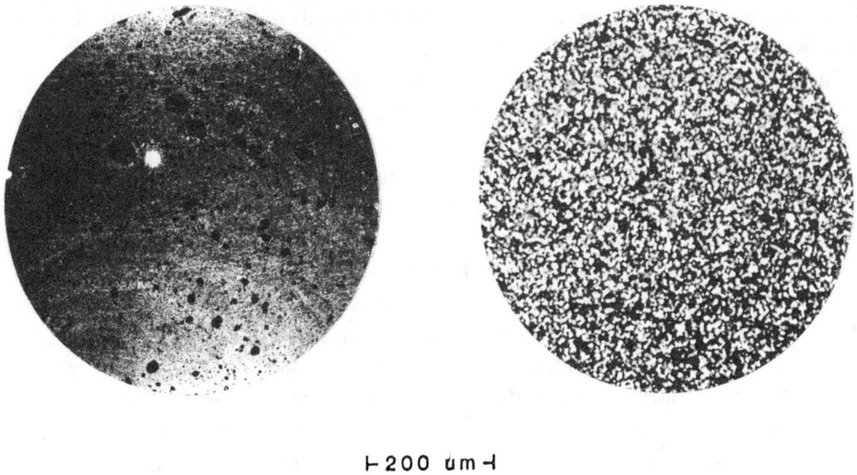

(a) (b)

Figure 1.8. Milling of magnetic oxide with and without dispersant.

thermal drying, was sufficient to reorient many of the particles during the short period between magnetic field interruption, solvent loss, and particle immobilization.

Dispersion may actually be employed to increase viscosity, if the particles are highly anisotropic, especially if needle shaped. Cosmetic and medical creams and pastes may be formulated at low viscosity by means of high shear stress with hectorite, halloysite, or attapulgite (palygorskite) minerals or organic structures of carboxymethyl cellulose and similar derivatives, then immobilized and thickened upon stress release. Such a system behaves most efficiently with highly dispersed components. The anisometric agents employed for this purpose commonly have thixotropic indices of 10 or above to prevent components in the system from settling or losing homogeneity.

Dispersion in public drinking water supplies is a curse. The goal of providing crystal-clear water from the public's tap is thwarted by ultrafine ($<0.1\ \mu m$) clay particles and organic debris. Municipal water processors exercise many techniques to flocculate and remove these impurities. Most often water resources, such as wells, lakes, reservoirs, etc., are pH adjusted to achieve a near-zero particle surface charge, causing aggregation and making the particulates larger and more readily filtered through fine sand beds.

Where this technique proves inadequate, aluminum sulfate solution is added followed by calcium hydroxide (lime) to form large gelatinous floccules of alumina hydrate which nucleate upon and precipitate the clays and other suspended solids, thereby forming much larger aggregates, suitable for filtration. Chemical quantities must be carefully adjusted, as their presence adds sharpness and tang to the water's taste, and produces calcium sulfate (gypsum), which builds up on faucets, shower nozzles, basins, and other plumbing equipment.

Human blood contains upwards of a hundred different substances, some dissolved, some suspended, which must be maintained in homogeneous dispersion to permit transit throughout the body's plumbing array and across cell walls to perform various biochemical functions. The body manufactures its own dispersants for many of these, aminohexoses, glucosamines, hyaluronic acid, etc., and wraps others in a protective, dispersant envelope. Most of these dispersions are extremely pH sensitive. Blood, nominally at pH 7.35 ± 0.05, can be affected severely with some components flocculating upon a pH shift of no more than 0.5 pH units. Death is the consequence!

Coffee and tea are colloidal dispersions of complex organic molecules including alkaloids. As either of these liquids age or reach higher concentrations, dispersive forces on the molecules begin to break down, either by decomposition, oxidation, or an inability to overcome strong van der Waals forces in close particle proximity, and the system commences to coagulate. This is seen first as opacity (haze), then bitterness, and finally precipitation and rancidity as the system's dispersant fails!

Wet grinding of slurries to obtain ultrafine particles for cosmetics and pharmaceutical applications requires both efficient and low toxicity dispersants. Facial cleansing masks often contain clays—kaolins, attapulgites, and bentonites. To give these products smoothness and creamy textures, they are milled or formulated with FDA-approved agents having dispersion effectiveness but topological inertness as well. The clays must go onto the skin, dry, and there absorb oils and defoliate dead skin flakes. Floccules do not function well on the skin in this capacity. The dispersant must be stable, bacterially insensitive, and harmless to skin, and must possess no offensive odor.

Pharmaceutical preparations may have dispersants employed to make solid–liquid systems stable, enabling homogeneous mixtures to be present, or produced with mild shaking, from the first spoonful to the last. Often these suspended systems are designed to flocculate deliberately in the stomach, allowing the active agent to curdle and localize along stomach and intestinal walls, there to adsorb an alien species or be assimilated by the body.

Food products, especially those with high weight costs, such as chocolate and exotic spices, are jet milled to reduce particle size. In applications these frequently contain an FDA-approved dispersant, often lecithin, castor oil esters, or similar natural polyhydric compounds. Upon entering the mouth, the expensive flavor components immediately disperse and allow a rapid response from taste buds. This is important inasmuch as residence time in the mouth for most food products is short. Employing less than 1% of a dispersant allows the chocolate content (an expensive commodity) to be reduced by as much as 25–30%. Not only are economies of production realized, but consumers find the product extremely tasty. It is this rate of change of suspension that the tongue experiences as high flavor intensity.

1.5 Limits of Dispersion and Disadvantages

All dispersed states have solids limits. These may result from volumetric limits of the liquid phase or from immobilization of the liquid phase through ion solvation. Commercially, the highest level of pigment solids is likely attained with pigmentary titanium dioxide shipped at about 80% solids. Here particles are all in the narrow range from about 0.15 to 0.35 μm. This size range represents a very narrow distribution bordering upon dilatancy. However, during production titania particles aggregate, forming dimers, trimers, tetramers, and even large agglomerates. It is this distribution of sizes that results in a quasi-log-normal distribution, permitting more nearly Newtonian flow. Were all of the aggregates broken down into a monodisperse system, solids level for processing and shipping would be reduced significantly.

Not all powders can be dispersed at such high solids because of surface character, particle morphology, and dispersant–surface affinity. The latter is evident from the dispersant shift to solution phase in Figure 1.6. Kaolinite shipped at 70–72% solids is a slurry with the consistency of heavy dairy cream. An increase of 1% solids changes this dispersed slurry to a rigid solid. Titanium dioxide, however, can be shipped readily at 75% solids because its higher density results in lower solid volume.

The processing of calcium carbonate for pigmentary applications may take two forms, one produced by grinding white marble to below 1 μm and the other by controlled precipitation of calcium carbonate to form particles less than 0.3 μm. The latter can yield very narrow size distributions, ideal for light scattering but dismal for high solids loadings of coating films. Wet grinding with dispersants, by contrast, can produce very nearly log-normal distributions, at least sufficiently so to effect high solids for coatings. These, however, are less efficient as photon scatterers.

However, the two modes of carbonate production have other differences than just distribution size breadth (sigma). Precipitated carbonate has highly perfected surfaces showing few defects and little stress. Ground surfaces are highly stressed for periods following dry grinding of as long as several days. These surface character differences often reveal themselves in dispersant demand change, adverse chemical reactivity with polymers, dispersant breakdown, and other detrimental effects. Such adverse results are magnified with dry grinding but often are attenuated with wet milling in the presence of dispersing agents.

Alkali inorganic anion salt dispersants (sodium polyphosphates, silicates, borates, etc.) increase their affinity for the solid surface as pH lowers. Oxide surfaces decrease their negative charge at low pH values and become positively charged, attracting negative dispersant ions. Thus, where flocculation is employed to increase solids for filtration, the equilibrium between solution and surface shifts for the dispersant species dramatically to the surface. In some instances, for example polyphosphates on alumina hydrate, a permanent compound can be established on the pigment surface.

During high solids slurry drying or particulate dehydroxylation stages, alkali dispersants can flux and cement particles together, sinter strength increasing with temperature. During the calcination of kaolinite, to effect whiter and less hydrophilic character, fluxing often creates rigid interparticle sintering. Figure 1.9 is a shadowed electron micrograph of a dispersed kaolinite feed to a calciner, while Figure 1.10 is an SEM of the calcined and pulverized product. Prior to calcination, feed particles were 90% −2 μm, and primarly as thin platelets. Particles following calcination show cementation from alkali dispersants with many of the fines adhering to coarse particles. Light scattering, which optimizes with particles between 0.2 and 0.5 μm, is drastically impaired by such sintering.

1.6 The Rheological Profile

As increasing amounts of dispersant are added to a flocculated system, viscosity drops dramatically, often by 3 to 4 orders of magnitude.[25] Optimum dispersant

Figure 1.9. Dispersed kaolinite slurry particles (calciner feed).

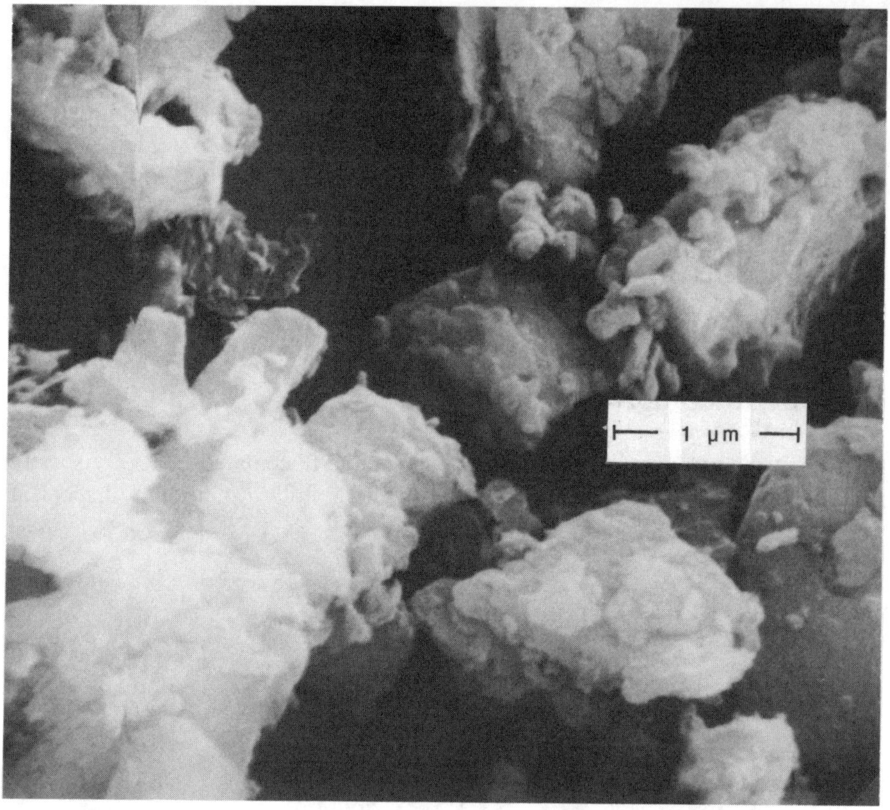

Figure 1.10. Thermochemical sintering of calcined kaolinite (product).

"demand" is attained at the minimum point on a plot of viscosity as a function of dispersant level. Continued additions of dispersant may cause the curve to increase, especially at elevated solids, in an almost mirror image of the descending curve. This "U"-shaped curve is essentially catenary in form, although as often as not it bears some asymmetry, skewed slightly to the higher dispersant dosage level.

Viscosity increase with overdispersion is often termed *chemical viscosity.* This phenomenon derives primarily from small-molecule, ionically charged, inorganic species of dispersants. Typical examples are shown in Figure 1.11. A brief overview for the curve's shape follows. The phenomena involved will be described in more detail pertaining to specific dispersants and interactions later in this book.

1. With zero dispersant the particulate system is highly agglomerated, disallowing individual particle movement (agglomerates are sheared off by the impeller blades). The entire agglomerate attempts to move in response to shearing

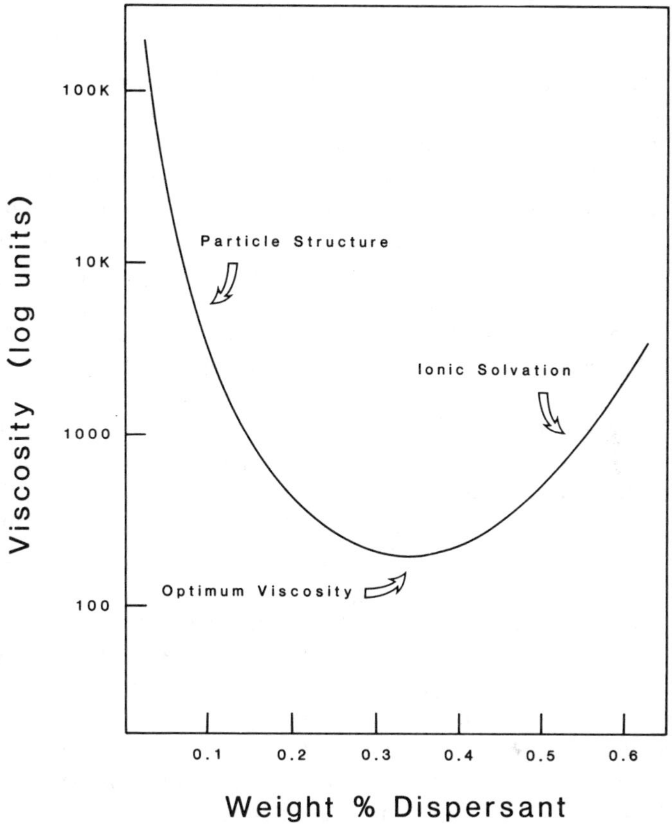

Figure 1.11. Catenary rheological curves for dispersant level.

activity and follow the turbulent flow in the container. Viscosity is very high due to interparticle bridging structures.

2. With low levels of dispersant and progressively more particles being fully removed from the assemblage, agglomerates become smaller, and discrete particles, having no mechanism to return to the agglomerate, other than diffusion, are able to move more freely in the system. The viscosity diminishes.
3. As the dispersant level increases, all particles are eventually dispersed, and the system attains its maximum mobility. Viscosity reaches a minimum. At high solids a significant portion of the liquid medium is adsorbed as shells about the dispersed particles, much of it immobilized through hydration of the counterions in the ionic dispersant itself.
4. Increasing the dispersant level beyond the minimum in viscosity shifts more ionic material into the solution phase, where it immobilizes additional liquid from the fluid phase. Viscosity rises because system fluidity is further reduced by ion–liquid solvation.

5. If sufficient dispersant is added, the aqueous phase will become completely hydrated, at which point the system forms into an immobile quasisolid.

If the particles dispersed are sufficiently fine, their surface will immobilize water by structural entrapment at lower solids. A paste or viscous cream form results. This is the origin of thixotropy or false bodying. A gel state will be formed at zero shear if weak intermolecular forces in the liquid phase are completely overridden by weak structuring forces created by the solid's highly polar surface and the dispersant molecules adsorbed thereon. At this point the liquid phase is fully immobilized, trapped in structural "cavities" created by the solid bridge network. Upon application of additional external force, or applied shear, the weakly structured network breaks down into a more mobile one. This point is termed the *yield point* (see earlier discussion). Yield points often are recyclable.

Where particle size distributions are narrow (i.e., where the value D_{75}/D_{25} is 2 or less), the fine and coarse ends of the Gaussian distribution are significantly diminished. Here, large numbers of internal voids of liquid exist and solids levels obtainable are less. In broader distributions, these voids would be filled with smaller particles and the number of all particles at the same solids would be reduced to balance the mass.

Where a narrow size distribution is dispersed, its particles can move if they are permitted time to displace neighbors mechanically. Adequate fluid is usually present at moderate solids to accomplish this, but only marginally. However, if this particle aggregation is forced to move rapidly, such as by high applied shear, a logjam is created with the appearance of temporary solidity. Upon removal of the shear force, particles can resume their relaxed positions, and the system reverts to fluidity. This mechanism defines dilatancy.

A common example of this type of dilatancy occurs with wet beach sand, which feels firm under foot during walking. Yet walking on adjacent dry sand results in immediate sand collapse. Ocean waves have classified to a very narrow breadth the size distribution of sand grains. When the same sand is encountered with traces of clay in a jungle, it is more popularly termed "quicksand" (meaning "live" sand).

Optimum fluidity is created with a log-normal distribution of particles,[26] for this assemblage has the greatest free liquid space for a given mass of solid. As shear is applied, flow rate increases, but resistance to flow is in equal proportion. The particle assemblage moves proportionally. Such a system, one where viscosity is constant with shear, can be dispersed at the highest solids level of all particle size distributions and will display the greatest fluidity at any specific solids level. This system is defined as Newtonian.

1.7 Types of Solids and Their Characteristics

Both the selection of dispersants and interactions between solid surfaces and the liquid phase are influenced by solid character, that is, its morphology (shape), particle size, crystal structure, surface chemistry, and types of chemical bonding. Solids may be divided into five general classes, four of which may be considered organized, based on atomic and molecular order and electron bond strengths.

Crystalline-covalent: compounds usually with high bonding energy ("hard"), typified by quartz, SiO_2, corundum, Al_2O_3, titanium dioxide, TiO_2, diamond, C, ferric oxide, Fe_2O_3, talc, and clays.

Crystalline-metallic: metal atoms (as Ag, Cu, Ni, Al, W) directly bonded together by a combination of covalent and ionic forces.

Crystalline-ionic: compounds with intermediate bonding energy, represented by calcite, $CaCO_3$, barite (barytes), $BaSO_4$, silver chloride, AgCl, and gypsum, $CaSO_4 \cdot 2H_2O$.

Crystalline-molecular: (also termed "van der Waals solids") compounds possessing weak intermolecular attraction and represented by sucrose (sugar), $C_{12}H_{22}O_{11}$, phthalocyanine blue, naphthalene, $C_{10}H_8$, and graphite, C.

Amorphous: solids with no apparent organizational structure and low binding energy, including cellulose, rubber, polyethylene, gels, and natural products (spices and herbs).

The first four categories are shown on the energy diagram in Figure 1.12.

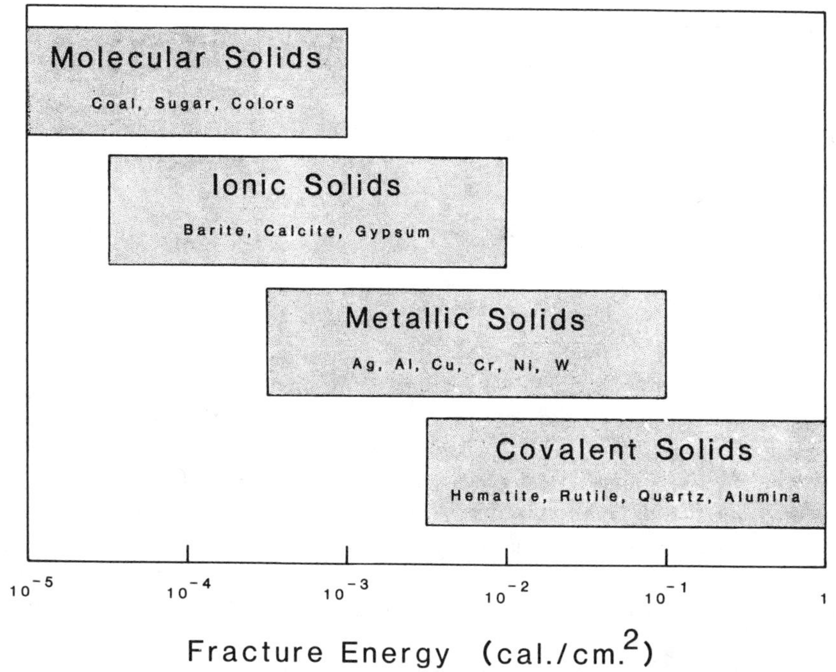

Figure 1.12. Bonding energy range for various solids.

Clearly evident is the relationship with both the Moh scale for hardness (e.g., diamond = 10, gypsum = 2) and the Bond Work Index (e.g., silica = 16, calcite = 10).[27] The ability to comprehend attachment of dispersant molecules to such varied surfaces, as appear throughout this book, can be aided considerably by molecular modeling software,[28] especially that with molecular docking capabilities.[29]

Within any of these four structure types can occur morphological differences which influence surface character (charge and structure), particle separation stability, and rheological characteristics (flow properties). Such structural types include:

Isometric: the equidimensionality of particles, that is, as displayed by spheres and cubes.
Anisometric: particles having at least one dimension significantly greater than the other(s).
Irregular: particles having no definable shape, usually cloudlike with rounded and convoluted exteriors.
Asperic: particles having sharp points or apices (asperities) over their surface.
Faults: dislocations, atomic replacements, inclusions, voids, twinning planes, or interstitial substitutions.

Clearly, the ability of a particle assemblage to flow in a liquid environment, especially one where particles are in close proximity, will be influenced by these morphological factors.

Dispersion systems are formulated on a weight basis while rheological properties derive from a volume basis (but not exclusively). Thus solids density is an important parameter in acquiring or assessing dispersancy. Figure 1.13 shows the proximity between particles (based on spheres) at various solids levels and densities. Hydrodynamic similarity studies[30] show the fluid wake of one particle to influence the motion of neighboring particles when separation is three diameters of less, the influence increasing with proximity. This level is achieved at less than 10% solids, irrespective of density. Other particle proximity studies based on Monte Carlo random traverse show that thixotropy derived from acicular particles and dilatancy derived from uniform size particles do not become significant until interparticle separation is about 1 particle diameter.[31]

At more practical levels of solids–liquid formulation (e.g., 50–60% solids), proximity is less than about 0.5 diameters. With more anisometric particles, the ability to flow is diminished dramatically because of their need to orient along laminar flow lines and because particle–particle separation decreases with anisometry. These systems commonly require reduced solids levels to obtain the free-flowing characteristics required in production or customer applications.

As will be discussed in more detail in later chapters, surface texture and chemical character may influence dramatically rheological behavior. Whereas grinding (comminution) commonly produces parallel Gaussian distributions, only finer in size,[32] stressed surfaces and altered morphologies may change surface change, hydration, and dispersant stability significantly.[33]

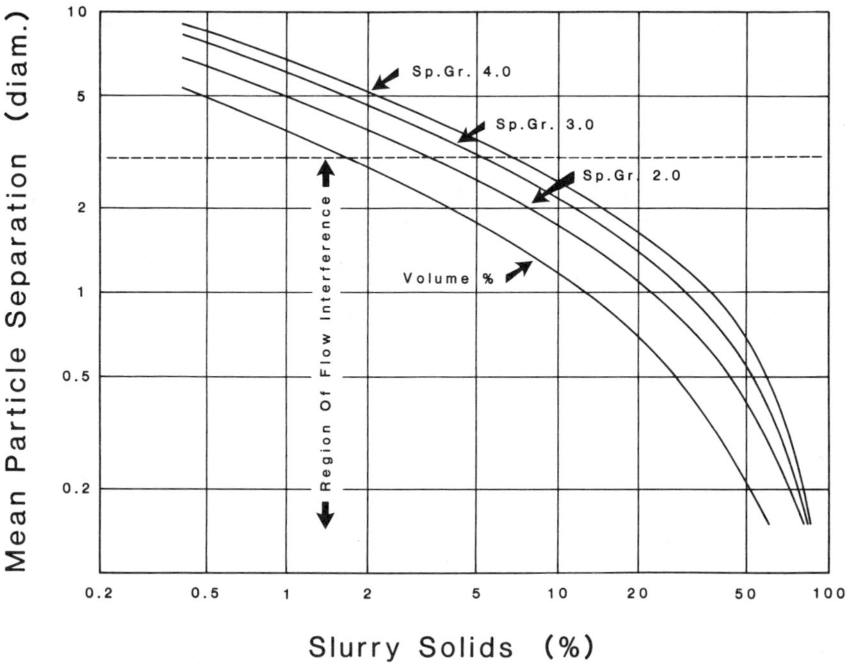

Figure 1.13. Solids level and particle proximity.

While many subtle factors play roles in the dispersive process, the dominant players and their character in the system are:

Liquid system: viscosity, polarity, acid–base character.
Solid surface: area, acid–base character, shape, stress.
Agent: affinity for solid, affinity for solution, stability.
Mechanical application: shear, energy (RPM and blade size).

Of these the latter two are most amenable to modification. Solid surfaces in instances can be modified chemically, but this is as often a theme variation on "agent" change as it is true structural modification. Only the liquid phase is a fixed parameter, for it commonly serves as a major factor in application dynamics or process control.

For any line of commercial products where rheological properties are important parameters, quality control and quality assurance of the products is mandatory to retain a customer base. As a consequence, the various plant variables, particle size control, dispersant dosage monitoring and stabilization and attention to application conditions are important considerations for the plant manager, the process control chemist, and the sales personnel.[34]

In the following chapters these categories will be discussed in detail, methods will be suggested for proper material and process selection, and a number of process examples will be given, all based on a knowledge of liquid and surface characteristics.

References

1. Bosse, D., ''Dispersing Fine-Particle Materials. Part I,'' *Dispersion of Pigments and Resins in Fluid Media*. Kent State University, Kent, OH (1981).
2. Moore, W. J., *Physical Chemistry*, Prentice Hall, Englewood Cliffs, NJ, 3rd. ed., p. 78 (1962).
3. Cropper, W., *J. Chem. Ed.* **48**, 182 (1972).
4. Nash, L., *J. Chem Ed.* **47**, 355 (1970).
5. Bruff, T., and Stillinger, T., *J. Phys. Chem.* **25**, 312 (1956).
6. Crowl, V., *Agriments*, Elsevier, Amsterdam, p. 192 (1967).
7. Boersman, W., Leven, J., and Stein, H., *A.I.Ch E. J.* **36**, 321 (1990).
8. Chechetkin, A. et al., VINITTI, p. 1696 (1974).
9. Parthasany, R., and Faeth, G., *J. Fluid Mech.* **220**, 515 (1990).
10. Allen, T., *Particle Size Measurement*, 3rd ed., Chapman & Hall, London, p. 136 (1981).
11. Pfragner, J., *Prog. Colloid Polymer Sci.* **77**, 177 (1988).
12. Mewis, J., *Adv. Org. Coat. Sci. Tech.* **10**, 65 (1988).
13. Zubarev, A., Katz, E., and Latkin, A., *Inzh. Fiz. Zh.* **58**, 721 (1990).
14. Cornell, R., Goodwin, J., and Ottewill, R., *J. Colloid Interface Sci.* **71**, 256 (1979).
15. Urev, N., and Akhterov, V., *Particle Sci. Tech.* **7**, 253 (1989).
16. Kovar, J., and Bohdaneck, M., *Col. Czech. Chem. Com.* **44**, 2085 (1979).
17. Quemada, D., *Rheo. Acta.* **17**, 643 (1978).
18. Barnes, C., *Proc. Royal Soc. London* **368**, 177 (1979).
19. ''More Solutions to Sticky Problems,'' Brookfield Eng. Labs, Stoughton, MA, p. 19 (1994).
20. Milichowsky, M., *Sb. Ved. Pr. Vys. Sk. Chem.* **51**, 149 (1988).
21. Conley, R, ''Cadam Kaolin Processing,'' Rept. #85-JP-178, Jaako Poyry, Sao Paulo, BR (March 5, 1985).
22. Block, H., and Kelly, J., *J. Applied Phys.* **21**, 1661 (1988).
23. Conley, R., ''Magnetic Materials and Magnetic Impulse Members,'' U.S. Patent #3,115,470 (1963).
24. Conley, R., ''Magnetic Materials and Methods of Making Same,'' U.S. Patent #3,399,142 (1968).
25. Conley, R., *J. Paint Tech.* **46**(594), 60 (1974).
26. Gray, W., *The Packing of Small Particles*, Chapman and Hall, Ltd., London (1968); Michaels, A., MIT communication, June, 1960.
27. Bond, H., *Trans. A.I.M.E.* **193**, 484 (1952).
28. Studt, T., *Res. & Dev. Magazine*, Feb. 1995, p. 77.

29. Biosym Technologies, Inc., Materials Science Software, San Diego, CA (1995).

30. Conley, R., *Powder Technol.* **3**, 102 (1970).

31. Andreev, V., and Lukyanov, A., *Kolloidn. Zh.* **51**, 748 (1989).

32. Gaudin, A., and Meloy, T., *Trans. A.I.M.E.* **223**(3), 243 (1962).

33. Croucher, M., ''Hydrodynamic Methods for Studying Colloidal Stability of Disperse Systems,'' 28th Annual Tech. Conf. Cleveland Soc. For Coatings Tech., Cleveland, OH, p. 7 (1985).

34. Polke, R., *Chem. Ing. Tol.* **62**, 813 (1990).

CHAPTER

2

How Dispersants Function

The obvious goal of all dispersants lies in stabilizing as discretely as possible particulate distributions in fluid media. While certain special applications may be an exception to this rule, it is a valid requirement for the bulk of slurry applications. Chemical agents functioning as dispersants must accomplish at least three purposes. First, the effective area of the particulate surface must be adequately covered by dispersant molecules (adsorption property). Second, that portion of the dispersant molecule directed outwards into the liquid phase must coordinate with or display similarity to that phase (semblance property). Third, a barrier must be created around each particle capable of preventing other particles from coming into direct contact (isolation property).

2.1 Energetics of Wetting and Adsorption

It is not the intent of this text to delve into the complex mathematics of surface phenomena. Many thorough texts in a variety of fields cover this fully.[1,2] However, to understand the competitiveness of the various gel–liquid–solid interactions that may influence dispersant selection and the mechanical means to achieve a dispersed state, it is necessary, at least on a qualitative mathematical basis, to review these at the molecular level.

The four primary processes involved in stable dispersion are: (1) displacement of adsorbed gas on the solid phase by the liquid phase; (2) formation on the solid surface of a protective boundary to prevent particle–particle adhesion; (3) mechanical separation of these particles to allow the liquid phase complete encapsulation of them; (4) complete and homogeneous redistribution of the particles throughout the liquid volume. These stages must be inspected in detail.

All solids at room temperature have atmospheric gases adsorbed upon them, primarily oxygen, nitrogen, water vapor, and carbon dioxide. Energies of adsorption for these gases are not great, of the order of 50–500 erg/cm^2, the energy increasing with polarity of the gas and the surface. This assumes chemisorption (true chemical bonding of the gas species to the surface atoms, e.g., oxygen on metallic nickel powders) does not take place. When a single solid, as a specific component within a powder assemblage, is added to a liquid for the purpose of establishing a slurry, that liquid must displace the gas film and adsorb directly. Adsorption of the liquid must involve energy components not possessed by simple gas molecules. If the energy of liquid adsorption is low, the particle will simply float on the liquid surface supported by its interfacial gaseous barrier. A common example is talcum powder sprinkled on tap water.

The diagram in Figure 2.1 illustrates a magnified view of a solid particle on which a drop of a liquid phase has just made contact. If energy of wetting (enthalpy of wetting) by the liquid exceeds that of the adsorbed gas, the liquid in the drop will move outward until the internal force holding the liquid molecules together (essentially the latent heat of vaporization) balances the energy of wetting provided by the surface. The latter is termed the *free surface energy*. At this point, in the absence of evaporation, the droplet will assume a constant configuration. The consequence of this activity is that work is being done on the droplet to increase its contact area.

More correctly, the surface wetting by the droplet is the result of *excess*

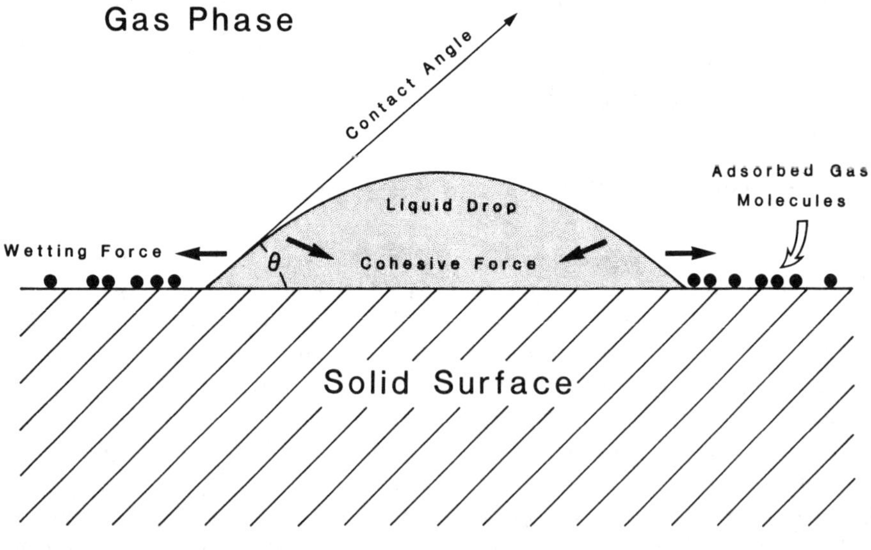

Figure 2.1. Forces operating during solid wetting.

surface free energy.[3] It is measured in energy units per unit area, erg/cm^2 or cal/cm.^{2} Surface molecules are thus being pulled in normally, toward the solid surface, a situation equivalent to a surface tension. The surface energy expended is not truly the total surface energy but that "left over" as a consequence of surface atoms or molecules being at the surface.

Were the liquid not to experience this free energy, the settled drop would have a negative contact angle, and its internal cohesive forces would overrise any spreading forces created by energy of wetting to the solid.

The force created by wetting energy in Figure 2.1 is shown as γ_{ls}, that for gas–solid by γ_{gs}, and the third force component, adsorption of gas on the liquid drop, by γ_{gl}. These are related mathematically by the Young equation*

$$\gamma_{gs} = \gamma_{ls} + \gamma_{gl} (\cos \theta) \tag{2.1}$$

Rearranging this equation gives

$$\cos \theta = (\gamma_{gs} - \gamma_{ls})/\gamma_{gl} \tag{2.2}$$

In this equation the contact angle of the liquid on the solid, θ, becomes a direct measure of the ability of the liquid to wet out the solid surface, or for the solid as a powder to disperse in the liquid as the bulk phase. Well-dispersed powders have contact angles approaching zero, poorly wet surfaces angles approaching 90 degrees or more (sessile conformation). The latter is illustrated by the perverse phenomenon of raindrop "beading" on freshly waxed automobile surfaces.

Where the wetting energy in Figure 2.1 only slightly exceeds the gas adsorption energy, the edge of the liquid contact of the droplet will slowly advance while the drop approaches a thin film. If the gas phase is the more energetic of the two, the contact edge will recede and the liquid droplet will assume the appearance of a squashed ball.

In some evaluation methods, rate of droplet spread is employed to measure wetting.[4] However, quality dispersants usually wet powders so rapidly that spreading velocity cannot be measured accurately. A variety of innovative techniques also have also employed, including measuring the liquid pressure involved in wetting a fixed, compressed volume of powder.[5]

These forces are significant, as demonstrated by the parlor trick of floating a clean, steel razor blade on pure water. The blade can be increased in thickness until its gravitational force exceeds the cohesive force of the water environment (in this instance essentially an infinite diameter drop), at which point the tension is broken and the blade sinks. Yet the thin blade can be made to sink instantly with but a drop of soap or detergent in the liquid. The modified liquid then possesses a greatly increased energy of wetting for the steel surface.

---

* Thomas Young, a brilliant mathematician born in England in 1773, formulated this now-famous equation around 1800 as part of a broad spectrum of investigations into scientific phenomena. He held doctorates in physics, medicine, archaeology, and astronomy and was considered a genius at linguistics, deciphering Egyptian hieroglyphics and other arcane ancient scripts. Young has been said to have been the last man on the face of the earth who truly knew everything!

Elegant means of assessing contact angles, and hence wetting energy, are available, some of which are described in more detail in Chapter 13. However, for the hurried chemist the simple approach of dusting a small portion of powder onto the warmed liquid surface usually suffices in a qualitative sense to assess wetting capacity.

Solid wetting also has ramifications in systems where one liquid must replace another (usually water) to enable a powder to wet out in the second liquid and retain mechanical integrity within the film (paint) or resin casting (plastics). Table 2.1 is a literature collection[6] of enthalpies of wetting for various particulates in water and hydrocarbon (usually heptane). As with isoelectric pH, the crystal structure of the solid has a pronounced effect on adsorption energies.

Competitive adsorption of a dispersant in a competitive liquid and for two competitive liquids has been analyzed mathematically by considering the two adsorption profiles.[7] For constant temperature adsorption (isotherms):

$$n^{\circ}\,\overline{X}_1 = n_1^s X_2 - n_2^s X_1 \qquad (2.3)$$
$$n^{\circ}\,\overline{X}_2 = n_2^s X_1 - n_1^s X_2$$

where n° is the total number of moles solvent and dispersant; $\overline{X}_1$ and $\overline{X}_2$ the change in mole fraction from adsorption; X_1 the mole fraction of solvent after adsorption; X_2 the mole fraction of dispersant after adsorption; n_1^s the number moles solvent adsorbed; and n_2^s the number moles dispersant adsorbed.

Table 2.1 Energies of Solids Wetting by Water and Hydrocarbons

Solid	ΔH(Water) (erg/cm^2)	ΔH(hc) (erg/cm^2)
Al_2O_3 (alpha)	−750	−150
Al_2O_3 (gamma)	−610	−85
Al_2O_3 (amorphous)	−454	−80
$BaSO_4$ (barite)	−490	−140
Bentonite	−500 to −1000[a]	−90
$CaSO_4 \cdot 2H_2O$	na	−120
Fe_2O_3 (alpha)	−400	−85
Fe_2O_3 (gamma)	−460	−85
Graphite	−50	−120
Kaolinite	−590	−90
MgO	na	−125
SiO_2 (quartz)	−820	−80
Silica (amorph)	−215	−65
Silica (fume)	−165	−118
SnO	−240	−70
Teflon (PTFE)	−6	−58
TiO_2 (rutile)	−550	−135
TiO_2 (anatase)	−490	−120
ZrO_2	−600	−190

[a] Dependent upon which cation, Na, K, Ca, is the dominant species.

Solution of Eq. (2.3) is tedious because each point on both adsorption isotherms must be measured or calculated to determine the competitiveness at each solids level in suspension. Most investigators prefer to assume a monolayer coverage based on relative energies of wetting.

Because dry powders consist of agglomerated assemblages of particles, the energy of wetting also must exceed the interparticle binding energy. Wetting energy dominates with agglomerates, is competitive for aggregates, but fails completely with sinters. However, even with agglomerates, the process is liquid diffusion limited. Addition of mechanical force is required, as a consequence, to separate these assemblages and permit dispersants to adsorb onto those unwet, excluded portions of the surface.

2.2 Stability of the Dispersant–Substrate Structure

Dispersion and flocculation are two sides of the same slurried coin. Their rate processes are similar, with the exception that flocculation most often occurs under quiescent, diffusion-limited conditions, whereas dispersion is accelerated by high-shear mixer blades and impatient formulators. The primary variables in each case are: (1) concentration (proximity) of particles; (2) viscosity of the slurry's liquid phase; (3) dipolar surface forces on the particles and in the liquid; and (4) temperature. Mechanical separation of particles produces an unstable system that quickly reverts to a reflocculated state. The Smoluchowski–Einstein[8] relationship provides a simple flocculation rate expression based on molecular kinetic energy for the time required, $t_{1/2}$, for half of the particles to assemble into floccules, assuming each collision is a "sticking" one:

$$t_{1/2} = (0.75\ \eta k T n_o) e^{W/RT} \tag{2.4}$$

where η is the liquid viscosity in poise; k the Boltzmann constant; T the absolute temperature; n_o the number of particles per cm^3 initially; W the energy barrier on the particles; and R the gas constant. In the absence of dispersant or inherent surface charge, the value of W is essentially zero in aqueous systems. With noncharged species (e.g., alkanolamines) adsorbed on the solid phase, it rises to 15–20RT and with charged species to 25–35RT.[9]

Were the particles to have no interactive capacity other than van der Waals forces, they would require collision contacts to form a network. At 50% volume solids (appr. 72% by weight at sp. gr. 2.6), the number of isometric particles per cubic centimeter with a mean diameter of 2 μm is about 53×10^9. At 25°C in water, the Smoluchowski–Einstein equation reduces to:

$$\begin{aligned} t_{1/2} &= 2 \times 10^{11}/53 \times 10^9 \text{ sec} \\ &= 3.8 \text{ sec} \end{aligned} \tag{2.5}$$

the half-life for single particles. For 10 half-lives, representing 99.9% of the particles flocculated into a network, the time rises to 38 sec. This reasonably approximates what is observed in poorly dispersed (by chemicals) but highly sheared kaolin suspensions of that size. Increasing the energy barrier to $30kT$ increases $t_{1/2}$ to a very long wait, 1.3 million years! In actuality, the time required rises to infinity because particle collisions no longer are "sticking" ones.

Studies with magnesium hydroxide sols[10] dispersed with sodium lignosulfonate using high-speed microcinematography showed the number of interparticle contacts at room temperature at critically dispersed levels (approaching flocculation) ranged from 10^9 to 10^{11} per second per square centimeter. The force of these ranged from 10^{-6} to 10^{-4} dynes per contact. This force is apparently just below the critical barrier for penetrating the charge–hydrate shell.

Kinetics of flocculation have been followed microscopically by employing polystyrene spheres of both mixed and single sizes.[11] The process is a steady state one thermodynamically. The reverse operation, redispersion, indicates certain agglomerate sizes have preferential lifetimes and that the lifetime (or half-life) increases slightly as the units progress from massive agglomerates down to doublets and finally singlets. This would indicate that either geometries or mass (van der Waals effect) render disproportionate stability to terminal floccules. Evidence also shows that singlet particles actually diffuse into, through, and out of floccules, where the binding energy of the floccule ("loose" floc) is not high.

Behavior of singlet particles and floccules, especially long-range forces influencing their stability, suggest that diffusion of dispersant into such agglomerates is the primary rate-determining process in dispersion[12] and the primary beneficial activity of high-shear mixing.

Electrical conductivity studies on carbon blacks dispersed in linseed oil[13] suggest black particles form an ever-changing network, even when fully dispersed, that contributes to the "incomplete" thixotropy of the system. Electrical conductivity of conductive blacks is a measure of network strength and single-particle population. Further, mechanical strength versus time curves (rate of structure formation or breakdown) show a definite restoration time requirement that is nearly proportional to applied deformation stress. Interestingly, continuous deformation–restoration cycling results in a slow degradation of permanent structural integrity. This suggests mechanical damage of the carbon black surface occurs. The solids levels for this effect ranged from 15% to 22%.

Similar phenomena are observed with ultrafine silica dispersions where the Si–OH group is the effective bridging mechanism.[14] In silica, however, reduction of either the dispersant or liquid phase results in a gel formation as interlocking silanol groups convert to chemical bonds.

In the submicrometer size range, dispersion becomes especially critical and difficult because of floccule strength and the quasisingle particle character of floccules in response to shear. Energies with which these are associated are greater than the van der Waals mass interactions would suggest. It has been proposed[15] that special, high-energy liquid bridges are responsible for this refractoriness towards shear dispersion.

2.2.1 Charge-Dependent Agents

Several means have been described previously by which a protective barrier can be established on a solid particle in a liquid phase. Other factors are also subtly involved which may alter functionality under special conditions. In an aqueous environment the anionic portion of a charge-type, dispersant molecule will most often have a lower solvation energy than bulk water molecules, nominally <9.6 kcal/mole, while the cationic portion has a higher energy. Second, the particulate surface (p) must possess a higher energy of adsorption for the dispersant molecule (d) than for the liquid phase (l) (true in both water and oleophilic liquids):

$$E(\text{ads.})_{(p\text{–}d)} > E(\text{ads.})_{(\text{p–l})} \tag{2.6}$$

For pure physical adsorption of the dispersant molecules onto a solid surface, the energy should equal or exceed twice the heat of vaporization of the adsorbed liquid film. Should chemisorption occur, this energy will be at least seven times the heat of vaporization. However, chemisorption usually requires a certain quantity of activation energy (heat, very high shear). The rate for chemisorption is more often slow compared to that for physical adsorption of the dispersant.

If the foregoing energetics conditions are not met, equilibria will be established in which only a small fraction of dispersant molecules are attached to the particulate surface. Thus two particles, both solvent enshrouded, could approach each other sufficiently close to form an incipient floccule. It will be seen that even when the above energy conditions are met, many systems will have an equilibrium established in which a significant portion of the dispersant appears in the liquid phase.

Those factors that dominate in dispersant adsorption include:

1. Shape and size of the dispersant molecule;
2. Orientation of the dispersant molecule on the surface;
3. Number of molecular layers possible (including micelles); and
4. Sufficient solubility of the dispersant in the liquid phase to reach the maximum adsorption capacity.

As shown in Figure 2.2 for a polar surface in water, two barriers are established, one a charge repulsion and the other a water insulation. Because dispersant anions are larger and more massive than their associated cations and are capable of greater charge (polyanion structures), they have greater adsorption potential for solid substrates. Anionic molecular mass provides increased van der Waals attraction for the surface. Charge, or more properly, charge density, is the significant parameter as it provides the ionic field strength for interparticle repulsion (equal to the product of the charges on both particles divided by the square of the distance of separation) and the adsorbed water barrier, whose effectiveness lies in its thickness, provides the insulation for dispersancy.

Studies on annealed silica[16] and sucrose[17] indicate the attractive (flocculation) force between two particles lies in the range of 12–13 nm corresponding to about

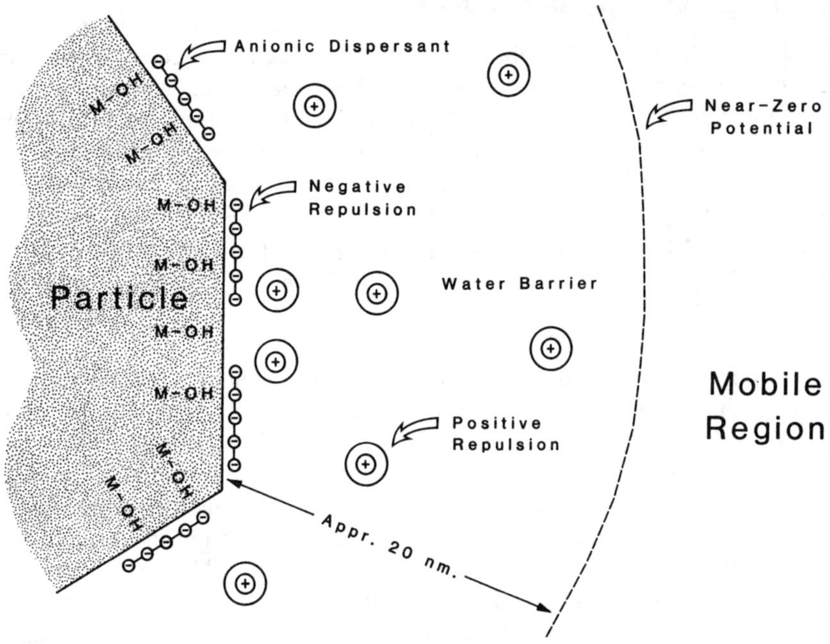

Figure 2.2. Barrier formation from dispersant adsorption.

40 water molecules (20 molecule thickness per particle). At best, the greater portion of these molecules are weakly associated with the solid surface. Gas adsorption studies with water vapor (BET technique) indicate that polar solids (such as oxides) adsorb only about three layers of water with any significant strength. The attractive force for such molecular adsorption falls off at a third power rate. Thus an adsorbed, charged dispersant must provide an electrostatic barrier spanning this distance (i.e., 12–13 nm) to provide a greater repulsive barrier than water alone.

Freshly ground crystalline silica appears to have an even greater interparticle attractive force because of the enhanced ionization of surface silanol molecules. Wet-ground material has a lower ionization (acid strength) by at least half a pK_a unit than dry-ground. Thus dispersant-ground quartz behaves considerably different in paints, for example, than dry-ground and freshly formulated quartz.

As a consequence of the two factors, aqueous bridging and electrostatic repulsion, a popular method of inducing charge-operative dispersion involves employment of polyvalent anions. Several of the more economical are shown in Table 2.2. These function most efficiently where inherent (as opposed to induced) surface charge of the solid surface is positive or only slightly negative.

Oxide and hydroxide solids, which include ferric oxide, alumina, zinc oxide, kaolinite, talc, chrome oxide, and titanium dioxide, all are influenced by system

Table 2.2 Common Inorganic Polyanion Dispersants

Anion Molecule	Chemical Structure	Popular Name
Pyrophosphate	$(P_2O_7)^{4-}$	TSPP
Hexametaphosphate	$(PO_3)^{1-}$	Calgon, SHMP
Tetraborate	$(B_4O_7)^{2-}$	Borax
Silicate	$(SiO_3)^{2-}$	Water glass
Carbonate	$(CO_3)^{2-}$	Soda ash

pH through surface ionization reactions. However, contrary to common chemical expectations, these solid surfaces are rarely unilaterally charged entities, that is, neither wholly acidic nor wholly basic. More frequently they occupy an intermediate position.

The surface of oxide or hydroxide solids may ionize via several routes:

Surface Reaction	Surface Category
(a) $M\text{-O-H} + H^+ \longrightarrow M^+ + \text{H-O-H}$	hydroxylic
(b) $M\text{-O-H} + OH^- \longrightarrow M\text{-O}^- + \text{H-O-H}$	protic
(c) $M\text{-O}^- + H^+ \longleftarrow M\text{-O-H} \longrightarrow M^+ + OH^-$	amphiprotic

That pH value at which the reactions represented by the combined equations in the forward and reverse surface reaction (c) are equal in magnitude is the point termed *isoelectric pH.* A series of isoelectric values for common pigments[18–20] are shown shown in Table 2.3 to illustrate the broad range of surface diversity among solids.

Note the two crystalline forms of titanium dioxide, rutile and anatase, have significantly different isoelectric values. The values quoted in Table 2.3 for these materials are for pure solids. Commercial titania often has coatings of silica gel and/or alumina gel to alter its surface character. This coating, as shown by Parfitt,[21,41] significantly alters isoelectric values. For example, the now obsolete sulfuric acid process, which left both residual acid and gypsum (calcium sulfate) on the titania pigment surface, often showed an isoelectric value of 3.0–3.5. Measurements by the author on alumina gel-treated titania for alumina-rich, alumina–silica gels) gave isoelectric values for anatase of between 7.0 and 8.0, depending upon the method of precipitation and the alumina–silica ratio. Pure alumina gel precipitated on rutile titania also can yield a pigment with a near-neutral isoelectric point (7.0).

Adsorption of any anionic gel coating onto a solid surface, if sufficient in volume, will alter wholesale the chemical character of that solid. This technology has been employed to alter many pigments and enable their compatibility with other pigments whose surface character may be reactive, of different surface charge, or form bridge bonds with dissimilar pigments within the formulation.

Tin oxide, SnO, has been employed for specialty applications in coatings because of its unusual electronic characteristics. The isoelectric pH, which is

Table 2.3 Isoelectric Values for Various Oxide Pigments

Pigment	Chemical Formula	Isoelectric pH
Antimony pentoxide	Sb_2O_5	0.3
Fumed silica	SiO_2	1.8
Ground silica	SiO_2	2.2
Rutile	TiO_2	4.7
Kaolinite	$Al_2Si_2O_5(OH)_4$	4.8
Vanadium pentoxide	V_2O_5	5.0
Talc	$Mg_3Si_4O_{10}(OH)_2$	5.1
Hematite (alpha)	Fe_2O_3	5.2
Anatase	TiO_2	6.2
Alumina hydrate	$Al(OH)_3$	6.4
Maghemite (gamma)	Fe_2O_3	6.7
Chrome green	Cr_2O_3	7.0
Mica (sericite)	$KAl_3(Mg)Si_3O_{10}(OH)_2$	7.9
Zinc oxide	ZnO	9.0
Litharge	PbO	10.3

Values for any of these oxides may change if foreign atoms partially substitute for the primary atom in the structure.

highly dependent upon the method of precipitation, averages about 9.8. However, even partial oxidation of the surface to stannic oxide, SnO_2, alters this value greatly, dropping it into the range 2.5–3.0. This alteration, from alkaline surface to acid, affects greatly the effectiveness of a dispersant selected for stannous oxide in formulations.

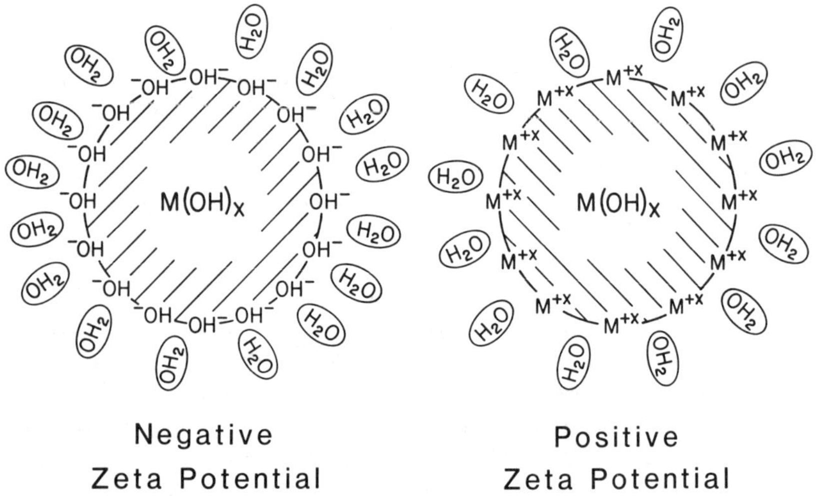

Figure 2.3. Water molecule orientations on charged surfaces.

Surface charge in sign and magnitude will have obvious effects upon molecules adsorbing upon that surface. Even water molecules have preferred orientations which influence dispersions, as shown schematically in Figure 2.3. Ionic attraction brings oppositely charged oxide particles together, such as kaolinite and anatase at pH 6.0, and dipole interaction between their adsorbed water "clouds" contributes additional attraction.

This fundamental solid surface charge clearly affects stability of adsorbed dispersant molecules having anionic charge. In the instance of polyphosphate dispersion of alumina or alumina hydrate, the system pH influences greatly adsorptivity of TSPP or SHMP. This in turn affects the equilibrium distribution of the dispersant between liquid phase and solid surface and the effective repulsive charge acquired by that surface.

This basic surface charge divided by the distance between the surface and the random solution-phase ionic atmosphere, or shear plane, is termed *zeta potential.* Two potentials, as sensed from a distance by a secondary particle, provide the electrostatic repulsion known as charge dispersion. Zeta potential becomes significant only at close proximity between particles[22] because of the inverse square force it creates, as defined by

$$\text{Zeta voltage} = \frac{Q}{S}$$

$$\text{Repulsive force} = k\frac{Q_1 Q_2}{S^2} \tag{2.7}$$

where Q_1 and Q_2 are the respective charges on two approaching particles and S is the separation distance between them. The term k represents the dielectric constant of the liquid medium (water = 78). That layer defined by the solid surface plane and one parallel and constructed through the centers of the adsorbed anions (or cations) is termed the *Helmholtz layer* (more accurately termed the "inner" Helmholtz layer).

However, it is not the adsorbed anions that repel one another upon close approach of two particles, but the counterions (cations associated with these anions on the dispersant molecule).

Liquid medium and dispersion (dielectric sensitivity) influence is illustrated in Figure 2.4 by a sedimentation particle size analysis of a clay, one dispersed in water ($K_e = 78$) with TSPP and the other two dispersed similarly but in water diluted slightly with acetone to provide K_e values of 61 and 50. Each will appear uniformly dispersed upon cursory visual inspection.[23] Particles below 0.5 μm, however, do not appear present in the acetone–water systems, and coarse particles appear slightly coarser. Because smaller particles have a higher surface-to-mass ratio, they are more susceptible to charge reduction than are coarser ones. Partial flocculation results in which fine particles adhere to coarser ones. This is confirmed by scanning electron micrographs for a variety of particulate systems.

Small reductions in the dielectric constant obviously have great effects upon

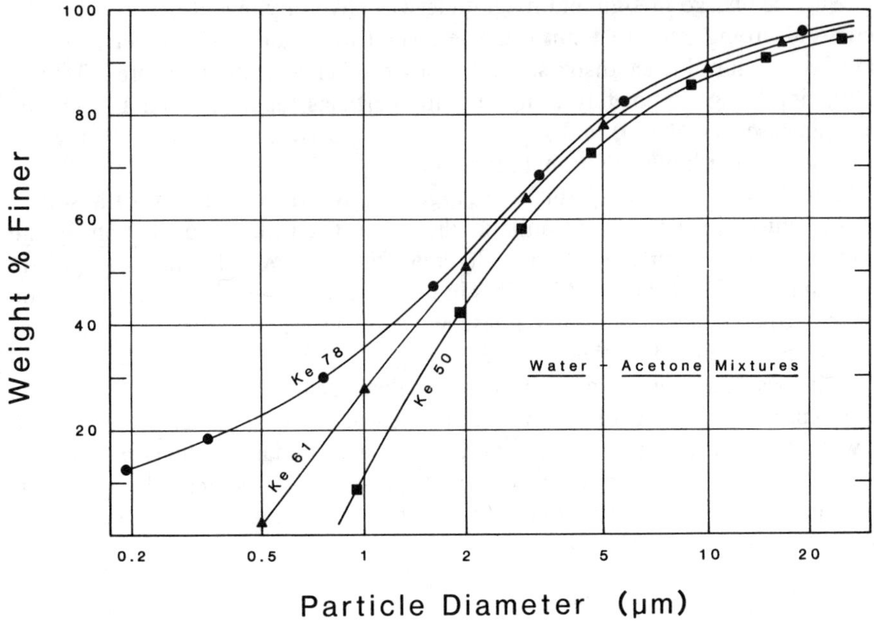

Figure 2.4. Sedimentation particle size in diverse dielectric media.

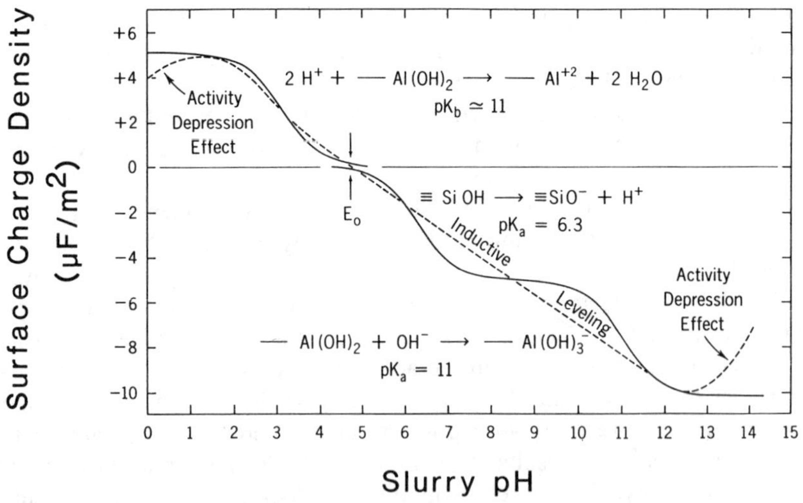

Figure 2.5. (a) Zeta potentials and reaction mechanics for kaolinite as a function of pH.

dispersion. Where K_e decreases greatly, such as with hydrocarbon liquids (K_e ~2), the charge effect, both attraction and repulsion, reduces dramatically and strong flocculation occurs through van der Waals interactions.

The zeta potential of dilute (0.1%) ultrafine silica dispersed in aqueous polyoxyethylene solutions gradually increases (more positive) as ethanol is added.[24] It is thought alcohol reduces dispersant adsorption on the silica that moves the counterion slippage plane toward the solution phase. Obvious limits exist for this phenomenon, however. As alcohol concentration increases, dielectric constant decreases, and charge repulsion falls off. Tests on a variety of fine silica dispersions suggest this does not occur until the alcohol concentration is at least 50%, however.

Where a surface can change polarity of ionization, as with metal oxides, the surface charge, as measured by zeta potential, will vary significantly with pH. This is shown by the complex curve for kaolinite in Figure 2.5(a). On the acidic side (pH 4.5 or below) the surface ionizes positively by way of the aluminum octahedron:

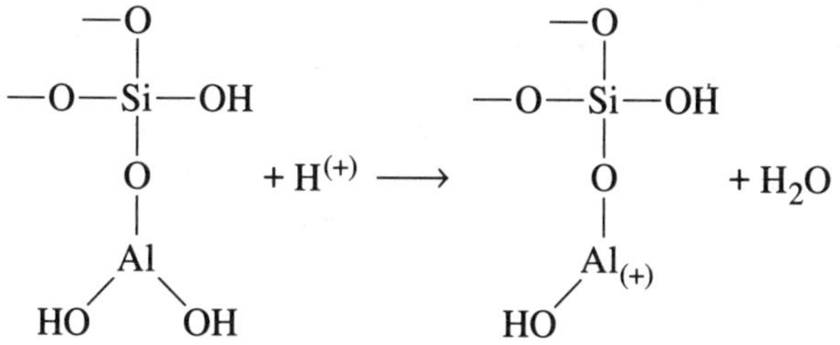

(2.8)

Population of the aluminum ion on the pigment surface increases with acidity, which in turn increases surface charge and zeta potential.

With alkalinity (pH >7) silicate surface potential becomes progressively more negative by the following reaction:

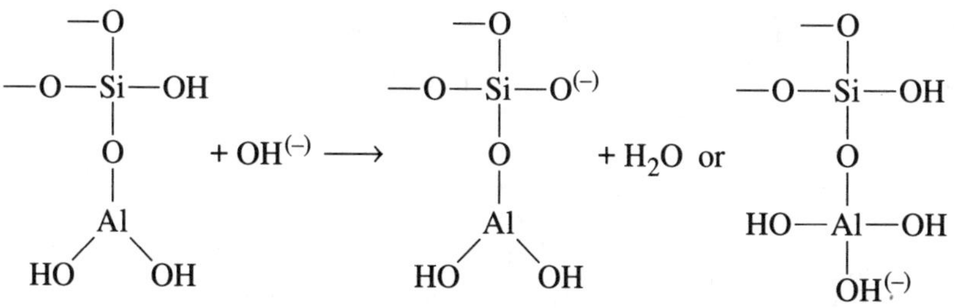

(2.9)

At very high values of pH, hydroxyl ion activity is no longer proportional to concentration, and the potential becomes slightly less negative. This effect is also present on the acidic–positive side of the zeta potential curve, but the acidity where it appears is great enough to dissolve partially the oxide particle. Both effects derive from chemical activity, not concentration.

Alunite, a hydroxylated alum mineral, with the chemical formula $KAl_3(SO_4)_2(OH)_6$, is negatively charged at pH levels above 2.5. Finely ground alunite flocculates readily in water, where it equilibrates to about pH 3. Addition of 2 millimoles per liter of sodium tripolyphosphate, however, enables slurries to be dispersed at all pH values from about 4 upward.[25]

Studies of dispersed states show a potential of about -30[26] to -35 mV (-7 μfaradays/m^2) is necessary to override van der Waals and solvent-bridging forces. Thus for kaolinite in water a pH of nearly 10 is required for a simple, dilute pH dispersion. This degree of alkalinity is generally much too high for commercial applications because of degradation of secondary system components. Eventually, it will degrade the solid (by partial silica solution).

Adsorption of a polyphosphate [Figure 2.5(b)] generates a surface potential of -30 mV at a pH of only 5.5 Note that orthophosphate also produces a negative charge, but significantly less than that of TSPP. The quantity of TSPP required to attain the critical charge on kaolinite at high solids is about 0.2–0.4 weight percent.

As noted previously concerning Table 2.1, crystal structure is a factor in establishing surface ionization. Rutile and anatase have identical chemical compositions but different crystal systems. Their zero-point potentials (isoelectric pH values) occur at significantly different pH values, 4.7 vs. 6.2. Insofar as the formulation pH (solvent, polymers, additives, etc.) is 6.3 or above, both pigments will behave as negative colloids. However, at pH 6.1 or below, in the absence of a charged dispersant, charge disparities between these two forms may exist. Similarly, alpha and gamma ferric oxides, also identical chemically, differ in their isoelectric points by 1.5 pH units.

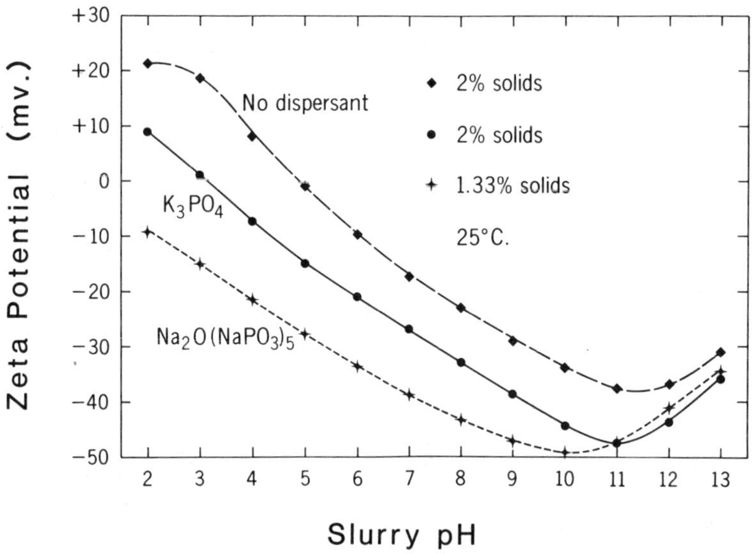

Figure 2.5. *(continued)* (b) Influence of phosphates on kaolinite's zeta potential.

Bulk solution counterion charges must exactly equal adsorbed anion charges, but not in the same locale. Counterions, being small and highly hydrated, tend to be swept off or diffuse away from the particle surface by energetic water molecules colliding with the dispersed particle. Thus the population of counterions around a given particle is always less than enough to balance out the charge. The net effect is shown in Figure 2.6. While zeta potential derives from a net negative charge, actual interparticle repulsion derives from their two cationic layer interactions, rather than the anionic. At great distances, as one particle diameter of separation (1 μm or 1000 nm), the net electrical field sensed by a neighboring particle is essentially zero.

Ionic repulsion between dispersed particles is an inverse second power force and becomes significant only upon close approach. Van der Waals forces, which attract particles due to mass alone, are inverse fourth power forces and do not become significant until particles are sufficiently close for hydration spheres to interact, (0.001–0.01 μm). As a consequence, the effective range required for dispersion by an ionic atmosphere about a particle, as calculated from high solids and high shear studies, is well below 1 μm and usually of the order of 0.02 μm (20 nm).

Measurements on fine silica[27] have confirmed this. These indicate attractive

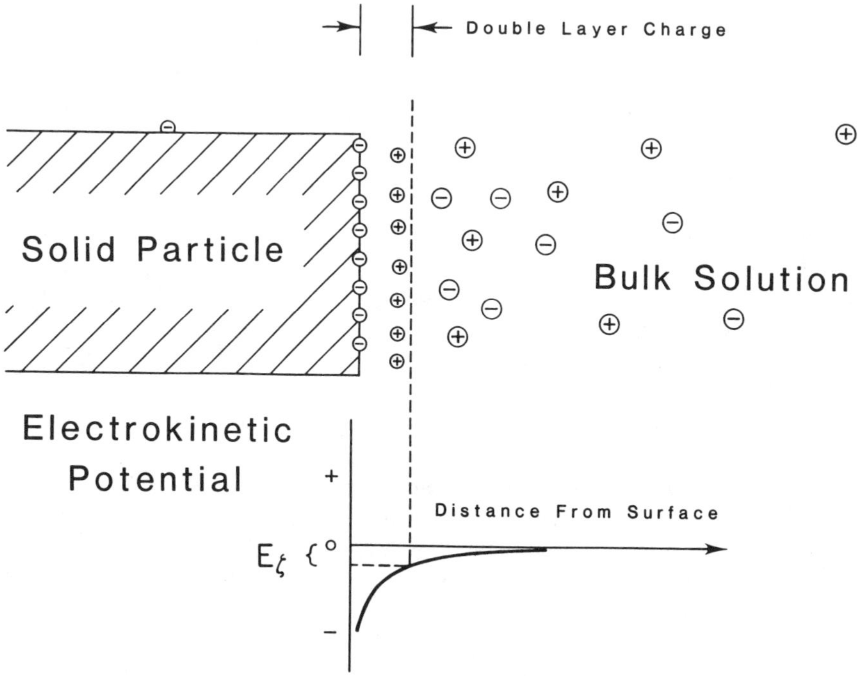

Figure 2.6. Ionic atmosphere about a dispersed particle.

forces become effective in the 20–60 nm range and are not overcome by dispersion forces (polyamines) until the silica particle's charged envelope thickness approaches 15 nm (app. 25–30 nm total separation). Similar studies have been made on mineral suspensions[28] that also show forces in this general range.

Surface ionization of the pigment, whether from pH or natural ionic dissociation, becomes an important factor in designing both single pigment–dispersant and mixed pigment–dispersant systems. Unfortunately, instances are rare in which a pigment surface can be sufficiently self-ionized to create a dispersant state. Most such surface groups, usually –*M*–O–H in structure (where *M* is the electropositive element), are ionized by only a few percent, providing insufficient charge to establish repulsive fields capable of overcoming van der Waals forces of attraction and floccule formation.

Because close-approach separation is of the order of several tens of liquid-phase molecular diameters, overcoming van der Waals forces (20 nm) through imposed charge is not always a necessary criterion for achieving dispersion. The creation of artificial water (or solvent) barriers around particles at dimensions exceeding approximately one-half of the van der Waals range can often produce a stabilized system.

Dispersants that function in the above manner fall into the category of dipolar or polypolar agents. Such agents are based either on hydrocarbon chains having relatively few carbon atoms with two terminal polar groups or on large, complex molecules with many dipoles (dominantly –OH or =O, secondarily $-NH_2$ or =NH) arranged along their structures. One or more portions of these molecules must adsorb upon particles with an attractive energy greater than that of water, $E_{ads} > 9.6$ kcal/mole. Other polar groups on the molecule are directed into the liquid phase. The outer (solution-directed) dipole group(s) hydrate and provide the barrier necessary to prevent close approach between particles. The molecular weight per active group with polar dispersants approximates those of ionic agents.

Because interdipole repulsion is significantly lower in magnitude than interion repulsion (inverse fourth-order force versus inverse square force), the number of molecules adsorbed on each particle surface must be significantly greater than for anionic molecules, or their physical extent must be much greater. Usually a compromise is reached between these two alternatives. Figure 2.7 represents a simplified structure for this type of dispersancy.

Because small molecules are involved (even though oxygenated) in anionic dispersants and large molecules in multiple-dipole dispersants, the latter will possess higher molecular weights. Coupled with the need to have greater numbers of the latter on a surface to effect a protective barrier, molecular weight increase dictates significantly greater weight requirements (solids basis) for this class of dispersants, usually by a factor of 2 to 3, often more, to achieve equivalent dispersion.

Polypolar agents may lie down fully upon a solid structure (train configuration), may be terminally coupled only (tail configuration), or may have any of a number of intermediate orientations (tortuous configuration). However, suffi-

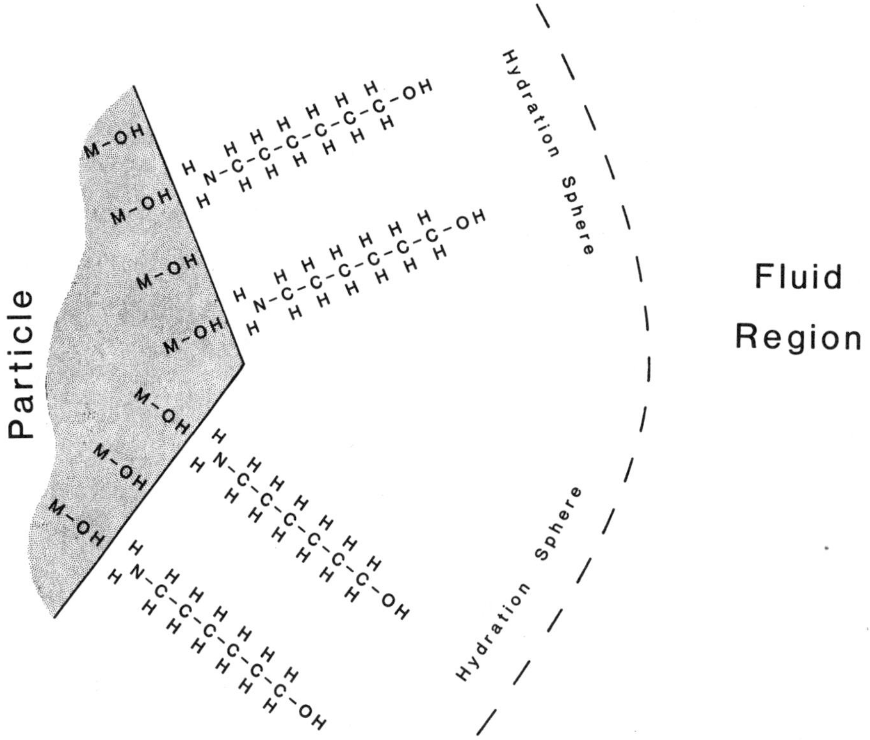

Figure 2.7. Dipole dispersed particle.

cient polar groups must be attached to surface polar sites, usually an emergent oxide or hydroxide (covalent solid) or coordinating atom (transition metal compounds), to form an enveloping shell about the particle.

Figure 2.8 contains Langmuir adsorption isotherms for a diamine on three kaolinites differing in surface area. Steepness (45°) in the initial profiles followed by flatness in the tops indicates a moderately strong and quantitative affixation of the dispersant molecules for the silicate surface. Table 2.4 gives the exchange capacity of the three kaolins and the calculated molecular adsorption of the diamine. Two of the samples (K-1 and K-2) have average aspect ratios of about 6:1, while that of the third is nearly 3:1, based on shadowed transmission electron microscopy work.

This method, adsorption of amines on acidic surfaces, coupled with BET nitrogen data, can provide information on the acidic site location or, as with kaolinite, the particle shape factor.[29]

Because of a closely competitive energy status with water, an equilibrium occurs in aqueous supension between adsorbed water and the polar agent for attachment sites. In some applications dispersant molecules or assemblages are

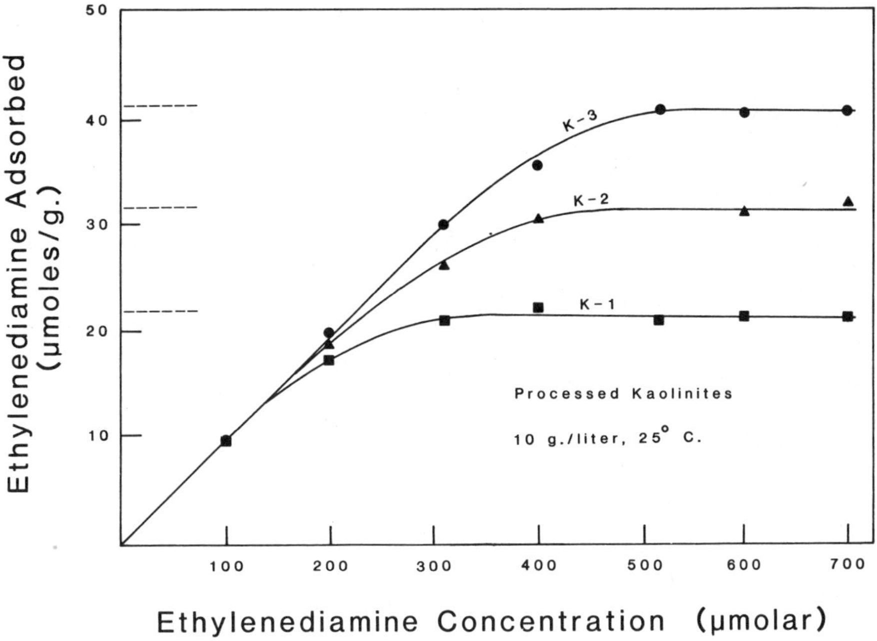

Figure 2.8. Adsorption isotherms of ethylene diamine on kaolinite.

sufficiently large to fully envelop a small particle. The state produced is termed a *protective colloid.*

In addition to the influence contributed by surface charge on the particulate solid, its structured configuration may be a factor in holding or stabilizing dispersant molecules. Many oxide solids are octahedrally coordinated (sp^3d^2 or d^2sp^3), either in a dioctahedral or trioctahedral state ($\frac{2}{3}$ or $\frac{3}{3}$ central atom occupied). The core within this octahedron is slightly too small for a central metal atom. Thus the adjacent oxygen–oxygen bond spacing in these octahedra expands to about 0.6–0.7 nm.

Table 2.4 Surface Properties of Adsorbate Kaolinite

Sample Designation	Cation Exchange Capacity[a] (μequiv/g)	Langmuir Monolayer Amine (μmole/g)
K-1	24	21.9
K-2	35	31.5
K-3	40	41.3

Determined colorimetrically by cobalt adsorption.

Polyphosphates, by contrast, are tetrahedrally coordinated (sp^3) with an oxygen–oxygen bond distance of about 0.52 nm. An almost identical spacing occurs with polysilicates. Thus two triangular tetrahedral faces of the dispersant molecule can fit neatly into a reentrant "cove" provided by two triangular faces of a solid's octahedral site. Because of the coordinated fit, this method of attachment of a dispersant to a solid surface has been termed *template adsorption* or *templature* and is a special case of selective van der Waals adsorption.

Further, a dispersant having double tetrahedra, as tetrasodium pyrophosphate (TSPP), exactly fits the double face presented by the reentrant locale, as illustrated by Figure 2.9(a) for a dioctahedral base (micas, clays, etc.). Longer chains of tetrahedra, 5 to 6 for sodium hexametaphosphate (SHMP) and about 4 for sodium silicate, must contort themselves around the crenelated oxide base structure to adapt to the surface. A sketch of such an arrangement is shown as Figure 2.9(b).

This sorbate-on-surface mechanism has been likened to a molecule in a potential box,[30] similar to the classical Schrödinger electron particle in an energy box. The dispersant can be considered as locking into a geometric lattice. Because of atomic proximities, the charge-compensating cations (usually hydrated) may be neglected with regard to the potential well on the lattice box.

The dispersant molecule is not usually held tightly to the surface if it possesses a charge. Langmuir adsorption studies of various phosphates upon an aluminum silicate pigment show a distinct equilibrium between surface and liquid phase (see Chapter 3). Initially all phosphates, including the ortho form, trisodium phosphate, are essentially quantitatively adsorbed. However, as the solution concentration of the dispersant increases, progressively more of the dispersant remains in the solution phase. At the level considered for optimum viscosity, approximately 0.25–0.3% solids basis, only about 30% of the dispersant species

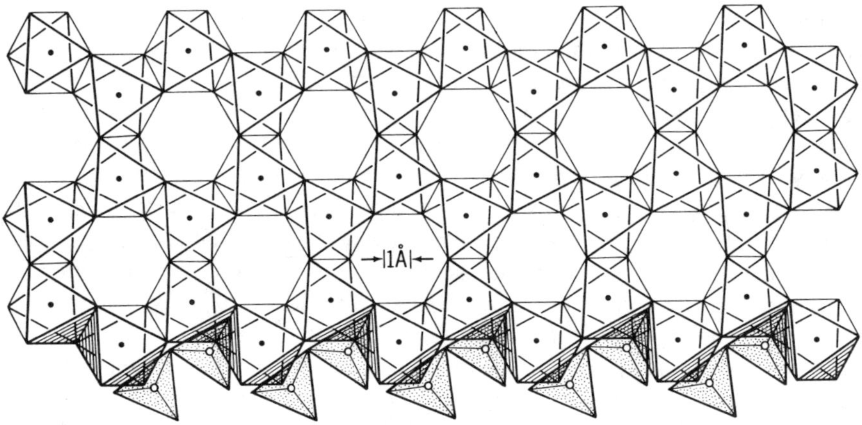

Figure 2.9. (a) Octahedral crystal with binary tetrahedra dispersant.

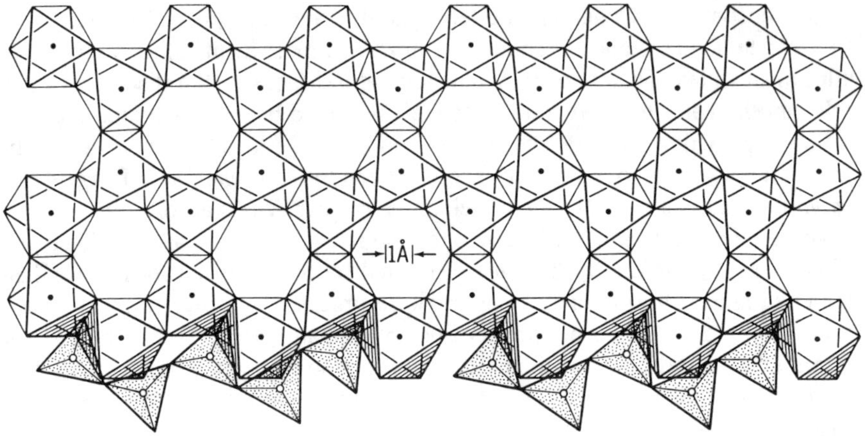

Figure 2.9. *(continued)* (b) Octahedral crystal with multiple tetrahedra dispersant.

remains attached to the solid surface. It is this large equilibrium shift to the solution phase that is responsible for high viscosity at high dispersant dosages.

Chain dispersants with multiple attachment groups may reorganize upon a surface with time, the result of entropy, enthalpy, and thermal diffusion. Studies on various pigments suggest that tail-to-train transitions are common for complex chain dispersants, but not with short-chain, terminal attachment species, as diamines. Such non-Langmuirian behavior is atypical of adsorption processes and can best be described in terms of structural reorganization.

The adsorption concentration where a surface configuration transition occurs will appear at slightly higher concentrations as the dispersant molecule increases in length. This reflects the influence of van der Waals forces on the adsorption process.

In the train configuration, anionic dispersant molecules represent good, but imperfect, fits between tetrahedra and octahedra. This element of misfit increases as chain length increases, and accumulates along the surface. Thus an optimum chain length should exist for surface coverage beyond which the molecule simply folds itself out as a partial tail into the solution phase. This can be shown by employment of a specialty glassy polyphosphate product made for water softening. The glassy material has a chain length in the 15–25 phosphate unit range. Its effectiveness as a viscosity reductant, however, is considerably poorer than shorter chains, as illustrated in the viscosity study in Table 2.5. The solubility of these hyperchain agents decreases dramatically, which also limits their use.

Peak values for polyanionic dispersant adsorption does not represent the lowest viscosities for the system nor the most discretely dispersed particulate assemblages. It represents, in fact, only about one-fifth to one-third the amount of agent necessary for optimum dispersion. A full monolayer is not obtained until much higher concentrations of dispersant are employed.

Table 2.5 Polyphosphate Chain Length Effect on Viscosity

Dispersant Type	Chemical Formula	Level Employed[a] (mmoles/g)	(Wt. %)	Brookfield Viscosity (10 RPM, cp)[b]
TSPP	$Na_4P_2O_7$	22.6	0.300	280
STPP	$Na_5P_3O_{10}$	22.9	0.281	260
SHMP	$(NaPO_3)_{5\text{–}6}$	22.7	0.260	320
Glassy	$(NaPO_3)_{15}$	22.3	0.237	930

[a] Based on number of active phosphate tetrahedra, anhydrous salt, and 70% solids.
[b] Selected because of sensitivity to floccule formation.

Optimum phosphate concentration, that is, the amount required for a viscosity minimum, lies between 0.25% and 0.3% for the first three species in the table.

On a kaolinite substrate dispersant chains must lie in parallel configurations, normal to the C crystalline axis, due to the laminar structure of the aluminum silicate. Because of this restriction, the entire alumina octahedral network likely can never attain full monolayer coverage. Calculations based on surface area and morphology of the solid suggest that monolayer coverage is not attained until well beyond 10 micromoles per gram. The adsorption profile does demonstrate that a tail configuration has a much lowered effectiveness in dispersant activity.

A moment of reflection upon tail geometry suggests that the extension of both anion backbone and cation counterions into the liquid phase produces no net apparent charge (zero additional zeta potential) because an equal number of cations and anions are introduced into the double layer around the particle. This orientation may in fact reduce net charge by allowing tails to intermingle, cation to anion, so to speak. At best a near-neutral field develops around such particles.

A second observation from the Langmuir study lies in the absorption constant C from Langmuir's equation:

$$C = e^{-k[\Delta H(\text{abs}) - \Delta H(\text{hyd})]} \tag{2.10}$$

which for polyphosphates on octahedral pigments lies in the range of 5–10. These values suggest a moderately low adsorption pressure for polyphosphate molecules exists on the aluminum silicate edge surface, even during the initial adsorption process, that is, below 0.1 mmole concentration. At optimum dispersion of around 0.3 mmolar phosphate, this equilibrium is yet farther shifted to the solution phase. These studies readily explain the phenomenon termed *chemical viscosity*.

Because of the low energy of adsorption and poor localization of inorganic anionic agents, any solution activity that might affect their chemical composition or structure will result in adverse effects upon viscosity of the system.

Most oxide pigments display a skewed catenary rheological profile as a function of pH. Two typifical profiles, for rutile titanium dioxide and kaolinite, are shown in Figure 2.10. Values for pH at viscosity minima are neither those for

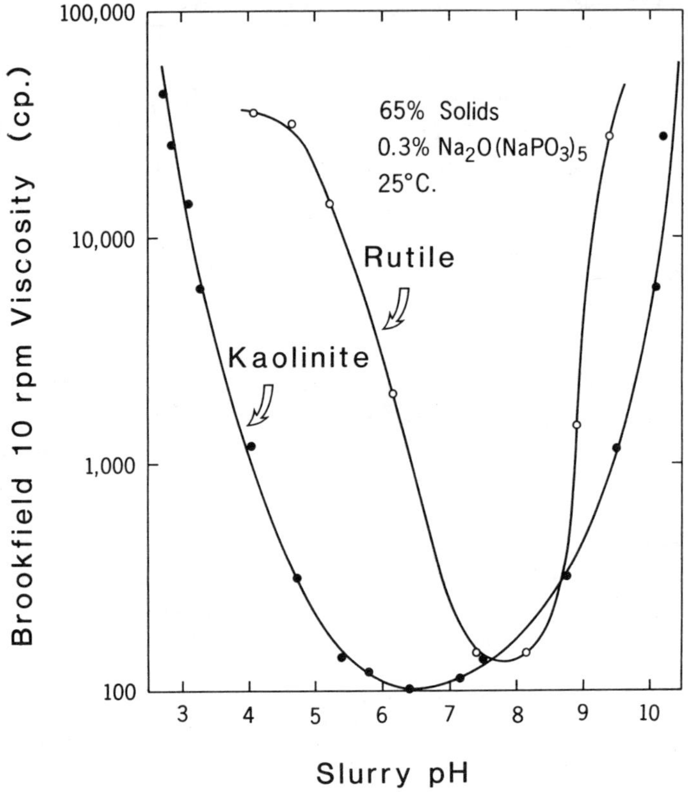

Figure 2.10. Effect of acidity on rheology of rutile TiO_2 and aluminum silicate.

the pigment in water (i.e., without dispersing agent) or for the dispersant in solution without pigment. Rather, they represent an interactive value for the dispersant adsorbed upon the active pigment surface.

The solid may reconfigure the dispersant and actually change its ionization:

$$Na_4P_2O_7^{4-} + H^+ \xrightarrow{\text{surface}} Na_3^+\ H\ P_2O_7^{3-} + Na^+ \tag{2.11}$$

With strong acid addition, this equation will shift sufficiently to the right that the dispersant becomes singly charged. The ejected cation(s) in the equation no longer occupies the counterion region but diffuses into the bulk liquid phase to immobilize additional liquid. Viscosity will rise dramatically during this acidification process as a consequence.

As progressive increments of alkali are added (the curve's left leg) to the point a minimum in viscosity has been established, alumina octahedra ionize negatively (see Figure 2.5 for kaolinite), electrostatically desorbing polyphosphate chains. Zeta potential for the solid is shifted to a greater negative value, which will eventually override the van der Waals adsorption energy and desorb the dispersant. While the limited negative octahedral change in itself may produce minor dispersancy, the particulate surface actually drops in charge density and in the ability to repel adequately other similar particles. Left on its own, the expelled dispersant transfers to solution and solvates the water phase. Both processes elevate viscosity. While the acid and alkaline processes have certain similarities, they are not identical, and the rheological profile always assumes a nonsymmetrical form, usually axially skewed towards alkalinity.

Optimum stability of a dispersed state, therefore, is a direct function of the fundamental surface chemistry of the solid onto which dispersant molecules must attach. As temperature elevates, equilibrium between particle surface and solution for the dispersant molecule shifts to the solution phase because of increased kinetic energy of solution molecules. Thus viscosity should actually rise from loss of surface-charge retention. However, because water viscosity diminishes with temperature, the charge reduction effect, depending upon its magnitude, may be masked.

In summation, with equilibrating charged dispersants, typified by polyphosphates and polyacrylates, the quantity of dispersant actually adsorbed does not increase appreciably with dispersant concentration in solution phase beyond a certain level, usually less than half that required to provide optimum viscosity.

2.2.2 Polar Dispersants

Were the adsorption energy by the substrate for the dispersant greater, the shift of dispersant to the liquid phase would not control the rheological impairment by overdosage. Such is the case with the families of dispersants that includes alkanolamines (also known as amino-alcohols) and polyamines. These agents are best described as dipole–dipole dispersants. They constitute a major family of nonionic materials for dispersing pigments and other powders. Their activity is best understood by considering the simplest members of the family.

Adsorption energy for the amino group onto an acidic silica group, or silanol structure, is of the order of 12–15 kcal/mole, as determined by isothermal degassing studies.[31] At this energy of attachment it becomes inherently stable as regards Langmuir adsorption and solution–surface equilibrium. Both diamines and alkanolamines demonstrate true Langmuirian profiles, as will be discussed in greater detail later. The difference in amine dispersancy compared to phosphates, as exemplified by aluminum silicates, derives from polyphosphate adsorbing at alumina octahedra, while amines adsorb along edge silica tetrahedra. The computed monolayer, from Langmuir data, corresponds moderately closely to the population of silanol groups on the silicate surface.

Hydroxyl groups on the dispersant molecule's end have essentially the same adsorption energy for the solid surface as water and cannot compete effectively with the amino group for adsorption. Thus all tails of an alkanolamine possess the same configuration when adsorbed, amine-to-solid, hydroxyl-to-solution. Diamines, by contrast, possess equal end probabilities for solid attachment.

Neither diamines nor alkanolamines impress a charge upon the solid pigment surface. Rather, they produce a buffer zone of terminally hydrated polar groups (uncharged). This appears at first glance little different from two "naked" but water-enshrouded particles coming into contact and forming floccules. However, if the dispersant chain is of sufficient length, the restricted approach between particles will be such that solution kinetic energy (which provides Brownian forces) will overcome weakened van der Waals forces. This has been shown previously to occur in the general range of 15 nm of molecular "thickness."

This distance would suggest a molecular length of at least 10 atoms plus hydration spheres, based on floccule formation with hydrated surfaces and the range of van der Waals flocculation forces. However, in practice a distance of this magnitude is not required. As the chain length of the dispersant increases, viscosity decreases and fluid movement between particles increases. Also the fluidity of the system becomes more Newtonian.

At 2.5 nm chain length, viscosities are attained which approach those of charge-functional dispersants. Because of the molecular weight increase of the agent and the need to cover the substrate fully, the mass dosage of alkanolamines is substantially greater than for inorganic anionic dispersants.

A broad survey of inorganic anionic dispersants and dual dipole dispersants is provided by the family of curves shown in Figure 2.11 for comparative purposes. Three important factors can be derived from studies with polar dispersants:

1. Dipole–dipole agents produce slightly higher viscosity minima than polyanionic agents;
2. The weight dosage of dipole–dipole dispersants is greater than with polyanionic agents; and
3. Overdosage by dipole–dipole dispersants causes only minimal increase in viscosity.

While the amino group has strong bonding potential for surface hydroxyl groups (those naturally forming on oxide pigments), alkaline oxides and salt pigments (e.g., calcium carbonate, barium sulfate, strontium chromate, etc.) have very low absorption energies for amines. As a consequence, little viscosity improvement occurs with these substances when diamines or alkanolamines are employed as dispersants.

Those oxide pigments with surfaces having greater than 50% acidic ionization, as exemplified by isoelectric values at pH 5 or below, respond well to alkanolamines as dispersants. Thus rutile titanium dioxide, for example, disperses satisfactorily while anatase does not.

Employing amino groups at both ends of a dispersant chain will produce a stronger terminally hydrated zone around the particle. However, when the chain length exceeds a certain value, logic would suggest that a chelated ("clawlike") structure might ensue that would destroy the hydrate barrier and make the particle considerably less hydrophilic and partially hydrophobic. Viscosity profiles

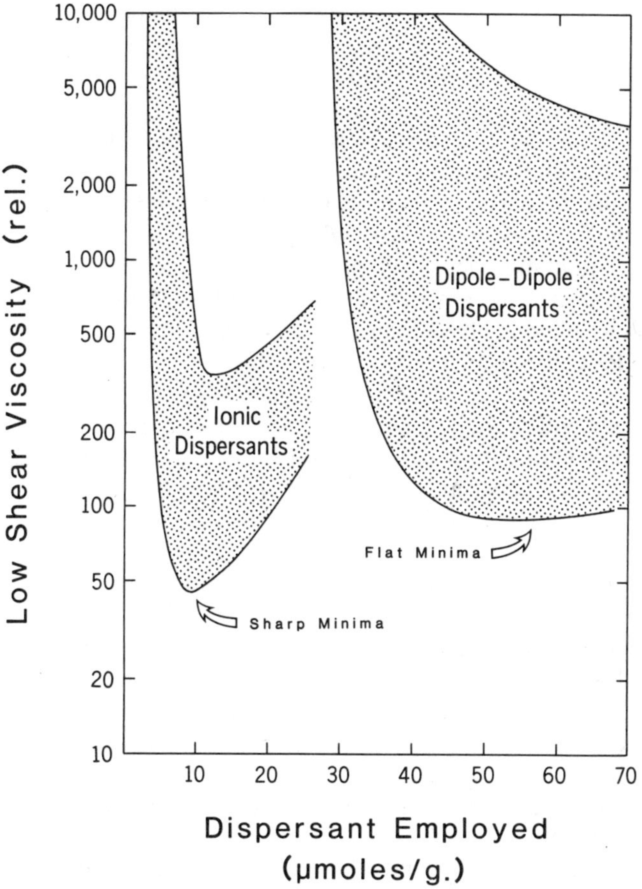

Figure 2.11. Comparative rheological profiles of dipolar vs. charge dispersants.

for most polyamines show precisely this chain length (or segment length) criticality for dispersancy.

For several years diamines as dispersants were employed for specialty pigment applications, primarily those involved in reacting pigmentary and filler silicates with aqueous emulsions of polyester and epoxy polymers. Excess polyamine, especially with lower molecular weight members, would be fugitive during drying processes. However, due to the toxicity of polyamines and the incompleteness of these reactions, their use has been greatly restricted.

Use of diols in place of diamines generates much the same characteristic, producing mildly flocculated structures with 4 or more carbon atoms in the hydrocarbon spine. In addition, the reduced adsorption energy of the alcohol group upon oxide surfaces, compared to diamines, causes a substantial shift of the diols into the solution phase. During any film-formation application (paints, inks, and coatings) diols slowly migrate to the surface. Thereafter they progressively evaporate or wash away through the agencies of surface moisture, cleaning, rain, etc.

While polyanionic species also migrate to the surface under humid conditions, they do not evaporate. Rather, these tie up surface moisture and form white deposits on the surface, a phenomenon termed *blooming*. Blooming is most noticeable on deep tone colors, particularly with external house paints in high-humidity and -rainfall regions. The phenomenon also occurs in molded plastics where filler pigments were inadequately washed free of dispersing agents. Thus amino alcohols and diols have long-term advantages in product quality through evaporative loss characteristics.

With the exception of ethylene glycol (ethanediol), most of these compounds, both amino alcohols and diols, are relatively nontoxic, finding their way into a variety of applications from paint to cosmetics. Chapter 5 describes specialty applications of diol dispersants in commercial products.

2.3 Interrelation of Viscosity and Dispersion

Viscosity is a mechanical characteristic of a slurry only partially related to dispersion. Ability to flow is a consequence of collective mobilities of a particulate assemblage in a liquid medium. Where all of the particles are of the same size (termed ''monodisperse''), about 55% of the total volume will be taken up by the solid phase at maximum packing. Thus about 45% of available space in the liquid–solid system is unused. Finer particles can be introduced into these interstices and elevate the solids. However, at equivalent solids the mixed-size particulate system will have more fluid space than the monodispersed one.

Mathematically it can be shown that the size–volume relationship maximizes when the particle size distribution fits a Gaussian distribution with a breadth at 1 sigma such that it falls on a 45° line[32,33] on log-size, probability function weight graph. Such particle size distributions are log normal in distribution and tend towards Newtonian behavior.

Grinding tends to produce log-normal distributions,[34] based on the theory of random distribution of energy imparted into solid particles. However, in processed materials (classification operations, coarse-size cuts, screening, desliming cuts, etc.) the size distribution may no longer be truly log normal.

An increase (or decrease) in one population within the size distribution, whether at the coarse, middle, or fine end, will affect the collective assemblage's ability to flow freely. Typical size skews having a common median size (50% by weight) and their projected rheological effects are presented in Figure 2.12 for comparison.

Increased fines in the distribution increase (1) the surface area, (2) the dispersant demand, and (3) the quantity of water or solvent from the fluid phase adsorbed upon the particles. The net effect is to reduce fluid mobility of the

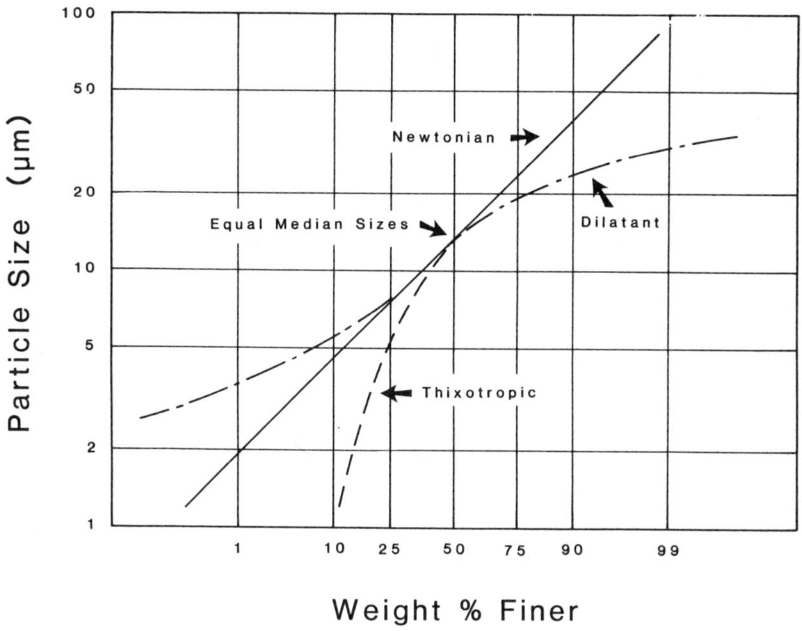

Figure 2.12. Rheological effects of size skews from log-normal particle size distributions.

system, especially under low-shear conditions. The result is most often thixotropy.

Increased coarse fractions, as has been described under the discussions of monosize distributions, impede mechanical fluidity, independent of dispersant level.

Increased levels of midsize components, both fines poor and coarse poor, occur more rarely. However, when produced, their rheological characteristics tend toward elevated viscosity at both moderate and high levels of shear, combining the worst characteristics of dilatancy for applications. Such an effect is observed with reflective paints where one pigment is fine, monosize glass beads. These are extremely difficult to spray from a thixotropic formulation (to prevent bead settling) onto road signs and highway lines, even with large-nozzle guns, often being brush applied because of the dilatancy.

In a detailed examination of viscosity and dispersion, the Kovar relationship has been developed to explain the complex interactions of particle assemblages in a two-phase system.[35] In this derivation the mathematical relationship between the observed system viscosity (which by measurement disturbs the system) and the concentration of discretely dispersed particles within the homogeneous liquid phase remains constant throughout the entire two-phase system. This is predicated on the solids concentration being expressed on a volume fraction basis (not weight fraction). It also assumes particles within the system interact with each other and the liquid phase solely in a hydrodynamic mode. For polypolar dispersed systems this is essentially true and for charged dispersants true except at very high solids concentrations.

The Kovar relationship holds true for monodispersed systems where congruous particles, that is, a collection that possesses the same shape but has differing size (as with phylloidal clays, acicular iron oxides, asperic quartz, etc.) will pack similarly in a highly loaded slurry and will have similar fluid transit. In the transformation from floccules to dispersion (or overdispersion), agglomerates form constant size and number distributions of subagglomerates, eventually diminishing to triplets, doublets, and finally to singlets. Each of these smaller species, prior to total dispersion, has a finite lifetime in the system, independent of shear or dispersant concentration. Similarly, a fixed mobility exists for single particles within these floccules. The process is not precisely reversible thermodynamically (a phenomenon termed *hysteresis*), viscosity usually being higher on the ''return trip'' to low shear and flocculation because size populations and rate processes are dissimilar.

The phenomena have been examined also in mixed dispersions, where intersurface effects become highly important.[36] Heterocoagulation is not only affected by these same phenomena, but by size ratio between the mixed pigments as well as their relative concentration. In such systems, with three independent variables, it is not difficult to understand the frustration of formulators attempting to produce stable dispersions for commercial applications.

The properties of these systems may be directly scaled by the Kovar principles.[37] For heterodisperse systems where particles have the same qualitative distribution of shapes and sizes, each particle in one element of the dispersed system can be scaled relatively to a corresponding particle in another element of the system by a factor that is constant for that material and liquid.

Using this relationship, viscosities can be predicted based on a knowledge of sizes and shapes of the particulate body. It also can be employed to explain the change from thixotropy or dilatancy to Newtonian character as agglomerates break down and become discrete particles during the dispersive process.

The Kovar principle has been demonstrated by observing flocculation kinetics and dispersion of monodisperse particles, including titanium dioxide, polyvinyl chloride emulsions, and precipitated calcium carbonate rhombs, all of which are steady-state transitions.

Thus far dispersion has been assumed where sufficient electrostatic or isolation barriers are established around particles. In a few specialized cases, this cannot be accomplished. Most illustrative is that of acicular magnetic pigments, both magnetite and maghemite. Because dispersive forces originate only at surfaces, magnetic behavior, which derives from the entire particle bulk and extends farther into the liquid phase than charge effects, usually overrides the dispersant. Even with the best mechanical and chemical efforts, particles will disperse then rapidly flocculate, often in ring or network structures, as illustrated by the specimen in Figure 2.13.

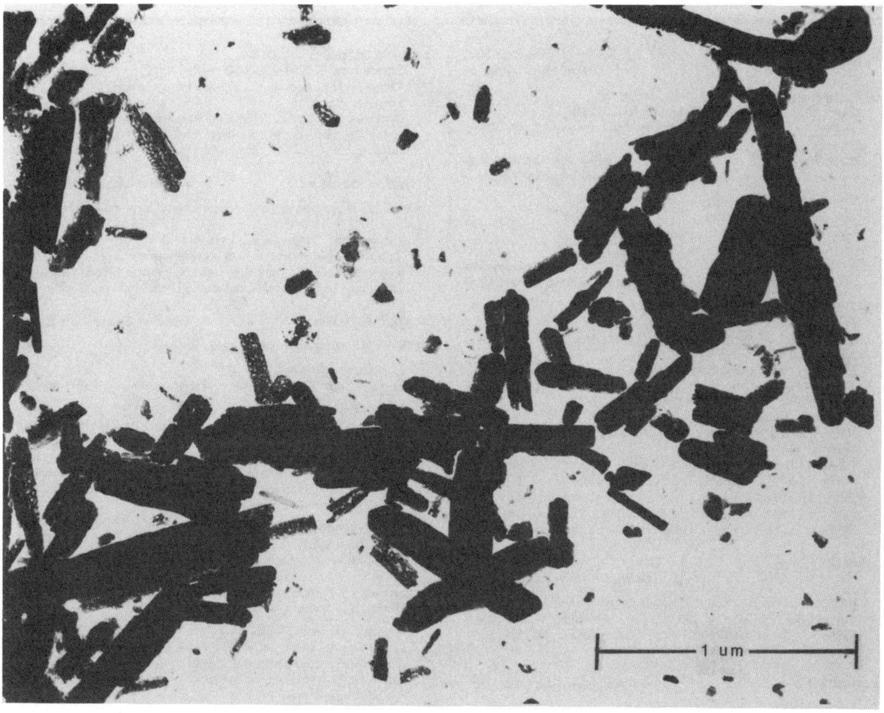

Figure 2.13. Magnetic flocculation overriding electrostatic dispersion.

2.4 The Hydrophilic–Lipophilic Balance

In dispersing organic materials in polar liquids and polar solids in nonpolar or reduced polarity liquids, a convenient effectiveness test has been developed termed the *hydrophilic–lipophilic balance* (water loving/oil loving) or HLB ratio. Although the concept undoubtedly was in use on a qualitative basis for many years, it was first published in 1949 by Griffin[38] to assist formulators of cosmetic lotions with emulsions and skin agents. Thereafter it was broadened and promoted by Atlas Industries,[39] a manufacturer of surfactants and chemical additives.

HLB ratios for many agents have been calculated by assessing the individual components of a dispersant molecule's structure, assigning them each a value and summing these. Component values range from about +40 (for a potassium alkyl sulfate grouping) down to about −0.5 for agents with major paraffinic hydrocarbon groupings. A polyhydric alcohol grouping, for example, is nearly zero (+0.5). However, exclusive chemicals, as heptane (wholly oleophilic) or trisodium phosphate (wholly hydrophilic), are not assigned HLB values because they are unable to function across polarity barriers.

Rycofax® 618

Rycofax® 618 — **Reilly-Whiteman Inc. Philadelphia Division**

Chemical Description: Fatty amido-amine salt
Form: liquid **Conc.%:** 98 **Type:** cationic
Remarks: Softener/debonder. Release agent for tissue manufacture.

Rycomid 2120 — **Reilly-Whiteman, Inc. Philadelphia Division**

Chemical Description: Fatty ester
Form: solid **Conc.%:** 100 **Type:** nonionic
Remarks: Foam stabilizer for detergent formulations

S-Maz® — **PPG Industries Specialty Chemical Bus. Unit**

S-Maz® 20
Chemical Description: Sorbitan monolaurate
Form: liquid **Conc.%:** 100 **Type:** nonionic **HLB:** 8.6
S-Maz® 65 K
Chemical Description: Sorbitan tristearate
Form: solid **Conc.%:** 100 **Type:** nonionic **HLB:** 2.1
S-Maz® 80
Chemical Description: Sorbitan monooleate
Form: liquid **Conc.%:** 100 **Type:** nonionic **HLB:** 4.3
S-Maz® 85
Chemical Description: Sorbitan trioleate
Form: liquid **Conc.%:** 100 **Type:** nonionic **HLB:** 1.8
Remarks: Polyol surfactants-fatty esters of sorbitol, lipophilic in character, dispersible in water and soluble in oils and solvents. Solubilizing emulsifying and dispersing properties.

S-Maz®-60K — **PPG Industries Specialty Chemicals Bus. Unit**

Chemical Description: Sorbitan monostearate, kosher
Form: flake **Conc.%:** 100 **Type:** nonionic **HLB:** 4.7
Remarks: For vegetable and dairy products, cakes, gloss aid in chocolate, non-dairy creamers.

Sanac — **Karishamns USA, Inc.**

Sanac C
Chemical Description: Complex carboxylated coco quaternary
Form: liquid **Conc.%:** 60 **Type:** amphoteric
Sanac S
Chemical Description: Complex carboxylated soya quaternary
Form: liquid **Conc.%:** 60 **Type:** amphoteric
Sanac T
Chemical Description: Complex carboxylated tallow quaternary
Form: liquid **Conc.%:** 60 **Type:** amphoteric
Remarks: Detergents, wetting and foaming agents, solubilizers of both organics and inorganics. Biodegradable.

Sandobet SC — **Sandoz Chemicals Corp.**

Chemical Description: Cocoamido propyl hydroxy sultaine
Form: liquid **Conc.%:** 50 **Type:** amphoteric
Remarks: For cosmetic and toiletries applications. High foaming, mild shampoo active with good viscosity properties, acid and alkali stable.

Sandopan B — **Sandoz Chemicals Corp.**

Chemical Description: Ethoxylated anionic complex
Form: liquid **Conc.%:** 40 **Type:** anionic **HLB:** 6.0
Remarks: Nonfoaming industrial cleaning ative, hydrotrope.

Sandopan DTC — **Sandoz Chemicals Corp.**

Sandopan DTC Acid
Chemical Description: Ethoxylated anionic complex
Form: liquid **Conc.%:** 90 **HLB:** 13.0
Remarks: Free acid form of DTC, oil and solvent soluble. Surfactant for industrial and personal care, conditioning shampoos, liquid soaps, household and industrial cleaners.

Sandopan DTC Linear Gel
Chemical Description: Ethoxylated anionic complex, Na salt
Form: gel **Conc.%:** 70 **Type:** anionic **HLB:** 16.0
Remarks: Alkaline and high temperature stable surfactant for industrial, personal care and household product use.

(cont.)

Sandopan DTC (cont)
Sandopan DTC-100, 100 MM
Chemical Description: Ethoxylated anionic complex
Form: liquid **HLB:** 17.0
Remarks: Surfactants for liquid detergents, all-purpose cleaners, germicidal cleaners; oil solubilizer for aqueous systems, acid bowl cleaners.

Sandopan KST — **Sandoz Chemicals Corp.**

Chemical Description: Cetyl alcohol ethoxylate anionic complex, Na salt
Form: solid **Type:** anionic **HLB:** 16.0
Remarks: Mild surfactant with good lime-soap dispersing properties for use in bar soaps, antiperspirants and other personal care products; inhibits sodium stearate crystal formation.

Sandoxylate — **Sandoz Chemicals Corp.**

Chemical Description: Alkoxylated alcohols

Sandoxylate SX 408
Form: liquid **Conc.%:** 100 **Type:** nonionic **HLB:** 10.0-12.0

Sandoxylate SX 412
Form: liquid **Conc.%:** 100 **Type:** nonionic **HLB:** 12.0-14.0
Remarks: Modified specialty nonionic wetting agents and dispersants for use in individual cleaning/treatment products and processes and personal care products. Low foaming hard surface cleaners. Stable in alkaline and acid media. Fragrance and oil solubilizers, low skin irritancy.

Sandoxylate SX 424
Form: semi-solid **Conc.%:** 100 **Type:** nonionic **HLB:** 16.0
Remarks: Modified specialty nonionic wetting agent, dispersant for use in individual cleaning/treatment products and processes and personal care products. Low foaming hard surface cleaner. Stable in alkaline and acid media.

Sandoxylate SX 602
Form: liquid **Conc.%:** 100 **Type:** nonionic **HLB:** 16.0
Remarks: Fragrance solubilizer, wetting agent, low/non-foaming surfactant for household and industrial products, for liquid laundry and hard surface cleaners.

Sandoxylate AD-6 — **Sandoz Chemicals Corp.**

Chemical Description: Alcohol ethoxylates
Form: liquid **Conc.%:** 100 **Type:** nonionic
Remarks: Modified specialty nonionic wetting agent, dispersant for use in individual cleaning/treatment products and processes. Non-foaming, stable in alkaline and acid media.

Santone — **Van den Bergh Food Ingredients**

Chemical Description: Polyglycerol esters of fatty acids

Santone 3-1-S XTR
Form: bead **Conc.%:** 100 **Type:** nonionic **HLB:** 7.2
Remarks: Used as a whipping agent in non-aqueous dairy, bakery and confectionery lipid systems.

Santone 3-1-SH
Form: plastic **Conc.%:** 100 **Type:** nonionic **HLB:** 7.2
Remarks: Replacement for polysorbates in icings and icing shortenings. Aerting agent in cakes and icings

Santone 8-1-O
Form: liquid **Conc.%:** 100 **Type:** nonionic **HLB:** 13.0
Remarks: Viscosity reducer in high protein systems, emulsion stabilizer, beverage clouding agent and replacement for polysorbates.

Santone 10-10-O
Form: liquid **Conc.%:** 100 **Type:** nonionic **HLB:** 2.0
Remarks: Emulsion stabilizer, dispersant, anti-foaming agent. Useful in frozen applications.

Schercamox C-AA — **Scher Chemicals, Inc.**

Chemical Description: Coco amido propyl dimethyl amine oxide
CAS: 68155-09-9
Form: liquid **Conc.%:** 30 **Type:** nonionic
Remarks: Detergent, wetting agent in household and institutional cleaners. Conditioner and viscosity builder in shampoos.

170 1994 / Emulsifiers & Detergents / North American

Figure 2.14. Representative page from McCutcheon's dispersant guide (Ref. 40).

The HLB value is in actuality a measure of the affinity a surfactant retains for the aqueous phase. It cannot be measured by physical experiment, but is derived mathematically from its known chemical structure. The more positive the HLB value, the more the agent will wet in water, while the more negative it is, the greater its probability of wetting in an oleophilic phase. The midpoint (equal proclivities) is not at the arithmetic average but in the vicinity of +6 to +8. As an example, sodium stearate, a common soap component, has a calculated HLB value of $17(-0.48) + 19 = +10.8$, giving it a hydrophilic edge.

Anionic or nonionic dispersants with HLB values around +8 have proven very effective in dispersing carbon black in polyoxyethylene nonylphenyl ether–water systems for cosmetic pigmentation (eyeshadow). A typical agent possessing this HLB characteristic is oleic acid.

HLB values provide a general guide to molecular affinities and as such serve a useful purpose in designing or comparing dispersants. They do not take into consideration molecular geometries, and as such may provide misleading information. For example, locating two hydroxyl groups in proximity near a long hydrocarbon tail end produces an agent with decidedly different surface chemical properties than one where each hydroxyl occupies a terminal position. HLB values are used less today, as the science of dispersion has become more quantitative, than it was at its inception a half century ago.

The most complete reference text on commercial dispersants is *McCutcheon's Emulsifiers and Detergents* (North American edition).[40] This informative tome provides the chemical names, structures and physical forms of thousands of commercial dispersants which are otherwise obscured by nontechnical and nondefinitive tradenames. Where such information has been provided by the manufacturer, the HLB value is included. The book, a page from which is shown in Figure 2.14, is updated annually.

References

1. Burke, J., Reed, N., and Weiss, V., *Surfaces and Interfaces*, Syracuse University Press, Syracuse, NY (1967).
2. Aveyard, R., and Haydon, D., *An Introduction to the Principles of Surface Chemistry*, Cambridge Univ. Press, London (1973).
3. Crowl, V., and Wooldridge, M., *Wetting*, SCI Monograph #25, p. 201 (1967).
4. Fowkes, F., and Sawyer, W., *J. Chem. Phys.* **20**, 1650 (1952).
5. Bartell, F., and Walton, C., *J. Phys. Chem.* **38**, 503 (1934).
6. Greg, S., and Singh, K., *Adsorption, Surface Area and Porosity*, Academic Press, New York, p. 301 (1967).
7. Greg, S., and Singh, K., *op cit.*, p. 278.
8. Smoluchowski, J., *Z. Phys. Chem.* **92**, 129 (1917).
9. Davies, J., and Rideal, E., *Interfacial Phenomena*, Academic Press, NY, p. 346 (1961).
10. Lukyananova, O., Pilinskaya, N., and Soloveva, Y., *Kolloidn. Zh.* **41**, 684 (1979).

11. Cornell, G., Goodwin, J., and Ottewill, J., *J. Colloid Interface Sci.* **71**, 254 (1979).
12. Sonntag, H., *Croat. Chem. Acta* **48**, 439 (1976).
13. Georgiev, G., *Khim Ind.* **51**, 441 (1979).
14. Rozenthal, O., and Mityakin, P., *Izv. Sib. Otd. Akad. Naak. USSR* **2**, 54 (1989).
15. Matsayama, T., *Inst. Ind. Sci. Univ., Funtai Kogaku K.* **28**, 188 (1991).
16. Estrela-Lopez, V., and Dudnik, V., ''Multiple Theory of Electrosurface Phenomena in Concentrated Disperse Systems,'' 5th Proc. Conf. Colloid Chem., Mem. E. Wolfram, p. 38 (1988).
17. Vaud, J., *Phys. Coll. Chem.* **52**, 314 (1948).
18. Parks, G., *Chem. Rev.* **65**, 177 (1965).
19. Somasundaran, P., and Agar G., *J. Colloid and Interface Sci.* **24**, 433 (1967).
20. Somasundaran, P., *J. Colloid and Interface Sci.* **27**, 659 (1968).
21. Parfitt, G., ''Titanium Dioxide Pigments,'' Lecture Series #5, *Dispersion of Pigments in Liquids*, Kent State Univ., Kent., OH (May, 1979).
22. 60 nm zeta potential range.
23. Conley, R., *Ceramic Industry*, p. 62, March (1963): also, Conley, R., *J. Am. Ceram. Soc.* **46**, 1 (1963).
24. Sergienko, Z., *VINNITI* **8784**, 527 (1979).
25. Hwang, J., SME-AIME Annual Meeting, New York, p. 85-11 (Feb. 24–28, 1985).
26. Parfitt, G., *Dispersion of Powders in Liquids*, Applied Science Publ., London (1973).
27. Estrella-Lopez, V., and Dudnik, V., 5th Proc. Conf. Colloid Chem., Wolfram Mem., p. 37 (1988).
28. Estrela-Lopez, V., et al., *Fiz. Khim. Mekh. Liofilnost Disp.* **19**, 17 (1988).
29. Conley, R., and Lloyd, M., *Clays and Clay Min.* **19**, 273 (1971).
30. Cannon, P., *J. Phys. Chem.* **63**, 1292 (1959).
31. Conley, R., and Lloyd, M., *Clays and Clay Min.* **18**, 37 (1970).
32. Gray, W., *The Packing of Solid Particles*, Chapman and Hall, London (1968).
33. Graton, L., and Fraser H., *J. Geol.* **43**, 785 (1935).
34. Conley, R., and Albus, F., *Size Reduction and Classification of Solids—Mechanics of Fracture*, Center for Professional Advancement, E. Brunswick, NJ, p. A-33 (1993).
35. First proposed by Janovich Kovar, celebrated Czechoslovakian colloid chemist.
36. Harding, R., *J. Colloid Interface Sci.* **40**, 114 (1972).
37. Kovar, J., and Bohdenecky, M., *Col. Czech. Chem. Com.* **44**, 2086 (1979).
38. Griffin, W., *J. Soc. Cosmetics Chem.* **1**, 311 (1949).
39. ''The Atlas HLB System,'' company brochure, Atlas Chemical Industries, Wilmington, DE (1962).
40. *McCutcheon's Emulsifiers and Detergents*, McCutcheons Publications, Glen Rock, NJ (1994).
41. Ottewill, R. and Tiffany, J., J. Oil Color Chem. Assoc., *50*, 844 (1967).

CHAPTER

3

Inorganic Acid Salt Dispersants

Chapter 2 provided background on the function of dispersants in creating isolation fields about particles to preserve entropy. This basic field, as noted, may be electrostatic or a solvent barrier, but both must provide impenetrable barriers to interparticle contact. All dispersants function within these two classes of activity. How charged species create this isolation barrier will now be reviewed.

3.1 Polymeric Anion Charge Generators

The most simple and economical of the dispersant species are chemical agents derived from mineral acids—common inorganic substances having complex anion structures with oxygen atoms surrounding central electropositive atoms. They include tetrahedral (sp^3), octahedral (sp^3d^2), and planar (sp^2) coordination species. The major families are:

1. Phosphate family;
2. Silicate family;
3. Aluminate family;
4. Borate family.

Dispersants in these families are salts and derivatives of moderately weak acids, that is, with terminal pK_a values greater than 4.*

* pK_a is the logarithm of the dissociation constant. High pK_a values indicate low molecular dissociation (ionization) and weak acid character, as well as highly alkaline salts of the acid. Strong acids possess pK_a values of 1 or less.

The chemical structure of a representative member of these families provides clues to their general activity.

Sodium hexametaphosphate (SHMP):

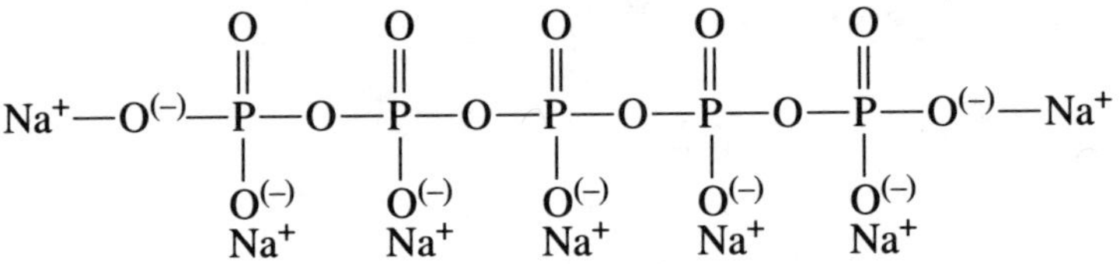

(3.1)

$(Na_2O)(NaPO_3)_x$

Sodium n-silicate:

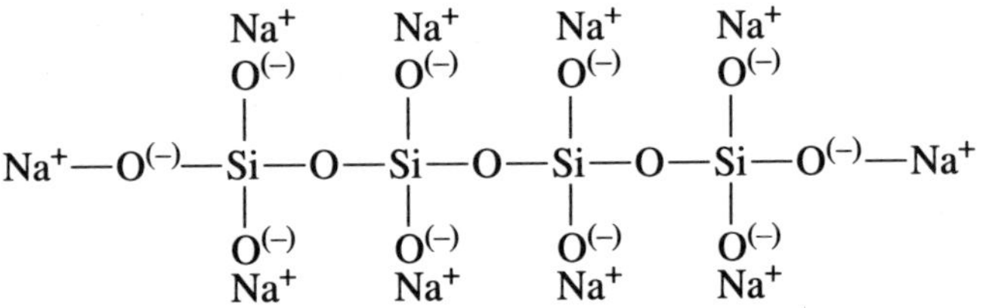

(3.2)

$(Na_2O)(Na_2SiO_3)_x$

Sodium polyaluminate (SPA):

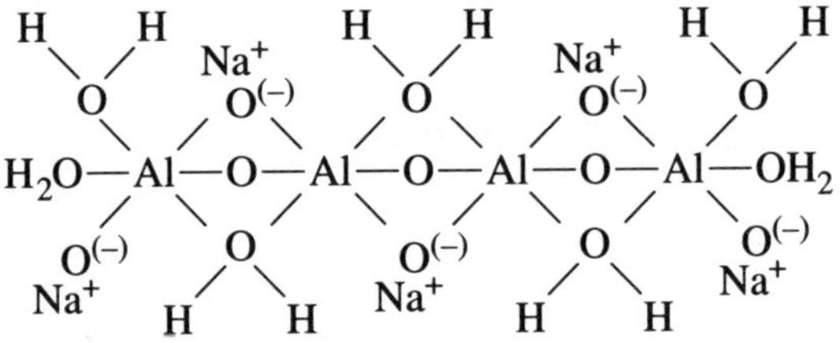

(3.3)

$(Na_2O)(NaAlO_2)_x$

Note: The structural components, $\overset{O}{H\quad H}$ and $\underset{O}{H\quad H}$, represent coordinated water of the overall species in solution.

Sodium tetraborate (borax):

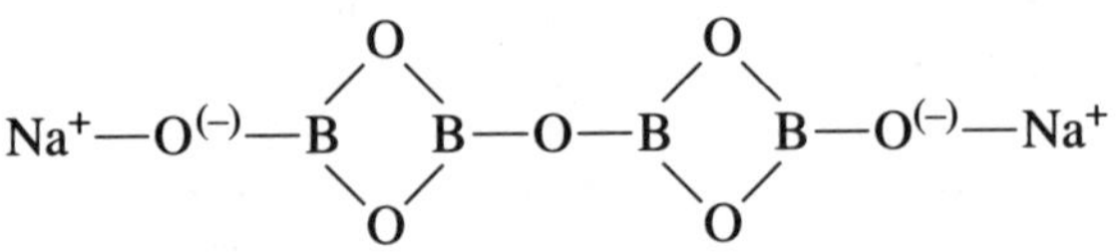

(3.4)

$Na_2B_4O_7$

These various species differ significantly in charge densities, that is, charge per anion group. With SHMP the value is 1.4 (unit avg. = 1), with sodium silicate 2.5 (unit avg. = 2), with SPA 1.25 (avg. = 1), and with borax 0.5 (unit avg. = 0.5). The first two species are sp^3 coordination hybrids having tetrahedral configurations, the third is sp^3d^2 octahedral, and the last sp^2 triangular planar.

Most of these families are available commercially with alternate alkali metal cations, including an ammonium species. As a broad generality, species stability decreases with electropositivity; that is,

$$K^+ > Na^+ > NH_4^+$$

During manufacture of longer-chain polyphosphates, a precise chain length is not generated. The commercial product SHMP, popularly termed "Calgon™," often averages closer to five phosphate groups than six, but includes members ranging from four up to eight. Polyphosphate chains can be produced whose length is great enough to be nearly water insoluble[1] ("Glass H" and "Aquafos").

Within the family of polyphosphates several species are also commercial, the most popular of which are tetrasodium pyrophosphate (TSPP) and sodium tripolyphosphate (STP), the former having two and the latter three phosphate groups.

Sodium silicate and sodium aluminate dispersants are also mixtures of chain lengths. The formulas shown are approximations of commercial products available and may vary between manufacturers.

Adsorption of these species onto an inorganic surface will increase with species molecular weight (van der Waals force) and increase in effectiveness with specific charge. For the polyphosphate family these values are shown in Table 3.1. Also included in the table for reference is trisodium phosphate (TSP), the popular household cleaner and wax stripper.

Because of limitations in the industrial methods employed, other polyinorganic anion dispersant families cannot be manufactured having the variety of

Table 3.1 Comparison of Species Characteristics for Inorganic Dispersants

Family Member	Molecular Group Charge	Average No. Groups	Average Group Charge	Molecular Weight[a]
SHMP	1	6	1.25	572
STP	1	3	1.67	368
TSPP	2	2	2.0	266
TSP	3	1	3.0	164
Sodium silicate	2	4	2.5	550
Sodium aluminate	1	4	1.5	390
Sodium borate	0.5	2	0.5	201

[a] Exclusive of hydration water.

the phosphate family, although sodium borate occurs as both the tetraborate, $Na_2B_4O_7$, and metaborate, $NaBO_2$.

All sodium (or potassium) ions in the complex phosphate polymers do not ionize equally and, as a consequence, do not contribute to forming a dispersive charge. Similarly, each of the species hydrolyzes differently in other environments (solution pH, companion solutes, and slightly covalent oxide surfaces). Employment of phosphate dispersants to add charge to a surface may produce highly variable results, depending upon the species selected, as shown in Table 3.2.

Specific selection of a member from the phosphate family to serve as dispersant may be influenced as much by alkalinity as by its charge imposition.

In foundry and ceramics applications, polyphosphates are employed to prepare casting slurries whose rheological properties are more complex than those of coatings.[2] For example, 60–65% slurries of mixed kaolin–bentonite increase in thixotropic character as longer dispersant chain lengths (low Na_2O/P_2O_5 ratios) are employed. This becomes an important factor where calcium ion, derived from gypsum casting molds, is present. Thixotropy prevents separation of components in the casting formulation and ensures uniformity and strength in the fired products.

Values in Table 3.2 have an important bearing upon dispersant stability. High-alkalinity dispersants possess higher pK_a values and a greater tendency for hydrolysis of their secondary and tertiary cations. Where pK_a exceeds 6 or 7, stability problems may arise with oxide-based pigments, Fe_2O_3, Cr_2O_3, SiO_2, TiO_2, $Al(OH)_3$, $Al_2Si_2O_5(OH)_4$.

Chrome green, chromium trioxide, when pure, is amphiprotic—its isoelectric pH is about 7.0. It is one of very few green pigments that demonstrates high light fastness. However, most commercial pigment grades involve coprecipitation processes designed to shift the pigment's hue slightly. These controlled cationic impurities also alter pigment surface chemistry and isoelectric point. The altered surface in certain cases results in pigments having unusual sensitivity to atmospheric sulfur dioxide.

For paint applications on highway equipment, bridges, toll booths, etc., the pervasive sulfur dioxide from both automotive exhausts and commercial oper-

Table 3.2 Phosphate Family Species Ionization and pH

Member	$pK_a(1)$	$pK_a(2)$	$pK_a(3)$	$pK_a(4)$	pH at 1% Solution Concentration
SHMP	<2	<2	<2	2.8	5.3
STP	<2	<2	<2	6.1	7.8
TSPP	0.85	1.96	6.68	9.4	8.0
TSP	2.15	7.1	12.4	—	9.0

$pK_a(n)$ indicates the hydrolysis constant logarithm for the *n*th member of the species, assuming all members below *n* to have ionized off in prior stages.

ations (refineries, smelters, coal burning utilities) can degrade the green color in a short period. To combat this chemical fading, polyphosphates or phosphonates with a small fraction of zinc polymetaphosphate are employed in dispersing the pigment.[3] The presence of the combined dispersant–zinc metaphosphate increases the fade life several years. By contrast, the use of zinc orthophosphate [$Zn_3(PO_4)_2$], while effective, is far inferior to the metaphosphate. It is believed that transfer and localization of the phosphate anion onto the chromium oxide surface is a major factor in enhancing color stability.

Grinding pigmentary chromium trioxide with a ball or hammer mill in the hot, dry state likewise increases its surface acidity, at least on a short-term basis. High-energy grinds of the pigment have been observed to raise the surface acidity sufficiently to bring about dispersant breakdown and flocculation of the color pigment within an hour. This phenomenon is described in more detail in Chapter 10.

For rutile titanium dioxide slurries 30% by volume (nominal 65% by weight), a viscosity minimum is displayed at 0.2% TSPP. If dispersant concentration is reduced below this level, bridge flocculation of the titania particles results.[4] As the concentration is increased, elevated viscosity results but with minimal thixotropy. For those conditions where thixotropic flow has advantages, as paints, mixing pumps, etc., slight underdispersion is preferred.

Although calcium carbonate is more commonly slurried with sodium polyacrylate (SOPA), SHMP produces certain rheological properties not found with SOPA.[5] A $-5\ \mu m$ sample of calcium carbonate slurried at 65% solids showed a Brookfield viscosity of 10,000 cp but was slightly dilatant, which impeded the makedown process for water-base paint systems. While this slurry, once formulated, diminishes in viscosity with slight aqueous dilution, it also becomes thixotropic.

Adsorption isotherms of polyphosphates on calcium carbonate show a Langmuir pattern with TSPP but display a slightly sloping plateau with SHMP. Unlike alumina and silicates, no template octahedra exist on the calcite surface. It is believed that phosphate tetrahedra spacing in the dispersant molecule, which is a poor match for the calcium–calcium ion spacing on calcite, makes longer-chain train configuration less probable. Phosphate adsorption density for TSPP calculated at the Langmuir plateau value was 2.66 mg per square meter of area.[6] This computes to about 6.11 molecules per square nanometer or 0.164 nm^2 per molecule. At this level of adsorption slightly less than 10% of the surface is being covered. The lowest viscosity is observed at about 80% of this "monolayer" value.

Sodium silicate and sodium aluminate dispersant species are yet more alkaline than phosphates, approaching pH 10 at 1% solution level. Sodium silicate has been employed for years as a deflocculant in kaolinite slurries for ceramic casting because of its inert effect upon molds. The yield stress value, important for producing high green strength for handling prior to firing, changes with both pH and the level of silicate addition. While the zero-potential pH for pure kaolinite is about 4.5, sodium silicate lowers it well below 3. At about 0.3% sodium silicate solids basis, the zeta potential drops to slightly over −40 mV while the pH rises to above 7. These conditions are ideal for general clay slip handing and casting.

Studies on 51% clay suspensions for casting show high initial values for rheological shear stress.[7] However, this changes with time, suggesting a reorganization of dispersant on the clay surface. The change also alters the elastic and plastic behavior of the slip. When sodium silicate dosages are increased to 0.4% or above, the above phenomena attenuate.

The sodium silicate phenomena are similarly observed with ground silica (quartz). When freshly ground fine quartz is suspended in alkaline dispersants, its surface character is observed to change with time. This suggests the formation of a surface gel, probably as the (H_2SiO_3) structure.[8] The zeta potential diminishes during this period as well as the general dispersibility (network formation). Oven heating of the quartz or allowing it to accommodate to room temperature for several weeks following grinding causes the effect to disappear. This is best explained by highly active silanol groups dissolving and precipitating on the surface in alkaline media.

The question may arise why a single cation–anion species, such as sodium chloride, does not function as a dispersant, as such compounds also would carry an average unit charge of 1. Their basic failure lies in van der Waals forces of surface attraction being too small for them (molecular weight 58.5 for NaCl) to overcome the slight repulsion between emergent electron clouds (arising from surface atoms on the solid) and the anion's negative charge. In some experiments with extremely large anions,[9] such as iodide, adsorption does occur, although not quantitatively, and a small amount of dispersancy is observed. Other more complex factors, such as template structure, also influence their ineffectiveness.

Additional special inorganics may fall into the classification of polyanionic dispersants, for example, orthorhombic phosphorus pentoxide and polyphosphonates. However, these have highly specific and limited applications and are most often highly expensive to manufacture. Thus their use in common industrial practice is economically restricted.

Orthorhombic P_2O_5 is a solid having approximately a sodium hexametaphosphate structure but with each of the sodium ions replaced with a hydrogen ion. The material is reasonably stable as a crystalline solid but decomposes to orthophosphate when in prolonged solution. It is quite acid.

Phosphonates are organic derivatives of polyphosphoric acid with an esterified group replacing one or more P–OH groups. One advantage of these structures lies in lower alkalinity and another in increased molecular weight. Most phosphonates have short to moderate hydrocarbon chain lengths, as typified by:

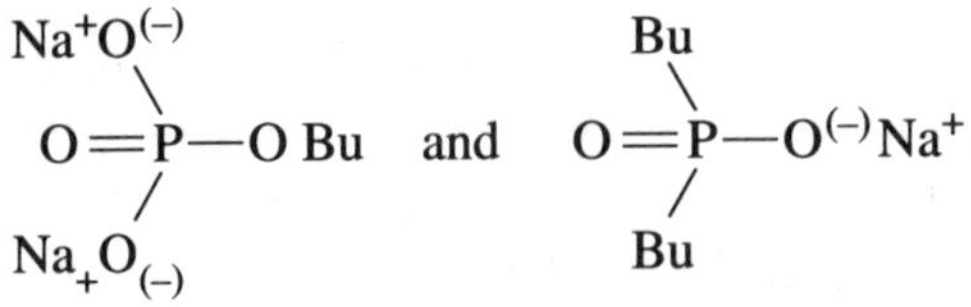

(3.5)

$Bu = C_4H_9—$

The use of inorganic species as dispersants does not appear to have a great effect on induced particle charge, as is noted by similar rheological profiles (see Figure 1.11). The three species, tetrasodium pyrophosphate, sodium N-silicate and sodium polyaluminate, have nearly equivalent viscosity minima. However, the number of dispersant unit moles required to attain these minima varies significantly, 9 for TSPP, 11 for sodium *n*-silicate, and 25 for sodium aluminate. The molecular weight per unit monomer differs very little for these three species. Hydrolysis must play a highly important role in the dispersion as the minima for the three differ by over 3.5 pH units, even with the buffering presence of the acidic kaolinite surface.

Borate and carbonate rheograms are likewise similar but produce minima at more elevated viscosity values.

Alkaline salt dispersants producing high-pH slurries are subject to hydrolytic interaction with the surface. This is especially true for the aluminate species.

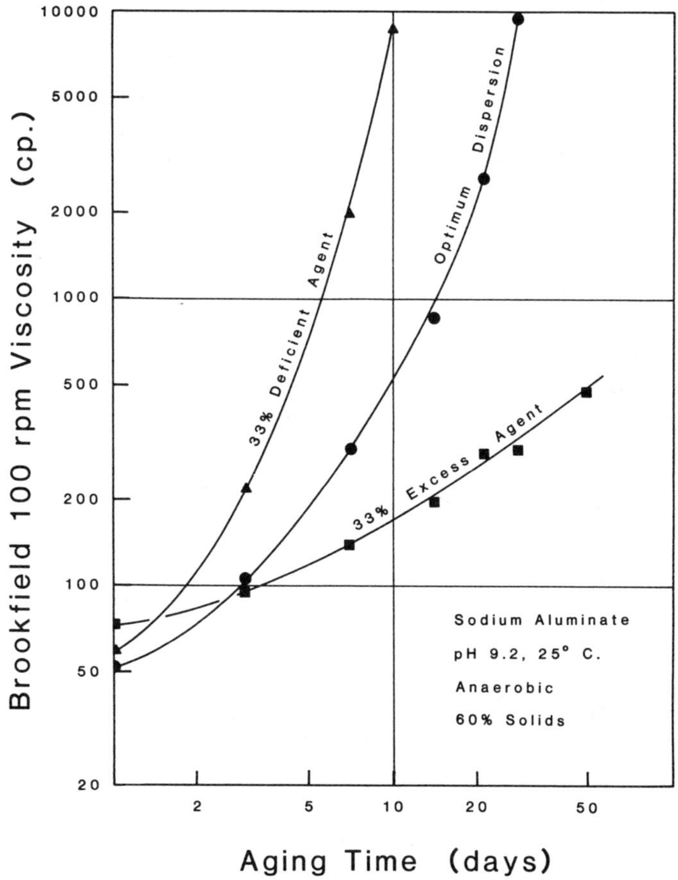

Figure 3.1. Time-dependent rheograms during hydrolysis.

Figure 3.1 shows comparative rheograms for time-dependent viscosity change as a consequence of hydrolysis. Although by application standards this is a relatively slow process, it may affect long-term slurry transport for commercial applications, a situation where both time and temperature influence the reaction.

Molecular orientation, especially with longer-chain inorganic species, will influence both surface-imposed charge and pigment solids level. Tail-to-train transition on templated substrates is prevalent, as demonstrated by the Langmuir data in Figure 3.2. Obviously, no sharp boudnary exists for this transition, and the surface is populated by both modes of attachment from initial addition to maximum dispersive effectiveness.

An excellent illustration of this occurs with sodium tripolyphosphate. At low levels of dispersant, the anion orients initially in the tail configuration. At about 50% of the effective level for obtaining a dispersion minimum (about 0.15% by weight on the silicate surface), solution pressure results in a slow transition to a combined tail–train configuration as depicted in Figure 3.3(a).

Adsorption of polyphosphates onto insoluble salt pigments (barium sulfate, silver chloride, calcium carbonate, strontium chromate, etc.) involves weaker forces because the van der Waals template effect is absent. With these substrates Langmuir patterns display no inversion effects but adsorb, apparently in a train configuration, as demonstrated by a gradual plateau reminiscent of trisodium phosphate. Salt substrate adsorption is illustrated in Figure 3.3(b).

The net imposed charge changes dramatically between tail and train orientations. In the tail configuration with tripolyphosphate, no more than a third of

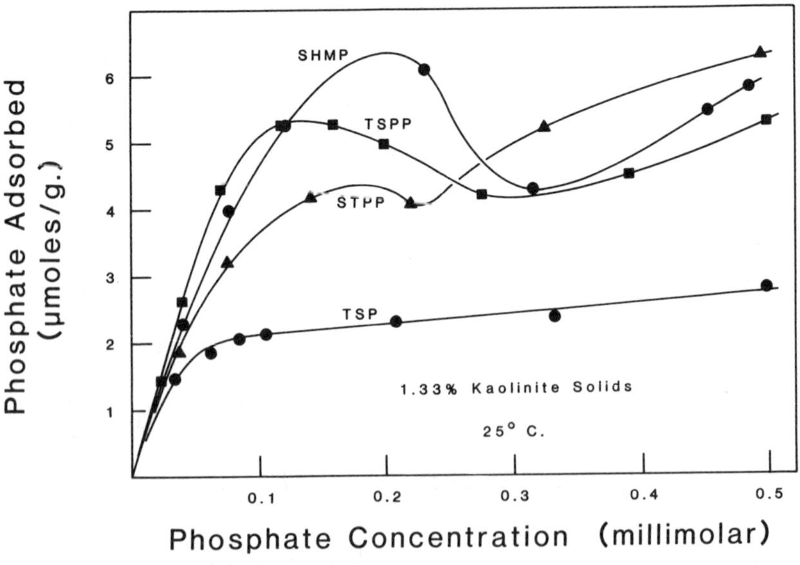

Figure 3.2. Langmuir adsorption profiles for phosphate on kaolinite.

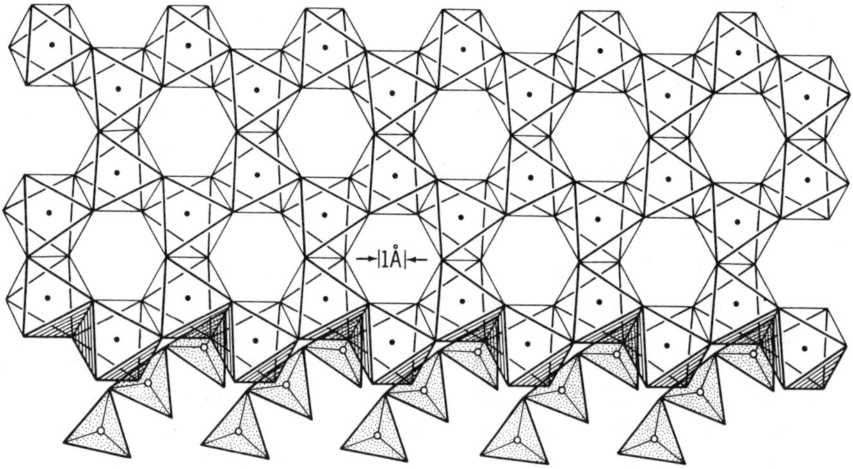

Figure 3.3. (a) Adsorption orientation of sodium tripolyphosphate on an octahedral pigment surface.

the surface can contain dispersant anions, based on simple arithmetic calculations from the Langmuir data of Figure 3.2. The two unit members of the phosphate chain directed into the liquid phase add essentially zero net charge to the surface as both cations and anions now occupy the Helmholtz layer, cancelling each other's charge effects.

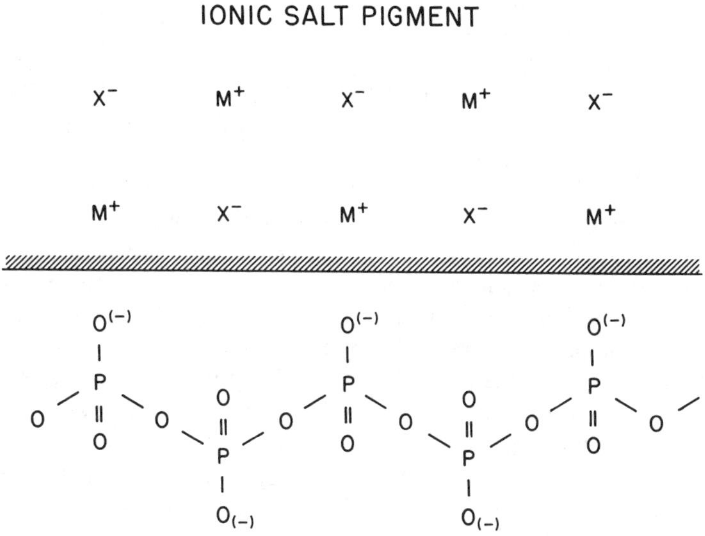

Figure 3.3. *(continued)* (b) Adsorption orientation of polyphosphate on salt pigment.

At the transition point where about $\frac{1}{3}$ of the minimum dispersant has been added, the viscosity is still very high. Addition of the remaining dispersant, approximately half, or 0.15 weight percent, contributes very little additional tripolyphosphate molecules to the surface, again based on Langmuir data. Yet viscosity drops at this dispersant level from above 3000 cp to less than 50 cp (100 rpm Brookfield). Only about 50% of the surface need be covered with tail-forming molecules (see Chapter 7) to achieve the result of full coverage, due to a phenomenon known as *kT* molecular spinning.

Based on the BET surface area of the tripolyphosphate pigment, 7.1 m^2/g, and a 6:1 plate morphology as established by scanning electron micrography,[10] the edge area, where dispersancy takes place, would represent about 1.2 m^2/g. In the tail configuration 18.9 micromoles per gram of 0.72 wt% would represent full edge coverage. In the train configuration these values reduce to exactly one-third, or 6.3 micromoles/g and 0.24 wt%. This is confirmed by the Langmuir curve, where extrapolation indicates that about 6.4 micromoles/g represent the adsorption monolayer coverage for tripolyphosphate, that is, 100% active site coverage.

On neither substrate type are inorganic salt dispersants quantitatively adsorbed, the equilibrium being shifted progressively to the liquid phase with higher levels of addition. Figure 3.4 for the phosphate family, including orthophosphate, clearly demonstrates this low energy of surface coordination.

Optimum (minimum) viscosity is attained at about 0.35 wt% tripolyphosphate. At this level only 20% is retained on the surface, or 0.07 wt%. Thus approximately one-third of the optimum is attached, and two-thirds are in the solution phase. Impression of charge actually reduces the need for dispersant (compared to the approximate 50% for long-chain dispersants). The differential, $\frac{1}{2}$ versus $\frac{1}{3}$, is the consequence of high ionic charge.

These calculations can also be made for SHMP and TSPP, whose factors are five and two, respectively, for tail versus train orientations. The curve in Figure 3.2 indicates a higher monolayer for SHMP and a lower one for TSPP. For TSPP tail-and-train formation would appear identical because the reentrant positions on the alumina octahedral surface of kaolinite expose exactly two octahedral faces. However, a definite transition occurs with TSPP in Figure 3.2 at about $\frac{1}{5}$ monolayer coverage (peak at 0.12 millimolar solution concentration).

Unfortunately such studies can never be as quantitative as research dispersant chemists desire because industrial chemists are concerned more about productivity than purity in family-structured dispersant species,[11] with the result that a given product comprises several different chain lengths.

Trisodium phosphate, the most alkaline of the family and the member with the highest charge, might appear to provide the best dispersancy. For years it was sold commercially as a household cleaner, especially for stripping kitchen floor waxes. However, when Langmuir adsorption isotherms are performed, the number of phosphate groups adsorbed, which for TSP represents the entire molecule, is almost exactly half that for TSPP.

The dispersion disparity between TSP and TSPP is easily rationalized as

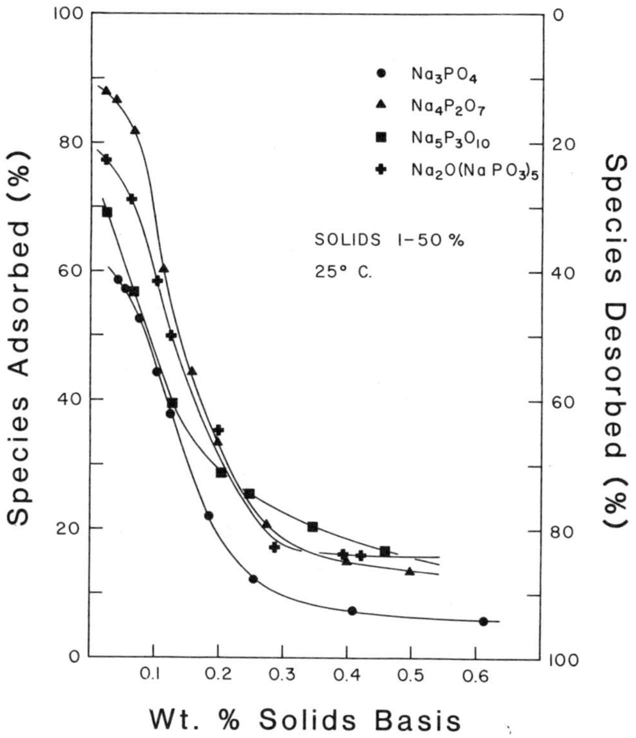

Figure 3.4. Equilibrium distribution of phosphate dispersant on kaolinite.

arising from a single molecule with the monomeric species being able to fit into the reentrant position on the alumina and, because of its high charge, to repel other members from the locale. One phosphate per reentrant site represents one phosphate per two total sites. Thus, even with all reentrant positions occupied, 50% of the alumina with its charge remains exposed to the liquid phase. Tail formation is not possible for TSP, which is borne out by the conventional Langmuir form in the adsorption study.

TSP is so ineffective a dispersant it is not employed, even as a diluent, in phosphate mixtures. Its second disadvantage lies in its alkalinity. Whereas at high clay solids the minimum viscosity pH for SHMP lies between 5.5 and 6.5 and for TSPP between 7.0 and 8.0, that for TSP is in the vicinity of 9.0. For many applications this is simply too alkaline a condition to be compatible with other system components.

Langmuir data indicate all polyphosphate dispersants undergo tail-to-train transitions at 50–75% of surface coverage. This results in part from surface pressure of the increased solution ionic atmospheres, a modified form of double-layer compression.

Other factors may influence the selection of a particular phosphate or other inorganic dispersant species for slurry application. Aqueous solubility for SHMP is many times that for TSPP, making the former more practical for liquid feeder use than the latter. Both sodium silicate and sodium aluminate have similar high solubilities.

3.2 Optical Properties of Dispersed Pigment Films

Dispersion of a pigment affects not only rheological properties, it also affects numerous optical properties. Polyanion salt dispersants are most commonly employed with paint and paper coating pigments and in ceramic casting slurries. With the first two applications film coating properties are of great importance. Where a particle assemblage is composed of spheres having a log-normal size distribution and homogeneously distributed as a dispersed phase, a film formed by drying such an assemblage will possess the ultimate in compactness, as shown in Figure 3.5(a). While film density would maximize, all optical properties would not. This can be illustrated by the interaction of light with particulate matter.

A photon of light entering the film, as shown by the broken line, is refracted

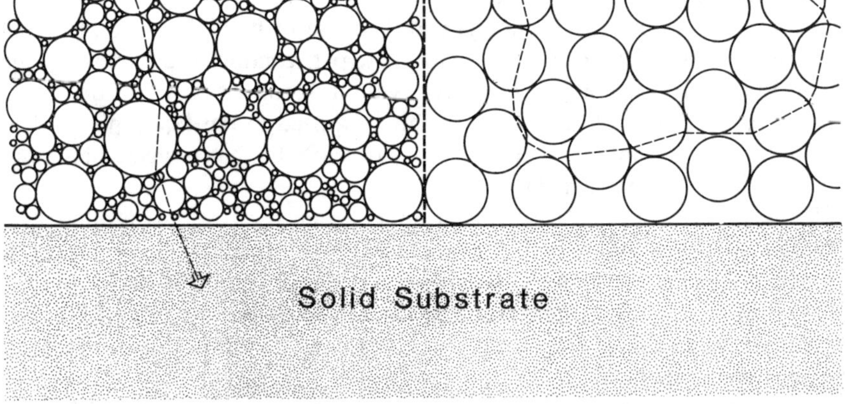

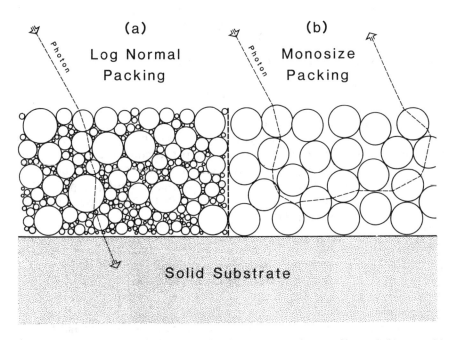

Figure 3.5. Light transmission through (a) a compact pigment film and (b) one with voids.

by an angle R such that:

$$\sin(I - R) = \sin I/n_R \tag{3.6}$$

where I is the incident angle of light entering the film or particle and n_R is the refractive index of that film or particle. The sphere performs as a type of lens—as the photon emerges from the sphere, it no longer proceeds in the same direction. However, because a semicontinuum of mass exists, the photon's probability of passing, albeit circuitously, through the bed of particles and eventually reaching the substrate is high. For each photon reaching the substrate and being adsorbed, the film will appear progressively darker (energy adsorption) at that wavelength. The object of most coating films, especially white ones, lies in preventing just this type of light interaction.

The film must allow photons to enter but refract them sufficiently that they are bent backwards, in the direction from which they came, and emerge from the film unadsorbed. A perfectly white material will have a relationship with a parallel beam of illumination:

$$dI_{(out)}/d\theta = I_{(in)}(K - L) \tag{3.7}$$

where K is a scattering efficiency constant and L represents the absorptive loss within the film. For all angles greater than the critical angle (that angle where 100% of the incident radiation is reflected), reflectivity, or the ratio of incident light to emergent light, is wholly independent of the angle of illumination or the wavelength. For most practical white materials, $dI/d\theta$ falls off slightly for values of θ less than 30°, but not with wavelength.

Addition of absorptive color body particles homogeneously distributed throughout the film will allow L to change with wavelength but not K. It can be shown mathematically that where either the primary diffractive pigment or the color bodies are not well dispersed throughout the film, a color variation will occur both regionally across the film and with angle of illumination. This shows up occasionally in poor automotive repainting as islands of color variation, especially when viewed at differing angles.

A second approach to increasing opacity (ratio of emergent to incident light intensity) is to compose the film with a monodisperse fraction of equivalent morphology particles, as shown in Figure 3.5(b). When this film dries, voids will occur that greatly diffract the illuminating beam. If, however, these voids are filled with a binder having a refractive index only slightly below that of the primary pigments (typically 1.50–1.55), essentially the same optical effect will be produced as by employing a log-normal distribution of primary pigment particles, as illustrated by Figure 3.5(a). To prevent photon transmission from film top to film base requires a significant modification of one or more components within the system:

1. A portion of the void space can be left empty ($n = 1.00$) using just enough binder to hold particles together at contact points.

2. A pigment with a refractive index significantly higher than the binder, >1.55, may be used in the film.
3. Particle morphology can be changed to provide more anisometry, greater random orientations, and fewer interparticle surface-to-surface contacts.

The obvious method of achieving opacity is to employ a partially flocculated state with anisometric (preferably platelike) particles. The greater the randomness of their orientation, the greater the probability of light scatter and photon return to the surface. Particle shape becomes important because of particle thickness. While equants have essentially a single interactive dimension with light and thus scatter is independent of their orientation, plate-shaped particles have three interactive dimensions with light waves. Two are large and one is small. While the large dimensions serve as "mirrors," only the smallest dimension interacts with light to produce scatter. From Mie scattering theory this should be about $\frac{1}{2}$ to $\frac{1}{3}$ the wavelength of the illumination source, nominally 200–300 nm. Figure 3.6 is a scatter function plot of aluminum silicate platelets averaging 6:1 morphology in a flocculated array formed on a glass slide. Seven fractions of very narrow distributions (sigma = 0.2) made from selective classification were employed for the data points with white light as illuminant. The particle size referenced is the mean plate diameter as determined by electron microscopy.

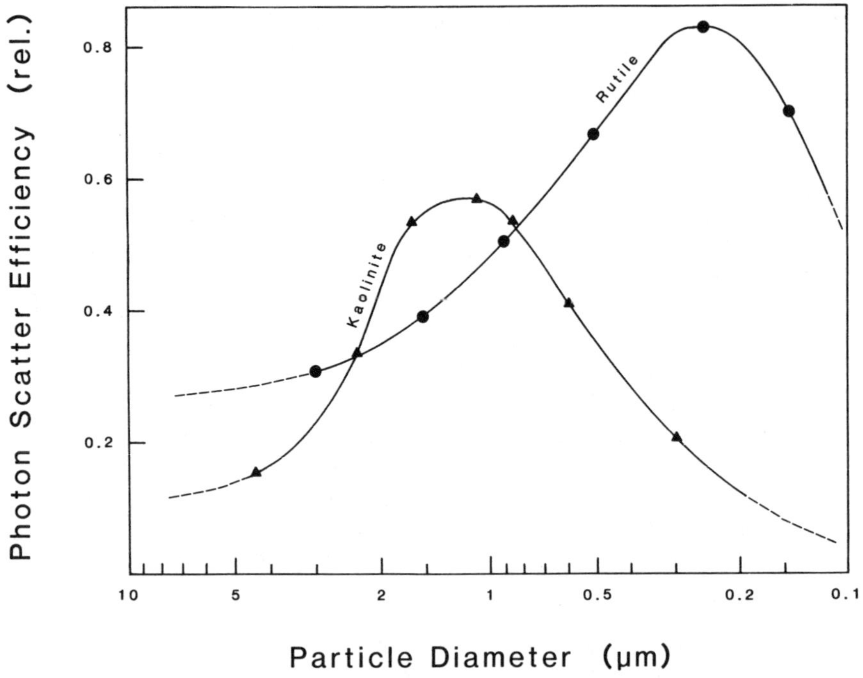

Figure 3.6. Scattering efficiency of differentially sized pigments.

The consequence of broad-wavelength illumination is a broad peak in the scatter function. Also in Figure 3.6 are the scatter data for near-monosize titanium dioxide occurring in a broad energy of aggregates ranging in size from 0.2 (fundamental size) up to about 3 μm. Optimum scatter for the phylloidal pigment occurs in the range 1500–1200 nm (1.5–1.2 μm) diameter, corresponding to an average 250–200 nm particle thickness, essentially equivalent to the most effective size for titanium dioxide.

Studies on paint and paper coating films substantiate this additionally. Figure 3.7 contains opacity data for kaolinite slurries,[12] both initial and aged, as a function of dispersant dosage. Film densities, obtained by film microtoming and measuring optical density and void space, were approximately equal, about 77%.

Note that optimal dispersion in both coatings, nominally 0.35–0.40% SHMP, produces the lowest hiding power. Further, worst-case dispersion, with highly flocculated assemblages (zero dispersant), produces peak capacity. Film density in such coatings usually increases as opacity falls.

Brightness or whiteness (in the instance of clay and calcite pigments) is similarly a measure of efficiency of photon return from the film. Figure 3.8 shows brightness measurements performed on films with calcium carbonate[13] and kaolinite in various states of dispersion. A similar relationship to opacity occurs. However, brightness and opacity are but two characteristics of these pigments

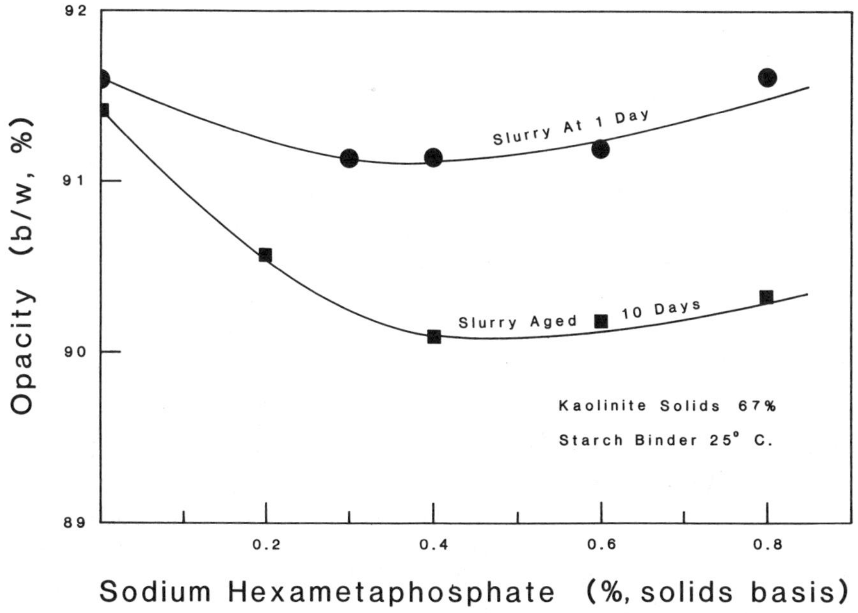

Figure 3.7. Opacity of pigmented films as a function of dispersant level (Refs. 13 and 26).

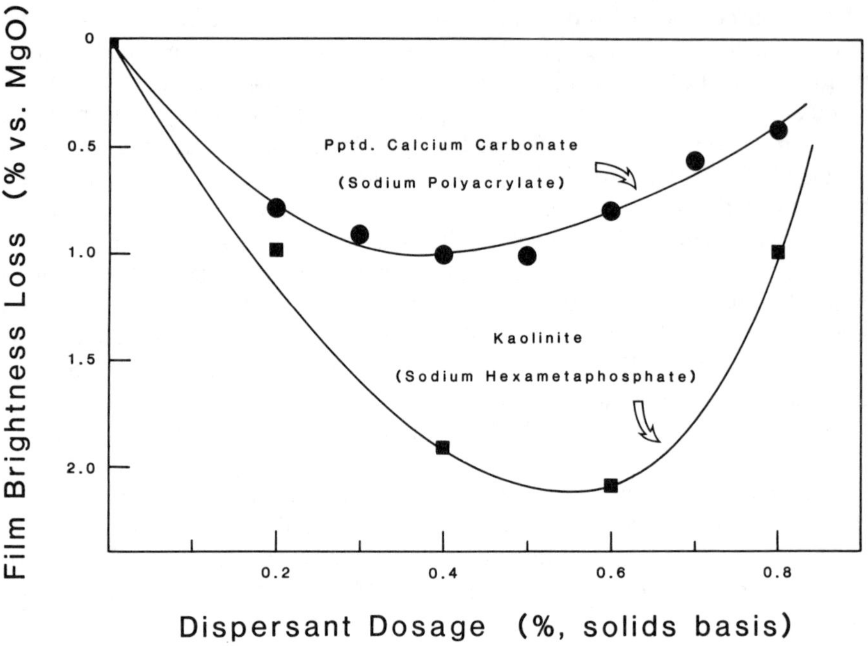

Figure 3.8. Film brightness dependency on state of dispersion (Ref. 26).

needed by a polymer-bound film for either paint or paper pigment coating. Viscosity and gloss are of equal importance.

Polyphosphates are employed both in industrial plant processing of pigments as well as in finished formulations. For water-base paint films the sodium salt of the dispersant is often replaced by the larger potassium or ammonium ion. Because paint films constantly cycle between wetness and dryness from weathering, moisture absorbs in the film then desorbs from the film through tiny apertures or pores. Paint technologists term this *film breathing*. Films containing flocculated particles breathe more readily than ones containing dispersed particulates.

Utilization of any pigment predispersed or subsequently dispersed with polyphosphates is subject to ionic transport as this water column moves into then out of an exterior paint film. Once on the surface, the dispersant salt dries and crystallizes, leaving a white superficial coating. Should calcium carbonate be present as an extender pigment, polyphosphates will sequester calcium ions and carry them to the surface. There they are subject to secondary reaction with atmospheric gases and can form calcium sulfite and sulfate and, as the polyphosphate degrades, calcium phosphate, all white, insoluble powders. This white film is barely noticeable on white painted surfaces, slightly on pastel finishes, but is offensive on deep tones. The common term for the phenomenon is *bloom-*

ing. Depending upon the calcium content in the film, the bloom may be removed by water or very dilute acid, usually acetic (vinegar) or hydrochloric (muriatic). Such paints are unpopular with homeowners because, once leaching commences, it is an annual exercise, in part accelerated by the acid cleaning!

Porosity inherent in the film also is opened up further by acid cleaning operations. When this porosity increases sufficiently, the film acquires a black mold (asperigellum niger) that roots nicely in the pores. Thereafter, the paint film must be chemically bleached, usually with acidic sodium hypochlorite. This also becomes an annual activity in regions where elevated humidity is common.

The annual ritual of scrubbing and bleaching soon deteriorates the film, damages color, and weakens the film bond to the substrate.

Well-dispersed pigment particles settle more freely and pack with higher densities in applied films than slightly flocculated ones. As a consequence, light radiation passing through a dispersed film of sedimented particles will encounter a more level stratum. This property is exhibited in gloss. Figure 3.9 contains data for a flocculated laminar silicate, consisting of particles essentially of 100% $< 5\ \mu m$ in diameter and near log normal in distribution. At the dispersant level where viscosity is minimum (0.35–0.40%), gloss is likewise minimum. Overdispersion, necessary to overcome residual chemical flocculants, improves gloss dramatically. However, gross overdispersion, which impairs particle mobility

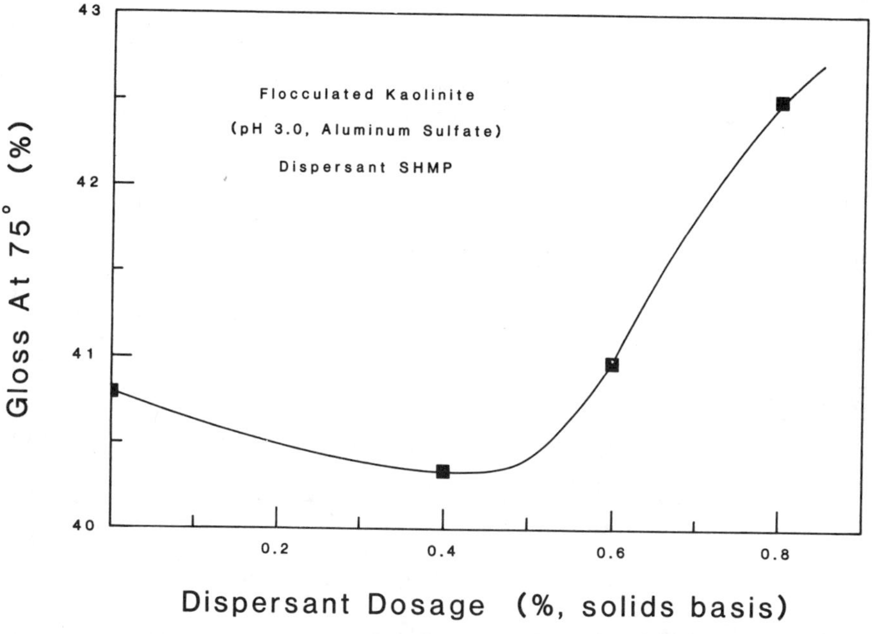

Figure 3.9. Film gloss dependency on dispersion (Refs. 13 and 26).

and adversely affects rheology, will degrade gloss characteristics in such coatings.

Optical brightness, measured by reflectance of incident radiation at 456 nm wavelength (blue-green), will diminish with better dispersion as particles pack closer than $\frac{1}{2}$ wavelength of light (300 nm). Such a film will scatter incident radiation and reflect it back in the direction of the source to a lesser extent than would slightly flocculated agglomerates, where mean interparticle spacing is greater. This is shown clearly for both titanium dioxide and kaolinite in starch coating formulations made on glass for comparative pigment studies in Figure 3.10. Employment of polyvalent cations as process flocculents, however, engenders brightness degradation, the extent increasing with level of addition. This derives from trivalent cations producing tight floccules which restrict diffusion and elutriation. These, in turn, prevent particles from becoming fully redispersed during coating formulation, where low-shear-energy equipment is often employed.

At optimum dispersion and with parallel alignment of anisometric particles or log-normal distributions of equants, light scattering reduction will permit more photons to pass through the film and reach the base layer, diminishing film opacity (demonstrated in Figure 3.7). Overdispersion, however, shows an increasing trend toward opacity gain. The opacity parameter appears to degrade

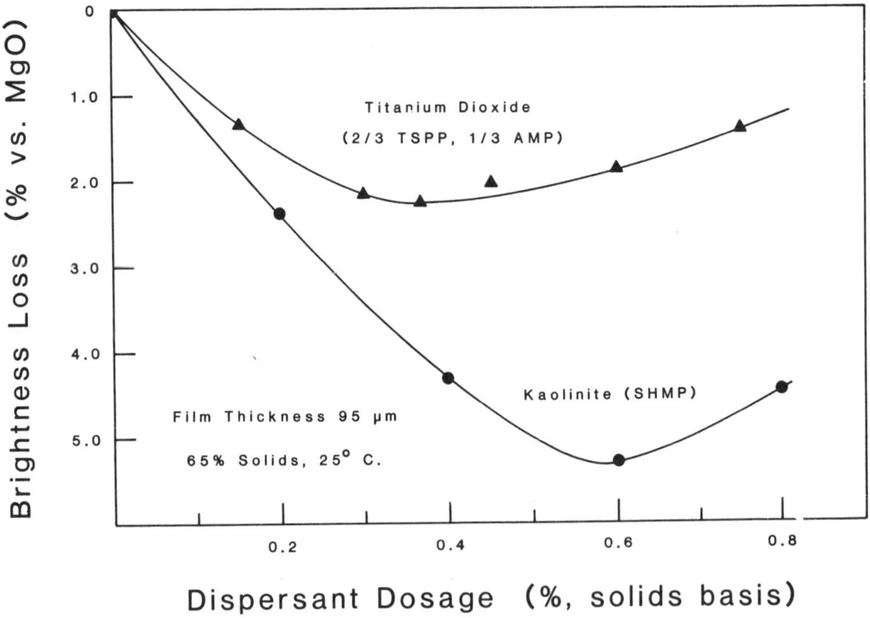

Figure 3.10. Titanium dioxide and kaolinite brightness dependence upon flocculation on glass coatings (Ref. 26).

with time, that is, slurry aging. In years past this was theorized to result from a slow release of fine particles which filled voids. However, present evidence suggests that dispersant decomposition and particle cementation is more likely the cause. Film porosity, as measured by pressure drop from an applied non-adsorbing gas (krypton), is an excellent method of assessing void porosity. Film packing density can be seen to be a direct consequence of degree of particle dispersion.

Certain formulations employ pH adjustment by alkaline dispersants (morpholine, amino alcohols, TSPP, sodium carbonate, etc.) to reach the critical point where one pigment is dispersed but the second is partially flocculated to provide improvement in opacity. Examples of this practice are discussed in Chapter 11.

Addition of low concentrations of flocculant tends not to affect the entire size population equally, but serves to attach the finest particles to coarsest ones. This was illustrated by particle size analyses in Figure 2.4. Figure 3.11 shows both the particle size change and the optical improvement during the initial stages of flocculation. Missing particles below 0.5 μm would appear to bridge coarse particles and improve the random scatter of photons. Flocculation size loss becomes significant below about 1.5 μm. This effect results from conservation of system charge; that is, the generated fine-to-coarse agglomeration represents the lowest-energy state of the assemblage.[14] With increasing additions of flocculent, progressively coarser fine particles attach to finer coarse ones until all

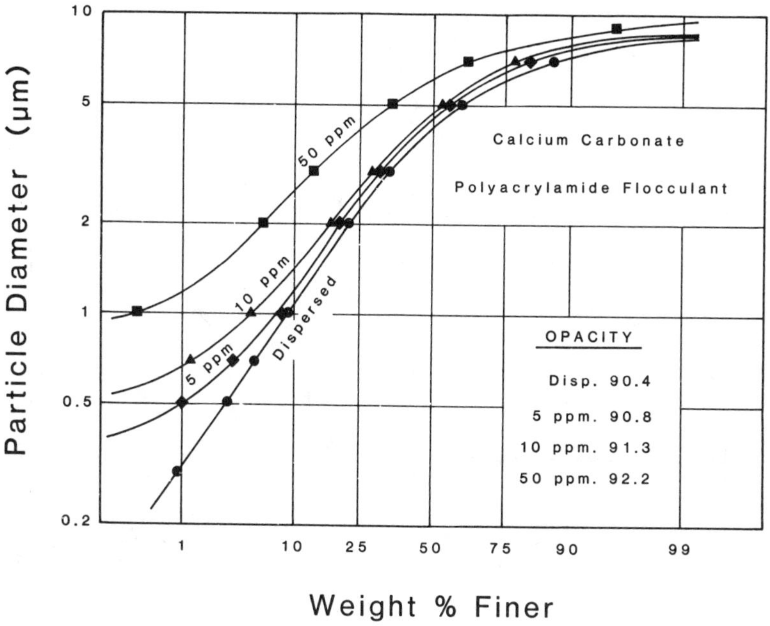

Figure 3.11. Differential flocculation of fine particles.

prefloccules generated approach a common size and a network assemblage develops.

Small amounts of flocculent, as a consequence, effectively change the particle size distribution. Viscosity measuring equipment generally does not provide sufficient shear to break apart these floccules when very small. Thus a log-normal particle distribution displaying Newtonian characteristics will slowly approach a dilatant state as flocculent is added, illustrated by the family of barite data in Figure 3.12. Log-normality for Newtonian character theoretically maximizes around a D_{75}/D_{25} ratio of 3.5 (Chapter 1), while dilatancy attains its maximum at a ratio of 1.0 (monosize distribution).

Also in the figure are the median sizes (D_{50}) for each value of log normality, ranging from about 1 μm for the dispersed state to 50 μm at 100 ppm. polyacrylamide. The recommended dosage of this flocculent is about 500 ppm. With high rates of shear, as provided by sophisticated viscometers, floccules may be broken down, or at least shear deformed, to conform to fluid flow. The apparent size distribution will then not be as narrow, and viscosity will appear less dilatant.

Should residual flocculent remain from processing operations as filtration, dewatering, etc., the resulting dried products will possess chemical sinters, especially where aluminum salts are employed. Aluminum ion hydrolyzes and cements particles together with aluminum hydroxide gels. Sinters become even more significant with flash drying and extrudate traveling bed driers. In these

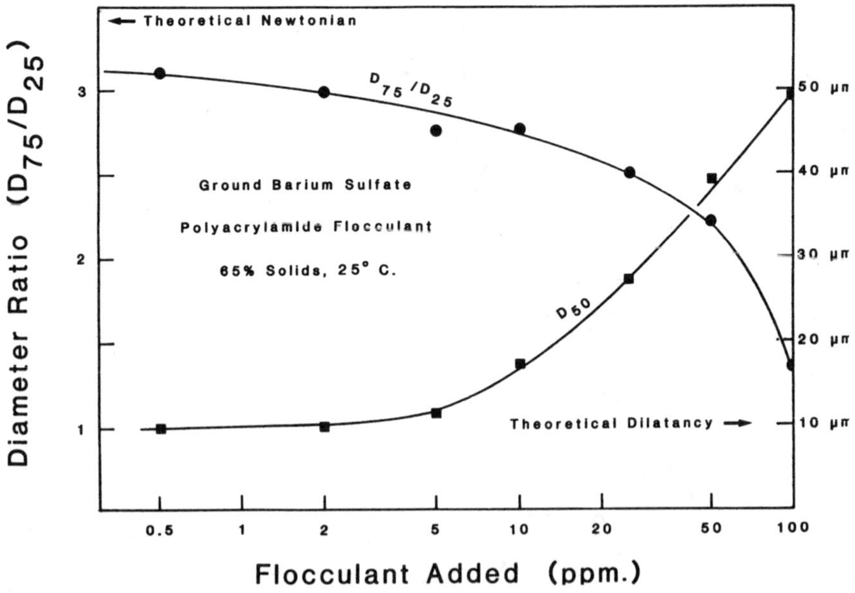

Figure 3.12. Newtonian-dilatancy transition with flocculation.

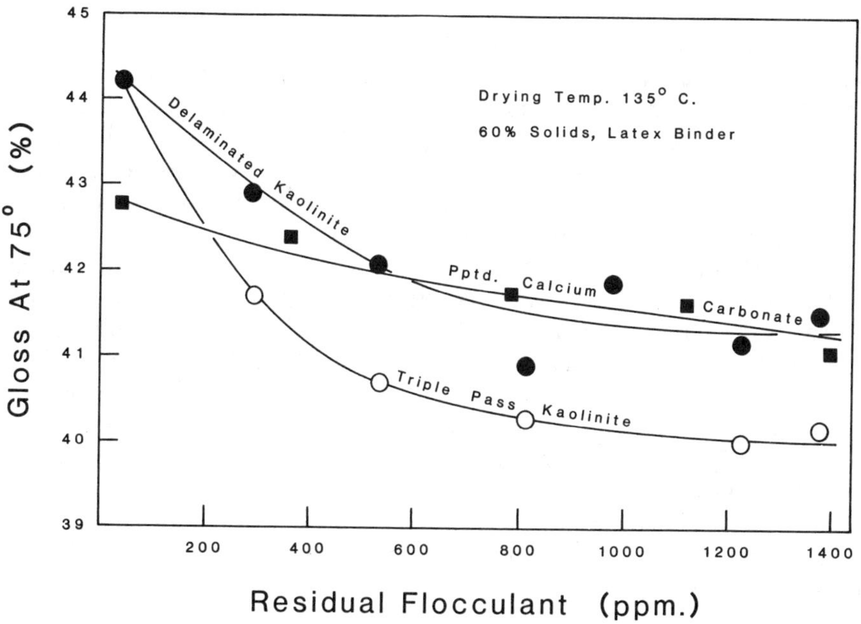

Figure 3.13. Gloss degradation from chemical sintering.

operations, dry material is blended into wet feed to achieve higher solids in the slurry. As a consequence of repeated chemical additions and recirculation, further sintering occurs. A 30% dry add-back to a 50% solids slurry will under steady-state conditions contain several percent of material with three or more passes through the drier. This is illustrated in Figure 3.13 for both delaminated kaolinite (aluminum sulfate flocculent) and precipitated calcium carbonate (organic flocculent), where the parameter gloss degrades from flocculant residuals. Calcium carbonate, having a much narrower size distribution, is less adversely affected than the broadly sized silicate with a single drier pass. However, gloss degrades yet further with multiple passes.

Pigments employed for the coating of paper often are wet processed by grinding, size classification, chemical bleaching, flotation, magnetic separation, and/or morphology alteration. Many of these unit operations must be conducted in the liquid (or slurry) state to be effective. The degree of process chemical removal will have pronounced effects on the coating film with regard to brightness, gloss, and opacity.

3.3 Dispersant Instability

Table 3.2 gave the ionization constants for the various phosphate members of the dispersant family. These values are directly related to the hydrolysis of these salts. The presence of any accessory acid having an ionization contant greater than that of a table member will cause an equilibrium shift of the salt to the un-ionized hydronium form. Consider as an example the silica structural component from freshly ground kaolin. Titration studies indicate a pK_a for the edge silanol ionization of about 6.2.[15] If the particle is stressed or broken by milling, this value diminishes. As shown in Chapter 2, proton mobility is slow on a silicate surface, often taking days to move one unit cell in distance. On kaolinite this takes the form depicted in Figure 3.14.[16]

The structure shown lies at the particle edge, and the proton migrates essentially parallel to the *c* axis. This is termed, somewhat erroneously, an ion-exchange reaction because acidified kaolins eventually release hydrolyzed aluminum ion by this mechanism. It is, in reality, acid decomposition of the aluminum silicate.

However, should a phosphate dispersant molecule be attached to the alumina octahedral edge site, proton transfer is more probable to that entity because it is loosely associated with the structure, not held by crystal field forces, as is the bonded alumina dihydroxide.

Proton mobility takes the general form noted in Figure 3.14 but establishes an equilibrium with the dispersant TSPP anion:

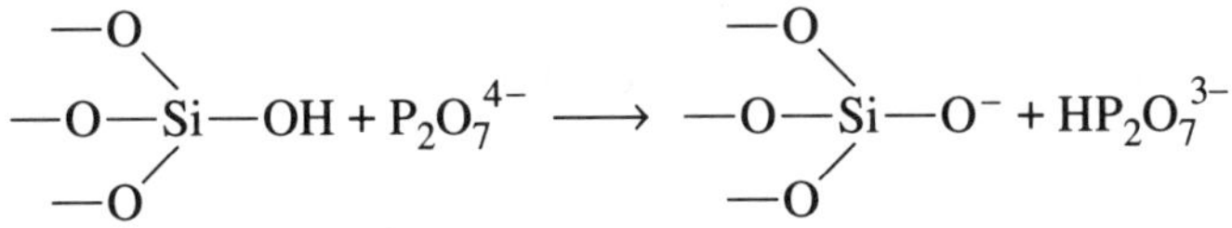

The reaction is simply a proton transfer between weak acids whose equilibrium constant can be expressed mathematically as:

$$K_{\text{hyd}} = \frac{(\text{SiO}^-)(\text{POH})}{(\text{SiOH})(\text{PO}^-)} \tag{3.9}$$

For any phosphate group with a *pK* in excess of 6.2, the equilibrium will shift such that greater than 50% of the proton population associates in time with phosphate dispersant molecules. The concentration of silanol groups on this silicate will equal the number of alumina sites [based on the formula $Al_2Si_2O_5(OH)_4$] and hence the number of potential phosphate groups assimilated in the template dispersant. The proton transfer reaction, consequently, has the potential for being quantitative.

Thus far in this scenario no net change in surface change has occurred nor any alteration of dispersancy. The problem develops not with charge but with dispersant molecular stability. Polyphosphate molecules are not highly stable. In acidic environments, they hydrolyze by scission, that is, by clipping off terminal tetrahedra.

For TSPP anions this results in total decomposition, as shown in the schematic representation of Figure 3.15.

The terminal species in Figure 3.15, $H_2PO_4^-$, dihydrogen phosphate, has no dispersant activity whatsoever. It actually flocculates the particle system slightly.

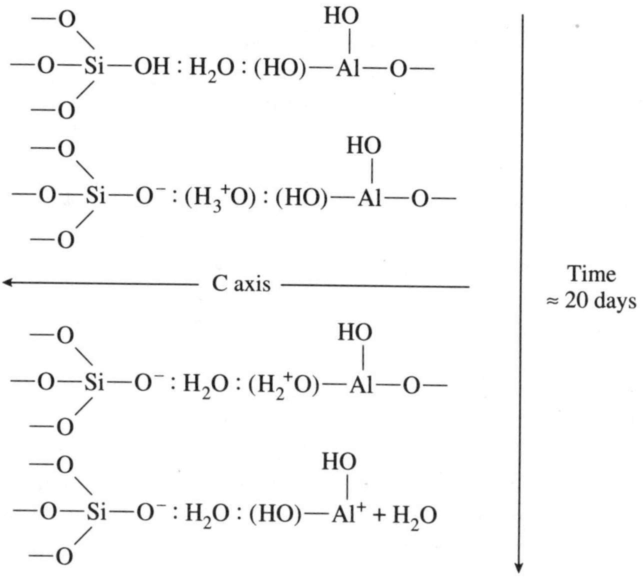

Figure 3.14. Proton migration on silicate surface.

Proton transfer in the decomposition reactions results in the chemical conversion of pyrophosphate to orthophosphate and mechanical transition from dispersed state to flocculated state. Part of the mechanism for this reaction lies in a catalytic effect provided by the solid surface. The template fit between sequential units in polyphosphate chains and the sequential alumina octahedra is not

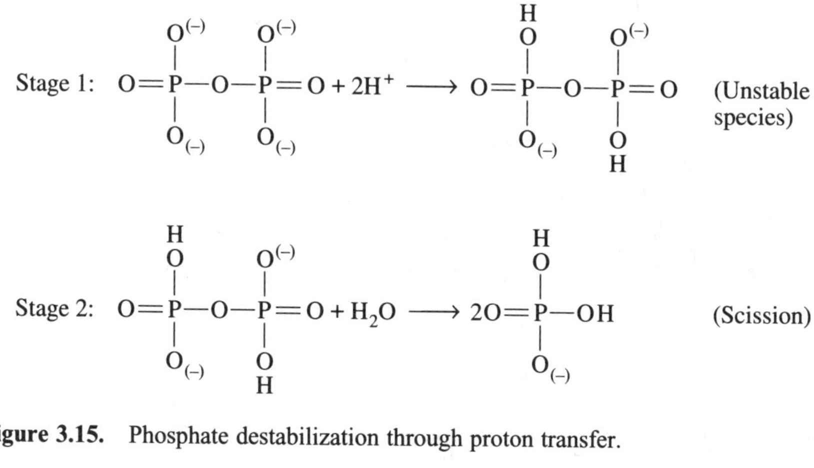

Figure 3.15. Phosphate destabilization through proton transfer.

exact. As a consequence complex phosphate molecules experience some coordination strain in the train configuration. Obviously, this mismatch and strain increase as the polyphosphate molecule lengthens. However, with longer molecules the loss of a terminal PO_4 tetrahedra is not as critical as with TSPP, even with the greater strain. Very long polyphosphates, where the chain consists of 20 or more monomers (e.g., Glass H™) show this effect to a lesser extent because they do not lie down fully in train configurations but possess significant tail orientations. Because of their poor dispersancy, including interparticle bridging and flocculation, the detection of decreasing dispersancy is difficult.

The rate of orthophosphate production, as the species sodium dihydrogen phosphate, would be expected to increase for a given surface as chain length and misfit of the dispersant molecule increase. In long-term studies[17] this is proven to occur, as is shown in Figure 3.16.

At ambient temperature, pyrophosphate is about 30% decomposed in about 300 hours, or 12 days. At elevated temperature (50°C) that time reduces to 30 hours. The rates of decomposition for SHMP and STPP are substantially greater than for TSPP because of their more strained adsorption configurations.

As conversion from polyphosphate to orthophosphate reduces the surface charge on the particle surface, dispersion would be expected to degrade at the

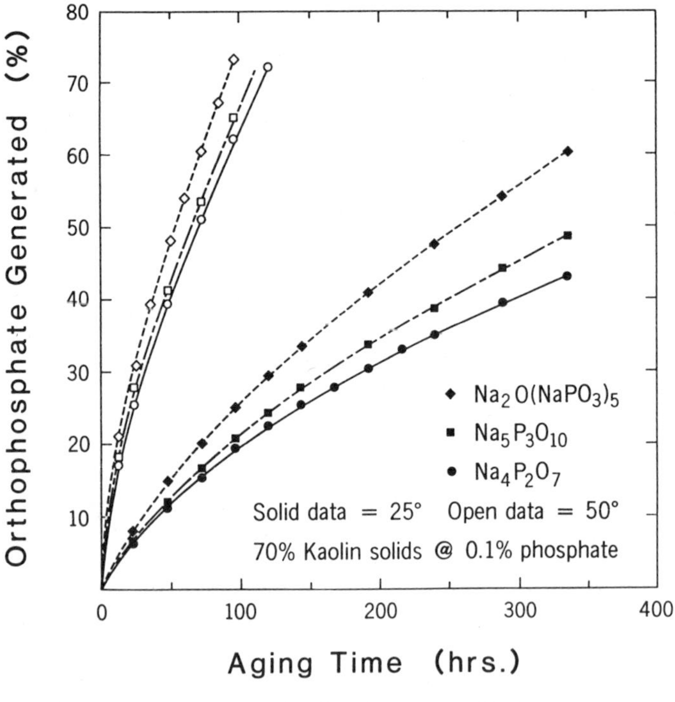

Figure 3.16. Dispersant decay to orthophosphate on kaolinite surface.

same rate. Figure 3.17 shows the viscosity increase during the same period as a consequence of this proton transfer reaction and surface catalysis. Because viscosities varied slightly in formulation, their relative change at time t, η_t/η_o, is employed for all studies.

Decomposition of dispersant can become a disaster for predispersed slurry systems, especially those in transport. Because proton transfer rates are slow, materials formulated at manufacturing sites may display excellent rheological characteristics, but change dramatically after time lapse. In years past, prior to the knowledge that dispersants were subject to surface catalytic decomposition, high-solids slurries would be plant formulated at 70–72% solids, pumped into tank cars and shipped transcontinentally. These transits could require 10–20 days, including siding storage. During warm summer months railcars, particularly those painted black, traveling through southwestern U.S. regions, could expect a temperature rise of 10 to 20°C during transit. A well-dispersed slurry with 0.35% polyphosphate might have a formulation viscosity between 1000 and 2000 cp at point of origin. After transit and 30% decomposition, the system would yet have 72% solids and have 0.35% phosphate present, but would have converted to a rigid solid, requiring air hammers to remove 50 tons of solidified pigment from a 10,000 gallon tank car!

To mitigate this problem, slurry pH must be elevated. Higher alkalinity in the solution phase reduces the free, or mobile, proton population on the pigment surface by reacting at the first stage of acid activity shown in Figure 3.14. In

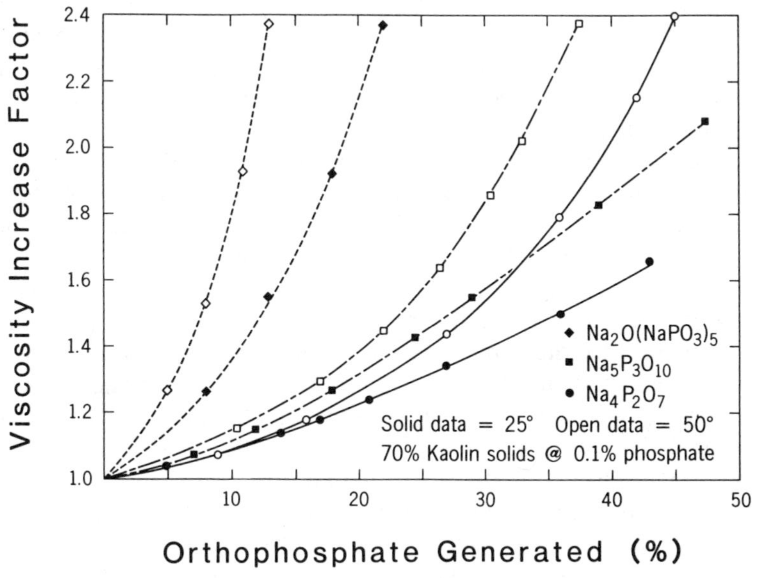

Figure 3.17. Rheology degradation from orthophosphate conversion.

certain applications, pH elevation may be restricted. With these, the superior dispersant of choice is TSPP because of its lower catalytic degradation rate. A second solution to the problem lies in shipping in white or aluminized tank cars. A third is to eliminate polyphosphates altogether!

The decomposition reactions above have been proposed to influence eutrophication, or excessive algae growth, in waterways and lakes by way of discharged polyphosphates from cleaning products. Living organisms are unable to metabolize polyphosphates. The decomposition is believed to be promoted by aluminum hydroxide gels derived from water purification (lime plus alum) and natural clays entrapped in them where the pH values fall to 6.5 or below.[18] The same mechanism has been extrapolated to natural colloidal materials on clay surfaces to account for polyphosphate degradation there. This, however, is at best questionable. Investigations of polyphosphate decomposition rates on chemically cleaned versus natural kaolinites show no relationship to natural colloidal bodies present.[19]

Commercial titanium dioxide pigments, however, some of which bear aluminum hydroxide gel coatings, may well accelerate dispersant breakdown where the system pH is on the acidic side.

Another problem develops simultaneously with catalytic phosphate degeneration. Bacteria-laden pigment slurries metabolize excessively in the presence of orthophosphate, but not polyphosphate[20]—a process equivalent to the eutrophication just noted. This process is most rapid in the common slurry range, pH 6.0–7.5. These bacteria lower pH as a consequence of their metabolizing phosphate, which accelerates additional phosphate breakdown. Alkali alone is insufficient to destroy bacterial activity. Slurries must also contain specific bacteriocides, compatible with the application products of the slurried pigment. Mercurials were once the common treatment, but these have more recently been replaced by brominated organics, phenol derivatives, and hydroxyquinoline compounds. Silicates, however, are not the only pigment species subject to problems of phosphate instability.

Alumina and alumina hydrate are manufactured in a variety of forms which may be dispersed in certain applications with polyphosphates. Because they have structures akin to kaolinite (and other octahedrally coordinated pigments), SHMP, STPP and TSPP serve as excellent dispersants.

Alumina ceramics employ mixtures of alumina and kaolinite in approximately a 9:1 ratio for casting slurries to be used in high-temperature bodies: furnace liners, crucibles, electrical-grade insulators, and sparkplugs. The slurry solids range from about 50% to 65%. For very fine, high-density firing bodies (e.g., sparkplugs), the viscosity minimum corresponds closely to the most negative zeta potential. This point occurs at about 0.2% STPP. If the dispersant level is increased to 0.5% STPP, fired body strength decreases.

For these high-density alumina bodies sodium silicate provides superior rheological properties. Its basic shortcoming as a dispersant lies in its poor long-term stability. Both sodium tripolyphosphate and sodium silicate can be employed to spray dry slurry for ceramic grog formation if the slurries are prepared fresh.

Alumina hydrate is manufactured in various morphologies, tabular (platelike), prismatic, and amorphous. Polyphosphates coordinate with these surfaces in a fashion similar to kaolinite. However, the isoelectric pH for these forms of alumina lies between 6.0 and 7.0.[21]

No protons are generated in these aluminum oxide compounds because active silanol groups are absent. However, formulation pH can become a critical factor

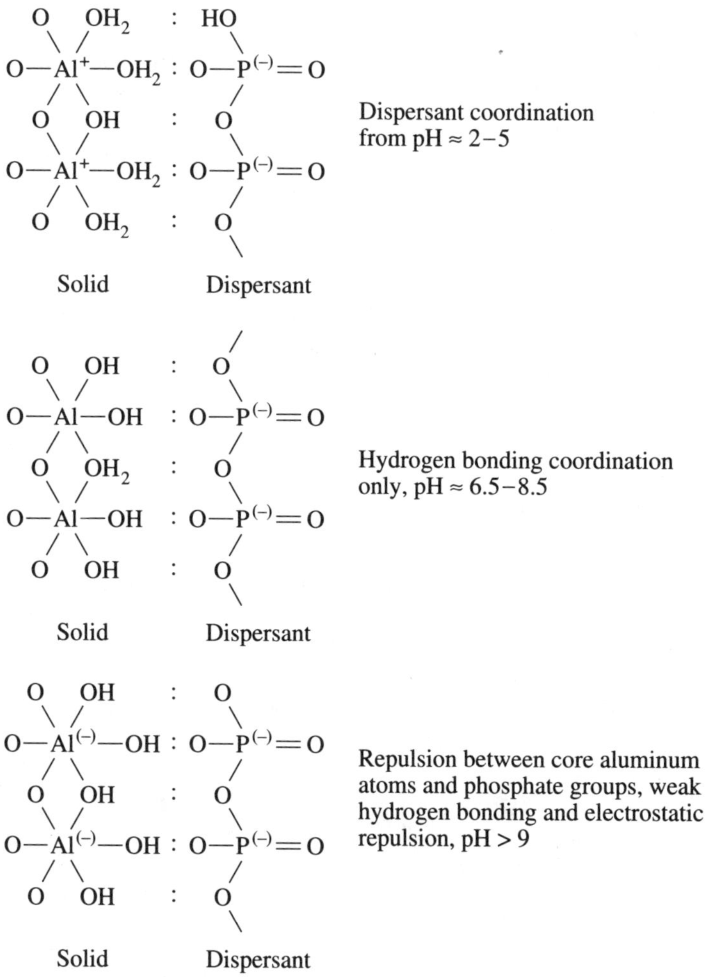

Figure 3.18. Phosphate destabilization on alumina hydrate surface.

in slurry stability. Figure 3.18 illustrates the three ranges of dispersant–pigment interaction as a function of the operation or application range.

No protonic degradation or scission of the dispersant occurs with alumina hydrate. However, prolonged association between the polyphosphates and active alumina hydrate below about pH 6.0, where electrostatic attraction pulls the dispersant molecules into the coordination spheres of aluminum atoms, results in slow chemical reaction between the two. This takes the form as depicted in Figure 3.19.

With longer-chain polyphosphates not only is dispersion lost, but new interparticle chemical bonds may be established and a quasisintered structure result. These same reactons, interestingly, in nature in the presence of copper ion are the basis for the formation of the semiprecious gemstone turquoise, $H[CuOH][Al(OH)_2]_6\,[HPO_4]_4$.

In certain ceramic applications members of the polyphosphate family with increased chain length have found use[22] because of their resistance to this type of short-term degradation observed with both TSPP and STPP. Chains having 13 to 21 members are most useful for operations where prolonged slurry states are required.

A typical rheological profile of a mixed-pigment system as a function of aging is shown in Figure 3.20. Both components of the pigment mix (rutile and kaolinite) possess acidic surfaces and promote phosphate dispersant breakdown. Note the general viscosity increase, the change of shape in the rheograms, and the displacement of viscosity minima towards higher dispersant dosage with

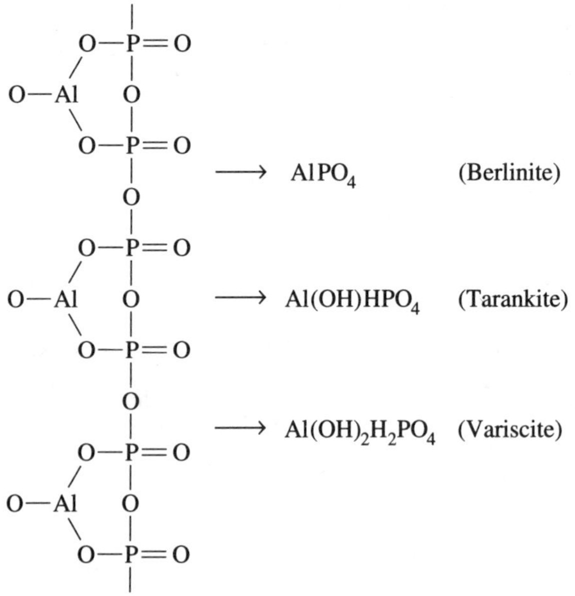

Figure 3.19. Phosphate interaction with alumina surface.

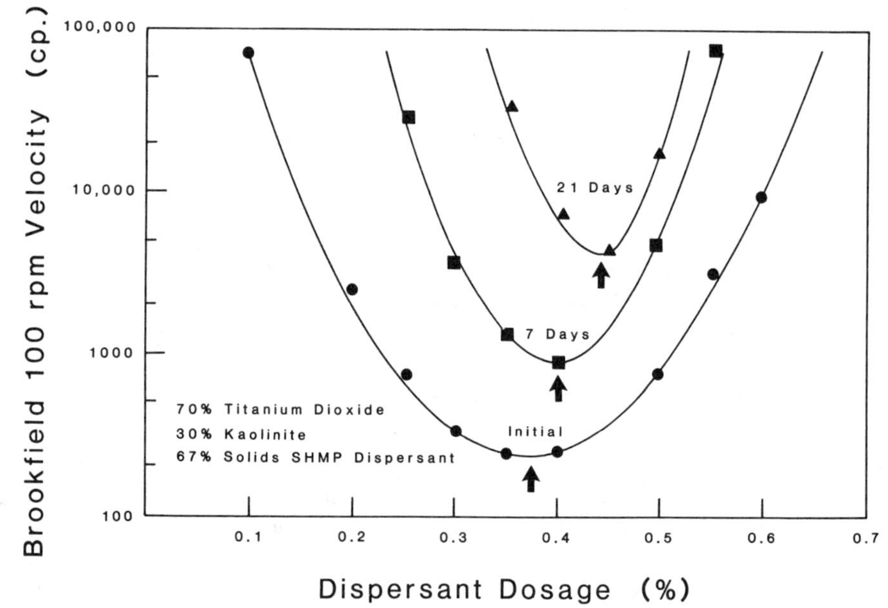

Figure 3.20. Rheology change of rutile–kaolinite pigment system with aging.

time. Rheogram shape also will be altered slightly by dispersant molecular length, adjutant pH control, and other factors that affect breakdown kinetics.

Ferric iron reacts similarly to alumina with polyphosphates. While ferric oxide is rarely slurried for pigment applications, a unique yellow "micaceous" pigment has been made in Japan that take advantage of this reaction.[23] Solutions of ferric salt, below the critical hydrolysis concentration, are acidified and TSPP or similar phosphate added and agitated. Thereafter, sufficient strong alkali is introduced to raise the pH above 7 and the entire system is autoclaved. The yellow flake morphology pigment has been employed as an economical gold leaf substitute in paints, inks, food products, etc.

In the manufacture and transportation of high-solids slurries of titanium dioxide (>70%), sodium hydroxide is added with SHMP to raise the pH as high as 10 and stabilize these slurries for long periods. Titania, in contrast to silica and silicates, is more resistant to chemical etching from alkali. By adding alkali following the dispersant makedown, less primary dispersant is required, and lower viscosities are obtained.[24] Brookfield viscosities at 100 RPM for 70% solids slurries have been formulated in the 600–700 cp region by this technique.

Several factors affect the kinetics of any dispersant decomposition on a solid surface: pH, bond strength, pK_a, slurry solids, temperature, and the insolubility of any interactive species.

Sodium polyaluminate is one of the more alkaline dispersing agents and in that respect should be more effective because of its strong zeta-potential effect, as shown in Figure 3.21. The molecular configuration and size are essentially exact fits for the alumina and other octahedra on mica, alumina hydrate, talc, kaolinite, ferric oxide, etc., as illustrated in Figure 3.22. However, bond strength in the anion complex is low, owing to a strong cation occupying the core position of the coordinated anion. Its octahedral anionic species (slightly puckered because of bond ionization), having a nominal four units per chain ($[AlO_2(H_2O)_4]^-$), can undergo both chain scission by acidic decomposition and configuration stress, in a fashion similar to polyphosphates. While the molecule is reasonably stable on many octahedral substrates, it suffers when employed on tetrahedral silica where the molecular misfit is over +0.5 nm.

Figure 3.23 contains time-dependent rheograms of finely ground silica (quartz) having a median diameter of 4.6 μm dispersed with sodium aluminate. The product was developed as a flatting pigment for paint applications.

Note in the figure the rapid rise in viscosity once chain scission begins. In this instance a complex surface molecular inversion reaction develops, as depicted in Figure 3.24, where the circled "W" denotes coordinated water molecules, three of which coordinate in each alumina octahedron.

A similar reaction occurs where sodium silicate is employed to disperse alumina or alumina hydrate. Calcining the alumina at elevated temperatures greatly

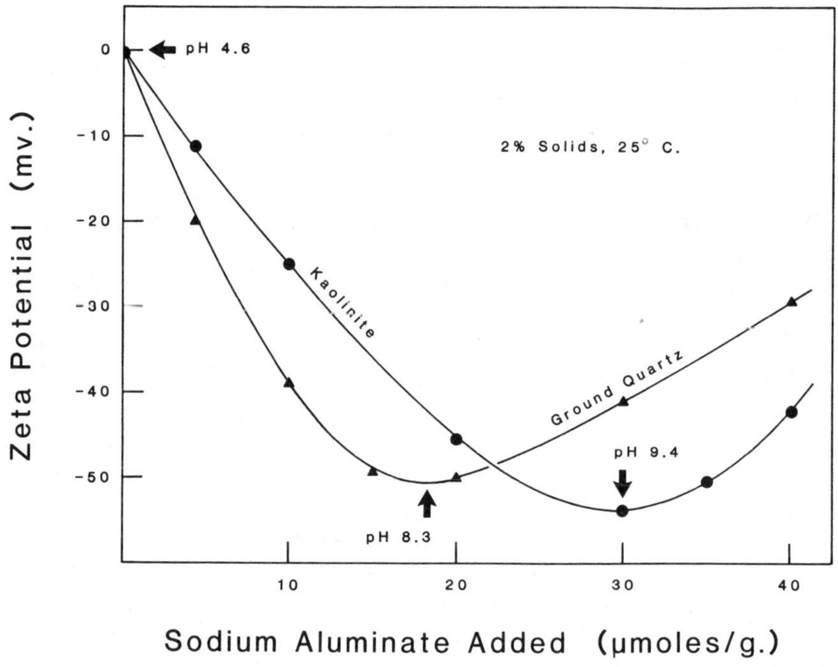

Figure 3.21. Zeta-potential plot for sodium aluminate on silicate.

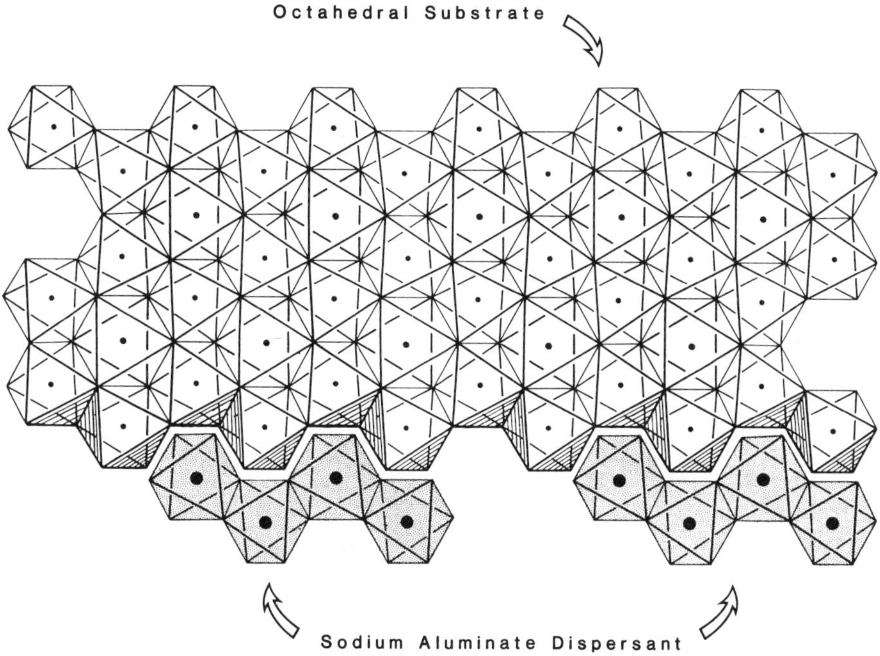

Figure 3.22. Structural coordination of sodium aluminate on octahedral substrates.

reduces this reactivity. As an example of the extent of this reaction, mixing solutions of the two dispersants, sodium aluminate and sodium silicate, causes an almost instant gel that slowly reforms to an open structure whose size and composition are dependent on relative concentrations, temperature, pH, etc. It is in crude form one of the methods employed for producing synthetic zeolites and petroleum cracking catalysts.

Sodium borate, having an electronegative element in the coordinate group, is moderately stable. Unfortunately, it suffers from having only a short chain (two groupings) and low charge density with but two terminal anions. It is less reactive with acidic pigment surfaces and has been employed for silica dispersions in certain applications.

Sodium silicate, though highly effective, is generally avoided as a dispersant because of its excessive alkalinity and tendency to polymerize irreversibly upon drying (the "water glass" effect). It possesses an advantage, however, as a codispersant with phosphates. Tetrasodium pyrophosphate in particular has a chemical tendency to dissolve zinc from brass bearings and structural elements of chemical processing equipment (a process termed *dezincification*) to which slurries are exposed (pumps, blendors, valves, etc.). The reaction is sufficiently rapid and strong to freeze or embrittle such equipment after short periods of

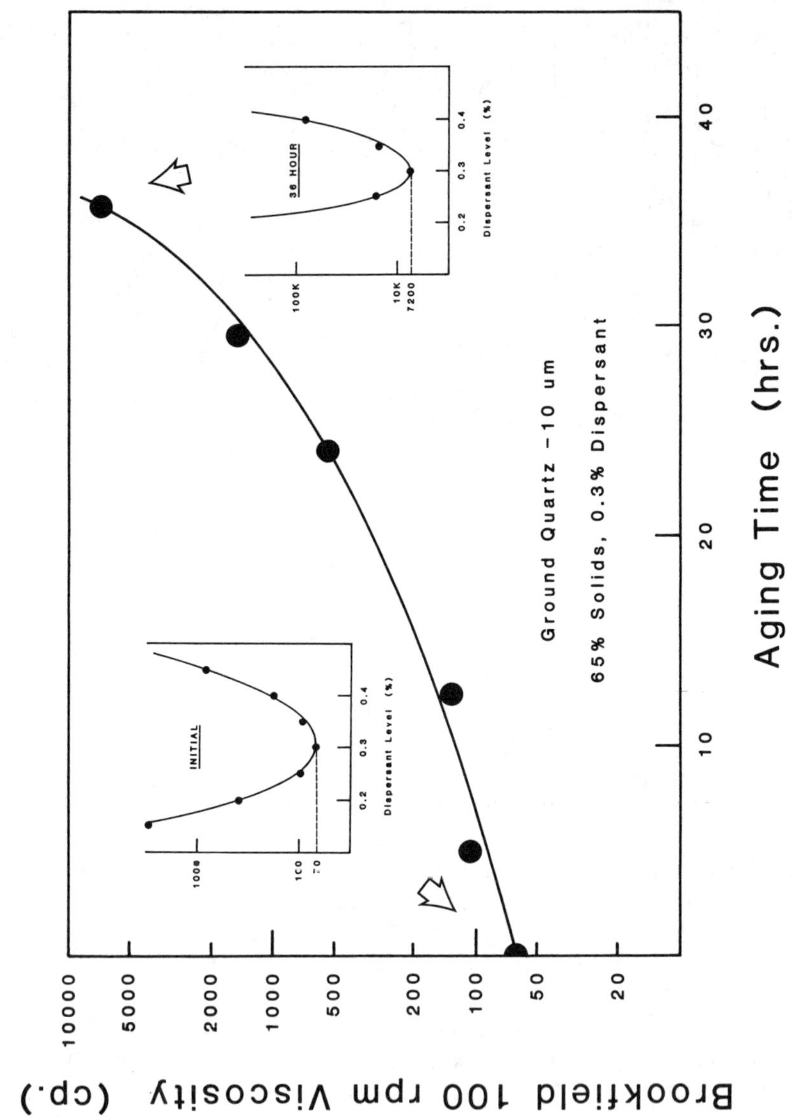

Figure 3.23. Ground silica rheograms with scissioning dispersant.

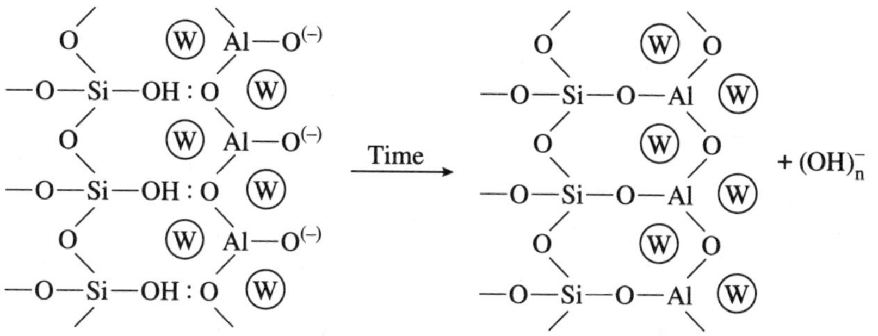

Figure 3.24. Interaction of aluminate anion dispersant on silica.

slurry exposure. The addition of but a few percent (relative basis) sodium silicate to the sodium polyphosphate produces a passivated surface. Passivation is assumed to result from zinc silicate formation on the brass element.

The broad spectrum of phosphate reactions noted will not occur with alkaline oxide particles, magnesium oxide, calcium carbonate, lead oxide, zinc oxide, etc. However, the potential for dispersant degradation, either through decomposition or surficial reaction, is substantial with acidic surface pigments. Alkaline oxide pigments do, however, have the potential for dispersant interaction where strongly alkaline dispersants are employed. This is especially true for sodium silicate, where calcium, magnesium, and zinc silicates can form and alter surface charge.

An example of long-term stability of polyphosphates on nonoxide pigments lies in the production of barium sulfate dispersions for X-ray and fluoroscopy applications. These slurries are pharmaceutical formulations which must retain stability for many months, even years, and, because of being internally consumed, cannot exhibit elevated pH. This is one of few applications where SHMP displays superior long-term stability to sodium polyacrylate. Because of the density of $BaSO_4$ (4.2 g/cm^3) and its proclivity to settle and cake during slurried storage, small quantities of hydroxymethyl cellulose (food grade) are added after the barium sulfate has been thoroughly dispersed. Most barite employed for this application is chemically precipitated, sulfate rich (to minimize the presence of toxic barium ion), and -50 μm in size.[25] An additional advantage of employing SHMP as a dispersant lies in its capacity to sequester barium ions, should any be present from manufacture or develop from application.

Obviously, stomach and intestinal acids will slowly degrade the SHMP. However, because residence time is short, most often 2 hours or less, this degradation is of minimal concern. High acidity along the intestinal wall does cause minor flocculation and attachment of particulates to the wall surface. Highly acidic ulcerous regions can be detected by just this phenomenon as such flocculi show up as light patches (high $BaSO_4$) on the X-ray screen or film.

Ionic sequestering phenomena, as discussed previously, are especially important with the use of polyphosphates on gypsum. There an exchange reaction occurs whereby structural calcium (likely in solution equilibrium at the surface) reacts with the polyphosphate to form sequestered calcium ions. Eventually, sufficient dispersant is complexed in this manner that its effectiveness for producing surface charge is lost and the slurry flocculates. This activity is far more pronounced with SHMP than with TSPP because of its long molecular configuration.

With more alkaline dispersants, especially polymeric ones, reactivity is both more rapid and more damaging upon acidic substrates. A limited number of exceptions occur. These include, for example, sodium *n*-silicate on ground silica (quartz) and SHMP and TSPP on calcium phosphate (rock phosphate). In both instances the equilibrium constant between acidic surface and phosphate grouping, noted earlier, must be 1.0 or less.

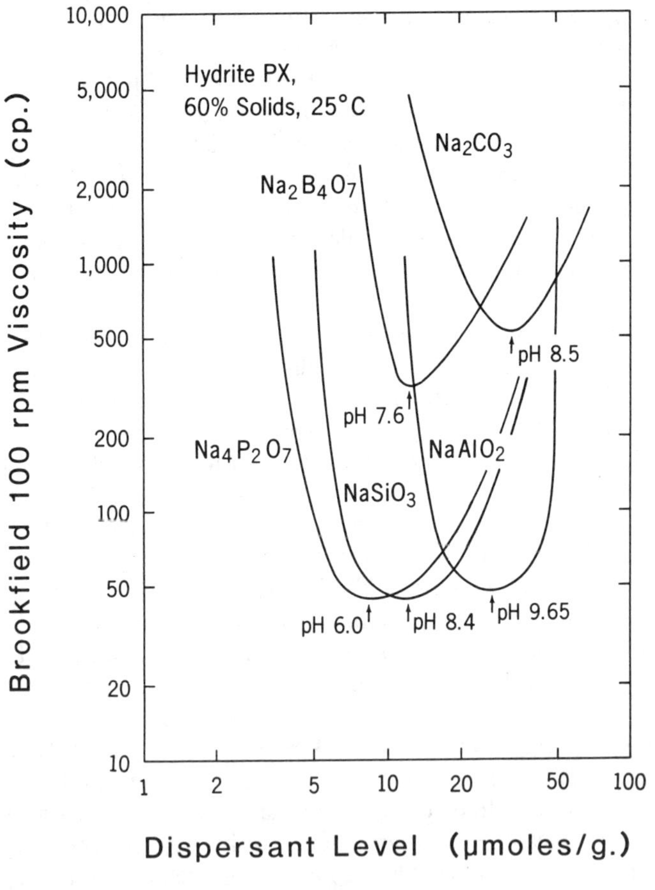

Figure 3.25. Inorganic salt dispersant family rheograms.

3.4 Family Rheograms

The family of inorganic salt dispersants share a general property in that they all experience a substantial shift in concentration to the aqueous phase at high solids dispersion. It is this phenomenon that generates the catenary rheological profiles common to each species.

Figure 3.25 is a short survey of these species on a moderate-solids, single-silicate pigment. Except for pH, aluminate, silicate, and pyrophosphate are virtually identical when the dosage is monitored on a molar basis rather than by weight. Both borate (dimer species) and carbonate (monomer species) have substantially higher viscosity minima (factor of 10) and a sharper rheological trough than the other members. SHMP produces a curve essentially identical to TSPP but possesses a pH at the viscosity minimum about 0.5 units lower.

References

1. "Stability of Complex Phosphates," Tech. Bul. 810-A, FMC Corp., (1981).
2. Manfredini, T., et al., *Appl. Clay Sci.* **5**, 192 (1990).
3. Dobson, D. "Modifying Chromium Oxide Pigments," Canadian Patent #1,090,503, April 18, 1977.
4. Kanno, T., and Umeya, K., *Zairyo* **30**, 83 (1981).
5. Leaf, C., and Liggett, L., *TAPPI* **39**, 142 (1956).
6. Matsumura, T., and Iwasaki, K., *J. Chem. Soc. Japan* **58**, 903 (1955).
7. Garrido, L. Gainza, J., and Periera, E., *Appl. Clay Sci.* **3**, 923 (1988).
8. Martynov, G., *Kolloidn. Zh.* **40**, 1110 (1978).
9. Both the iodine molecule (from hydrocarbon solvents) and iodide ion (from aqueous solvent) are known to absorb nearly quantitatively on many solid surfaces. This alteration of surface charge and mobility is the basis for an early method of assessing surface area, especially on very fine powders. See: Shimoiisaka, J., et al., *Mineralogical Soc.* **3**, 12 (1962).
10. Conley, R., *J. Clay and Clay Min.* **14**, 317 (1966).
11. Van Wazer, J., *J. Am. Chem. Soc.* **72**, 647 (1951).
12. Bundy, W., *Chem. Eng. Prog.* **63**, 57 (1967).
13. Mathur, K., Pfizer MPM Div., private communication (1978).
14. Verwey, E., and Overbeek, J., *Theory of Stability of Lyophobic Colloids*, Elsevier, NY (1948).
15. Conley, R., and Althoff, A., *J. Colloid and Interface Sci.* **37**, 186 (1971).
16. Fripiat, J., and Toussaint, F., *J. Phys. Chem.* **67**, 30 (1963).
17. Shepherd, J., FMC private communication (1972).
18. Clesari, N., and Lee, G., *J. Air & Water Poll.* **9**, 723 (1965).
19. Shepherd, J., *op cit.*
20. Stumm, W., and Morgan, J., *Aquatic Chemistry*, Wiley-Interscience (1970).

21. Parks, G., *Chem. Rev.* **65**, 177 (1965). (Also Alcoa publ.)

22. Haber, R., *Ceram. Eng. Sci, Proc.* **12**, 106 (1991).

23. A.I.S.T., ''Precipitating Ferric Oxide as Mica for Pigment,'' Japan Patent #80,154,318, Dec. 1, 1980.

24. Story, P., ''Manufacture of High Solids Titanium Dioxide Slurries,'' Intl. Patent Appl. #WO-90-13606, Nov. 15, 1990.

25. Wang, H., and Hu, X., *Yaoxue Xuebao* **16**, 610 (1981).

26. Bundy, W., and Murray, H., *Clays and Clay Min.* **21**, 295 (1974).

CHAPTER

4

Organic Polyacid Salt Dispersants

Techniques for imparting an isopolar charge to a solid surface and thereby creating a repulsive field are not limited to inorganic species, although until the 1950s these were the most popular agents. By employing an organic chain as the polymeric species (almost always an anionic surfactant), a much greater variety in dispersant species, group spacing, chain length, and other variables can be induced than with the more simple inorganic anionic chains. These materials are not without their drawbacks, however, cost being uppermost. Organic charge dispersants must be synthesized from petroleum or animal byproducts or manufactured by complex polymerization chemistry from natural products, which limits their supply. Two of the family's major assets lie in their low alkalinity and anticorrosive character. These assets greatly increase their value when incorporated into organic application systems (paints, plastics, pharmaceuticals, and cosmetics).

4.1 The Acrylate Family

One of dispersant technology's great advancements came about in the 1960s when polyanionic salts of organic acids (chiefly alkali metal salts) were introduced into the pigment and milling industries. For years sodium citrate had been employed[1] as a general grinding aid in the processing of cement clinker, limestone, and red iron oxide. But it was not until technology derived from manufacturing polyacrylates for military aircraft windows matured and such materials entered the public domain that chemical derivatives of these family became available and their functionality understood.

Salts of organic polyacids (SOPAs) display a number of remarkable properties. Coincidentally, at the time of their development enrivonmentalists discovered phosphates were the direct cause of eutrophication (algae buildup in lakes, ponds, and streams). Thus problems created by flushing waste polyphosphates from metal cleaning and laundry operations and from incorporating dispersed pigments in systems that would later be flushed out (e.g., paper coating operations) motivated serious interest in this new dispersant family.

Many of these neutralized organic compounds were made by polymerizing acrylic and maleic acids and mixtures thereof. Those two compounds are similar, producing polymers that differ only by the separation between acid groups. The monomer structures and their polymerization reactions are depicted in Figure 4.1.

Later variations on the structure came about through substitution of methacrylic for acrylic acid, which introduces a methyl group for one of the alpha hydrogen atoms, and includes an olefin that increases the carboxylate spacing.

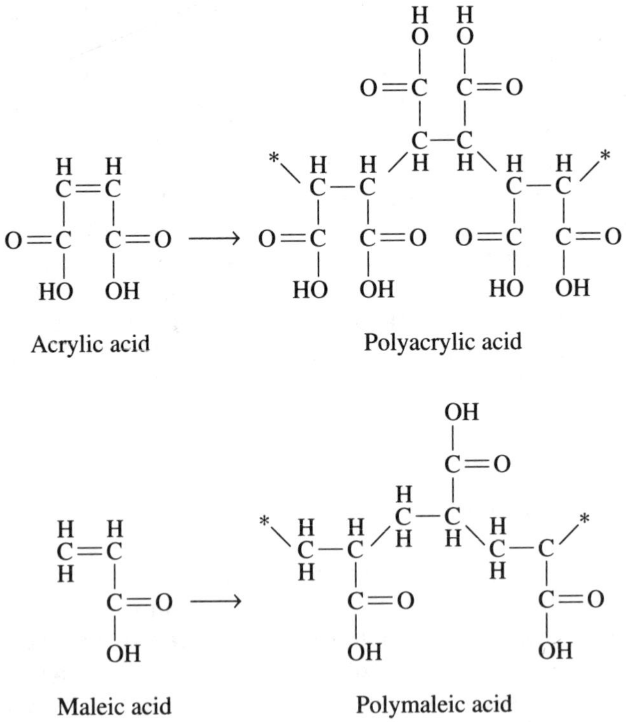

Figure 4.1. Polyacrylate production reactions.

Olefin insertion usually increases the acidity (pK_a) of the nearest –COOH grouping and increases the van der Waals attractive force through molecular mass increase. Except where specific differences in chemical behavior exist, the term *polyacrylates* will be employed generically in this section to designate these polymers and their copolymers collectively.

A second advantage to these materials lies in the stability of their polyorganic acid structures. They show little or no catalytic degradation from acid-active pigment surfaces, especially upon long contact, and display no bacterial decomposition in the solution phase. The pK_a values for all *R*–COOH groups on the polyacrylate chain are 5 or below, greatly limiting the proton exchange with silicates and other pigment surface acid donors.

Third, although salts of weak acids and strong bases, the products have a significantly lower alkalizing effect upon slurries compared, for example, to TSPP. Their structures, resembling an accordian, have the capability to expand or contract, which allows individual (or alternate) acid (or salt) groupings to configure to a broad variety of solid surfaces and coordination structures. Whereas phosphates and silicates fit tetrahedral dispersant groups into octahedral niches, polyacrylates fit triangles into those same reentrant positions. Mechanical modeling suggests the triangles lie planar to the octahedral array in compounds like iron oxide, titanium dioxide, aluminum silicates, etc.

Fourth, many SOPAs can be manufactured as 40–45% solutions, which is not possible with TSPP and other phosphates. With such high water solubility concentrated dispersants can be metered directly as solution additions to slurries without requiring large additive tanks.

The molecular structure of the SOPAs (Figure 4.2) compares favorably with that of the well known citrate except for molecular length.

In the polyacrylate molecule an independent hydrocarbon chain serves as a backbone, threading its way along the entire molecule. Alternate carbon atoms have branched sp^2 planar active groups, $–COO^{(-)}Na^+$. This molecule has a significant difference and advantage over polyphosphate molecules. Its unit charge density is the same, 1.0, but spacing between charge active groups, which can coordinate with a solid surface, is nearly twice that of polyphosphates. Functionally, alternate carboxylate groups coordinate with positively charged (ionic) or octahedral (covalent) sites on the solid surface, while intermediate carboxylate groups extend into the solution phase to provide molecular charge and a double layer. The primary advantage of this structure has been noted—flexibility of the independent hydrocarbon spine allowing the molecule considerably greater latitude in expansion or contraction. This enables adaptability to charge and structural sites on solid surfaces without inducing bond stress into the molecule.

Although lower in molecular weight per active group, 71 for acrylate versus 87 for phosphate (TSPP), which reduces the unit van der Waals attractive force for the solid surface, the net molecular weight can be much higher, typically 1500 to 5000. Studies show polyacrylates often to be more stable than polyphosphates.[2]

As with polyphosphates, an increasing chain length of polyacrylates results

Sodium citrate

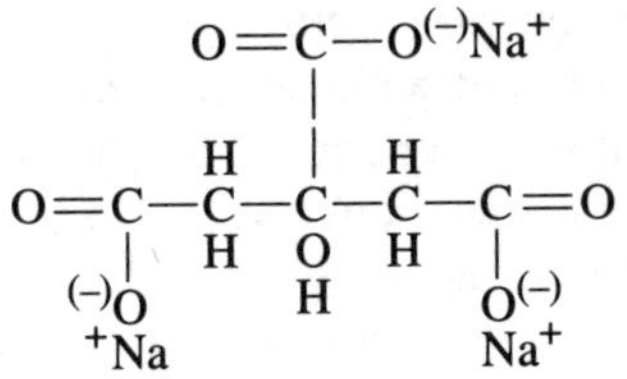

Sodium polyacrylate

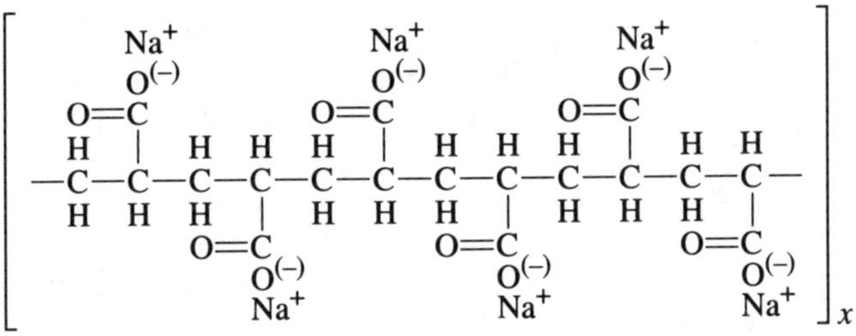

Figure 4.2. Molecular similarity between citrate and polyacrylate.

in decreased aqueous solubility. However, because of increased molecular freedom at each adsoption juncture, polyacrylate's configuration upon solid surfaces alters with length, a characteristic also of polyphosphates but to a much more limited extent. Beyond a molecular weight of about 3000–5000, representing 20 to 30 active groups (10 to 15 site active groups), dispersancy actually diminishes, except with special cases.[3] A case study for calcium carbonate is shown in Figure 4.3, where very long chains apparently bridge particles and serve to flocculate particle assemblages. Polyacrylates with molecular weights 100,000 or greater are utilized as mild flocculants for the same materials that short-chain polymers serve as dispersants.

Also like polyphosphates (especially SHMP), commercial polyacrylate dispersant products contain mixtures of polymers varying about ±25% of the molecular weight for the shorter members and even more with the longer species.

Polymeric salts of all polyorganic acids do not necessarily serve well as dispersants. In a study on kaolinite processing,[4] the sodium salt of a suite of organic acids was employed to reduce viscosity of a 65% slurry of the pigment in water. Results are shown in Table 4.1.

Inspection of the tabular data reveals polymeric acrylate to be the superior dispersant, with sodium citrate following in second place. The acrylate polymer possesses the highest molecular weight, ca. 2000, and the highest charge density.

Figure 4.4 shows the accordianlike configuration of a polyacrylate molecule

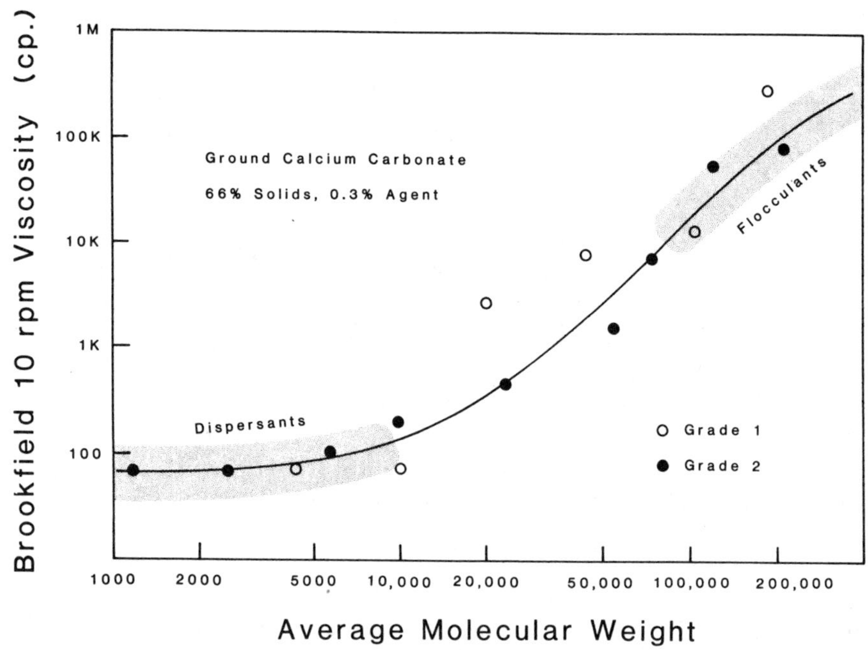

Figure 4.3. Polyacrylate chain length influence on viscosity.

on an ionic solid, barium sulfate (barite, often termed ''barytes''). Barite has one of the highest cation–cation spacings among common pigments, approximately 0.70 nm.[5] At this distance a polyacylate molecule must compress and extend its alternate charged groups well out into the solution phase.

A quick assessment of this structure would suggest, especially based on the slightly positive zeta potential of barite, that certain facets of the crystal could have sulfate ions dominating as the more protuberant species along the surface.

Table 4.1 Organic Polyacid Salts as Viscosity Modifiers

Species	Anion Formula	pH	Viscosity (cp)
No dispersant	—	6.0	>200,000
Sodium malonate	Na OOC—CH_2—COO Na	8.1	31,000
Sodium succinate	Na OOC—C_2H_4—COO Na	8.9	38,000
Sodium malate	Na OOC—CHOH—CH_2—COO Na	8.5	13,000
Sodium glutarate	Na OOC—C_3H_6—COO Na	9.3	40,000
Sodium glutamate	Na OOC—C_2H_4—$CHNH_2$—COO Na	8.5	46,000
Sodium citrate	[Na OOC—$CH_2]_2$—COH—COO Na	8.3	800
Sodium acrylate	[Na OOC—CH_2—$]_{20}$	8.0	240

Note: 65% kaolinite solids, 25°C, 30 μmoles/g level.

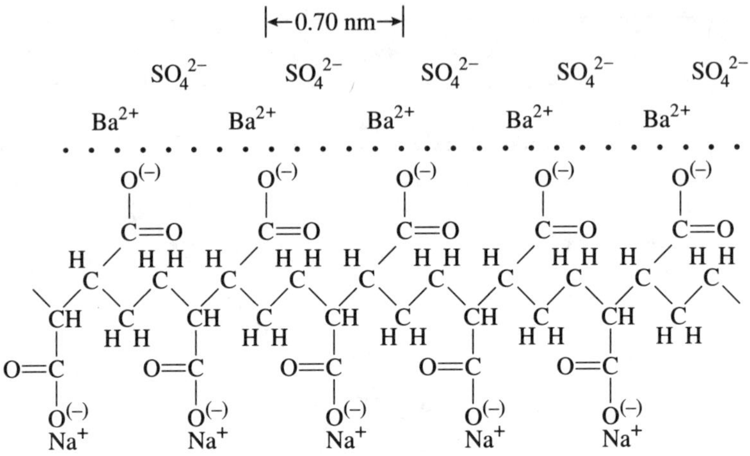

Figure 4.4. Coordination of polyacrylate anion on barium sulfate.

On these dominantly negative faces, polyacrylate ions would be repelled. This apparently occurs because the quantity of dispersant required to provide maximum electrophoretic mobility, as well as minimum viscosity at high solids, is well below that calculated based on molecular sizes and weights and powder surface area. This same reduction of dispersant has been observed with calcium carbonate and other white pigments.[6]

Water adsorbed on the barium sulfate surface is believed to enhance the adsorption of polyacrylate due to thermodynamic effects. Because of the large size of both the barium and sulfate ions, these do not hydrate during the crystallization process. Adsorption studies on the solid show that entropy decreases during the water adsorption process, that is, prior to the addition of dispersant. It is believed water molecules site themselves preferentially on the protruding sulfate ion facets. This preferential blanketing of the surface may direct the sodium polyacrylate molecules to the dominating barium faces or locales. The enthalpy of polyacrylate adsorption (heat of adsorption) decreases with the amount of water adsorbed prior to dispersant addition.[7] This factor in itself may account for the reduced level of dispersant required to achieve minimum viscosity.

Polyacrylate on ionic species, for example barite and calcite, like polyphosphates on covalent species, establishes an equilibrium with the solution phase, as deduced from the quantities required for minimum viscosity and surface areas of the solids, as well as their non-Langmuirian adsorption profiles.

Polyacrylates are used to deflocculate kaolinite slurries, both for production purposes and in application (paint and paper pigments). In these applications a close correspondence exists between the minimum amount to accomplish viscosity reduction (left arm approach on the rheological profile), as determined by sedimentation packing volume (see Chapter 13), and extrapolation of the near-

Langmuir adsorption isotherm.[8] Unlike polyphosphates, polyacrylates show much less tendency to undergo a tail-to-train reconfiguration during the course of the adsorption process, as illustrated by Figure 4.5.

Calculations based on the monolayer and the surface area of the silicate pigments indicate that polyacrylates, like phosphates, adsorb only along edge faces of the crystals (approximately 16% of the particle's total area). However, because the adsorption energy is greater (deduced from the greater isotherm flatness), the level required to disperse the pigment is less (less material being equilibrated into the solution phase).

Polyacrylates can be manufactured with sodium, potassium, and ammonium cations. Of these, sodium is the most popular because of its lower cost. In ceramic casting slips, however, sodium creates problems because dispersant cations migrate into the gypsum mold and slowly disintegrate it by solution weakening. For this reason potassium salts are found to provide superior clay casting, release, and mold life in the ceramics industry.[9] The economics here are simple. The dispersant amounts to approximately 0.1 pound with a differential cation cost of about $0.25 (1992 data). Gypsum molds can run several hundred dollars. Thus an increase in production of 800 pieces is required for break-even economics, far more than most common molds produce. However, the production

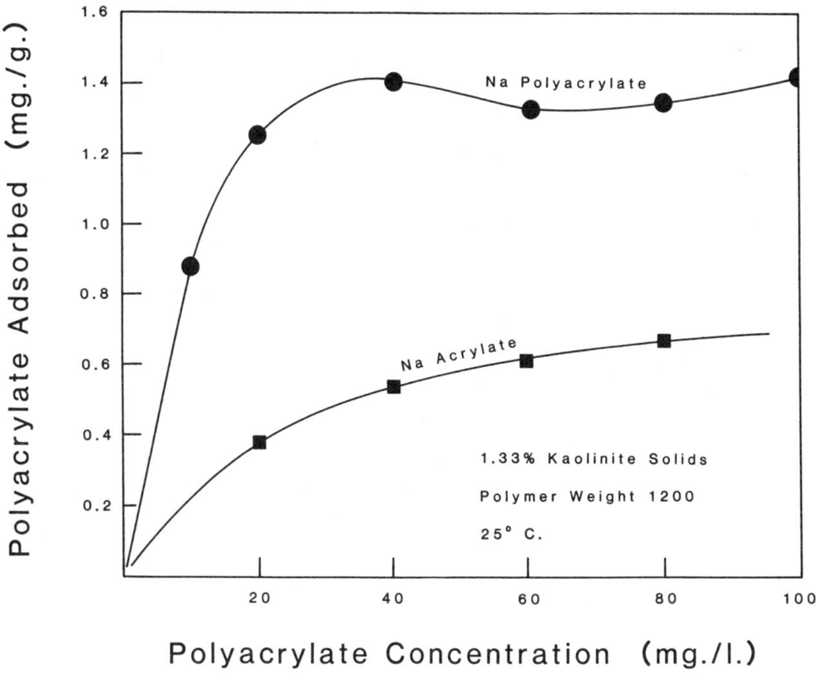

Figure 4.5. Adsorption isotherms for polyacrylates.

down time to remove and setup new molds more than outweights the cost differential between dispersants.

Some ceramic casting operations improve mold release and mold stability by employing a trace (1:100) of polyvinyl amine or long-chain alkanolamine as codispersant with the potassium polyacrylate.

Where the casting slip contains fibrous silicates (nonasbestos) to impart high impact strength, their high surface area (>10 M^2/g) will increase dispersant demand to 1.0–1.5%, solids basis. Palygorskite (attapulgite), sepiolite, halloysite, hectorite, and smectite replace in part regular casting clay (kaolin) to provide these enhanced mechanical properties. Sodium polyacrylate will reduce slightly the firing temperature because of high alkali loading, while potassium and ammonium species show little decrease. The use of polyacrylate for this purpose results not only in increased strength but in higher color intensity where the dispersant is also employed in glaze colors added to the casting.[10]

Even though adsorption of polyacrylate on clays and ceramic bodies is more Langmuirian than are polyphosphates, some equilibrium is established with the liquid phase. The consequence of this equilibrium is nonsymmetry in the caternary rheological profile. Figure 4.6 shows a comparison of viscosity of a 65% solids casting clay with TSPP and with a medium-molecular-weight polyacrylate, both sodium salts. Note the high dispersant arm of the curve is much more

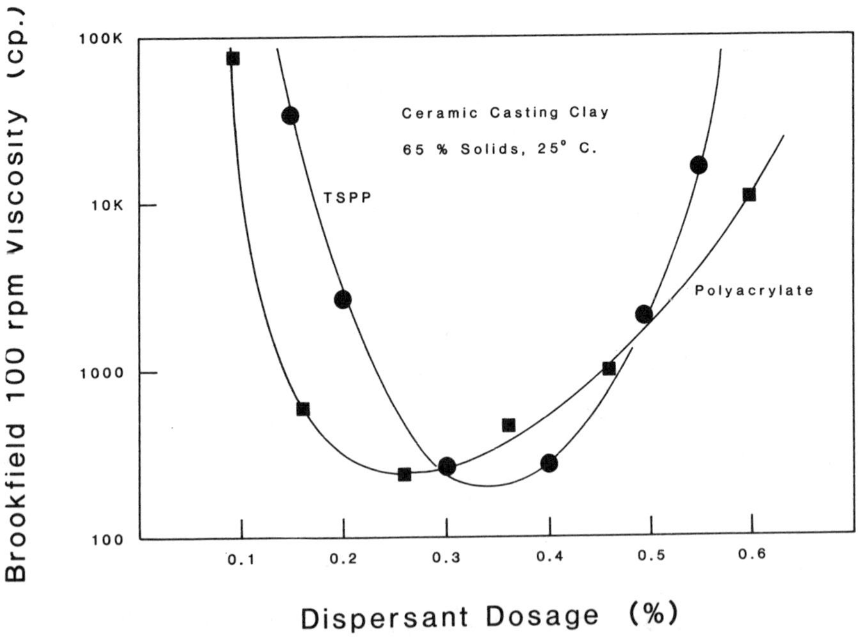

Figure 4.6. Rheological profile of polyacrylate on casting clay.

shallow with the organic agent than with the inorganic, reflecting the difference in equilibrium shift to the solution phase.

The minimum in viscosity is both a function of the clay's surface character (crystallinity, smectite content, surface area, morphology, etc.) and the molecular weight of the dispersant, as shown by the curve on the left side of the graph. The curve both rises and narrows as molecular weight increases until the value reaches 100,000 or more. Studies on casting slips for specialty ceramics have shown that destabilization of the dispersion and the inability to produce consistent and homogeneous castings occurs between 100,000 and 300,000 molecular weight units.[11]

For high kaolin based sanitary ware, as bathroom fixtures, sodium citrate can substitute for sodium polyacrylate as casting dispersant up to the latter's molecular weight of about 25,000 units and attain the same viscosity minimum and profile width. However, the gravimetric weight requirements for citrate increase significantly. Sodium citrate also has a detrimental effect on gypsum molds at high levels.

Because of the superior dispersant properties and better mold release provided by polyacrylates, casting rate dramatically increases with its use, typically 35–40% when compared to sodium silicate and sodium silicate–sodium tannate or humate mixtures.

Kaolinite with a better rheology and phylloidal morphology can use polyacrylates for the coating of paper. There, slurry solids are as high as rheological properties dictate to minimize water removal and sheet drying. High water contents of the coating formulation reduce the strength of the paper by maintaining it in a wet state longer and allowing sizing compounds (binders) to slowly break down or migrate.

Polyacrylates appear less sensitive to polyvalent ions, especially those present in commercial water supplies derived from purification treatment. Polymers which provide superior characteristics are higher-molecular-weight members approaching the destabilization, or flocculation, range. Studies confirm styrene–maleic anhydride copolymer neutralized with sodium carbonate can be improved upon more by esterifying the copolymer after manufacture with short-to-medium chain aliphatic alcohols (C-2 to C-15). Advantages of esters over sodium salts lie in reduced susceptibility to both acids and bases[12] (in coating and printing formulations) and higher resistance to flocculation from calcium and aluminum ion. This is illustrated in Figure 4.7 for TSPP compared to standard sodium polyacrylate (MW = 2500) and an estified form of the basic polymer.

In paint applications mention already has been made of the phenomenon of "blooming" from sodium polyphosphates. Blooming arises partly from the mobile and highly hydrated sodium ion and partly from phosphate desorption through dispersant breakdown and shift to the solution phase. Where polyacrylate substitutes for polyphosphate, this shift is much subdued, making the sodium ion far less mobile within the paint film and blooming less prevalent.

A second method of reducing blooming and obtaining better dispersion can derive from the reduction in the quantity of polyacrylate actually employed.

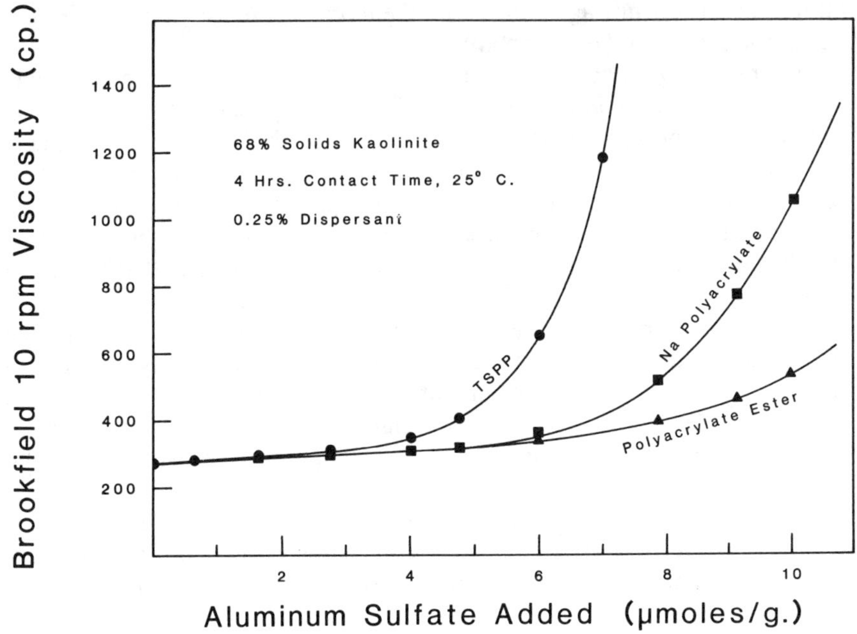

Figure 4.7. Influence of aluminum ion on acrylate dispersed pigment.

Figure 4.8 depicts the rheological profiles for precipitated calcium carbonate in a pigmentary premix at optimum dispersant level over varying aging times at a moderately elevated temperature.[2] Even though calcium carbonate contains virtually no acidic groups to break down phosphate chains, some deterioration occurs, likely the result of solution-phase (sequestration) chemistry.

Polyphosphate results in a threefold viscosity increase in just 10 hours, rising nearly linearly with time. Polyacrylate, by contrast, approximately trebles up to about 40 hours, after which it remains essentially unchanged. At higher levels of shear, as provided by a Hercules viscometer, both systems display a drop in rpm shear with time, the result of partial flocculation. Precipitated calcium carbonate is manufactured deliberately with a very narrow particle size level, approximately 0.2 μm. At high shear, shown in Figure 4.9, this system exhibits considerably dilatancy, with the SHMP dispersed material reaching rigidity, the consequence of this very narrow size distribution as well as dispersant decay.[2] However, some flocculation also develops with the polyacrylate dispersion, but its rheology is yet acceptable after 40 hours for coating applications.

A stabilized system for kaolinite slurries in paint applications may await neutralization until the polyacrylic acid complexes calcium, iron, and other viscosity-influencing ions. Citric acid becomes an important component as it is superior to acrylate as a complexer. A typical formulation contains 15% citric acid, 85% polyacrylic acid, agitated then neutralized to pH 7.1 by sodium carbonate. The

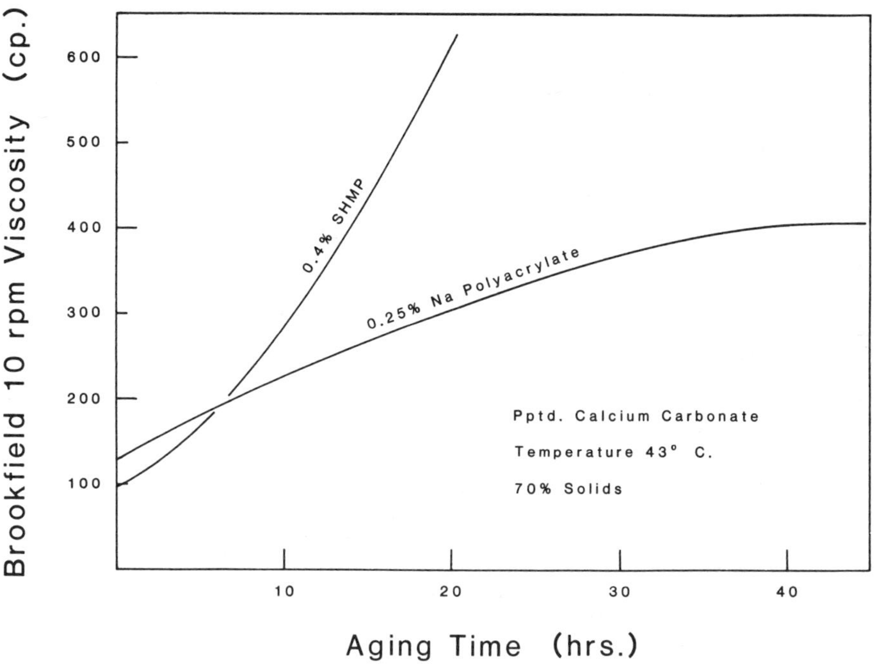

Figure 4.8. Low-shear rheological profile for elevated temperature aging of calcium carbonate (Ref. 2).

final combined dispersant dosage is 0.2%. This is purported to yield a 70% solids slurry with substantially increased long-term stability than is obtained by adding the sodium salts initially.[13]

Most applications of sodium polyacrylates alone show the optimum molecular weight to lie in the range 2000 to 10,000 MW units, even though some dispersion can be obtained with higher-molecular-weight polymers. Langmuir adsorption studies with progressively higher molecular weights of polyacrylate dispersants on calcium carbonate indicate the steepness of the curve's initial rise increases almost in proportion to the dispersant's molecular weight, a direct consequence of van der Waals forces. However, time-dependent studies at high solids with acidulating pigments suggest that adsorption changes progressively from train to tail along the molecular structure as the chain length increases. This comes about in part from lower solids being obtained at a given viscosity with progressively higher-molecular-weight members of the family.

While tail enmeshment is not a likelihood because of the charge-bearing elements within the dangling chains, increased solvation by water and the potential for bridging unquestionably increases with chain length. This has been demonstrated by direct comparison between long-chain sodium polyacrylates and equivalent chains of polyoxyethylene nonionic dispersants.[14] The latter have

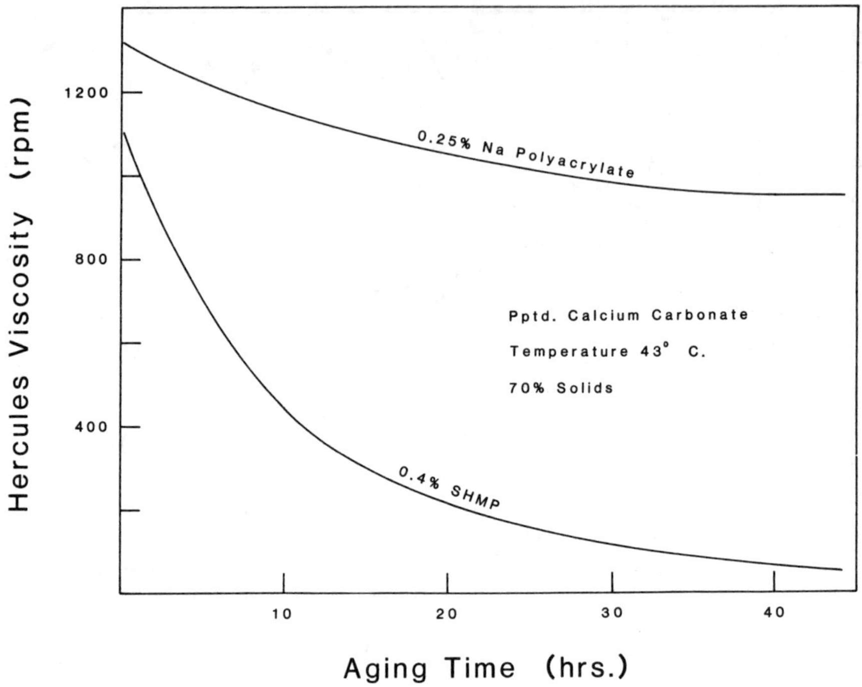

Figure 4.9. High-shear rheological profile for elevated temperature aging of carbonate (Ref. 2).

considerably reduced adsorption energy on calcium carbonate surfaces. Interparticle bridging by long-chain dispersants increases in probability with solids in suspension due to particle–particle proximity. At solids below 10% it essentially disappears.

Calcium carbonate appears to be a pigment moderately sensitive to polyacrylate structure variations. Studies on the alkali salts of copolymers of acrylic acid with isobutylene, styrene, and maleic anhydride show a broad variety of rheological properties in the molecular weight range from 2000 to 8000.[15] While all of these agents can provide a workable viscosity for calcium carbonate alone, addition of other pigments, especially those with low isoelectric pH values, can alter the application properties (brushing and spraying with paints, trailing blade properties with paper coating, extrusion, pumping, etc.). For example, styrene copolymers with acrylic acid are superior with calcium carbonate alone but not with titanium dioxide–calcium carbonate mixtures.

Certain solids appear to exhibit high sensitivity to chain length of polyacrylates with regard to dispersion, perhaps due to their coordination, their adsorption energy for the organic molecule or their poor aqueous wetting characteristics. One of these is talc (hydrated magnesium silicate), which requires a molecular weight for conventional sodium polyacrylate between 5,000 and 10,000. Species outside the range display a definite rheological deterioration.

Kaolinite, which disperses relatively well with most acrylate polymers, shows a special adsorption preference for the ammonium salt of vinyl ester copolymers with maleic anhydride having a molecular weight of only 1100 to 1200. This molecular arrangement apparently provides a superior conformation to the alumina octahedra on the pigment's surface.

Other solids appear to react with the polyacrylate molecule. Magnesium oxide derived from precipitation reactions can be dispersed readily at about 40–45% solids for applications in high-temperature, high-strength ceramics. When freshly formulated with 1% sodium polyacrylate (solids basis), viscosities in the range of 100 cp (Brookfield 10 rpm) are observed.[16] However, upon sitting unattended for 24 hours, slurry viscosity rises to 1500–2000 cp. Although studies have not been made on the structures involved, replacement of sodium by magnesium in the dispersant molecule may result, paralleling the long-term reaction of polyphosphates on alumina hydrate surfaces.

Precipitated magnesium hydroxide is employed in detergent matrices, partly as a suspending agent and partly as a mild abrasive. Sodium polyacrylates having short-chain-length species commonly are employed as dispersants. However, prolonged shelf storage has shown a slow breakdown of the dispersiveness (elevated viscosity) and a change of pH, attributable to interaction of the hydroxide with the acrylate chain.[17] No solution apparently exists for this combination except to minimize storage time.

One of the more difficult pigments to produce, transport, and handle is satinwhite, a cross precipitate, or copigment, between calcium sulfate and aluminum hydroxide. The pigment is manufactured by adding aluminum sulfate solution in stoichiometric proportions to concentrated calcium hydroxide.[18] The material is very white, has an acicular crystal form, and carries the approximate composition:

$$(CaO)_3\ Al_2O_3\ (CaSO_4)_3\ (H_2O)_{32} \tag{4.1}$$

Its advantage as a paper pigment lies not in its refractive index, which is poor at best, but in its ability to produce voids in a polymer paper coating. The voids develop from water loss by the pigment upon elevated temperature drying. Voids, although filled with air, are negative refractive index centers and have the same effect on optical diffraction as a pigment particle of the same size having a refractive index of 2.00. Air voids also are very light in weight—much lighter than refractive index 2.00 particles!

It is this property of water loss that makes satinwhite's manufacture difficult. It cannot be dried for shipment but is transported as a high-solids paste. Polyphosphates were employed to reduce the material's viscosity but also were found to react slowly with both the calcium and the aluminum[19] in the copigment with prolonged storage.

Another attempt to improve upon the shipping–storage instability involved dispersing with a combination inorganic and organophosphate, using a comixture of satinwhite and kaolin,[20] and spray drying the combination quickly to prevent hydration water from releasing. A transmission electron micrograph of the spray-dried, dispersant-containing pigment is shown as Figure 4.10. Individual needles of the pigment resolve readily. Upon immersion in water, the thermally stable dispersant immediately allows the spherules to break apart rapidly and reassume their acicular, water-bearing structure.

This technique yields an interesting product. Satinwhite's needles are in the range of 0.5–1.5 μm long with an aspect ratio of from 8 to 12. Spray drying, depending upon feed solids, temperature, and line pressure, produces spheres in the range of 3–5 μm with the needles aggregated in concentric brushpiles throughout the ball. This drying technique allows some free moisture, several percent, to remain in the spheres until use, assuming the material is kept cool and at moderate humidity.

A more recent improvment in stability enhancement is based on blending satinwhite with ultrafine (<1 μm) calcium carbonate, dispersing with a long-chain, coagent sodium poly(acrylate–maleate) having a molecular weight over 10,000 and spray drying. During spherule formation from the drier mist, acrylate–maleate polymer, being noncrystalline (compared to phosphorates), acts as a cementing and barrier agent for the acicular particles near the spherule's sur-

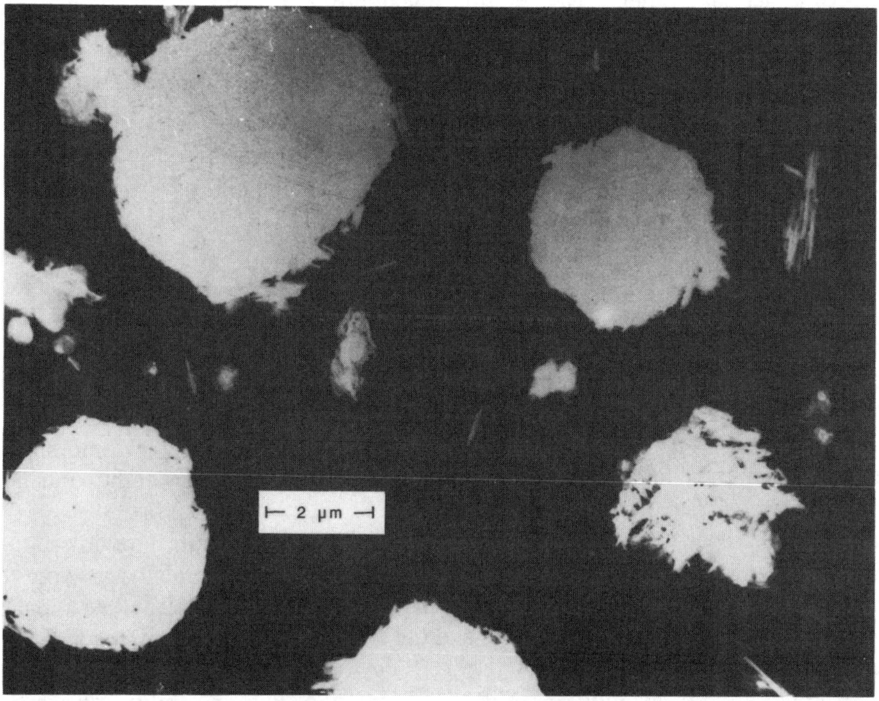

Figure 4.10. Spray-dried satinwhite with thermally stable dispersant.

face to trap moisture inside. This increases the stability greatly, especially towards warm storage.

Upon adding this satinwhite to a coating formulation, the polymer serves again as dispersant to maintain lowered viscosity as pigment is coated onto the cellulose sheet. In addition to improved long-term stability (in drums) and reduced viscosity (in applications), the dispersant provides better spreadability onto magazine stock paper. Polyacrylate's advantage lies in not creating dispersion problems by formation of competitive chemical products with the mobile calcium or aluminum ions in the pigment.

By modifying polymaleate with propylene during the polymerization reaction, the active group spacing can be lengthened further, to the point where coordination takes place with alternate cations on a nonoxide surface. This ability to control group spacing is important in minimizing chain compression, as has been noted previously for barite. Although the number of net surface contact points is lessened per unit length along a solid surface, sufficient tie points exist to stabilize the polymer for dispersancy. In action reminiscent of satinwhite drying, calcium carbonate dispersed in this manner binds strongly to cellulose fibers[21] when the carbonate pigment is employed in paper coating formulations.

A similar chain alteration can be made by introducing butyl acrylate into the acrylic acid polymerization reaction and neutralizing the copolypolymer with sodium carbonate. This material also has specific advantages for dispersing ultrafine calcium carbonate.

Where thermoplastics are molded using ultrafine calcium carbonate as a coloring or opacifying pigment, the finished casting gloss is a true measure of the dispersion by polyacrylate. Glossy-surfaced items are more easily marketed and more easily cleaned, two major sales advantages. Studies on 0.1–0.2 μm precipitated calcium carbonate in polypropylene at 50 parts per 100 loading (33%), dispersed with the butyl acrylate copolymer described at a 1.5–2.0% level (solids basis), show the dispersant not only improves casting gloss, it also improves color and impact strength.[22] The latter physical property is almost always a good test for dispersion in the polymer after molding.

Acrylate polymer salts may be amended to enhance calcium carbonate coatings by employing codispersants. The secondary agent is most often selected for its association with either copigments or to enhance substrate acceptance of the coating system. This is the case with paper coating where polyethyleneimine is employed to coat cellulose fibers[23] and allow the carbonate to flow more freely across the surface to develop a smoother texture and gloss. Because the fiber coating involves a greater surface function than the carbonate surface (where particle diameter is $\sim$2 μm), demand for a coagent will exceed that for the primary dispersant, in this instance by about 4:1.

As a consequence of potential for surface chemical damage to alumina hydrate from phosphate interaction (see Chapter 3), especially during extended storage, sodium polyacrylate and its allied compounds render a highly valuable service where alumina hydrate is employed. Tabular alumina hydrate has a morphology virtually identical to kaolinite, but the material's surface chemistry and color are superior in many applications. One of these is as a fire retardant pigment in water-base applications and in polymer resins.

Polyacrylate, as well as the maleic acid and fumaric acid copolymers, adsorb over the entire surface (all facets) of the hexagonal platelets of the hydroxide particles, unlike on kaolinite, which increases its weight demand in high solids coatings. Langmuir adsorption isotherms show a two-stage profile, the first representing about 25% dispersant and the second the remaining 75%. Dispersant demand for 10μm diameter particles is 0.3 to 0.5%[24] compared to about $\frac{1}{3}$ that amount for an equivalent particle size kaolinite.

Because of the broad dispersing applications of polyacrylate, it has come to be assumed it will disperse any inorganic powder. However, in binary mixtures, where isoelectric points vary considerably, a single polyacrylate species may not serve all species well. Studies show that by employing styrene ($C_6H_5CH{=}CH_2$) during the polymerization reaction, a benzene ring is introduced two or more carbons away from the acid group, increasing its pK_a due to electron-withdrawal characteristics of the ring. In white paint, where a mixture of calcium carbonate (isoelectric pH = 9.4), kaolinite (isoelectric pH = 4.5), and anatase titanium dioxide (isoelectric pH = 6.2) are employed as pigments, the ammonium salt of this polyacrylic acid is superiorior to any of the conventional sodium polyacrylates.[25] Electron withdrawal by the benzene rings in the dispersant molecule stabilizes the ammonium salt toward hydrolysis.

Barium titanate ferroelectric powders provide an example where chain geometry is not as determinant a factor in obtaining good dispersancy as is cation character on the chain. Earlier it was mentioned that polyacrylate chains are compressed in lying down upon barium sulfate, but are too short to span two barium ions (about 1.4 nm). Where the substrate is the salt of a strong base and a very weak acid, as barium titanate (H_2TiO_3, pK_a above 6), the salt has a net alkaline character. As such, a less alkaline dispersant, ammonium polyacrylate, rather than sodium will show a greater affinity for the surface. While this same phenomenon occurs with calcium carbonate, the solid is such a low-cost commodity (compared to barium titanate) that increased dispersant cost (nearly 50%) cannot be justified, whereas it can with high-end products. For water-base applications of dielectrics, barium titanate has been dispersed very well with short-to-medium chains of quaternary ammonium hydroxide salts of polyacrylic acid.[26] Where the organic chain length increases to above C-12, the products can be dried and redispersed in organic vehicles.

Advantageous characteristics of the polyacrylate family often lie in secondary application properties. For example, because of reduced alkalinity over KTPP, alkanolamines, and similar dispersing agents, bubble stability in an applied film is significantly less, an important factor in paints, whether applied by amateurs or professionals! Studies show polyacrylates also improve leveling, flow control, and reduce reagglommeration and film tearing. These characteristics appear to be general properties independent of chain structure, density of active groups, cation nature, or molecular weight.

4.2 Polyether Organic Salts

In polyacrylates the solid-seeking groups, as the neutralized acidic members might be termed, are the sole members coordinating with the surface. By reducing greatly the number of acid groups, often to only one, and by altering the remainder of the chain to include the dipolar alcohol or ether groups, coordinating units are provided for oxide surfaces and reduced charge sites. Such a molecular system ties up less water through ion solvation in high-solids systems. This often becomes very important in application systems where high charge density is a disadvantage.

This category of dispersants contains compounds that, while not containing as frequent acidic groups along the spinal chain as acrylates, have periodic active groups with terminal groups at the tail. Such structures are produced by copolymerization of other organic acids with polyols and polyethers. Their production results not as much from molecular engineering in active group spacing as from the utilization of raw materials containing mixtures of hydrophilic groups and organic acids.

The rheological profile for polyacrylates is less steep on the high-dispersant-level arm than is that for polyphosphates having the same weight dosage because of reduced cation density. And, as will be shown in detail in Chapter 5 for nonionic agents, this arm drops almost to a horizontal line when charge completely disappears. It is this overdispersion sensitivity that has brought about the polyether, polyol organic salt dispersant family.

A simple agent can be singly branched or doubly branched, depending upon the use of maleic acid versus acrylic acid, as illustrated in Figure 4.11.

Because of the complexity in anionic bases and the variations in polyether linkages, the number of such agents is limited only by an organic chemist's imagination and budget! As an example, the pigment red oxide of iron (alpha ferric oxide) has been commercially dispersed in polyacrylic latex formulations employing a dispersant having the structure:

$$[\mathrm{CH_2{-}CH{-}O}R]_n\,[\mathrm{CH(COO^-\,Na^+){-}CH}\begin{matrix}\diagup\mathrm{COO}R\\ \diagdown\mathrm{COO}R^*\end{matrix} \qquad (4.2)$$

where R is a C-6 to C-40 alkyl group and R^* a polyether, polyol group.

It is apparent that mutual particle repulsion evolves from the small charge adsorbed as well as portions of one or more of the three hydrophilic tails which may extend into the aqueous phase. The ether groups, possessing a moderate dipole moment (1.1×10^{-18} esu cm compared to water with 1.85×10^{-18} esu cm), can expand or contract to a limited extent and coordinate with the hydroxyl groups associated with the surface ferric atoms. In instances the spacing may be sufficient that alternate hydroxyl groups are the coordinating sites (as noted earlier with barium compounds).

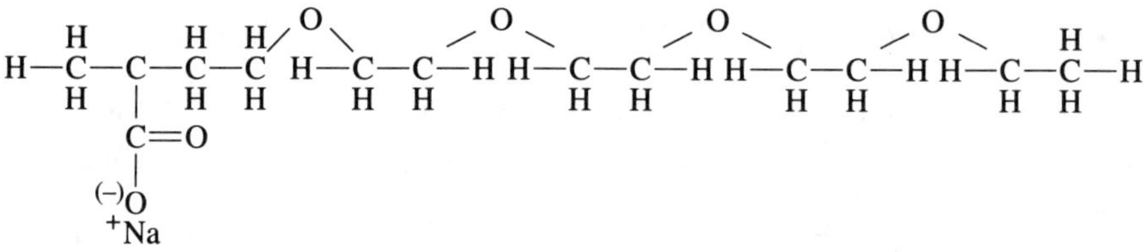

Mono-polyether substituted sodium maleate

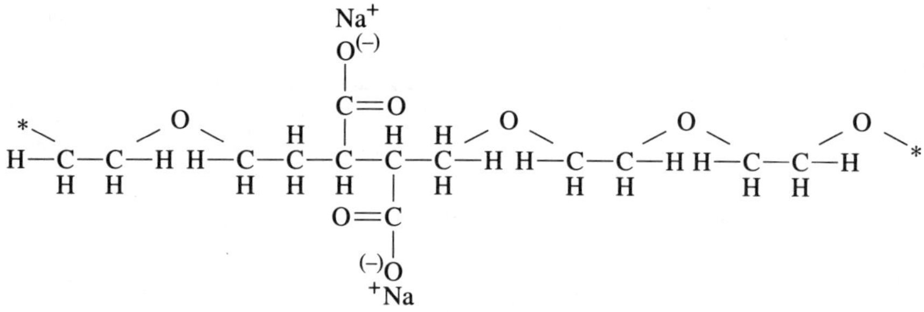

Di-polyether substituted sodium maleate

Figure 4.11. Dipole modified organic anionic dispersants.

While no shortage exists for calcium carbonate dispersants, a slight variation on the above structure has been made for this pigment employing a two-part mixture.[27] The first component is a polymerized fatty acid vinyl acetate ester, where the fatty acid has 10 carbon atoms or less, and the second, an olefin–maleic anyhydride copolymer, where the olefin group is from 4 to 10 carbon atoms. Distribution in the mixture is about 5% of the former and 95% of the latter. The technological object is to obtain very high calcium carbonate loadings for latex emulsion paints without incorporating highly ionic species that restrict pigment mobility in the aqueous phase.

Calcium carbonate slurries designed for transportation, especially in trans-oceanic ships where long-term stability is a major concern, use a mixture of polyacrylate–polymaleate sodium salts in combination with a copolymer between para-styrene sulfonic acid and vinyl alcohol.[28] These alcohol groups are assumed to provide low-energy interparticle bridging to prevent dilatant cake sedimentation from prolonged and compacted (from vibration) storage.

Other reactive systems include a broad variety of substituted maleate salts where the substituents include polyoxyalkalene ester salts and vinyl amides. Some of these show high specificity for certain pigments as clays,[29] and others appear generally applicable to water emulsions requiring low electrolyte.

Some of these compounds have interesting cationic functions. As noted earlier, for insoluble alkaline salts of weak acids, replacement of the cation in the polyether compounds changes not only the induced charge, but the affinity of the ionic component for the specific oxide. In detergents designed for high-efficiency iron stain removal without bleaching, the calcium salt of the vinyl copolymer with acrylic acid, dodecyl-substituted alcohols (*R*—OH) and mercaptans (*R*—SH groups) have been used with good results. These combinations function better than common sodium sulfonate–sodium polyphosphate–sodium carbonate agents and better than conventional sodium polyacrylates.[30]

Employment of magnesium oxide in cleaning compounds, both as a suspension stabilizing agent and as a mild abrasive, incurs serious problems with polyacrylates under prolonged slurry or formulation storage. This can largely be averted by using sodium alkylbenzene sulfonate in conjunction with a short-chain ester, as glyceryl triacetate or methyl diacetyl amine. Such compounds are often blended with a strong inorganic dispersant and a bleaching agent, as sodium perborate. Where the solid surface being dispersed contains acidic oxides, the triamine dominates in the formulation. Where the surface is alkaline (as with calcium carbonate), the glyceryl ester dominates. A number of liquid industrial cleaners have employed this technology for cleaning cement on airport runways and limestone buildings.[31]

Chemical variations on side-chain modifications of the acrylic core are legion. However, a few are interesting because of their specific, tailoring nature. For example, if the side chains contain alcohol groups (polyols), these can be converted to amines by nitration and zinc reduction reactions. Amino groups have a strong coordination capacity with numerous elements possessing open *d* orbitals in their electronic structure. These include numerous metal orbitals, for example, compounds of nickel, copper, cobalt, and zinc. Eastman Kodak has taken advantage of this special form of dispersant attachment in their high-quality films.[32] For the best compromise between high resolution (sharpness) and light sensitivity (speed) in photographic films, particles of silver chloride and bromide must be thoroughly dispersed in the aqueous gelatine base. This is directly analogous to hiding power in a pigmented paint film. Maleic anhydride–styrene condensation products converted to side chains with amino groups make excellent dispersants for the silver halide particles. Further, they do not crystallize and compete with photon adsorption or cause photon scatter (haloing). Finally, they wash free from the gelatin base when the halide crystals are converted to silver metal.

This approach is employable also with zinc oxide and other pigments of the transition metal family (e.g., cobalt blue) in the formulation of specialty paints. Again, the problems of economics become the governing factors, not chemical efficiency. Silver salts and photographic film have an unusually high cost per weight basis, and the additional costs incurred from specialty chemical alteration can be borne readily. Common pigment compounds rarely can afford this luxury.

Sodium polyacrylate condensation polymers, while effective for titanium dioxide, require a high molecular compression to fit by templature into each octahedral reentrant. By increasing the intercarboxyl group spacing slightly, the

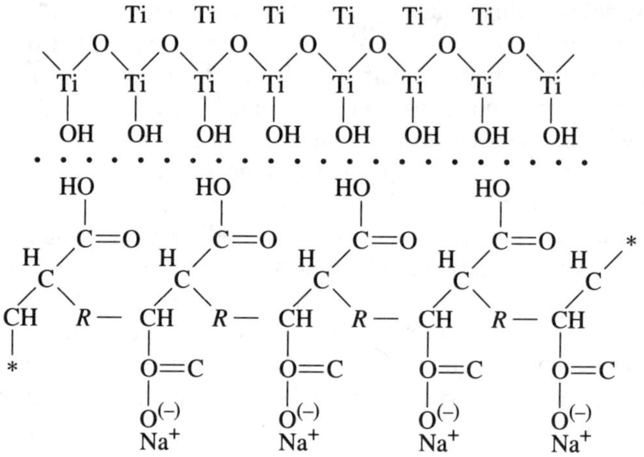

Figure 4.12. Half-neutralized polyacrylate on titanium dioxide.

molecule can adapt more readily through coordination with alternate octahedra. This is accomplished by introducing a hydrocarbon unit into the chain during polymerization. One of the substituents found to be highly effective is diisobutylene. Following polymerization only 50% of the carboxylate groups are neutralized with sodium hydroxide. When molecularly modeled, the structure appears to coordinate to the titania surface via the carboxylate's –OH grouping to the titanol groups at the pigment's surface, as sketched in Figure 4.12, where *R* represents the inserted diisobutylene molecule.

An examination of the infrared spectrum for the adsorbed species suggests half of the titanol groups are hydrogen bonded and half are not (split peak). All of the carboxylic acid groups appear to be hydrogen bonded. While other structures may be proposed for this linkage, the one above meets all of the dispersion properties observed by the combination.[33]

4.3 Other Anionic Organic Dispersants

An entire group of dispersants came about from employing pulpwood bleaching extracts, dominately lignins, in anionically converted forms. These materials, developed and promoted by the wood and paper industries, have as properties an amber to brown color, a pleasant bakerylike odor, and numerous interesting complexing characteristics. When clays are dispersed with a sodium lignosulfonate, for example, soluble iron on the surface is complexed and can be removed, a form of non-oxidative–reductive bleaching.

Kaolins have been processed with a somewhat more complex system that includes, in addition to the sodium lignosulfonate, a sodium naphthylene sulfonate–formaldehyde complex. This dispersant has high stability where protein binders (e.g., casein) are employed in kaolinite coatings. The dispersant serves both as antioxidant and hardener. Viscosities obtained generally are higher than with conventional polyacrylates, and a small component of polyacrylate may be required where rheological constraints are present.[34]

Because of the very large molecular weight of the lignin complex, these molecules are adsorbed nearly quantitatively upon kaolinites and other oxide pigments. Viscosity reduction is almost directly proportional to the quantity added, a strong indication of true Langmuirian activity. In some applications the lignosulfonate agents render the pigment UV photo-oxidation resistant.[46]

Commercial calcium carbonate slurry shipments have been made at solids above 70% by employing polyvinyl alcohol extended acrylates and maleates as well as the sulfonated alkylnaphthalene compounds already discussed. Some complex dispersants use sulfonated creosote oil in lieu of the alkylnaphthalene compounds because of its high bacteriocidal and fungicidal activity.[35] Very low viscosities on -2 μm calcium carbonate pigment have been obtained, less than 100 cp at 10 rpm Brookfield.

Many of these dispersants are designed to utilize waste chemicals and byproducts from petroleum and petrochemical operations. The pyrolysis or cracking compound byproducts are sulfonated with sulfuric acid and neutralized, usually with sodium hydroxide or sodium carbonate. Reactions generally proceed at reasonable temperatures, 120–140°C, quickly, often less than 10 min. Neutralized products are then combined with maleic or acrylic acid salts of the same alkali. These mixed dispersants and detergents enable waste and hazardous chemical materials to be converted into useful and salable products, thereby reducing a refinery's overhead and disposal costs.

One typical product is manufactured by reacting naphthalene and cresol (from wood distillation) with sodium sulfite followed by formaldehyde. Condensation is complex and must be neutralized after the reaction has been completed. The end product serves as a satisfactory bacteriocidal handwash detergent and has been commercialized for use in food preparation and medical preparation situations.[36]

Where reducing capabilities remain with a molecule once it has been converted into a suitable dispersant, the material possesses a strong antioxidant activity for inks, especially where atmospheric oxidation can damage severely the products prior to use (metallic powders, pressure sensitive leuco inks, etc.). Even *dl* ascorbic acid, common vitamin C, has been added as an antioxidant to these organic systems to increase operating lifetimes.

Carbon black with a particle size below 0.1 μm can be made into an ink for felt-tip pens by being dispersed under very high shear with a polyether-substituted sodium or ammonium carboxylate:

$$R\text{–O–}(CH_2\text{–}CH_2\text{–O})_n\text{–}CH_2\text{–}COO^{(-)}M^+ \tag{4.3}$$

It is imperative that the carbon black be thoroughly dispersed, otherwise floccules will build up in the felt that shorten pen life, long before ink is fully consumed. These systems are usually overdispersed to prevent felt blockage. As a consequence of the high carbon black surface area and overdispersion, dispersant demand is much higher than with common pigments. A typical formulation for such an ink is given in Table 4.2.

Similar formulations have been made substituting sodium and ammonium hydroxyphenyl stearates for polyether carboxylate and phthalocyanines (blue) and quinacridones (red) for carbon black.

Thus far in this chapter the simplest organic anionic agents have been omitted, alkali soaps and sulfonates. Most applications of straight-chain sodium fatty acid compounds tend to make the surface of inorganic oxides hydrophobic by orienting with the charged head to the solid surface and the hydrocarbon tail extending into the liquid phase (see Chapter 7). However, where that tail is short or furcated, where the acid is dicarboxylic, and in conjunction with other dispersants, a synergistic effect may be observed.

Sodium adipate, $Na^{+(-)}OOC–(CH_2)_4–COO^{(-)}Na^+$, mixed with SHMP provides a superior wetting agent for mill breakdown of titanium dioxide clusters from production. Sodium adipate appears to be less migratory than polyphosphate, once incorporated into a film. In the fashion of the isobutylene substituted polyacrylate shown in Figure 4.12, this material also may coordinate on alternate titania octahedra.

Sodium caprylate (octanoate soap) appears to represent the limiting length for a hydrocarbon tail adsorbing on an oxide surface and not generating hydrophobicity. It may represent the bridge species capable of providing useful dispersion in both water and hydrocarbon solvents. Kaolinite for paper coating also has been dispersed with the soap (or salt).[37] Magnetite, Fe_3O_4, black magnetic oxide, has been dispersed with the octanoate in both xylene and water for magnetic recording media (audio-video tape, check ink, and computer disks).

Mineral flotation is an example where a slightly hydrophobic tail with an ionic head has value in an aqueous system. To enable air bubble attachment to a specific mineral component in a mixed mineral system, one surface must be more hydrophobic than the remaining components. Solid particle size, bubble size, and bubble affinity are critical for flotation separation. A typical grind size is in the 16 to 20 mesh range (1000–840 μm). A surfactant is added that is

Table 4.2 Overdispersed Felt-Tip Pen Ink

Component	Weight %	Function
Carbon black	16	oxidized pigment
Polyether carboxylate	4.5	dispersant
Propylene glycol	23	viscosity agent
Triethylamine	4	anticlogging agent
Benzotriazole	1.5	corrosion inhibitor
Water	51	diluent
	100.0	

selective for the component to be floated. To separate hematite from siliceous host material or impurities, either a dodecyl anionic or cationic agent may be employed. However, the choice must be based on how the surface has been altered through pH change.

With $Na^+SO_4^-C_{12}H_{25}$, recovery increases as pH decreases below hematite's isoelectric point (5.2) for nearly two full units where the surface potential reaches +35 mV. Acidifying further does not increase separation yield. Correspondingly, with $Cl^-NH_3^+C_{12}H_{25}$, recovery increases as pH rises above 5.2, maximizing at about pH 7.5–8.0[38] with a surface potential of −30 mV. At the isoelectric point, hematite will not float with either agent. At high positive surface charge, the anionic agent is tightly adsorbed onto hematite, rendering the surface sufficiently hydrophobic to permit air bubble attachment. At high negative charge the cationic agent functions in identical fashion. These phenomena are illustrated by the split molecular docking sketches in Figure 4.13.

Because silica and silicates have isoelectric values at lower pH units than hematite (quartz at 2.4 and kaolinite at 4.5), separation is more efficient with the anionic agent because of lowered competition. Quartz has been floated by a similar procedure,[39] but the process is difficult because of the activity required.

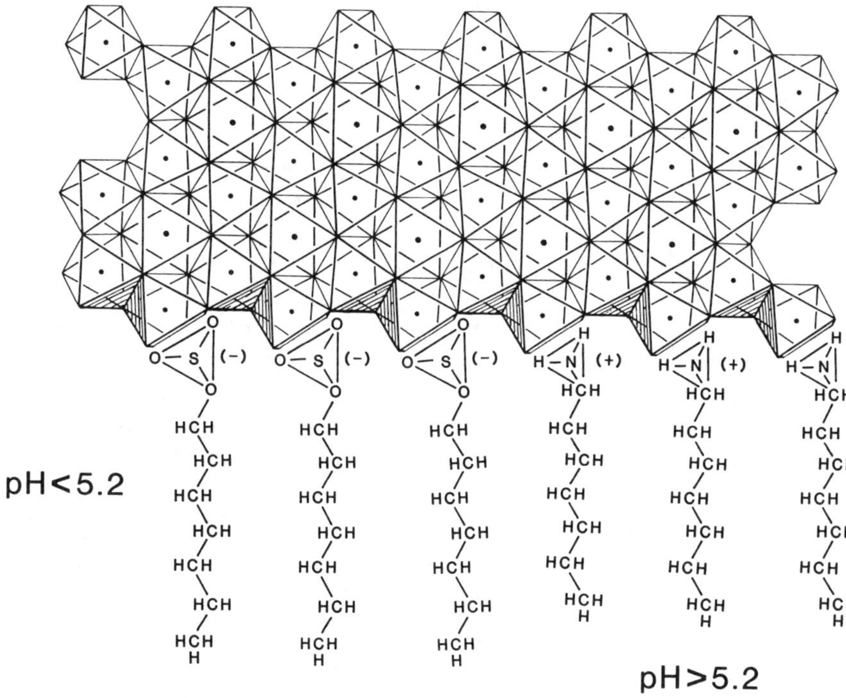

Figure 4.13. Cationic and anionic agents adsorbing onto hematite for flotation separation.

High-quality glass sand is removed from kaolin in a Texas plant employing cresylic acid by such a procedure. The kaolin is then calcined and sold as a medium-quality paint pigment.

Other analogous dispersant structures and their applications include the following:

1. The compound $(RR'NZSO_3)_nM$, where M is any divalent transition metal ion or lead, R is an alkanol group with a carbon chain length from 8 to 40 atoms, R' is a similar alkyl group without alcohol linkage, and Z is an alkalene group up to 6 carbon atoms. This agent is employed for carbon blacks in automotive[40] high-gloss finishes.
2. Where biodegradability is important, as with filled and dyed paper, cosmetic lotions, soap formulations, etc., the role of the dispersant takes on even greater importance. Its molecular structure must contain one or more alcohol groups strategically located throughout the hydrocarbon chain to permit bacterial action and chain breakdown. Typical of these agents is the structure:

$$C_nH_{(2n+1)}(CH—OH—CH_2)_x(C_mH_{(2m+1)})COO\ M \tag{4.4}$$

where n, m = 6 to 20; x = 1 to 3; and M = Na, NH_4, and H. All three of the cation species give agents that are mild to the skin, provide a creamy texture, and decompose quickly in waste disposal sites. This series of substances has largely replaced sodium dodecyl sulfonate (SDS) in many soap and cosmetic lotions.[41]
3. Coloring and opacifying aluminum chloride aqueous solutions for leather treating, fluxes, and other highly specific applications requires perhaps the ultimate in dispersing power because of the great propensity of the trivalent aluminum ion to floc charged dispersed powders incorporated for color. The three most practical agents employed in these applications are oxygenated fatty acids (including polyether substitutions), sodium dodecyl sulfonate (anionic agent), and lauryl pyridinium chloride (cationic agent). The polyether acid is poorly ionized and, as a consequence, functions better than the other two. Of the ionics, the anionic is superior because it has little competition on the titanium dioxide surface ($\frac{1}{3}$ alkaline groups) and because the cationic is swamped by the highly competitive aluminum ions. For the SDS molecule to disperse titanium dioxide, a double-layer inversion must occur on the titania surface.[42]

$$TiO_2{:}^{(-)}SO_3\text{–}R{:}R\text{–}SO_3^{(-)}Na^+ \tag{4.5}$$

4. Silica colloids made from hydrolyzing silicic acid are useful in paints as viscosity agents and as colorless wood fillers. Because of high surface area and sol structure, they are difficult to disperse and stabilize. They have a great propensity to coagulate by chain formation, gel, and render the liquid system unusable. A hybrid anionic organophosphonate containing a silicon grouping[43] greatly stabilizes these systems even under the thermal excursions

in freeze-thaw and acidity down to pH 4.5. A typical structure containing this inorganic–organic complex salt is

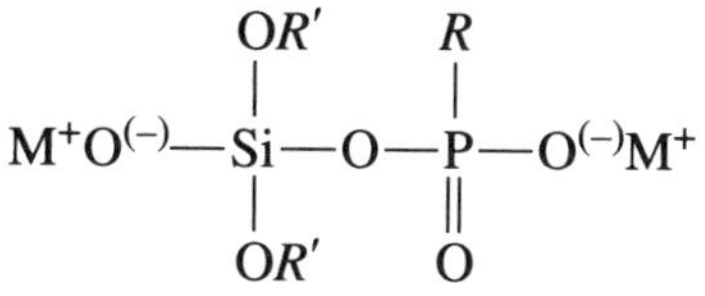

(4.6)

where R is a C-1 to C-7 chlorinated alcohol group; R' is CH_3 or C_2H_5; and $M = Na^+, K^+, Li^+$. What role the chlorinated group in R plays is uncertain, but its existence is apparently justified by improved dispersive power.

When R is replaced by a hydrocarbon tail and M by an organoquaternary ammonium ion, the silica colloid may be transferred to an oleophilic medium by a flushing (high-shear emulsification followed by rapid emulsion breaking) operation. Traces of water, however, carry over giving a slight haze to the organic liquid.

5. Ultrafine colloidal dispersions of pigments ($<0.1\ \mu m$) have been reportedly dispersed by a quaternary ammonium cation, organosulfate anion salt.[44] Included in the applicable sols are iron oxide, titanium dioxide, and alumina (probably as the hydrate). However, it is the anion in this system that is somewhat unique to cationic dispersants and that clearly alters its role of adsorption on oxide–hydroxide surfaces. The structure employed is

$$[R_4N]^{(+)}\ [CH_3\text{–}SO_4]^{(-)} \tag{4.7}$$

The organo grouping R may be any combination of one methyl, ethyl, or benzyl group and the remainder C-8 to C-18 alkyl tails.

6. One of pharmaceutical industry's more ubiquitous dispersants is cetyl pyridinium chloride, $C_{16}H_{33}\ [C_5H_5N]H^+Cl^-$, a cationic agent showing great affinity for proteinaceous matter, as well as being a mild bacteriocide and fungicide. The agent is believed to attach with its alkyl chain, as well as the polar aromatic pyridine structure, to hydrolyzed protein and as such frees food particles from lodged locales, such as between teeth, as well as releases bacterially degraded mucous from both mouth and throat. As a consequence of this activity and the material's innocuousness in the human body,[45] it is a frequent ingredient in many mouthwashes, gargles, and throat lozenges, as well as medical scrubbing preparations.

7. Hospital staff and other medical personnel whose duties require frequent hand washing employ a dispersant formulation designed to minimize skin cracking and other forms of dermatitis. This contains the ammonium rather than sodium salt of the anionic agent lauryl sulfate. A typical FDA-approved formulation[47] is shown in Table 4.3.

The first three components all serve as anionic or nonionic dispersants (surfactants), while the lanolin compound is an emollient. The hydroxysultain compound is a foaming agent enhanced by the lauramide and employed for both functional and aesthetic uses.

Table 4.3 General Medical Hand Wash Formulation

Component	Weight %
Ammonium lauryl sulfate	12
Ammonium cocyl isethionate	5
Lauramide DEA	3
Acetylated lanolin	6
Cocamidopropyl hydroxysultain	1
Fragrance and preservative	~1
Water	72
	100

Where more rigorous antimicrobial action is required, as for surgical procedures, a strong bacteriostat may be incorporated and the ionic agent replaced by a nonionic. This will be described in more detail under polyoxyethylene derivatives in Chapter 6.

4.4 Grafted Organic Ionic Agents

Large molecular species, which include color pigments, dyes, and pharmaceuticals, can be aqueously dispersed by chemically grafting organic ionic species onto the structure at one or more locales. Grafted species include sulfonic acid, $-SO_3H$, carboxylic acid, $-COOH$, and amine, $-NH_2$. These are then converted into salts with appropriate alkali for the first two species and HCl for the third.[48] With some color pigments, locating the solubilizing species directly adjacent to, or within a few atoms of, the chromophore group (color adsorption bond) causes a hue shift, making the water soluble products slightly different in color than the oil soluble ones. These differences, though subtle, create marketing problems with water-base and oil-base lines where color matching is critical (e.g., on walls and woodwork). To minimize this effect, a short hydrocarbon chain can be introduced between the fundamental molecule and the grafted ionic species, such as $-C_4H_8-SO_3^-Na^+$, which then serves as an electronic buffer.

Certain molecular configurations may contain a structurally inherent amine group that can be acidified directly. Such procedures more commonly are employed with pharmaceutical agents than pigmentary ones because of the electron-pair-withdrawal character on the nitrogen atom that results.

Where a charge-grafted color pigment will be employed in conjunction with inorganic, oxide-based pigments (titanium dioxide, zinc oxide, kaolinite, etc.) in water systems, the carboxylate and sulfonate are preferable modifications as they produce anionic pigmentary charge compatible with the oxide surface species. The amine hydrochloride modification, which produces cationic species, will flocculate onto oxides severely.

References

1. Rose, H., *Ball and Tube Milling*, Constable Press, London, p. 245 (1958).
2. Hagemeyer, R., TAPPI Monograph No. 38, p. 56, Pulp & Paper Inst., Appleton, WI (1974).
3. Where very high surface area particles, i.e., >20 m^2/g, or very fine acicular particles, are employed to effect thixotropy, longer polyacrylate chains appear to yield higher low-shear viscosities, usually a measure of dispersion quality.
4. Conley, R., *J. Paint Tech.* **46**, 51 (1974).
5. Pauling, L., *Nature of the Chemical Bond*, Cornell Univ. Press, Ithaca, NY (1960).
6. TAPPI Monograph No. 28, p. 108, Pulp & Paper Inst., Appleton, WI (1964).
7. Wu, Y., and Copeland, L., *Solid Surfaces and the Gas, Solid Interface*, American Chemical Soc., Washington, DC p. 366 (1961).
8. Nagarian, M., *Surfactants and Deterg.* **28**, 230 (1991).
9. Chiotoshi, K., Kazuhiko, H., and Yasuo, Y., *Bull. Res. Inst. Ceram.* **5**, 71 (1951).
10. Norris, J., Europatent #368,507, May 16 (1990).
11. Lapin, V., *Pap. Europe Symp., Ceramics*, p. 80 (1978).
12. Dikler, Y., et al., U.S.S.R. Patent #903,438, Feb. 7 (1982).
13. Hoyt, H., U.S. Patent #4,309,222, Jan. 5 (1982).
14. Oleinik, I., Polishchuk, R., and Dzyubenko, E., *Ukr. Khim. Zh.* **48**, 709 (1982).
15. Shimada, H., Japan Patent #6,272,762, April 3, (1987).
16. Falcione, R., McManus, R. and Aufman, J., U.S. Patent #4,230,610 Oct. 28 (1980).
17. Zini, P., and Coubot, Y., *Comun. Jorn. Com. Esp. Deterg.* **21**, 79 (1990).
18. Conley, R., Catherwood, B., and Lloyd, M., U.S. Patent #3,625,725 Dec. 7 (1971).
19. Catherwood, B., Conley, R., and Lloyd, M., U.S. Patent #3,854,971 Dec. 17 (1974).
20. Conley, R., Lloyd, M., and Catherwood, B., U.S. Patent #3,876,443 April 8 (1975).
21. Schuelde, F., and Kulisch, V., German Patent #2,700,444, Jan. 7 (1977).
22. Kao Soap Company, Japan Patent #165,960, Dec. 24 (1980).
23. Husband, J., Bown, R., and Drage, P., British Patent #PCT Intl. WO 8,341, June 13 (1991).
24. Takehara, H., and Okubo, A., Japan Patent #144,498 and 144,499, Dec. 15 (1978).
25. A general review of alteration of the hydrocarbon spine and its effect upon the dispersing properties of polyacrylates can be found in: Kulkarni, M., and Potniss, S., *J. Colour Sci.* **17**, 8 (1979).
26. Wilber, W., Europatent #260,577, Dec. 4 (1985).
27. Fukushima, Y., Miyazaka, H., and Maruyama, H., Japan Patent #2,298,331, Dec. 10 (1990).
28. Shioji, H., et al., Japan Patents #63,233,012, Sept. 28 (1988) and #63,248,718, Oct. 17 (1988).
29. Buerge, T., et al., Europatent #403,563, Dec. 19 (1990).
30. Yamada, I., and Takahashi, H., Japan Patent #1,310,730, Dec. 14 (1989).
31. Donker, C., and Mohammed, M., Intl. Patent #WO-12,313, Aug. 22 (1991).
32. Kodak (Gr. Brit.), British Patent #524,966, Aug. 20 (1990).

33. Routledge, A., Golightly, D., and Haughton, M., Australian Patent #499,791, May 3 (1979).
34. Nemeh, S., Sennett, P., and Sleptys, R., U.S. Patent #4,772,332 Sept. 20 (1988).
35. Fukushima, Y., Miyazaki, H., and Maruyama, H., Japan Patent #2,261,523, Oct 24. (1990).
36. Vanc, V., Thorovsky, Z., and Smetana, K., Czech Patent #244,709 Aug. 14 (1987).
37. Friberg, S., and Roberts, K., U.S. Patent #4,234,437, Nov. 18 (1980).
38. Iwasaki, I., Cooke, S., and Choi, H., *Trans. AIME* **217**, 137 (1960).
39. Furstenau, D., *Mining Eng.* **9**, 1365 (1957).
40. Nippon Paint Co., Japan Patents #65,050, June 2 (1981) and #67,532, June 6 (1981). Note that the Japanese patent office issues nearly 2500 patents in only 4 days, an indication of the ease of patenting and lack of critical technical review in that country.
41. Kamata, T., Gama, Y., and Ishigama, Y., Japan Patent #1,107,837 April 25 (1989).
42. Kawashima, N., Hiratu, O., and Meguro, K., Cong. FATIPEC, II, No. 15, p. 133 (1980).
43. Pleudemann, E., German Patent #2,912,430, Oct. 4 (1979).
44. Sato, N., Kimura, H., and Bando, T., Japan Patent #1,194,937, Aug 4 (1989).
45. Physicians Desk Reference—Non-Prescription Drugs, Vol. 47, ''Cetylpyridinium Chloride'' (1993).
46. Blaisdell, J., Marathron Chemical Co., private communication (1965).
47. Formulation courtesy Ballard Medical Products, Draper, UT (1995) (materials taken from Cosmetics Ingredients Dictionary, published by Cosmetics, Toiletries and Fragrances Assoc., Washington, DC, 1995).
48. Vernardakis, T., *Coatings Technical Handbook*, Chap. 61, Marcel Dekker, NY, p. 529 (1991).

CHAPTER

5

Bipolar Nonionic Dispersants

Nonionic molecules rely on hydrogen bonding primarily and van der Waals forces secondarily to attach themselves to powder surfaces and serve as dispersants. Except for exotic molecular structures, the following organic chemical groups, ordinarily bonded to simple organic molecules, serve this purpose.

1. *R*—$\ddot{\text{O}}$H (alcohol group)

2. 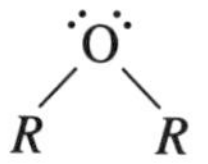 (ether group)

3. 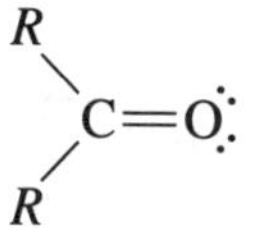 (ketone group)

4. 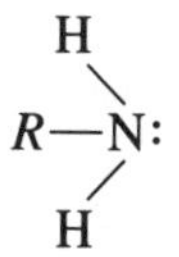 (primary amine group)

5. 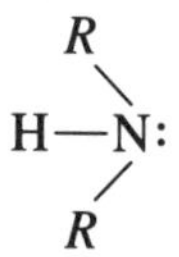 (secondary amine group)

6. 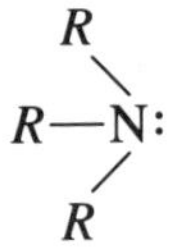 (tertiary amine group)

7. 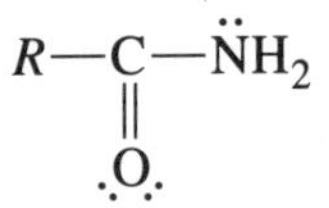 (amide group)

The double dot in the above formulas denotes a *p* orbital cloud extending into space. As a general guide, the farther this electron pair cloud extends outward from the central atom, the greater will be its potential of serving as a Lewis base and forming tight substrate–dispersant linkages. The organo-acid group, *R*–COOH, has been omitted deliberately from the series because of the descriptions given in Chapters 4 and 7 on its interactions, most of which are acid–base reactions rather than surface dipole coordination ones.

5.1 The Oxygen Dipole Family of Organic Dispersants

Within the oxygenated species, the molecular dipole moment is a strong inducement to associate with an oxide surface of a solid. For simple organic species the dipole moments are given in Table 5.1.

The term *dipole moment* used in Table 5.1 to express attraction between molecules and surfaces has the units of charge and distance. The unit 10^{-18} esu acting at 1 centimeter distance is termed a Debye, to honor the great physical chemist Peter Debye.

Oxygenated organics, as typified by the alcohols, must introduce dispersancy by competing with water upon surfaces. Because water has two hydrogen bonds

Table 5.1 Molecular Polarity for Oxygenated Organic Species

Organic Molecule	Formula	Dipole Moment (10^{-18} esu cm)
Methanol	CH_3OH	1.70
Propanol	C_3H_7OH	1.69
Phenol	C_6H_5OH	1.45
Ethylene glycol	HOC_2H_4OH	2.28 (1.14 per –OH)
Methyl ether	CH_3OCH_3	1.30
Ethyl ether	$C_2H_5OC_2H_5$	1.15
Acetone	CH_3COCH_3	2.88
Furan	C_4H_4O	0.66
Acetaldehyde	CH_3CHO	2.72
Hexane	C_6H_{14}	0.00
Benzene	C_6H_6	0.00
Water	H_2O	1.87

directed at approximately 105° and two electron pair orbitals directed into space, its enthalpy of wetting is high, about 8.5 to 9.5 kcal/mole, on solid oxide surfaces. Alcohol groups display a slightly lower energy, usually 6.0 to 8.5 kcal/mole.[1] Although an equilibrium concentration of any alcohol will be present on these surfaces, it will compete poorly with water and serve poorly as a dispersant in aqueous phase, unless agencies in effect on the solid surface intercede.

Silica and the silicates will adsorb an equilibrium concentration of both methanol and ethanol to a limited extent (approximately 0.1 to 0.2%) without giving evidence of flocculation. Use of polyvinyl alcohol, however, even at low concentration levels, introduces flocculation almost immediately. The proposed molecular spacing of the species

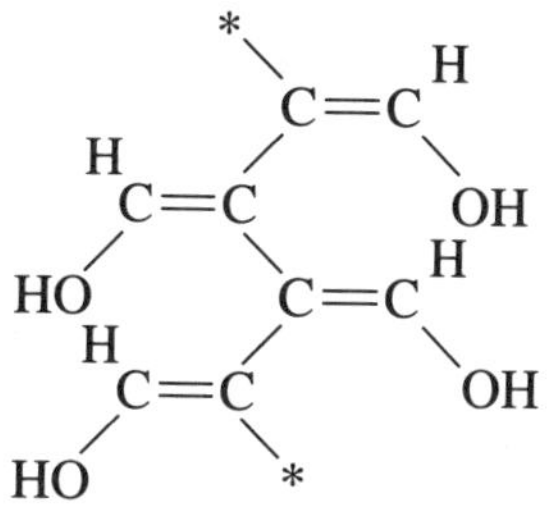

(5.1)

would suggest generation of a hydrophilic coating by directing half of the hydroxyl groups toward the surface and half outward. However, all alcohol groups appear to sit down on the silica surface, rendering the surface hydrophobic, based on infrared spectra of the adsorbed species. Sediment volume studies (see Chapter 13) indicate a very loosely associated, voluminous floc structure.

Studies on unmodified rutile (non-gel coated) show a preponderance of titanyl groups on the pigment's surface. In heats of wetting experiments[2] the enthalpies observed varied in the order:

isopropyl alcohol > water > acetone

indicating other factors than simple dipole moment are involved in the adsorption process (see Table 5.1). Obviously, adsorption of the alcohol does not serve to engender dispersancy in water because of the hydrophobic tail, albeit short, extending away from the titania's surface. The pigment so treated becomes slightly easier to disperse in polar organic solvents serving as predecessors to lacquer formulation. Alcohols of any greater molecular length than isopropyl induce mild flocculation quickly.

Studies with linear chain dihydric alcohols, especially *n* hexanediol and beyond, indicate longer members of these species actually create bridging between particles of inorganic oxides at high solids or close interparticle proximity and create a mildly flocculated structure. This effect can be measured readily, as shown for hematite in Figure 5.1, by terminal sediment volume (TSV) increase (see Chapter 13).

Several theories have been advanced to explain these trends, which vary

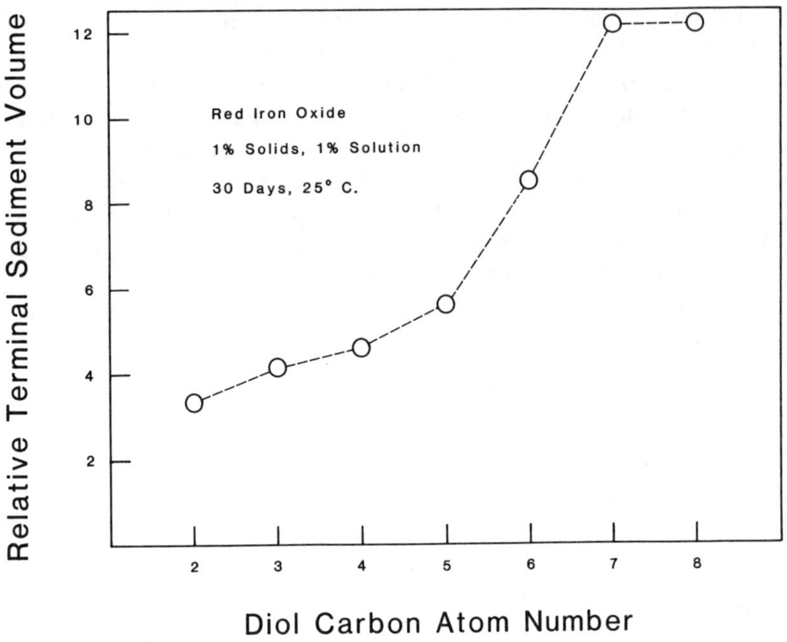

Figure 5.1. Iron oxide dispersibility with dihydric alcohols (Ref. 20).

somewhat among solids. Dipole constants for the two alcohol groups are identical because the carbon chain is too long to permit inductive effects. Second, a chain length of six carbon atoms is great enough to curve around the surface, enabling both alcohol groups to attach to alternate oxide or silanol groups (if the structure is silica or a silicate) and produce a chelation-type structure. Third, the attractive force for the surface docking sites is identical for each end of the molecule. The resulting configuration, if both ends of the molecules were attached to the same surface, would render the surface slightly hydrophobic. However, hydrophobic aggregation and particle bridge flocculation are significantly different with respect to sediment packing volume, with the latter dominating. The high relative terminal sediment volume values for C-7 and higher diols in Figure 5.1 strongly suggests interparticle bridging as a primary mechanism in these interactions.

Justifications for this behavior lie in probabilities of interparticle attachment. At 6 carbon atom distances particles are too far separated (see Chapter 2) for this interaction to be highly probable. Also, sedimentation volume in heptane should be comparable to that in water, which would indicate some interparticle structure, but it is not. Were a chelation structure the dominant form of diol attachment, particle terminal sediment volume (TSV) in heptane should approach a value of about 2, much like a dispersed clay in water, inasmuch as the surface should be rendered at least partially hydrophobic. However, TSV values are always much larger, often 6 or above.

Evidence does exist from terminal sediment volume work for interparticle bonding by long chains, that is, where the carbon chain exceeds 12 members. Again this evidence is not overwhelming. However, the sediment volume in water and heptane are similar.

Diols, through not highly effective as dispersants, have several advantages in commercial processing. Their water solubility is infinite (at least up to C-6). Most have pleasant odors, are easy to handle, and produce little skin irritation. Boiling points are high enough to prevent evaporation during slurry processing, yet low enough to be fugitive upon drying (200–250°C). Fugitivity commonly produces ''soft'' powders upon drying having high flowability and ease of secondary dispersion (customer makedown). Finally, diols impede sintering and ''tight'' agglomerate formation.

Because of their boiling points and low corrosivity, diols can be condensed from the vapor phase, recovered, and returned to the processing stream for process economy.

Most diols are reasonably inactive with regard to the surface chemistry or impurities in the pigment or solid structure and are colorless, for applications where whiteness is a critical factor.

5.2 The Nitrogen Dipole Family of Organic Dispersants

In the nitrogen-bearing family groups, dipole character follows a similar pattern to that of the oxygen family, if not more accentuated, as shown in Table 5.2.

Although a direct relationship might be expected between alkalinity in water and the molecule's dipole moment, any relationship is rather tenuous, as shown by the following alkalinity pattern:

Secondary amine > primary amine > tertiary amine > amide

Employing amine groups in place of alcohol groups changes dramatically the

Table 5.2 Molecular Polarity for Nitrogen-Bearing Organic Species

Organic Molecule	Formula	Dipole Moment (10^{-18} esu cm)
Methyl amine	CH_3NH_2	1.26
Dimethyl amine	$(CH_3)_2NH$	1.03
Propyl amine	$C_3H_7NH_2$	1.17
Trimethylamine	$(CH_3)_3N$	0.67
Aniline	$C_6H_5NH_2$	1.53
Phenylenediamine	$C_6H_4(NH_2)_2$	1.53 (0.76 per NH_2)
Acetamide	CH_3CONH_2	3.60
Ethylenediamine	$H_2NC_2H_4NH_2$	1.99 (1.0 per NH_2)

influence of these dipolar agents on particle dispersion. Monoamines, like single alcohols, impart a hydrophobicity to the solid surface onto which they are employed. However, the energy of adsorption for amines usually exceeds 10 kcal/mole and will most often displace resident water molecules. Part of this energy derives from the dipole effect but a greater part from multiple hydrogen bonds, as shown in Figure 5.2.

The most elementary symmetric species in the oxygen family, a double dipole hydroxyl compound, would be represented by hydrogen peroxide, HO:OH. This agent has virtually no dispersancy on oxide surfaces whatever. The nitrogen analog by contrast, hydrazine, $H_2N{:}NH_2$, does display a limited amount of dispersancy. The molecule has been shown to diffuse directly into kaolinite and other silicate interplanar regions, coordinate with silanol (or siloxy) groups on one unit cell of the silicate, then expand the crystal structure slightly.[3]

Unfortunately, a test of this principle of dispersancy by a diol or diamine molecule bearing a single carbon core cannot be made as such compounds do not exist. The smallest organic molecule capable of carrying dual groups is the ethane base (ethylene diamine and glycol).

Ethylene diamine, though a toxic agent, has been utilized as a dispersant for acidic oxides, especially where the diamine character may be later employed as a reactive adduct for secondary organic polymerization, as with epoxy and polyester compounds.

Figure 5.3 contains rheograms of pigmentary-grade kaolin processed with a spectrum of diamines and one triamine.[4] The data clearly indicate a dramatic dispersive activity up to a spine of about four carbon atoms.

Whereas the enthalpy of wetting for the alcohol group is less than water, that of the amine group on most acidic surfaces is higher than water. Enthalpy of wetting (adsorption) for ethylene diamine on ground silica is 10.5 kcal/mole, indicating that species would replace water at 9.2 kcal/mole. Surface titration curves for both silica and kaolinite are given in Figure 5.4. The stoichiometric point is clearly evident on both materials, though more so with the more acidic quartz surface. Calculations based on surface area and crystal structure suggest a quantitative adsorption mechanism. For competitive heats of wetting, however, chain length becomes a critical factor in surface adsorption. Figure 5.5 depicts

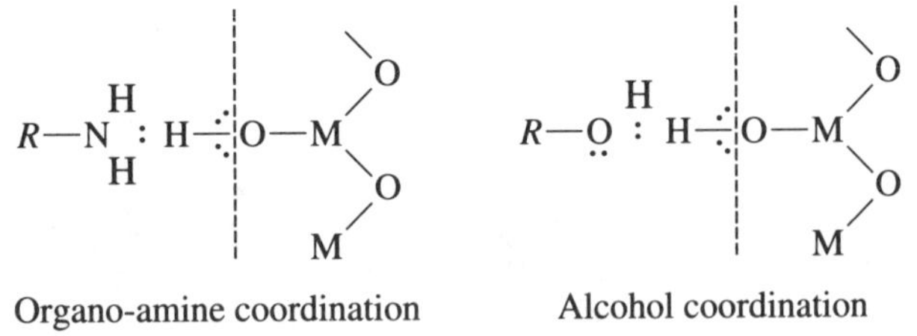

Figure 5.2. Comparative attachment mechanisms for amines and alcohols.

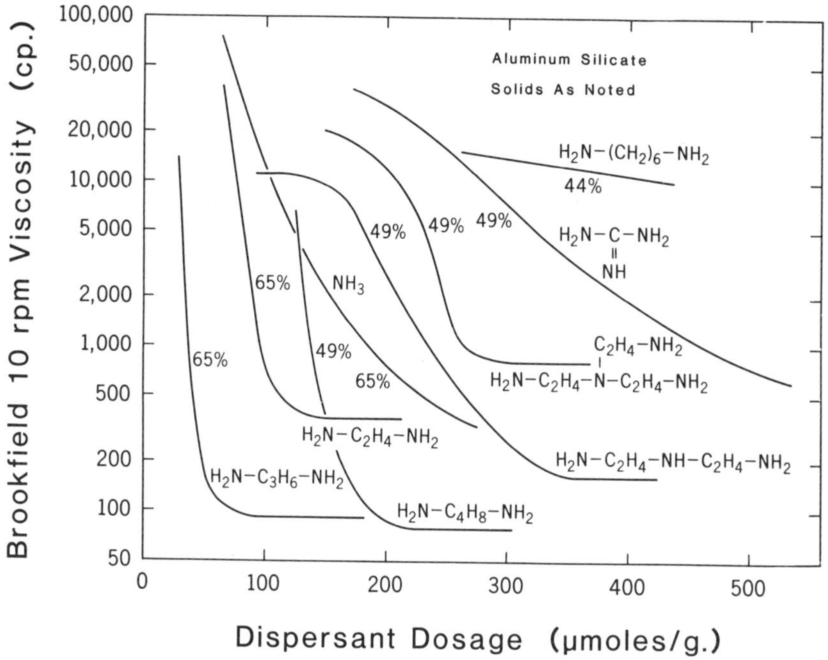

Figure 5.3. Diamine effects on aluminum silicate rheology.

two modes of diamine activity on the silica surface, the chelate configuration in (a) and a tail configuration in (b).

Chelation chemistry predicts five-membered rings represent a minimum stability configuration for a chelation complex. However, this principle becomes more complicated than simple ring chemistry suggests because two adjacent silanol (Si–OH) groups replace the central metal atom of conventional ligand or organic ring chemistry. Thus molecular span distance governs the limiting molecular size for the ligand species, not ring membership. Rheological profiles demonstrate this quite clearly. Dispersion data in Figure 5.3 shows a carbon spine of about 5 atoms represents the critical diamine molecular length for tail configuration.

The negative slope in the left arm of the rheogram of Figure 5.3 with hexylene diamine displays some reduction of viscosity with increasing dispersant level. Thus the study suggests equilibrium exists between the two modes of attachment illustrated in Figure 5.5, a train-plus-tail mixture.

Although diamines have excellent dispersive capabilities, several of their drawbacks can make their use impractical. In addition to rheological limitations, longer-chain diamines become highly colored (amber) and have more offensive odors (dead fish). All diamines have toxicological characteristics, skin irritation, lung irritation, some evidence of sterility, and carcinogenicity. They cannot be

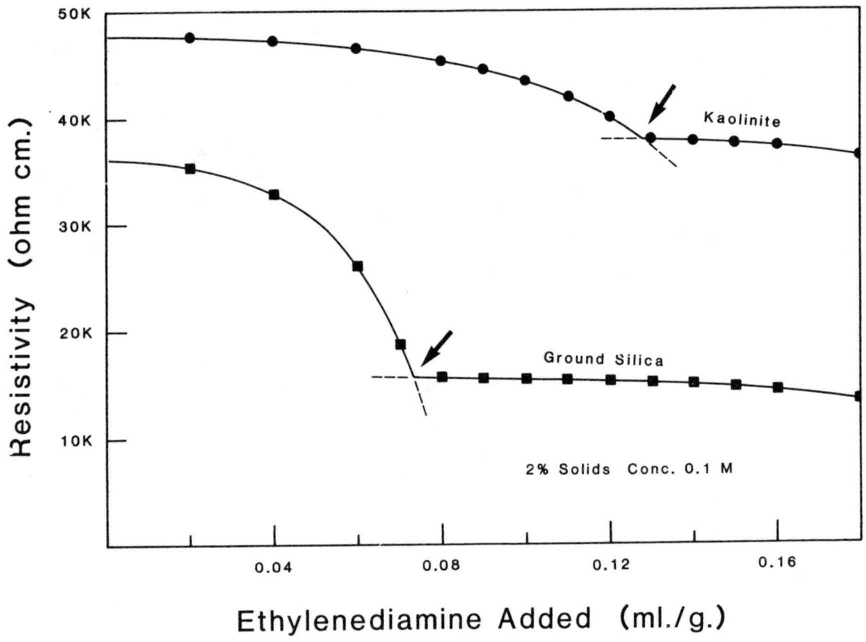

Figure 5.4. Surface acidity titration by ethylene diamine.

employed without the species eventually being fully reacted or removed. OSHA regulations limits their presence in worker air to 100 ppm.

Because diols are ineffective and diamines are toxic, a compromise in molecular structure can be realized wherein the molecular arrangement consists of a primary amine occupying one end and an alcohol group the other of an organic chain. Such compounds go by dual names, alkanolamines and amino alcohols. Their dispersive character was noted in the 1930s, with several reaction compounds between polyalkanolamines and fatty acids.[5] No commercialization appeared until the introduction of water-based emulsion paints.[8] Like polyhydric alcohols, the compounds are colorless and highly water soluble, but like polyamines they have an ammoniacal or fishlike odor. Although alkanolamines are alkaline and cause minor skin irritation in concentrated form, several have been approved by the U.S. Food and Drug Administration (FDA) and the Cosmetics, Fragrances and Toiletries Association (CFTA) for use as dispersants in cosmetic preparations at low concentrations.

Energy of adsorption for the aliphatic primary amine tail will be strongly dependent upon the acidity of the oxide solid surface, for here more than simple coordination chemistry is involved. Because the nitrogen molecule is a strong electron donor group (Lewis base) compared to its alcohol counterpart (a very weak Lewis acid), an amino alcohol has significantly reduced tendency for che-

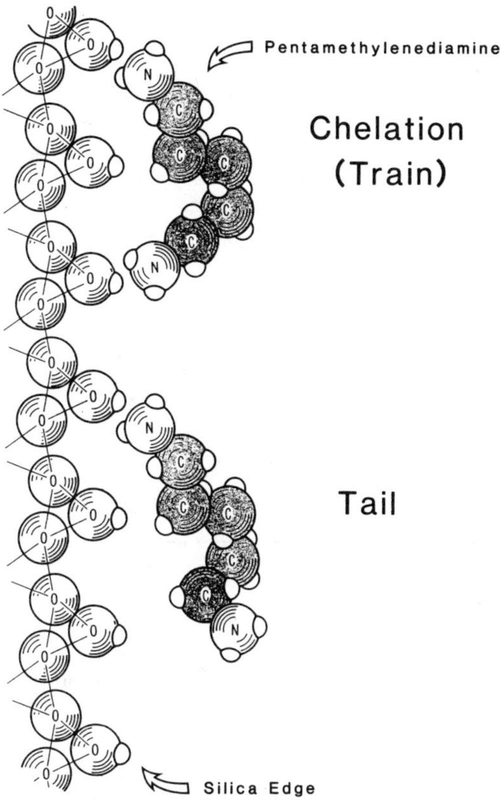

Figure 5.5. Adsorption modes of diamines on silica surface.

lation, most often being structured in a tail configuration. This is shown from heats of adsorption, which are greater for amines on acidic surfaces than alcohols. The molecular arrangement most likely is one where each surface acid group (hydroxyl) coordinates with an amino group and the alcohol group are solvent (water) directed. This is demonstrated further by Langmuir adsorption plots in Figure 5.6 for a selection of amino alcohols adsorbed on ground kaolinite. Note the aryl species, hydroxyaniline (also termed aminophenol), adsorbs poorly by contrast with the alkyl members. Hydroxyaniline has a pK_b above 9, while the others fall in the range of 5 to 7.

The extrapolated monolayer for the three aminoalcohols is approximately 42 μmoles per gram. This constitutes only about 20% of the powder's surface area, as determined by BET methodology. Thus the term *monolayer* with this silicate has a different meaning than with chemisorption or even nitrogen gas adsorption. The exchange sites, or silanol groupings, are distributed dominantly along the edge facets, and locales for amine fixation are about 0.32 nm^2 in area. Based on

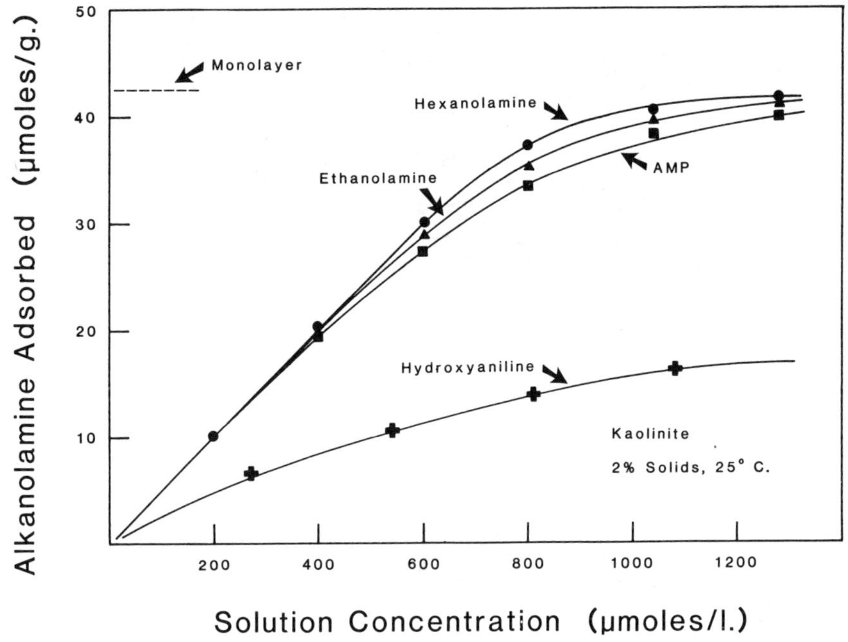

Figure 5.6. Langmuir adsorption of alkanolamines on kaolinite.

the surface area for this material, the amine adsorption operation constitutes about 18% utilization of the total area, a value essentially corresponding to the edge facets of the crystals. Note 2-amino-2-methyl-1-propanol (AMP), with its minor steric hindrance, adsorbs with lower energy than straight-chain alkanolamines.

Chain length is not a critical factor for alkanolamines because of the reduced potential for chelation with the differentially ended molecules. While the amino group has higher adsorption energy for the oxide (or hydroxide) surface than water, the hydroxyl end cannot compete as effectively (energetically) and, irrespective of length, dangles freely in the aqueous phase. Because dispersancy arises from creation of an insular barrier of water molecules, the longer this dispersant molecule, the greater should be this barrier and its attendant dispersant effect.

Figure 5.7 is a detailed study[7] of kaolinite viscosity as a function of molecular length of the amino alcohol. Because molecule weight increases with carbon chain length, dosage must be compared on a molar equivalency, rather than by percentage weight. Two effects are immediately obvious from the study. First, viscosity consistently decreases with chain length, even with a six-carbon chain. Second, the difference between 10 rpm Brookfield viscosity and 100 rpm viscosity also steadily decreases (more Newtonian) with chain length. Thus the

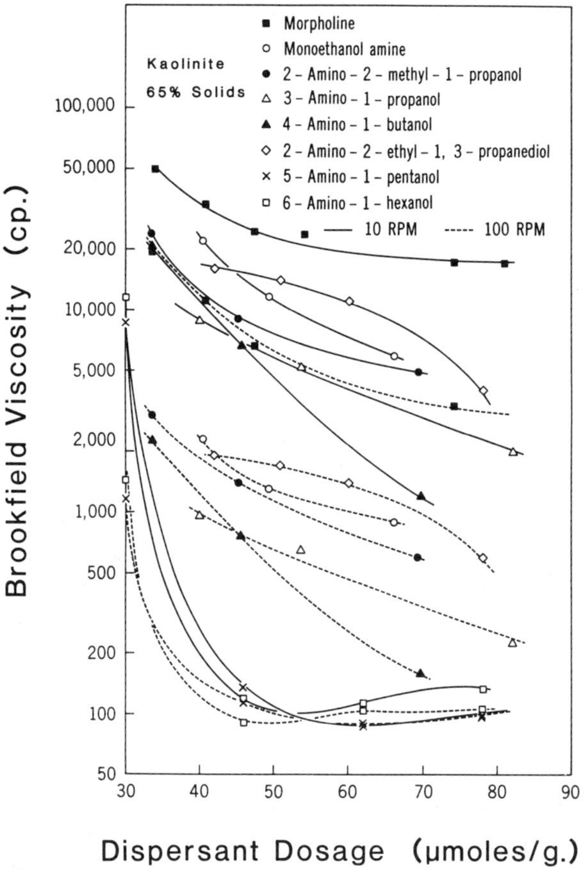

Figure 5.7. Alkanolamine molecular length effect on aluminum silicate rheology.

high-solids system progresses from a mildly thixotropic slurry for monoethanolamine (Thixotropic Index appr. 7) to an essentially Newtonian one (T.I. ~ 1) with hexanolamine. The optimum weight dosage for the hexanolamine is about 0.55% (solids basis), while the TSPP dosage for an equivalent solids is about 0.25%.

The study also includes morpholine, which is not a true alkanolamine. However, in solution morpholine hydrates strongly to give a pseudohydroxyl group:

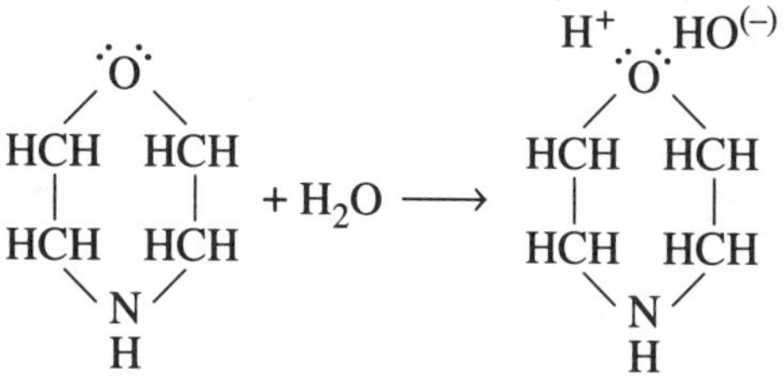

(5.2)

Water from the solvent phase that alkanolamine molecules could immobilize to form the barrier layer can be calculated by examining a long-molecule alkanolamine dispersant. Hexanolamine molecules are approximately 1.33 nm in length. While data on kaolin are useful and extensive, the pigment's surface activity is essentially restricted to edge faces, which makes such barrier-layer calculations difficult. Ground silica has particles ovoid in appearance with an aspect ratio of about 1.24. Active acidic silanol groups completely enshroud the particles. Silica employed for the Langmuir data in Figure 5.6, having a surface area of 2.23 M^2/g (mean size 0.67 μm), was selected for this calculation.

Immobilized volume is calculated by area times film thickness:

$$\begin{aligned} \text{Liquid volume} &= 2.23 \times 10^4 \text{ cm}^2 + 1.33 \times 10^{-7} \text{ cm/g} \\ &= 2.97 \times 10^{-3} \text{ cm}^3\text{/g (2.97 mg/g)} \\ &= 0.30\% \text{ (by weight)} \end{aligned}$$

Thus, at 60% solids, where the immobilized liquid/total liquid fraction is 0.445%, no critical rheological problems arise. Even at 70% solids the film immobilizes less than 0.7% of the available liquid phase. At 70% solids and a density of 2.64 g/cm, average interparticle separation is about $\frac{1}{4}$ diameter, which for a 0.67 μm particle is 0.25 μm or 250 nm, about five times the separation at which van der Waals forces become significant.

It is the effective length of the molecule that controls dispersion. The hexanolamine molecule at 1.33 nm is far short of the field required to overcome van der Waals attraction (20–25 nm), even if it is assumed the terminal hydroxyl group is hydrated heavily. This suggests that the enshrouded water film shields the particle sufficiently to cause interparticle contacts, water film against water film, to be sufficiently elastic to prevent ''sticking'' collisions of the Einstein–Smoluchowski type (Chapter 2). Water molecules are not monomeric species in the liquid phase, having a composition averaging about 16 molecules per molecular ''ball'' at ambient temperature. An association of such agglomerated species with the surface seems likely to increase the surface separation distance. Aligned, adsorbed polar molecules on the solid surfaces having a great thickness may also shield the direct and induced dielectric effects of one particle upon another. A parallel lies in a thin steel wall reducing dramatically the influence between two magnets in close proximity when inserted between them. While these concepts are conjectural, the ability of alkanolamines to disperse is not. They are highly effective.

Titanium dioxide without gel treatment can be dispersed to approximately 75% solids with AMP and at slightly less solids with morpholine, monoethanolamine, and triethanolamine. The molecular adsorption of the latter molecule is substantially less than for monoethanolamine, most probably because it extends three *R*–OH groups into the solution phase tetrahedrally and occupies a much greater effective surface area. This is almost exactly compensated because the effective weight level for the two is almost identical: 0.1% at 60% titania solids.

Unlike diamines, the amino alcohol family can be produced in nearly water-clear condition, important again for white and pastel pigments. When the dispersed material dries, even at elevated temperatures, discoloring is absent, and the agent is virtually quantitatively removed. Even if traces remain in the polymer casting or applied film, they do not form a white surface deposit following migration to the surface under humid conditions.

In contrast to polyphosphates, the chemical species not only supplies no nutrient materials for mold eutrophication and colony expansion, it also serves as a fungicide. One of its better applications relies heavily upon that property. In highly humid regions of the country, paint films eventually spawn mildew, usually a species termed *asperigellum niger*, a black tenacious mold observed to grow in a patchwork configuration across exterior painted surfaces. While the recommended antidote is to bleach, employing dilute, slightly acidified (with vinegar) sodium hypochlorite solutions, these eventually take their toll in color loss, film embrittlement, substrate breakdown, and peeling. Alkanolamines, especially 2-amino-2-methyl-1-propanol (AMP), as 2% solutions not only serve as a dispersant to remove mold, they kill it. Unlike bleach application, the painted surface does not require subsequent wash down with water. Residual alkanolamine on the surface retards subsequent mold growth. Retardency is somewhat short lived because the alkanolamine slowly volatilizes from the surface. The material's advantages in this category are lower toxicity, ability to clean a given surface repetitively without color and film deterioration, less physical work, longer "bucket life," more acceptable odor, and easier cleanup.

Amino alcohols have yet another series of properties that influence dispersancy. Tail configuration on a surface allows a greater number of molecules to adsorb than train configuration. This assumes each molecular protrusion from the surface (silica tetrahedra, alumina, iron or titania octahedra, etc.) represent attachment sites (acidic locales). Thus a typical 3- to 5-carbon amino alcohol with a molecular weight of 75 to 90 roughly corresponds to less than half that of a tetrasodium pyrophosphate molecule. Yet the optimum (minimum-viscosity) dispersant dosage for a pigmentary grade of kaolinite requires approximately three times as much butanolamine as TSPP. Also the minimum in viscosity is slightly higher for the amine than the phosphate as shown in Figure 5.8.

Although higher dosage levels incur higher costs, a major character advantage with a bipolar molecule appears in overdosing. Because the –OH group and the $-NH_2$ group do not differ greatly in solvation energies from water, their presence as free molecules in the dispersing liquid phase does not tie up significant additional water molecules by solvation. By contrast, sodium atoms at dilute levels coordinate 16 molecules of water per ion in their hydration spheres.[8] Polyphosphate anions, while solvating less, still consume at least half that. Solvation diminishes free liquid mobility, and by induction, individual particle mobility, thereby increasing viscosity. Because of low solvation energy and similitude between dispersant and solvent, amino alcohol molecules likely float nearly free in the liquid phase. As a consequence very little "chemical" viscosity arises from overdispersion. The nearly horizontal minima in Figures 5.3 and 5.7 are

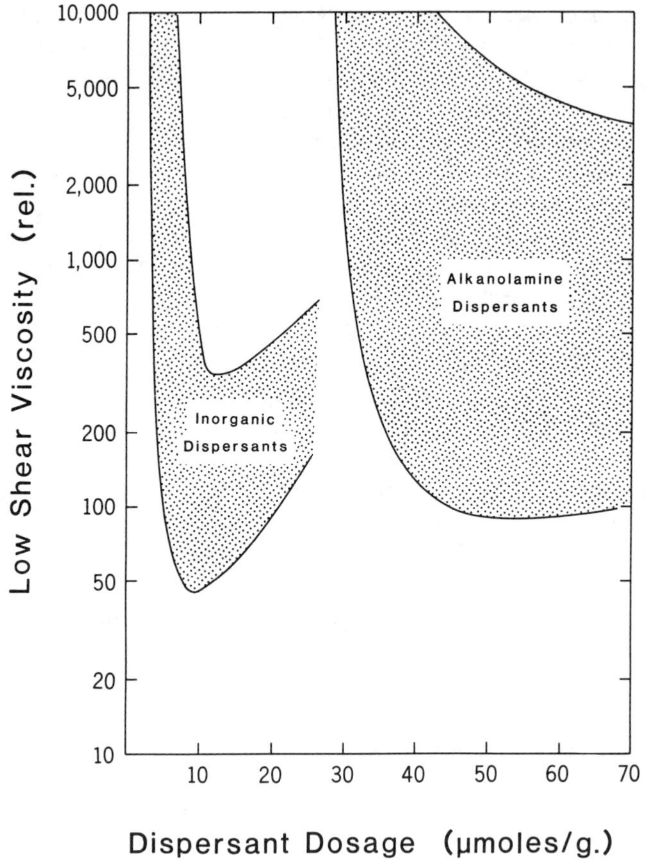

Figure 5.8. Rheological profiles for alkanolamines and inorganic ionic agents.

clear proof of reduced solvolysis effects with bipolar molecular dispersants. This can be of great advantage in grinding applications where high solids exist and surface area, with its attendant dispersant demand, increases constantly.

More detailed investigations of amino alcohols in other pigment applications show characteristics similar to those of kaolinite. For rutile without gel coating a slightly shorter list of amino alcohol dispersants has been evaluated, as shown in Figure 5.9. Again note that increased carbon chain length alters the system's rheological status from thixotropic to Newtonian.

A similar study on anatase, but with the single dispersant AMP, shows much poorer viscosity reduction. The influence of surface acidity is clearly a dominating effect on dispersancy for alkanolamines. This is shown dramatically by a series of rheograms for a broad spectrum of pigments, both white and colored, in Figure 5.10. Those solids having a preponderance of acidic hydroxyls on their

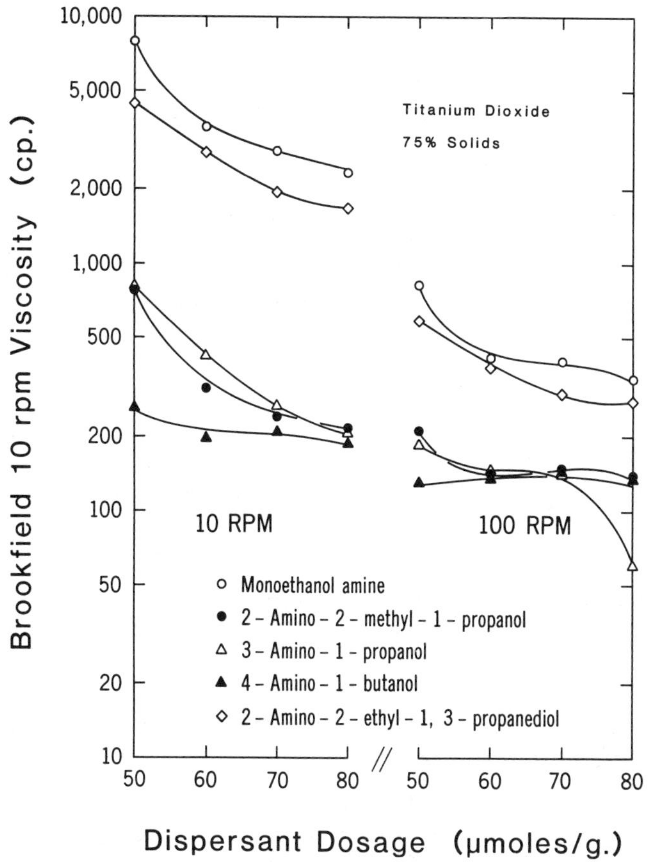

Figure 5.9. Rheological profiles for rutile with alkanolamines.

surface (rutile, kaolinite, talc, alpha ferric oxide, silica, and calcined Utah kaolin) display dramatic viscosity reduction with typical levels of dispersant, 0.2 to 0.5% by weight. Those materials having neutral or alkaline surfaces (barite, calcium carbonate, and strontium chromate) show little or no dispersancy at any level. Only zinc oxide displays minor dispersancy, a special case that will be treated later in this chapter.

Where critical viscosity values exist, alkanolamines have been employed in conjunction with polyphosphates. This combination enables a viscosity lower than with alkanolamines alone and broadens the rheological profile attainable with polyphosphates alone. For calcium carbonate a 1 : 1 molar mixture of short-chain sodium polyphosphate and butanolamine has provided excellent rheological dispersions of fine calcium carbonates and reduced the dispersion costs.[9]

While the usual application for alkanolamines lies in water-base systems, molecules may be tailored to effect different compatibilities for a broad variety

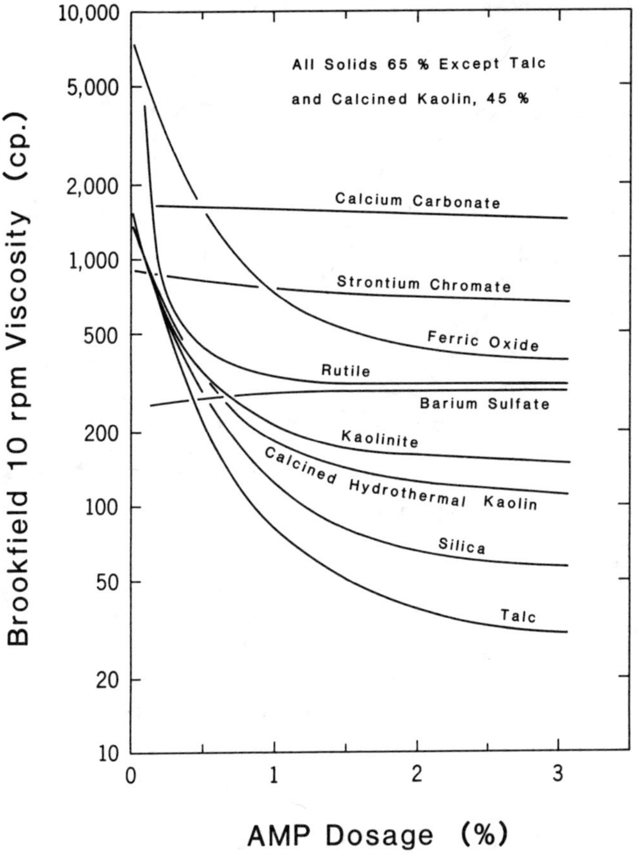

Figure 5.10. Rheological profiles for various pigments with AMP.

of media. For example, copper phthalocyanine is a pigment commonly about 0.2 μm in size. Dispersing this material in printing inks is difficult because of the carrier medium's high viscosity. The pigment can be more appropriately dispersed by modifying standard ethanolamine to the tertiary diethyl ethanolamine,

$$HO{-}C_2H_4{-}N\begin{matrix} \diagup C_2H_5 \\ \diagdown C_2H_5 \end{matrix} \qquad (5.3)$$

The addition of two ethyl groups bridges the hydrophilic–lipophilic boundary, making that end of the molecule more adaptable to the hydrocarbon ink carrier fluid. In this instance the hydroxyl group is believed to coordinate with the copper phthalocyanine pigment particles.

Fumed silica has an exceptionally high surface area and is employed as a viscosity additive to induce thixotropy into paints and plastics. Because the particles are extremely small (<10 nm) and the surface hydrophilic, dispersion incurs difficulties in oleophilic systems. One solution involves a flushing operation. Fumed silica is first dispersed in a long-chain alcohol (six carbons or greater) to which is added a paraffinic hydrocarbon, octane to dodecane.[10] It is uncertain whether the moisture-covered silica acts as an emulsion or whether a complex structure embodying four tiers of activity forms:

$$\text{silica: —Si—OH : } \overset{\text{H}}{\underset{\text{H}}{\diagdown\!\!\text{N}\!\!\diagup}}\text{—}C_2H_4\text{—OH : HO—}C_6H_{13} : C_{10}H_{22} \text{ : oleophilic resin}$$

The weak linkage in such a complex chain lies in the alcohol–alcohol hydrogen bond, about 8 kcal/mole. As a consequence the filled resin cannot be heated to any extent during polymerization or the network will break down. This dispersive technique is used in lieu of treating the silica with lactic acid and bypassing the long-chain alcohol intermediate. The lactic acid methodology gives poorer dispersion but greater thermal stability.

Polyester oligomers for water-based enamel applications may employ a moderately narrow molecular weight distribution to produce more consistent curing and gloss characteristics. To obtain the most thorough cross linking and film integrity (for high weather resistance enamels), the oligomers are dispersed with either dimethyl ethanolamine or a mixture of that substance and ethanolamine.[11] These agents can reduce the system viscosity by about half, making brushing easier, leveling better, gloss higher, and weather resistance greater. Triethanolamine shortens formulation shelf life and directly substituted amines react with the polyester interfering with cross polymerization.

Titanium dioxide has been dispersed with a broad variety of amino alcohol complexes, including amino glycol with alkyl-substituted glycols. In this instance the amino group is surface directed while the several alcohol groups are solution directed. The tenacity of adsorption of the single amino group permits several alcohol groups from the same molecule to be liquid-phase directed.

Where high-temperature applications occur, as in hot melt systems, the stability of the short-chain alkanolamine is endangered. For applications of finely ground, rounded (attrition milled) silica partcles,[12] the aminoalcohol may be added first and the system then dispersed in alcohol. This dispersion is added to about half again the silica weight of polyethylene glycol. Alcohol solvent is evaporated under vacuum at about 90°C and recondensed for secondary use. The resulting material is a molten mass at elevated temperatures with about 40% solids, which solidifies at temperatures below about 100°C.

The alkalene diamines can be substituted and complex duel alkanolamine molecules produced which may have special affinities for certain oxide surfaces. These can be employed with titanium dioxide, extender clays, silicas, ferric oxides, and chrome oxides in formulating ink pastes. Such molecular configurations are more difficult to predict because of stereospecific siting (docking) of such molecules on oxide surfaces. When these adsorb energetically, they produce a readily dispersible system for printing applications, including silk screening, because of their high thixotropy.

High-temperature aluminas have a minimal amount of their surfaces occupied with hydroxyl groups. When aliphatic amines or alkanolamines adsorb, a moderately tight association develops with the oxide surfaces. As the surface degenerates from hydrolysis or introduction of trace levels of water, the absorption energy of the alkanolamine or aliphatic amine diminishes. The two dispersants enable formulation of alumina for polish applications in either a water-base or petroleum-base formulation.

Where aluminum oxide has experienced long exposures to very high temperatures, the surface acidity rises.[13] If the alumina surface has had "active" exposure to water (hot water rinsing, etc.), this acidity decreases.[14] Infrared studies of alkanolamines adsorbed onto the acidic form of the surface show a slow transformation of the R-NH_2 tail to an amide structure that slowly releases from the surface because of reduced dipolarity:[15]

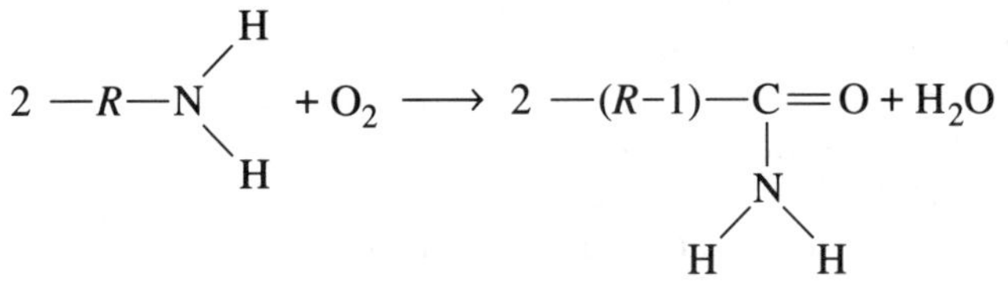

(5.4)

Alcohol terminal group, though apparently not adsorbed onto the alumina surface, are catalytically oxidized. Similar reactions are observed with olefinic amines. The energetics of this reaction as a function of varied levels of dispersant adsorption are shown for butanolamine in Figure 5.11. It is the initial approximately 2% of the alkanolamine adsorption that is susceptible to catalytic degradation. However, as the amide product is slowly desorbed because of its attenuated dipole moment, the remainder of the alkanolamine will become available to catalytic activity, much in the fashion of long-chain polyphosphate scission. This effect of surface will be described in more detail in Chapter 10.

Aluminum silicates are employed as fillers in caprolactam (nylon) polymers for a variety of reasons, including color and extension. However, certain nylon and polyester polymers can be given fiber surface texture by controlling particle size and surface character of the aluminum silicate filler (e.g., Qiana™). This method attempts to mimic the technique silkworms employ in extruding silk fibers. The worm's orifice "vibrates" during extrusion, giving silk a surface texture similar to a flexible vacuum cleaner hose but less accentuated. To obtain this character with a filler requires special dispersion. One technique involves a dispersant manufactured by a somewhat complex and circuitous route for this purpose. N-dodecyl 1-3 propanediamine can be reacted with caprolactam and an epoxide (oxirane)[16] to produce a slightly water-soluble material. This agent

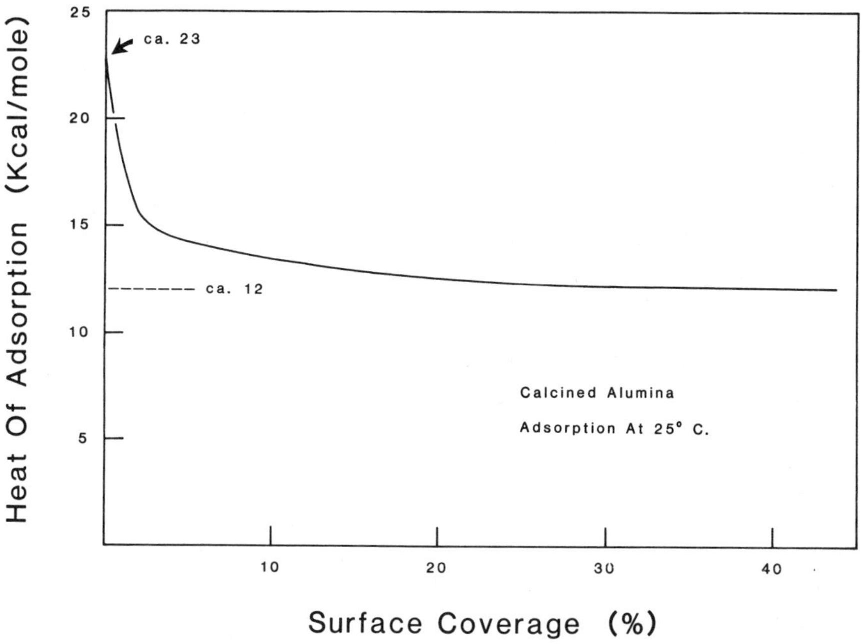

Figure 5.11. Energy of adsorption of butanolamine on alumina.

will disperse partially calcined aluminum silicate sufficiently that 40% solids can be introduced into the polymer with solvent. Upon extruding the polymer and allowing solvent to volatilize, a surface texture develops which upon twisting into multiple-strand threads and weaving, bears a remarkable similarity to the silkworm's product.

These fibers can be colored both by dyeing and by incorporating color pigments into the fiber prior to extrusion. The latter process provides both the highest light and laundry fastness.

Both rutile and anatase titania pigments may be coated with alumina gels to provide a more consistent and reactive surface. This coating may embody only a few weight percent but can cover the entire surface of the particles. By reacting alkanolamines with polyprotic acids (oxalic, succinic, acrylic, etc.), organo salts can be formed with dispersive properties. These will have pH values in the near-neutral range, 6–8, with only slight dissociation (pK_a greater than 4.4). Dispersants made by this method, for example, triethanolamine succinate,[17] stabilize and disperse alumina gel-coated titania particles for water-base emulsion polymer paints.

Another method of dispersing the alumina gel-treated titanium dioxide pigments is by cross products between organophosphorous compounds (see Chapter 3). Because the alumina gel has little acidity, AMP and similar compounds are ineffective. To assist AMP in dispersing acidic extender copigments (silica and silicates), coagent condensation products can be made from a 2-phosphono-1,2,4 butane tricarboxylic acid.[18] When neutralized with sodium hydroxide, the pH is about 9, eminently compatible with AMP and other aminoalcohols for multipigment dispersion. If neutralized with ammonia, the dispersant's pH drops to about 8.2, but its attachment onto the alumina gel is much weaker. Also, attempts to produce a triethanolamine reaction product do not yield a material that disperses the surface-modified titanium dioxide. It appears to function better on an untreated pigment.

Transition metal compounds have a peculiar association with alkanolamines. Because of the partially filled *d* orbitals in members of the series, amino groups can coordinate by contributing their electron pairs. Those elements most susceptible to this ligand formation are chromium, cobalt, nickel, copper, zinc, silver, cadmium, and platinum, as metal powders, oxides, or compounds. For example, cobalt blue, a mixed oxide with alumina having the approximate formula $CoAl_2O_4$, will coordinate amino groups and alter its color. The same phenomenon holds for copper silicate (lime green changing to deep blue). Because the amino alcohol molecules are pulled into the bonding orbitals, their dispersant property is greatly impaired due to the alcohol groups' restricted ability to extend their tails away from the surface and into the liquid phase. Metal coordination of alkanolamines results in larger quantities being required to effect a surface boundary layer of protective water. Only the surface and near-surface metal atom component of oxides and hydroxide groups can coordinate these molecules, as indicated by the color changes, which, when produced, are extremely shallow in the particles.

This same adsorption phenomenon also provides a mechanism by which metals can be dispersed with alkanolamines (see Chapter 12). It demonstrates, too, why zinc oxide, a highly alkaline material, can adsorb the amino group and disperse, even if viscosity reduction is not great.

This latter point is the basis of a masonry product developed for yet another purpose. Zinc oxide is incorporated into paint as a white pigment even though inferior to other primary white pigments in refractive index (2.00 compared to rutile at 2.76) and isoelectric pH (9.5–10 compared to 4.7 for rutile). However, the oxide has one outstanding and unique property—it is an excellent mildewcide. In regions where humidity is high and temperatures are warm, titania-based paints mildew heavily. Partial or whole replacement of the titanium dioxide by ultrafine zinc oxide is one effective alternative.

Bathroom tile grouting falls under the same conditions for generating mildew. Better grade grouts often contain a small amount of chlorinated phenol or brominated organics to resist mildew growth. However, after 2 years of heavy bathing activity, these agents leach out and mildew commences. Thereafter, cleaning is a frequent ritual. Grout cleaners are often formulated acidic to destroy mold, which slowly degrades the cement character, following which the grout begins to crack and slough out. This can be eliminated by employing zinc oxide in partial replacement of titanium dioxide as the white color body in the grout. As abrasion wears the surface slowly away, fresh zinc oxide is presented for continuing mildewcidal activity.

Some grout formulations are cement based (calcium aluminosilicates) and some are epoxy, acrylic, or PVC based. The latter are emulsifiable varieties much like the water-based epoxy resins for paints. Zinc oxide readily disperses in the amine phase of the two-part system, while the epoxy (oxyirane) component is in alcohol or emulsified. Zinc oxide disperses very poorly in that phase and earlier formulations in which the two were combined yielded a flocculated system tending to lump on prolonged contact, making the masonry activity of smoothing and troweling a tedious and time-consuming one.

However, when ZnO and water-based amine are blended first, the two-part system reacts to give a well-dispersed, tightly bonded pigment that in turn tightly bonds to the ceramic tile (acidic silicate surface). Cleanup of waste and smeared grout can be effected with water for a short time after application, and by alcohol if cleaning is delayed. Once polymerized, the new grout is much more acid, alkali, and scrub resistant than cement-based grouts and easier to keep clean.

5.3 Molecular Structure Effects

Molecular structure effects are involved in alkanolamine dispersancy. Reviewing the effect of AMP on kaolinite and rutile and comparing this with monoethanolamine, whose chain has the same length, shows the protruding side methyl groups to have an influence on coordinating water along the pigmentary surface. Consider the two dispersant rheograms in Figure 5.7. Redrawing these, the second can be seen to be identical with the first with regard to chain length, differing by the two methyl groups substituents on the chain:

Monoethanolamine (MEA)

2-amino-2-methyl-1-propanol (AMP)

With regard to development of an insular shield of coordinated water, the two should perform similarly. Comparison of the rheograms shows this to be true, both at high shear and low shear. Figure 5.7 is constructed on a equimolar basis, not weight percent. The two curves very nearly overlap. At high shear equivalent viscosities are attained at an AMP value about 10 micromoles per gram lower than with monoethanolamine and slightly less at low shear. This differential is the result of laterally extended chemical groups on the AMP molecule.

The effective area occupied at the surface contact point (silanol grouping) is identical for the two species based on Langmuir data. However, at the solvent facing end (alcohol group) the molecule sweeps out a greater protective area as AMP than as MEA. Reduction of 10 μmoles per gram clay at 60 μmoles, the estimated optimum for MEA, represents a savings of 16% while requiring an increase of nearly 46% due to molecular weight increase by the AMP (89.14 vs. 61.14). Thus, for this material, AMP has no dispersancy advantage over MEA.

Other simple alkanolamines which have been commercialized as dispersants include 2-amino-1-butanol, 2-amino-2-methyl-1,3-propanediol (AMPD), 2-amino-2-ethyl-1,3 propanediol (AEPD), and tris-(hydroxy-methyl)-amino-methane (THAM). The latter has unusual properties derived from its multiple alcohol groups:

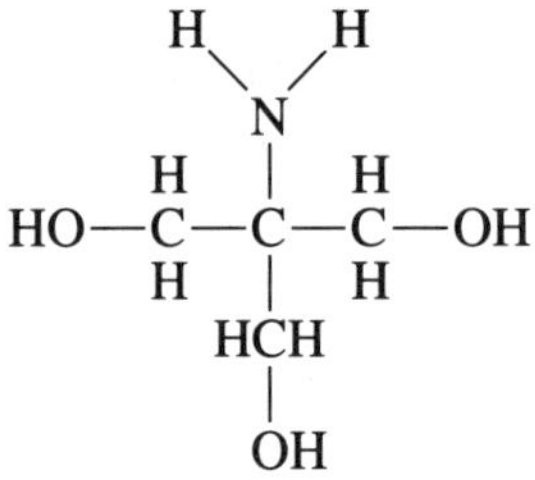

(5.5)

This molecule exerts a much stronger polarizing field, once adsorbed as a dispersant upon an acidic surface, which attracts and holds barrier water layers more tenaciously than do the single alcohol groups. Further, the effective area of THAM is approximately four times that of the MEA molecule. True mass savings result with THAM because the mass gain is only about 36%. It has the highest boiling temperature of the commercial amino alcohol compounds, 200°C, and is retained almost quantitatively upon drying at 125°C.

Several of these species show stereospecificity for certain oxide surfaces independent of the dipole moment and basicity of the amino group (which changes very little among the species). Silica and the silicates respond well to the linear molcules, giving the lowest viscosity as the chain length reaches five carbon atoms. As the structure becomes more compound, especially with multiple alcohol groups on the dispersant, site structure and molecular spacing often dictate which species displays the lowest rheology.

AMP has been found to provide superior stability to pigments in melamine and urea formaldehyde formulations, imparting, as a consequence, improved gloss and better color to the molded products. THAM (also called tris-amino) is highly effective in alkyd resins, bonding the polymers and their pigments more tightly, which increases both weather resistance and gloss retention,[19] advantageous properties in exterior application paints and plastic materials.

References

1. Zettlemoyer, A., *Chemistry and Physics of Interfaces*, Chap. 12, Am. Chem. Soc., Washington, DC, p. 137 (1965).

2. Munvera, G., and Stone, F., *Disc. Farady Soc.* **52**, 205 (1971).
3. Grim, R., *Clay Mineralogy*, 2nd ed., McGraw-Hill, NY, p. 472 (1968).
4. Conley, R., *J. Paint Tech.* **46**, 61 (1974).
5. Epstein, A., and Katzman, M., U.S. Patent #2,239,997, April 29 (1941).
6. Kingsbury, F., U.S. Patent #2,744,029, May 1 (1956).
7. This technical study represents portions of a joint research effort between IMC's Commercial Solvents Div. under Robert Purcell and the author at the Georgia Kaolin Co.
8. Jones, H., *Hydration in Aqueous Solution*, Carnegie Inst., Washington, DC, p. 20 (1907).
9. Robinson, G., U.S. Patent #4,345,945, Aug. 24 (1982).
10. Makarov, A., and Polishchuk, N., *Uk. Khim. Zh.* **46**, 1070, (1980).
11. Williams, R., *Proc. 15th Water-Borne, High Solids Coat. Symp.*, p. 478 (1988).
12. Baron, W., Knapp, M., and Marquard, K., German Patent #3,919,940, (1990).
13. MacIver, D., Tobin, H., and Barth, R., *J. Catalysis* **2**, p. 485 (1963).
14. Peri, J., *J. Phys. Chem.* **69**, p. 231 (1965).
15. Whalley, L., Doctoral Dissertation, Bristol Univ., Great Britain, (1964).
16. Andree, H., et al., U.S. Patent #4,264,480, April 28 (1981).
17. Story, P., U.S. Patent #4,752,340, June 21 (1988).
18. Teichman, G., Woditsch, P., and Linde, G., German Patent #2,725,210, Dec. 14 (1978).
19. "Chemistry and Use of Aminohydroxy Compounds," Tech. Bull. #TDS-10 and NP-6910, IMC Commercial Solvents Corp., pp. 2 and 8, (1969).
20. Initial test work in this study contributed by H. M. Johnson, Inc., and IM&C, Commercial Solvents Div.

CHAPTER

6

Polypolar Nonionic Dispersants

The general class of nonionic dispersants containing multiple polar groups functions on solid surfaces in manners little different from those ionic chain configurations described in Chapter 4 and the more simple alcohol-amine configurations in Chapter 5. Without ionized active groups, however, they display several general characteristics which either dictate their employment or aid in it.

Because these dispersant molecules are not ionized, solvation is less strong and percentage dosing less critical. In this respect the family resembles the diols, diamines, and alkanolamines. Unlike the alkanolamines, molecular structures are large, providing increased van der Waals attraction. Some are straight chained and some branched chained, but all contain numerous periodic polar groups along these chains to render them, at least in part, hydrophilic.

Molecular structures in this class most often contain polar groups at or near their termination(s). However, the central portion of the molecule may be polar or be devoid of polar groups. The former structures are useful for dispersing polar solids in polar media (including alcohols, esters, ketones, and aldehydes), while the latter function for nonpolar materials in polar media. A third class also exists wherein the structure allows polar adsorption but presents an oleophilic tail for dispersing in nonpolar media. This subgroup will be taken up as a member of a much broader class in Chapter 7.

This general class of appendant dispersants is characterized by polymerized ethylene oxide structures having large numbers of ether or alcohol groups:

$$R\text{—O(—CH}_2\text{—CH}_2\text{—O—)}_n\text{H} \tag{6.1}$$

where n lies in the range 2–20 and usually consits of mixtures. These are often defined as "polyethylene oxide" or "polyoxyethylene" derivates. Alkyl chain length, designated by *R*, can alter dispersant properties from dipolar to hydrophobic character, the transitional carbon number being around 4–5. Ethylene oxide groups absorb at polar sites of the particulate surface, while the tail extends into the liquid phase, either aqueous or oleophilic.

Short-chain polyethylene oxide derivatives with short *R* chains are often incorporated into cosmetic applications. There they contribute a soft gel-like, lubricating character but dry quickly, leaving skin with a satinlike texture. Many of these compounds have been cleared by the Federal Drug Administration (FDA) and the Cosmetics, Perfumers and Toiletries Association (CPTA) for topical applications and a few for internal use.

The hydrophilic–lipophilic balance (HLB) in these molecules is a direct function of alkyl chain length and ethylene oxide polymer number. Figure 6.1 is a graph of the theoretically calculated HLB values for a family of polymers having a range of alkyl chain lengths. Solid crosses represent a selection of commercial nonionic agents with a polymer average in the general range of the calculated lines. The graph is subdivided into four broad application bands in order of decreasing HLB value, (1) detergents, (2) oil-in-water emulsions, (3) water-in-oil emulsions, and (4) polar solids in oil, based on manufacturer and customer application success.[1] These bands do not exhibit sharp demarcations, as the

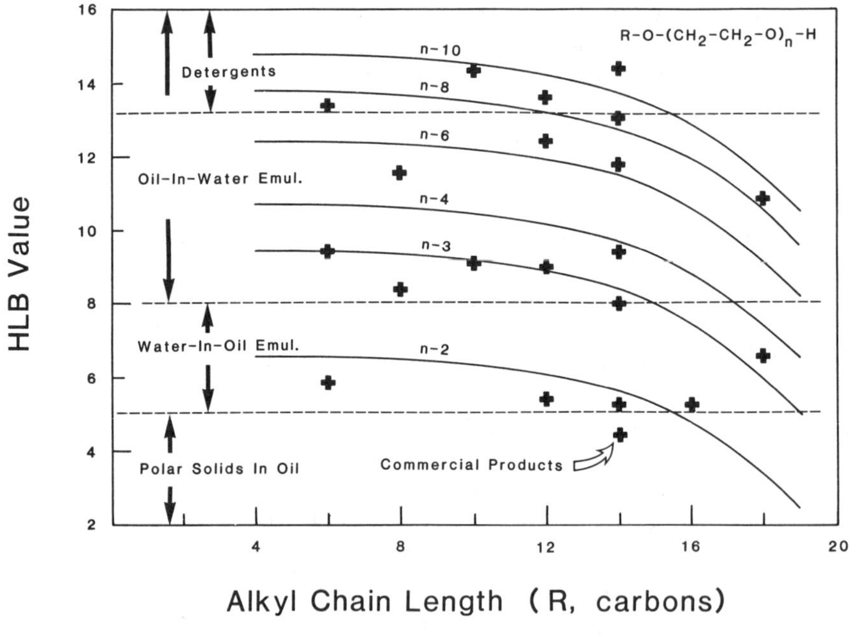

Figure 6.1. Application characteristics and HLB values for various nonionic dispersants.

figure may imply, and a marginal agent may well function satisfactorily in more than one application.

With longer hydrocarbon tails, nonionic dispersants serve for wetting organic materials, for example, oils, greases, dirt, etc., without the need for a charged molecular head to effect dispersancy. Rather, they rely upon multipolar groups, much in the fashion of dipole–dipole groups described earlier (Chapter 5). Such oil-in-water stabilized dispersions have the advantages of (1) reduced critical dosage and (2) avoidance of problems created by ionic agents, such as electrical conductivity and human topical sensitivity. Baby shampoos, where ionic components could severely burn delicate eye tissues, often employ nonionic dispersants. While they display reduced effectiveness in dispersant load, their compatibility with the skin and eyes overrides this limitation.

Molecular weights for the polypolar structures range from several hundred to a few thousand, and in special instances to 10,000 or slightly greater. Those same functional groups are operative as were described with the bipolar family: alcohols (polyols), ethers (polyethers), amines (primary, secondary, or tertiary), and amide groups. Because of increased molecular weight and the presence of multiple attachment groups, polypolar dispersants increase their affinity for polar surfaces via (1) numerical dipole site attraction, (2) acid–base interaction, and (3) elevated van der Waals forces.

Molecular chain structure, whether linear of branched, presents a more favorable conformation to the solid surface in a train configuration. Infrared spectra of the species' multifunctional groups (most often polyglycols or polyethers) are insufficiently definitive to determine whether part-train, part-tail configurations, as occur in transitional zones with polyanionic species, can occur on solid surfaces. With certain molecular structures these may be inferred, either by molecular modeling or by simple reasoning. Unlike alkanolamines, molecular chain length does not dramatically affect rheological properties. In the sense of increasing van der Waals forces and surface affinity, chain length does show small improvements in viscosity. Character shifts, as from thixotropic to Newtonian viscosity, are in little evidence. Most of these molecules either are sufficiently near the limit of aqueous solubility or their chain length sufficiently approaches the potential for interparticle bridging (similar to polyacrylates) that additional dispersant molecule size (and weight) may introduce additional properties disadvantageous to the system.

Nonionic agents as a class have certain special properties. Unlike ionic agents (polyphosphates, polyacrylates, etc.), nonionics exhibit a negative thermal coefficient of solubility. Their micelle structure increases in size (and molecular population) as the water temperature rises, contrary to what kinetic energy might imply. As the temperature continues to rise, these micelles approach colloid size where light scattering commences, generally in the range of 0.1–0.3 μm. At this point, which is not sharp, their solutions become hazy. The temperature at which haze can just be detected visually is termed the *cloud point*, a value often given in application literature for commercialized agents. Large nonionic molecular clusters are more prone to break and affix themselves to solid particles, cellulose

Table 6.1 Polyoxyethylene-Based Surgical Hand Scrubbing Formulation

Component	Weight %
Polyoxyethylene–polyoxypropylene copolymer	15
Lauryl dimethylamine oxide	3
Lanolin (polyethylene glycol 75)	5
Chlorhexadine gluconate (100% basis)	4
Isopropyl alcohol	4
Fragrance	~1
Water	68
	100

fibers, or other attractive surfaces at temperatures in the cloud point range because of thermodynamic stress in the micelle. Typical cloud points range from about 30°C for the medium alkyl–medium polymer chained molecules to near 100°C for the longer alkyl–larger polymer ones. The presence of dilute ionic species (salts, other dispersants) in solution usually lowers the cloud point by 20–30°C for the high-temperature species.

Surgical hand scrubs frequently are based on polyoxyethylene derivatives because of their reduced propensity towards dermatitis. A moderately strong bacteriocide or antimicrobial agent is incorporated that is assisted by the reduced alkalinity provided by the polyoxyethylene component compared to a sodium anionic agent. A representative formulation[2] is given in Table 6.1.

The primary dispersancy function is provided by the polyoxyalkene component. Lauryl dimethyl amine oxide also serves a dispersant function but provides high foaming activity in addition, important in floating skin debris to the surface for aqueous flushing. Anionic agents are deliberately omitted as they tend to inactivate the chlorhexadine ester, a broad spectrum microbial agent, and also cause more severe skin cracking with frequent usage.

6.1 Alcohol and Ether Macromolecules

This class of dispersants is dominated by derivatives of ethylene oxide because of that material's ready availability from petrochemical operations (est. 5 million tons annually), its ease of reaction (with water to form glycols, with alcohols to form polyethers, with amines to give ethanolamines, and with acids to give polyesters), and its versatility in producing a broad variety of dispersants.

Dispersant formation follows the general reaction scheme shown in Figure 6.2.

The *R* group in the dispersant can be highly variable. It may include any of the following organic groups: phenyl, tolyl, phenoxy, alkyl, alcohol, ester, and, as will be discussed in the section immediately following on, amines. It also may be sulfonated or carboxylated to produce anionic molecules.

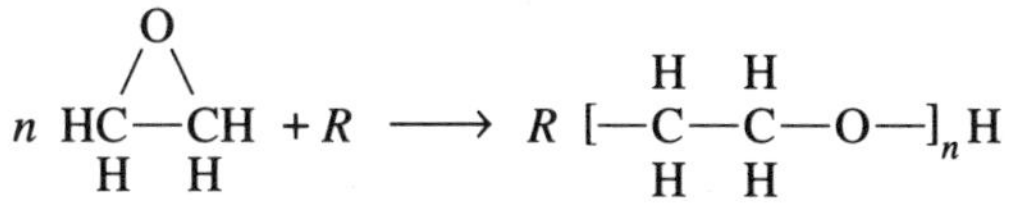

Figure 6.2. Nonionic dispersant formation via ethylene oxide.

Where the polyoxy organic molecule is sufficiently long and dipole groups moderately close (2–3 carbon atoms of separation), the train configuration prevents or significantly reduces the probability of interparticle bridging (a condition leading to flocculation), either in a partial mode (fines on coarse) or complete mode (total particle reticular network). A typical example, based on a glycol condensation reaction between ethylene glycol and stearyl glycol, is employed for carbon black dispersion in butanol solvent-based inks and polyester resins:

$$\mathrm{HO{-}\overset{H}{\underset{H}{C}}{-}\overset{H}{\underset{H}{C}}{-}O{-}C_{18}H_{36}{-}O{-}\overset{H}{\underset{H}{C}}{-}\overset{H}{\underset{H}{C}}{-}OH} \tag{6.2}$$

The molecular weight for this species is 374, and the number of surface attachment groups is four. However, the vibrational infrared spectra for the terminal hydroxyls suggest these remain free of surface coordination. This information leads to the conclusion that the molecule's dispersant function derives from the "basin" structure sketched in Figure 6.3.

These conclusions can be affirmed by observing the black's viscosity in butanol with variation in terminal, or leg, group chain length (C-2 to C-6) compared to the central, or spinal, section variation (C-4 to C-18), as shown in Figure 6.4. For the specific carbon black, a significant lowering in viscosity occurs with

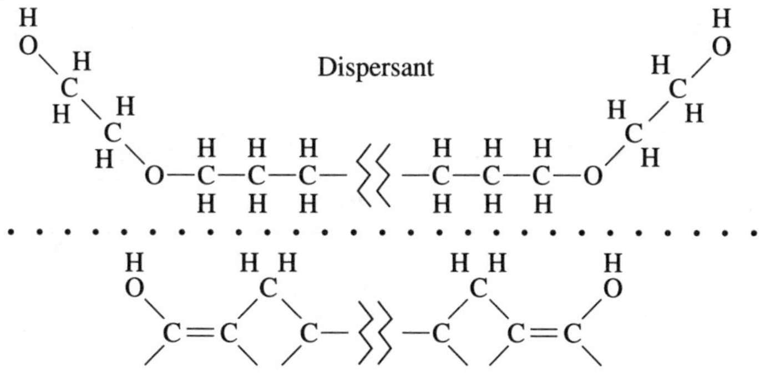

Figure 6.3. Polyol dispersant adsorbed onto partially oxidized carbon black.

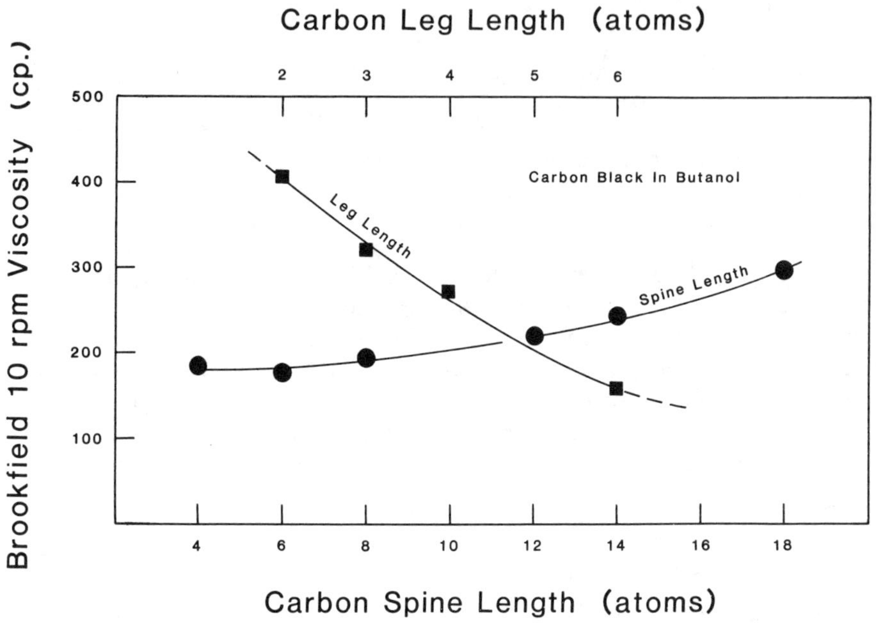

Figure 6.4. Effect of leg and spine length of polyether on carbon black viscosity in butanol.

increased leg length (isolation of the oleophilic surface from the polar alcohol), but only a small rise results with spinal length (decrease in leg population).

The degree of adaptation of the spinal molecular component is believed to be directly related to the level of oxidation[3] on the carbon surface. This spinal segment has the ability to flex and adapt and may configure variously with different blacks.

With a shorter central chain, this dispersant structure can be employed for water-based jet printing inks where carbon black is replaced by a fine (0.5–3

Table 6.2 Water-Based Jet Printing Ink with Nonionic Dispersant

Component	Function	Weight %
Quinophthalone black	Pigment	10
Ethylene glycol	Carrier liquid	20
Tributylene glycol (cond.)	Dispersant	2
Bentone 27	Thixotrope	1
AF-93 silicone	Defoamer	0.01
Water	Solvent	67
		~100

μm) quinophthalone dye. A typical formulation for this system is given in Table 6.2.

The dispersant's role in the table is actually twofold—first, to maintain a homogeneous distribution of dye particles throughout the ink, and second, to enable the ink to wet cellulose fibers slightly, but not sufficiently to create a wicking effect and cause ink spread.[4] Alkanolamines serving in this role enable ink to penetrate deeper into the paper print stock for a more smudge-proof printing, but also cause lateral ink spread and reduction of print resolution.

In this class is one of industry's most ubiquitous dispersants, Triton X-100™, a nonionic with the general formula

$$C_8H_{17}(C_6H_4)—O—(C_2H_4—O)_9—H \tag{6.3}$$

which employs an aromatic ring midway in the molecular chain. In certain instances this material serves as a more thorough dispersant (particle disaggregation) for silicas and silicates than the polyphosphates (see Figure 10.12).

Terminal alcohol groups may be esterified with a medium-to-long-chain fatty acid to provide a broad range of dispersancy, from water through polyester to alkyd resins, by changing the fatty acid chain length. By copolymerizing ethylene, propylene, and a C-9 chain length hydrocarbon and reacting this with butyric or stearic acid, an agent is formed capable of dispersing ground calcium carbonate (top size ~3 μm in either water-based acrylic emulsion or mineral spirits–alkyd resin systems.[5] Because of the dispersant's high molecular weight and the pigment's small particle size, approximately 5% of the carbonate weight is required for good dispersancy.

Shortening the fatty acid to two carbons (acetic) appears to degrade the dispersancy effect greatly in either formulation, even in water itself.

6.2 Protective Hydrophilic Colloids

As the chain length for these compounds increases and hydroxyl groups are added to the spine, they approach natural hydrocolloids, that is, polysaccharides, gums, and the like. Such hydroxylated agents are believed to adsorb at a level approaching quantitative surface coverage (a 5% level is required in the referenced Czech work, Ref. 5). Essentially, these substances wrap around the particle and extend hydroxyl groups both toward the polar solid surface and into the solution phase. In doing so they form what is termed a *protective colloid*, illustrated by Figure 6.5.

The colloid thickness on the solid surface is not critical except to provide an adequate number of exterior hydroxyl groups for the aqueous phase. Typical levels for medium-fineness pigments are in the 2–7% range. For the calcium carbonate pigment, 5% by weight represents a full surface coverage with a 10–15 nm thickness. This hydrocolloid layer plus its immobilized, coordinated water layer certainly extends to the 20 nm needed for interparticle repulsion.

Such coatings possess a close resemblance to long-chain glycols, known to

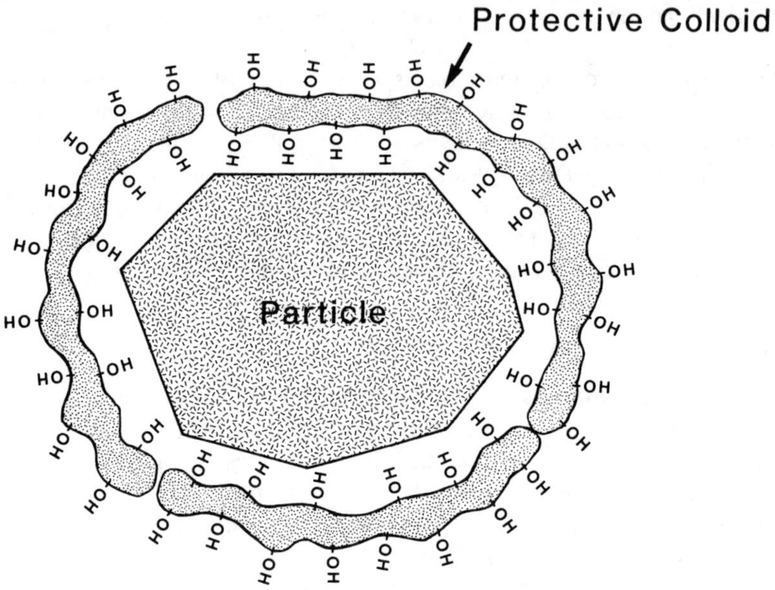

Figure 6.5. Polhydroxylated macromolecule encapsulation of particles.

flocculate oxide surfaces loosely (see Figure 5.1). However, at low to moderate levels of application, polysaccharides carry with them a strongly adherent layer of water molecules that maintains particle separation.[6] Significant overdoses of such agents, however, will result in bridging and loose flocculation.

Because of their high mass and van der Waals forces, polyhydric gums are neither selective in their site coordination on a given surface nor between different polar surfaces. Some gums, as arabic, pectin, and algin, contain small amounts of organic acid groups (*R*–COOH) situated periodically along their chains. These agents display a higher selectivity for metal oxides (Al_2O_3 and Fe_2O_3) and alkaline surfaces ($CaCO_3$ and ZnO) and may be employed for mineral separation via selective dispersion at very low dosages.

Studies of protective colloid adsorption and stabilization indicate very nearly full surface coverage must occur, even where an anisotactic surface is present, that is, one where charge or dipole character exists on a limited portion or selective facets of that structure. Similar phenomena exist for gums (polyhydric cellulose chains, proteins (poly-zwitterion condensates), and complex sugars. The key to effectiveness lies in molecular weight (van der Waals effect) and molecular flexibility (surface adaptability effect).

Where a structure as that shown in Figure 6.2 occurs, the appendant portion becomes more significant if the surface contains charge or possesses a significant dipole moment (as occurs with hydroxylated surfaces with 1.15 Debye per grouping). Where an ionic surface can exchange ions with the solution phase (barium sulfate, strontium chromate, etc.), the protective hydrophilic colloid has less opportunity to block the surface because it is adsorbed less tenaciously upon the surface. Ions can simply diffuse through the colloid layer. The mechanics of interparticle attraction will remain the same, with or without the colloid—only the kinetics change.

Protective colloid applications are specific areas of use of nonionic agents where the dispersant can be considered a macromolecule capable of fully enveloping individual particles. Depending upon macromolecule thickness, the weight demand for such agents can be moderately high, usually several percent.

One of the oldest agents employed for such dispersive activity is lecithin, a gelatinous complex derived from soybeans and other plants. Lecithin is a glycerol triester, where two of the alcohols are esterified with fatty acids and one terminal group is an organophosphatide with a terminal tertiary amine.

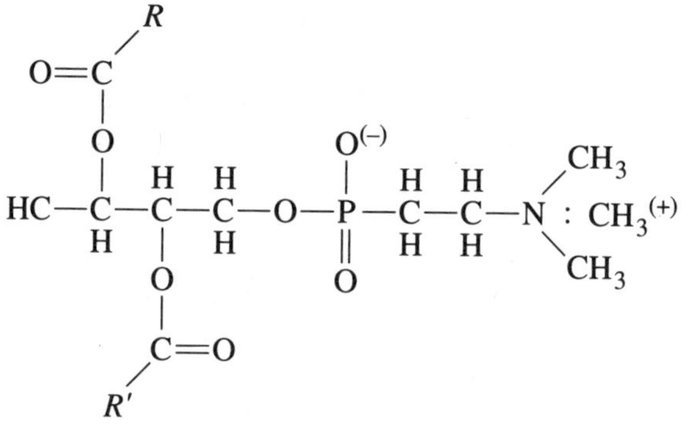

(6.4)

Groups *R* and *R′* are fatty acids. The molecule forms a semicontinuous colloid covering of slippery and emulsifiable character. On hydrophobic surfaces both alkyl chains adsorb, orienting the molecule's polar ester and amine groups outward into the aqueous solution phase. Lecithins have been employed to emulsify linseed oil for water-base coatings. They also function well on nonoxidized blacks, graphite, and a variety of organic, nonpolar pigments as well as on metal sulfides (as molybdenum disulfide) for specialty applications.

Lecithin covers most if not all of the surface of pigments on which it is employed. If applied directly to a pigment surface having approximately 2 M^2/g of area prior to its incorporation, the dispersant is more efficiently utilized, requiring only 0.5%. However, if the lecithin is added to the water-based latex formulation along with the hydrophobic pigment, the dispersant requirement is significantly higher, 4 to 10%.[7] Similar differences occur with polar pigments in nonpolar formulations (barium sulfate in alkyd resin). Because of this disparity, it is apparent that lecithin functions in a manner allowing its hydrophobic surface to interact with other molecules, much in the fashion of sodium stearate micelles. The viscous character of the latex formulation prevents these quasimicelles from being broken apart by shear long enough to permit the molecule's polar portion to find and adsorb onto appropriate pigment surface locales.

Lecithins are found in all animal and vegetable cells and compose part of the cell wall structure itself. They are believed to provide the dispersancy allowing oleophilic materials ready aqueous transport through these walls to assist in cell metabolism.[8] As cells age, the decreased ability to disperse colloids and transfer them through the wall is believed to result from lecithin deterioration and to be a major factor in aging and catabolism (cell breakdown).

A similar family structure, aminohexoses, compounds with long-chain polysaccharide backbones and periodic amine groups, are known to coat the surface of red blood cells and thereby provide their dispersion in the blood stream. Basic blood groups (A, B, AB, and O) vary in the self-dispersive composition derived from these compounds.

Lignins also fall in the natural macromolecule class, but with a phenyl propylene polymer structure derived from wood pulp, where they serve to bind cellulose fibers. Most commercial lignins have been sulfonated to strip them from cellulose bundles, rendering the bundles amenable to bleaching for brightness improvement in paper making.

Lignin sulfonates contain extensive hydroxyl groups capable of adsorbing onto polar solids and holding a layer of adsorbed water on their exterior, much like amino alcohols. Most lignins have two serious disadvantages. They are amber colored, partly the result of iron tied up in quasichelate form and partly the result of aromatic organic complexes. Unless highly refined, which is not often justified economically, high-molecular-weight sugars (pentoses and hexoses) are present which solvate water and diminish solid capabilities.

Straight lignins and lignosulfonates can be chemically altered to improve both their color and functionality as dispersants. Treatment with formadehyde converts the structures to hydroxyl methylated forms having phenolic –OH groups. The population of these, controlled both by lignin source and extent of treatment, determine the dispersivity of the molecule. Kraft lignins, derived for long-fiber pulp, are generally twice as effective as lignosulfonates, gauged by the quantity necessary to achieve a set viscosity. Modified lignins with these hydroxyl groups having molecular weights in the region of 4000 have been found effective as dispersants for acidic surface oxides, for example, rutile titanium dioxide, silica, kaolinite, and ferric oxide.[9] Solids having limited aqueous solubilities are dispersed better with larger molecules, those having molecular weights 11,000 or greater. Organic dye pigments, as typified by Disperse Yellow #54, require even larger molecular weights, 20,000–25,000.

Dispersants have been manufactured by condensing lignosulfonates (from spent sulfite pulp liquors) with formaldehyde (HCHO) and phenol derivatives in alkaline solution. Typical derivatives are the mono-, di-, and tri-methanol adjuncts of phenol and mixtures thereof. A typical manufacturing formula for this subgroup of dispersants is given in Table 6.3. This mixture is heated just below boiling for 5 hours, cooled, then the pH reduced to 9.0–9.5 with sulfuric acid.

Formaldehyde-modified lignins also have another property useful in dispersion. The molecules serve as sequestrants for a number of multivalent ions such

Table 6.3 Lignosulfonate Dispersant Formulation

Component	Weight %
Spent sulfite liquor	85.0
Sodium hydroxide	0.7
Formaldehyde solution (40%)	2.3
Phenol derivative	12.0
	100.0

as Ca^{2+}, Mg^{2+}, Fe^{2+}, and Al^{3+}. In the bleaching reaction of kaolins, iron from hydrous highly colored oxides is reduced from the ferric to ferrous state. Lignin molecules complex the reduced solubilized species and prevent or retard reoxidation and secondary color degradation. This is an important function for pigments where white color is a salient parameter.

Lignins have a binder property employed to cement particles, cellulose fibers, and other components together when a slurry is dried down. Because of their nontoxicity, they are widely employed in the food industry. There they disperse oleophilic components in hydrophilic systems and prevent recrystallization of water-soluble components (sugars, etc.), as well as water itself, in such products as cake icing, gravies, sauces, and ice cream. This property imparts a creamy texture to the product during prolonged storage or at low temperatures where recrystallization is the thermodynamically preferred state.

Gum arabic, mannogalactin, glucomannin, tragacanth, as well as common corn and potato starch, belong to this broad class of hydrophilic protective colloids. They are composed of highly polymerized sugar complexes (usually hexose rings) containing hydroxyl and ether linkages and produced by plants in seed endosperms (as jack-in-the-pulpit and calla lily plants). They contain both hydroxyl and ether groups and possess broad molecular weight ranges. Shorter chains serve as protective colloid dispersants in small quantities, as shown by the sketch in Figure 6.5, and as flocculants at higher levels,[10] where interparticle bridging can occur.

Because these colloids require low levels, less than monolayer adsorption occurs at maximum dispersivity. As agent concentration increases, molecule–molecule interaction becomes highly competitive with molecule–surface interaction. Hydrophilic colloids are believed to form linear linkages and bridge particles at high solids. Most of these materials are highly pH sensitive and require a narrow pH range in which to operate, commonly 6 to 8. Outside this range dispersivity falls off quickly and the system, solid plus agent, flocculates as a voluminous agglomerate.

Certain solids are more pH sensitive than others. A potential application arises for the separation of two components with such a selective dispersant.[11] One example is the grind dispersion of molybdenum disulfide, employed as space vehicle lubricant (lubricating in the pure vacuum of outer space where oils, greases, and graphite fail). Natural molybdenum disulfide, which bears a remarkable similarity in color and texture to graphite, commonly is combined in nature with quartz. The two may be separated quantitatively by grinding at equilibrium pH employing a mannogalactin at about 0.02%. This allows the silica to remain dispersed but the sulfide to settle out rapidly.

This technique also has been employed to separate two silicates having the same general composition but varied surface chemistry: poorly crystallized clay from well-crystallized clay,[12] bentonites from kaolinites, and silica from aluminum silicates. Extremely low levels are often required to differentiate disparate solids, as Figure 6.6 illustrates. Continued dosage beyond the critical level only succeeds in a general mutual flocculation of all particles, irrespective of species or surface character.

The feedstock, kaolinite, for the differentiation study in Figure 6.6 displayed a higher than practical viscosity, derived from poorly crystalline components and colloidal debris. At 25 ppm the bulk of these undesirable impurities are flocculated out, leaving the superior viscosity clay in the supernatant. Approximately 35% of the feedstock is lost via flocculation to accomplish the viscosity improvement. Material loss, not selective dispersant cost, is the determinant factor in the process. As mannogalactan levels are increased, progressively more quality material is brought down through interparticle bridging and viscosity is actually degraded by the operation.

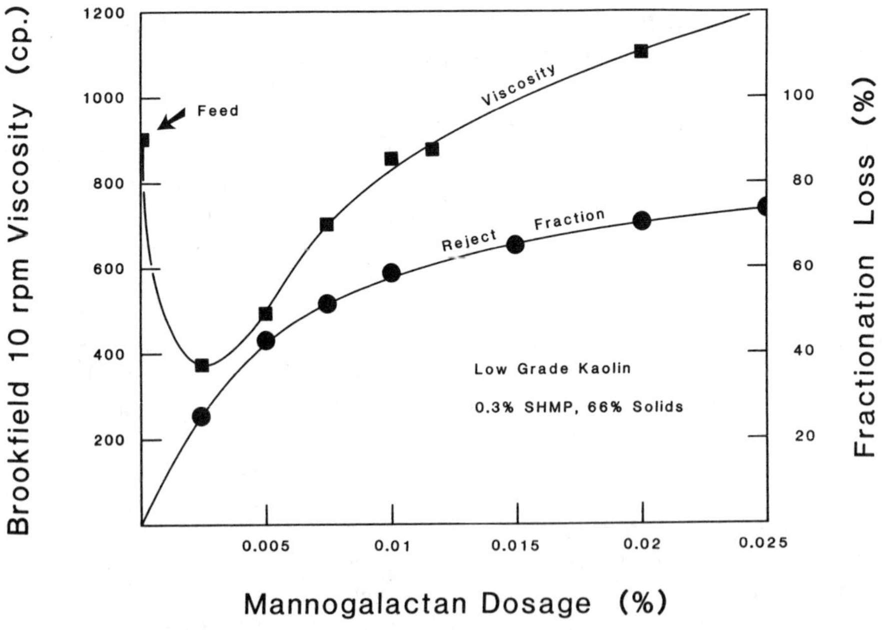

Figure 6.6. Mannogalactin influence on clay viscosity (Ref. 10).

At dosage levels in the range of 0.01% (100 ppm), gum arabic acts as a protective colloid dispersant for rutile titanium dioxide in formulations where ionic salt dispersants create problems, especially in electrical grade polymers for low-loss electronic applications.[13] These polymers possess a slight polarity and arabic-coated rutile disperses well for both quality color development and improved voltage breakdown resistance.

Polyhydric (and monohydric) acid molecules, which may not coordinate efficiently with certain surfaces, can be converted into dispersants by a two-stage treatment. The pigment is first coated with alumina hydrate through hydrolyzation, either thermally or chemically. The fresh surface is then reacted with gluconic acid, glycenic acid, or similar materials to attach the acid firmly to the intermediate coating [$Al(OH)_3$] and extend the hydrophilic hydroxyls into the aqueous solution phase. This technique is highly effective with surfaces possessing marginal polarity. It is less effective, though still employed, for surfaces predominantly hydrophobic in character. Levels of alumina hydrate are generally 0.5–2%, because monolayers are not formed in such procedures but aggregative patches.

One of the very early methods of employing protective colloids to coat pigments for nonaqueous applications involved alkyl esters, especially those with hydroxyl side groups. These esters, in conjunction with large complex sulfonated petroleum residues, form protective colloid coatings on iron oxide, lithopone, and graphite for dispersion in oil-based polymer paints[14] and molded phenolic–formaldehyde resins (Bakelite). Encapsulation enabled particles to remain well dispersed during film and polymer curing and to be permanently surrounded in the finished formulations with a water-resistant shell that improved weather and marine exposure resistance.

6.3 Nitrogen-Containing Nonionic Macromolecules

Ethylene and propylene oxides react with glycerol to form structural additions on each end of the glycerol molecule:

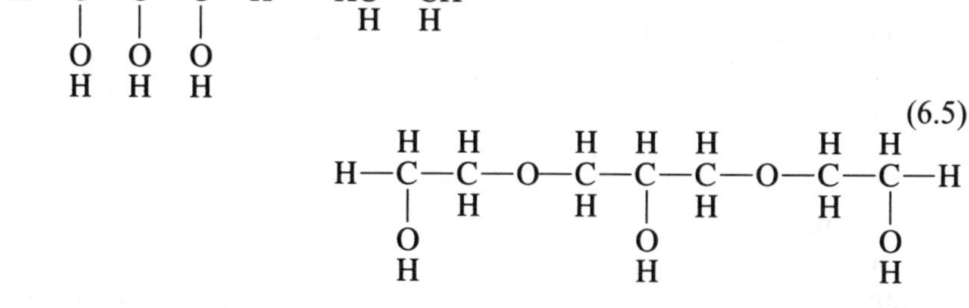

(6.5)

While the number of hydroxyl groups does not change (3), two ether groups are added, and the molecular weight nearly doubles (92 vs. 180). This structure will adapt to titanium dioxide. When the structure above is cross linked with toluene diisocyanate, the —N=C=O group polymerizes further to give a short urethane molecule. Molecular spacing of the polar groups in this structure is such that the large molecule adapts well to the rutile surface. The nitrogen group coordinates with the titania surface, leaving the hydroxyl groups of the dispersant directed toward the aqueous phase, which provides a protective colloid shell resisting interparticle flocculation.

Reaction of the ethylene oxide with a primary amine produces a triply branched structure with hydrophilic legs:

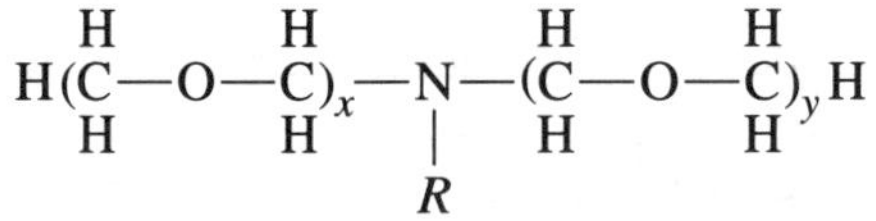

(6.6)

The grouping R is a short to medium hydrocarbon chain, and x and y range from 1 to 10, typically about 5. This material, similar to polyglycols, can be employed to disperse slightly oxidized carbon black in glycerol-based formulations and in benzyl butyl ethers for special printing ink operations.

One of the more interesting species from a structural standpoint is a subgroup of dispersants predicated upon the triazine molecule, $C_3N_3H_3$. This molecule, an analog of benzene, has a highly resonant structure with strong dipoles associated with each of the nitrogen groups. As each nitrogen atom in the ring has an sp^2 coordination, the lone electron pairs extend outward radially, in the plane of the molecule. Derivatives of this family are best known as melamine plastics precursors and herbicides. The general structure of the basic molecule is:

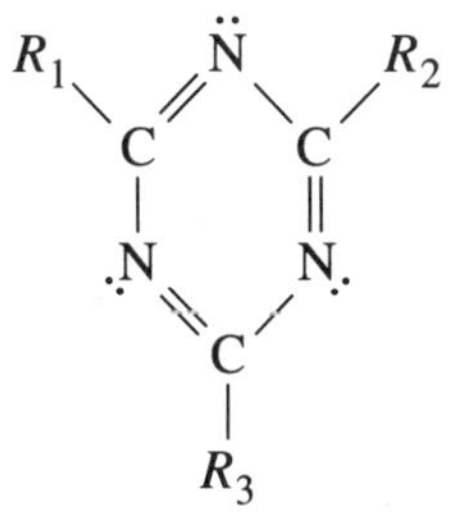

(6.7)

The substituents R_1, R_2, and R_3 are all hydrocarbon groups, four carbons long or greater, alkyl or aryl. The molecule disperses by sitting with its ring flat on a polar solid surface and extending oleophilic tails into the liquid phase. Where two or more of these tails contain polar groups (amines, sulfonated structures, alcohols, etc.), the molecule provides a thick hydrous barrier about polar particles. In yet another application one R group is bonded to a dye molecule, and the remaining substituents determine that annexed structure's hydrophilic or hydrophobic dispersancy potential.

Another class of complex nitrogen molecules serving as dispersants is that of aminimides. One such typical agent is:

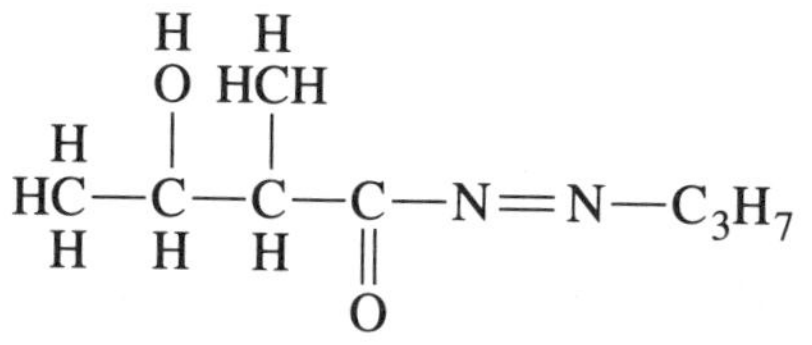

(6.8)

The structure functions as an exotic alkanolamine, but with reduced alkalinity. Increasing the terminal propyl group length, for example to C-8, shifts the system progressively toward lower polarity (lower HLB).

Copper phthalocyanine, as has been frequently noted, is a very fine pigment having the porphyrin structure. To make this water dispersible, manufacturers may bond directly to it an ionic structure (sulfonate, quaternary ammonium ion, etc.). This often shifts color slightly. Dispersants resembling a sector of the prophyrin structure have been employed[15] to assist in distribution of the pigment in glycols and other partially polar media.

One of these quasiporphyrins lies in the family of amides:

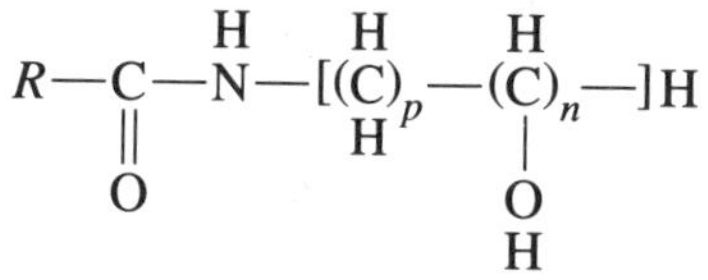

(6.9)

The terminal hydrocarbon *R*, falls in the lauryl to stearyl range (or greater), *p* from 2 to 4, and *n* is exactly half of *p*, 1 to 2; that is, alternate carbon atoms are alcohol groups, polyglycols. As with the previous family, *R* controls the degree of hydrophobic–hydrophilic character, while *n* and *p* control the surface affinity for polar solids.

Copper phthalocyanine, whose discrete particle size is typically about 0.05 μm, is somewhat more simply dispersed by condensing directly diethanolamine with C-10 to C-22 fatty acids onto the pigment surface. Because the molecular weight of this adduct is large, the dispersant requirement will run on the order of 5%. Even with these exotic dispersants, micrographic particle sizes in final formulations are rarely finer than 1 μm. This inefficiency in particle utilization reflects more on machinery limitations than dispersant activity.

Polyamines can be converted to polyamide structures by reaction with ethylene or propylene oxide, in which each secondary amine reacts to form the extension:

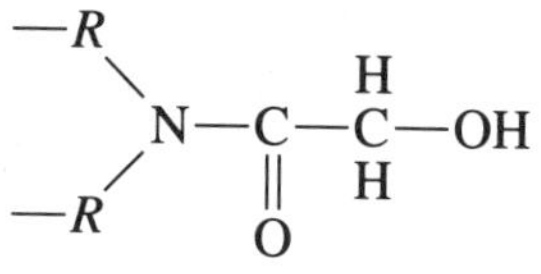

(Ethylene oxide reactant) (6.10)

This agent can be employed to disperse very fine inorganic, polyhydroxyl surface, inorganics, and low-nitrogen-containing organics.[16] An agent manufactured by reaction of triethylene tetramine with ethylene oxide is employed to disperse Pigment Red #14 at about 50% solids in water–glycol carriers for printing ink, both standard and that employed in colored ink jet printers.

Work along similar product lines involved modification of the ethylene oxide addition with polyether linkages.[17] This agent is slightly less deliquescent and aids in rapid drying on printing paper under moderate-to-high-humidity printing conditions.

A somewhat simplified molecular structure can be made with a single alcohol and amine group by reacting a C-16 to C-18 alkene oxide with methyl amine (CH_3NH_2) in alcohol or an alcohol–short-chain mineral spirits blend. The structure formed is:

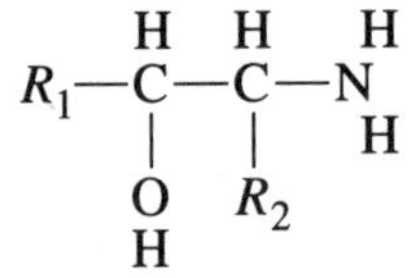

, $R_1 = R_2 =$ C-6 to C-22 (6.11)

Again the chain lengths of R_1 and R_2 control the liquid affinity. For the shorter C-6 chain,[18] the molecule disperses titanium dioxide well in slightly polar vehicles (alcohol and ketone solvent systems).

Many inert oxide–hydroxide pigments respond to these agents. However, not all polar solids are functionally dispersed with amide–amine–glycol–ether polymers. Certain amides serve better to disperse the silver halides (AgCl, AgBr, and AgI) in photographic gelatin than do the amines. The latter enter into complex solution reactions, which not only alter silver halide stability in film or paper but adversely change its photoreceptivity. Further, cationic agents destabilize the halides and cause a solution–recrystallization reaction, greatly reducing their effectiveness as photoreceptors. Studies show the nitrogen complex adsorbs onto the halide, especially the bromide, and permits the formation of the secondary complex, $AgBr_2^-$. It is this mechanism that accelerates recrystallization. Thus all alkyl, aryl, primary, secondary, and tertiary amine-containing agents should be avoided in photographic applications, in particular those where very fine particle size (high photographic resolution) is important.

With polyglycol amides or polyether amides, the number of amide groups must be reduced to a minimum for photographic applications.[19]

A chemical relative of this group, the amide derivatives of polyacrylic acid, can have an improved efficiency in aqueous systems. With short chains they can function as dispersants, much in the fashion of the acid salt (alternate amide groups associated with oxide and hydroxide groups on the solid surface). Intermediate amide groups, however, are solution directed, which, like amino alcohol groups, coordinate water and provide a dipole cushion between particles. However, the amide group is nearly twice as polar as the amine group and holds more tenaciously solvent water molecules at the surface. Also because of this high polarity, the structures lie in train configurations across the solid surface.

Both polyacrylamide chain length and concentration, as well as particle size of the polar pigment, are influential factors in effecting dispersion. For example, as shown in Figure 6.7, a −2 μm particle size yellow hydrated iron oxide pigment having 0.15% of a polyacrylamide with a molecular weight of 1000 is deflocculated. Upon increasing either the molecular weight or the application weight, viscosity is elevated and results in eventual flocculation. Decreasing significantly the particle size (which decreases interparticle separation) will also increase the probability of flocculation.[20] With a −5 μm product the viscosity is less, even though the quantity of polyacrylamide per unit area increases.

These same characteristics hold for rutile titanium dioxide. In low-solids slurries (20%) of that pigment, polyacrylamide concentrations up to 0.05% (solids basis) steadily decrease viscosity and yield coating formulations with progressively smoother surface texture and more uniform coating distribution.[21] At levels of polyacrylamide of 0.12%, filamentary micromolecules of the polyacrylamide form. These in turn develop into threadlike bridges between the titanium dioxide particles. At yet higher concentrations, more dense floccules form, reaching a peak aggregation with about 0.6% polyacrylamide (solids basis).

Polyacrylamide flocculated systems tend to be weakly associated. Such high-volume structures which readily break down under shear, quickly reform upon shear removal.

Fine alumina hydrate (0.1–2 μm), by contrast with the titanium dioxide above, can be dispersed at 50% solids, or higher, with 0.25% (solids basis)

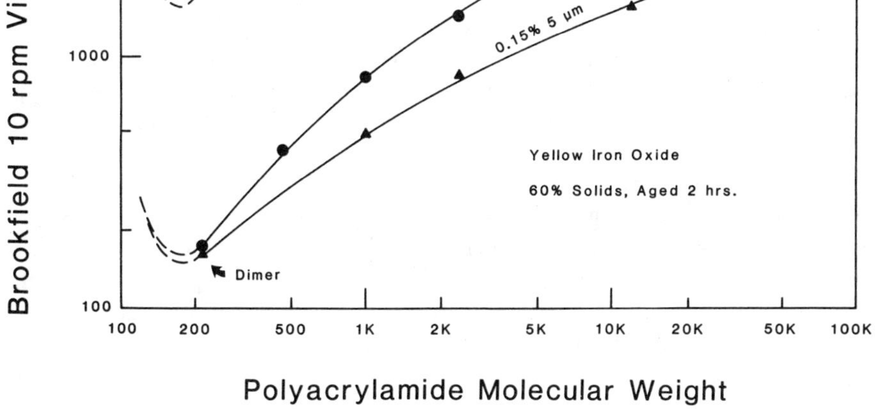

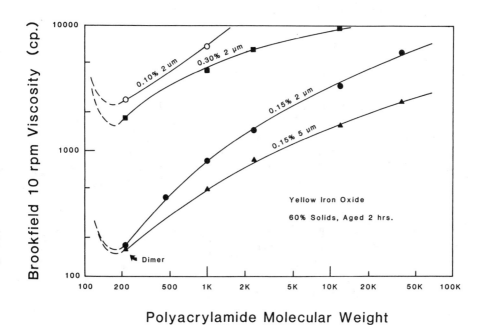

Figure 6.7. Effect of polyacrylamide chain length, dosage level, and particle size on iron oxide dispersion.

polyacrylamide for employment in paper coating applications.[22] To break down the dried pigment properly, dispersion is performed with a stirred bead mill employing 12 mesh (nominal 1.5 mm) glass media. The alumina hydrate pigment is an order of magnitude coarser than the foregoing titanium dioxide (nominal 2 μm).

An obvious difference in surface affinity exists between titania and alumina hydrate, just the reverse of what surface area might predict. One primary difference lies in the surface acidity, rutile being about 66% acid and 34% alkaline, while the surface of the hydrate is distributed close to a 50–50 acid–base ratio.

Acid amines of the general structure

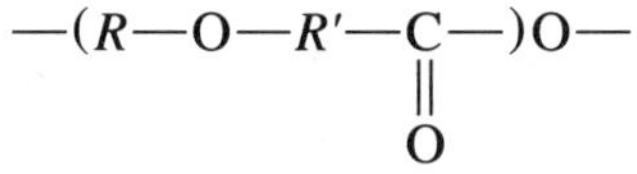

(6.12)

where R is an aromatic group and R' is aliphatic, when reacted with secondary aliphatic amines have been reported to have a fairly broad dispersant application for pigments.[23] In several of these applications the dispersant is formed in situ. With more acidic surface pigments, indications are the dispersant is more energetically attached to those surfaces.

In Chapter 4 a short discussion was given on organophosphates, particularly the alkyl phosphates. A dispersant that bridges both the two classifications in that chapter and the ones in this is the long-chain complex, polyoxyethylene dihydrogen phosphate,[24] whose chemical structure is:

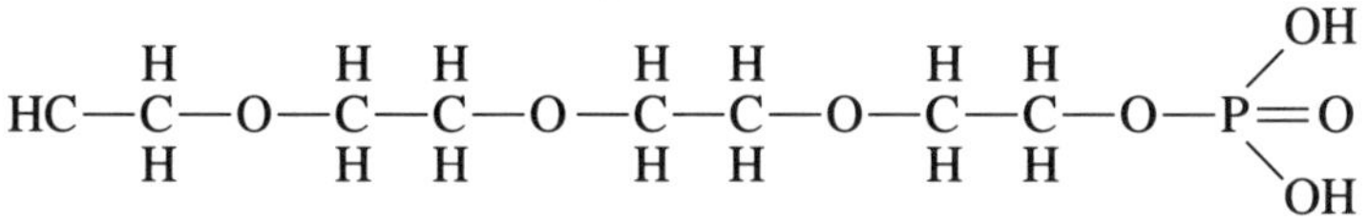

The molecule is especially applicable as a dispersant for pigments which display a slight solubility in water (gypsum, calcium carbonate, and low-temperature magnesium oxide). Both the dihydrogen phosphate head (un-ionized) and the polar organic tail are adsorbed onto these surfaces. Depending on the polyoxyethylene chain length, the dispersant's configuration may vary from train to tail, or display mixtures intermediate, as suggested by infrared spectra of the C–O bonds.

Nonionic structures can be developed that contain a highly unusual and specific grouping resembling closely the specific pigment for which the dispersant is to be employed. These are termed *tailored dispersants*. While many organic pigments have a self-dispersive chemical group intentionally grafted chemically (e.g., quaternary amine or sodium sulfonates), this addition may alter slightly the color of the organic pigment, as discussed for copper phthalocyanine, by shifting electrons to or withdrawing electrons from the color-sensitive chromophore group.

Other advantages of the tailored dispersant molecule lie in its weight reduction of the parent pigment by eliminating heavy anionic or cationic grafts. Perhaps most important, the dispersant, whose weight comprises at most 1–2% of the pigment, can be tailored for specific applications to alcohols, ketones, alkyd resins, polyesters, etc. It is this specificity that justifies the complex structures found in these applications. As an example of an exceptionally complex core structure for a dispersant molecule, the weather-resistant red pigment for melamine automotive and marine finishes, Irgazin DPP red, has a structure tailored for specific formulations, as shown in Figure 6.8. The terminal group, as well as the side chain, can be varied to include polyether linkages (via ethylene oxide), sulfonation group, tertiary amines, or terminal alcohol groups, depending upon system acidity–basicity and polarity.[25]

Terminal butyls in the dispersant structure may be replaced by more polar groups or groups which will chemically bond (as aldehydes) to the melamine polymer for which the dispersant was designed. A structure of this complexity is expensive to build chemically and can be justified in cost only because of its ability to conform efficiently onto the dye surface and effectively distribute the expensive pigment throughout the resin. Organic pigments, unlike titanium dioxide, kaolinite, and calcite, function optically not by reflecting (scattering) photons but by selectively absorbing them. The more thorough the dispersion, the more efficient will be this absorption. Earlier a study on copper phthalocyanine showed a primary particle size to be 0.05 μm while the effective film size was 1–2 μm. Clearly, core members of this assemblage never interact with photons, and far more pigment is required to effect the color depth required.

Specific tailoring is not restricted to organic dye pigments. Silica may be dispersed through the hydrolysis of tetraethyl silicate in ammonia–alcohol solution. Such a system enables the silica to be employed in polystyrene through this chemical grafting operation.[26]

Complex amines are often reacted with acid anhydrides to produce nonionic polymers with dispersant activity. One common agent so made is the trimethylamine neutralization product with maleic anhydride-styrene copolymer. While this molecule bears a certain similarity to the sodium polyacrylate family discussed in Chapter 4, the molecule is polar but not ionic, enabling its application

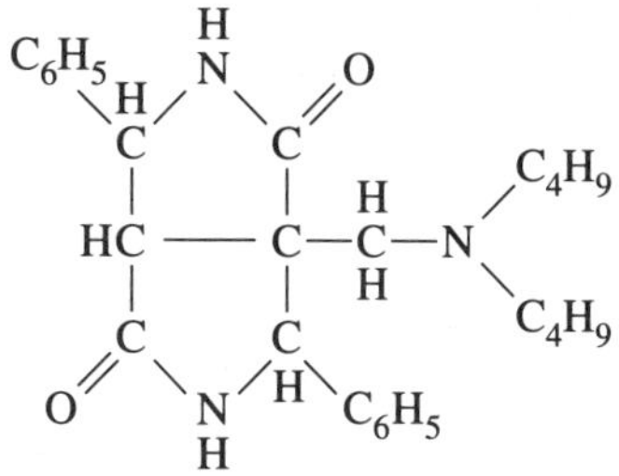

Figure 6.8. Tailored dispersant for specific pigment use.

in slightly polar polymers. Agents of this type are useful with a broad range of pigment surfaces ranging from copper phthalocyanine through barium sulfate.

Even more complex and greater in length is the salt of triethylene tetramine and myristic acid (*n*-tetradecanoic acid) employed for carbon black[27] in low-polarity carrier liquids.

Finally, the life process of reproduction itself relies heavily upon a polyamine dispersant. Human spermatazoa have a slightly negatively charged head (~5 μm in diameter). The tail of this genetic carrier, by contrast, carries little charge. The head is known to have a chemically active skin or membrane structure, containing various amino acids, possibly as zwitterions. At high pH sperm cells ionize and possess a negative charge, which maintains cell separation and activity as they flagellate along their course. Alkalinity for their motility is provided by the prostate carrier fluid, which contains spermine,[28] a dipropylene, butylene tetramine having the formula,

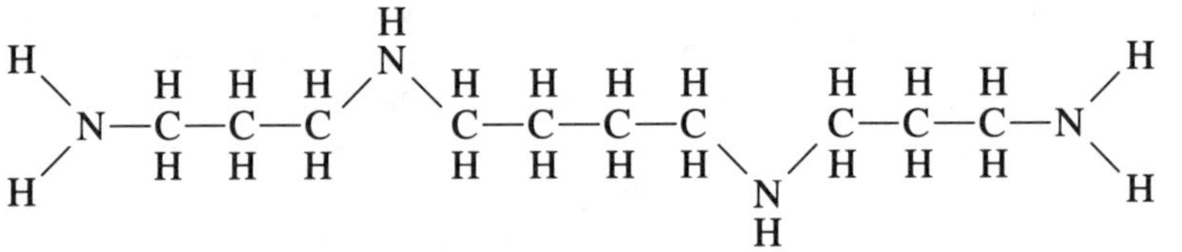

(6.13)

and a pH near 11. In addition to the head's slight surface charge, the tetramine molecule fixes itself onto the head and serves as dispersing agent. Each spermatozoa does not bear exactly the same charge, and those displaying higher charge appear to be more motile and, hence, more probable of reaching the waiting egg. Electrophoresis studies[29] suggest the sex pairing chromosone, XX versus XY, affects this charge also, the latter having slightly higher negative charge. This may account in part for the 6% higher conception rate of males (some studies indicate 10–20%) compared to females.

Spermine also may adsorb on the flagellum as both elements of the gamete (cell) effectively repel other nearby spermatazoa.[30] Gamete proximity in seminal fluid is about 40 μm—for a cell head of 5 μm and tail length of 60–70 μm. At this distance surface effects and need for dispersive forces are present early in the cell's activity (before dilution in the female tract), while pH effects become important later in its activity. Orientation of the spermine molecule onto the protein sheath (aminoacid structures) of the spermatozoa is not fully known. While four modes of attachment are statistically possible, from full-tail to full-train, a single (two-point) propylene diamine affixation seems most probable,[31] leaving the remaining two diamine groups in the solution phase extending at least 2 nm.

When multiple spermatozoa are forced together by cell manipulation techniques, they will not adhere,[32] as will most other human cells in suspension, indicating a highly effective dispersant shell. Figure 6.9 is a high-resolution, high-speed micrograph of active cells swimming at a typical 100 μm/sec and being forced together.[33] Note their closest proximity remains about one head diameter (5 μm or 5000 nm).

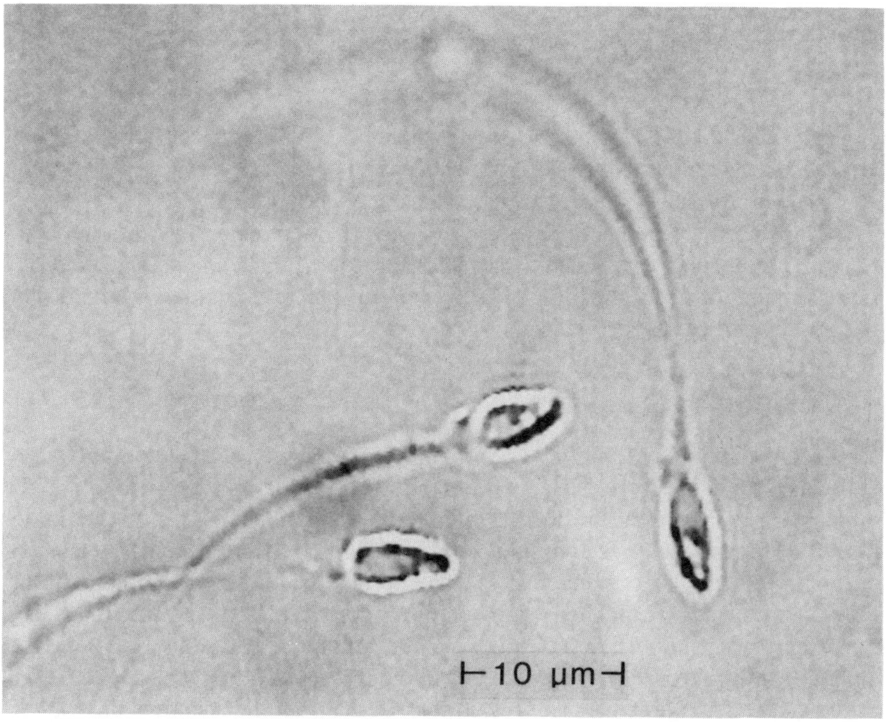

Figure 6.9. Human spermatozoa carrying natural polyamine dispersant (Ref. 33).

While the tetramine molecular attachment structure on sperm cells may appear academic at casual glance, submarine designers discovered in post–World War II era work that these vessels could travel faster for the same energy expenditure when the surface was wet out by chemical agents. In this configuration, which is also observed with the gamete in Figure 6.9, a thin film of liquid phase moves with the dispersed species, whether military submarine or its reproductive counterpart.

As pH drops, so does sperm motility, until the solution is slightly acid, pH 6 or below, at which point the cells flocculate and slowly die. Spermine not only provides the surface dispersive polarity to maintain cell activity (swimming!), it neutralizes the acidic walls (pH 3.5 to 4.5) of the female reproductive tract and maintains a functional and accommodative pH. Finally, it fluidizes (lowers the viscosity of) the mucous linings, allowing for more rapid transport.

Accompanying spermine is an enzyme (diamine oxidase) that slowly breaks down the tetramine molecule to two (or more) amine aldehydes. These compounds appear to raise the pH slightly more, apparently for the benefit of spermatozoan stragglers who have fallen behind in the reproductive olympics! It is this enzyme oxidation that gives rise to the strong fishy amine odor associated with dead spermatozoa.

Without a dispersing agent to assist them, the reproductive cells would be dead on arrival, as would the human race!

6.4 Other Applications

Thus far in this chapter only the dispersancy characteristics of the polypolar, nonionic molecules have been discussed. Because of their unusual characteristics, they find other applications that in many instances coincide with their dispersive properties. One of these is as defoaming agents.

6.4.1 Defoaming Activity

Dispersants having hydrophilic–hydrophobic components, such as sodium lauryl sulfonate, form a film at the air–water interface, with the hydrocarbon tails directed out of solution into the gas phase and the cations down into the liquid phase. Their presence as a continuous molecular film reduces the surface tension at this interface. Gas bubbles rising to the surface from mixing entrainment or other agitation operations penetrate this film but carry part of it at their surface, the result of the molecular elasticity of the anionic dispersant. Bubbles will emerge with films of water, usually only a few molecules thick, and a monolayer of the dispersant. New dispersant from the internal liquid phase then diffuses to the surface to fill the breaches created by the gas bubbles. This process is called *foaming*. Because surface tension is attenuated by the dispersant, bubble stability is greatly enhanced and bubbles are not broken without physical rupture or water evaporation, neither of which is practical or expeditious in commercial processes.

Where a polyethylene glycol is added to the liquid phase, it causes a similar alignment at the gas–liquid interface. However, the film these molecules form is highly inelastic. Under these conditions gas bubbles reach the surface, but must burst to penetrate this polypolar barrier. The two processes are shown in Figures 6.10(a) and (b). Typical antifoaming agents are the polyoxyethylene carboxylic acids, amines, esters, and sulfonates.

6.4.2 Freeze–Thaw Stabilization

Water-based formulations such as emulsions, latex paints, sprays, etc., may be highly stable within a fairly narrow temperature range, usually 10° to 40°C. Outside this range thermal effects may disrupt the stability or effectiveness of dispersants. This is especially true in freeze–thaw conditions. In earlier years, water-based latex paints, especially butadiene–styrene formulations, if stored in a garage during the winter, broke their emulsification and could not be reconstituted to the original, or even a workable, consistency. Paints undergoing this situation were simply thrown away, much to the dismay of purchasers. However,

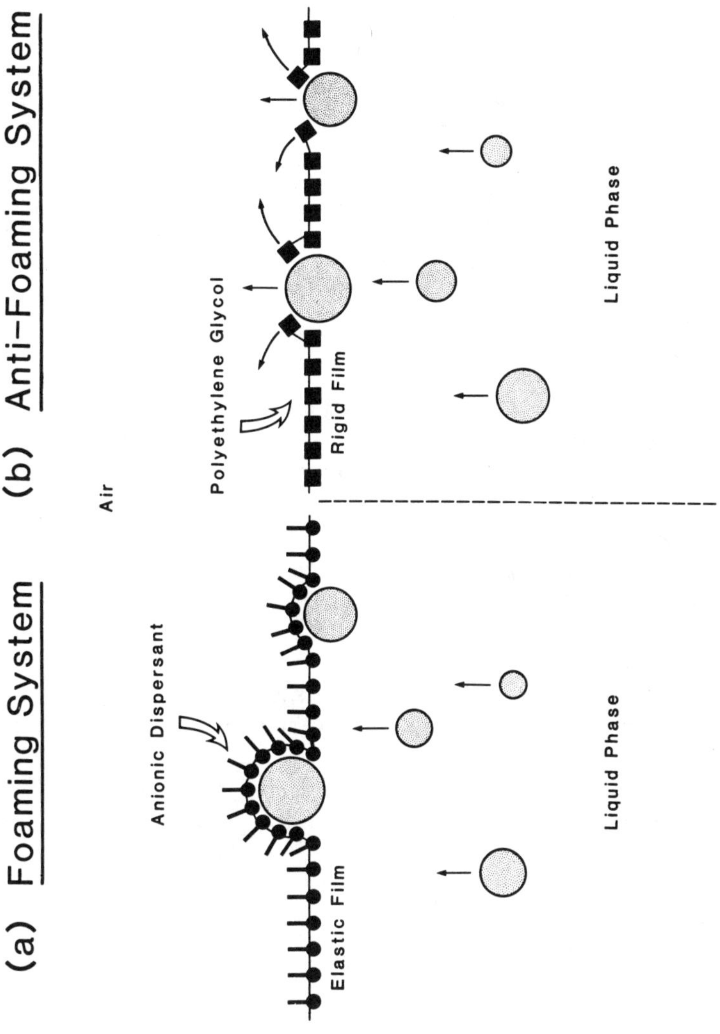

Figure 6.10. (a) Effect of rigid surface film (foaming) on bubble evolution. (b) Effect of antifoaming agent on bubble evolution.

the introduction of long-chain polyether or polyglycol agents greatly reduces emulsion breakdown from freeze–thaw conditions.[34] As a consequence, where temperatures do not drop significantly below freezing, a stored paint can be reused simply with the introduction of high shear once warmed. Such freeze–thaw inhibitor compounds include polyethylene oxides having terminal alcohol, oleate, and carboxylate groups.

6.4.3 Gloss Enhancement

Where gloss enhancement is an important product parameter, as in paint finishes and molding resins, polyalcohol groupings have great value in the polyethylene oxide chain. These adsorb upon the pigment particles and allow a slow leveling operation to proceed within paint films and a portion of the pure polymer to rise to the surface while the solvent slowly evaporates. Good dispersion in casting polymers also allows the polymer to flow to the mold contact surface and produce a smooth texture to the polymer during the curing stage.

6.4.4 Lubricant Properties

Machinery lubricants are most often based on hydrocarbons, graphite, and combinations thereof. Where machinery must produce food, cosmetic, pharmaceutical, or other products destined for human contact or internal consumption, such petroleum-based lubricants present health problems. Polyethylene glycols and natural oil-substituted derivatives (commonly castor oil adducts) provide quality lubrication and, should they accidentally become incorporated in the product, minimal health hazards.

A broad series of personal products, including intimate lubricants, are sterilized, polyethylene glycol-based, aqueous solutions formulated with aloe vera, nonoxynol-9, triethanolamine, and other dispersants or thixotropes.[35] They also may be employed to suspend a variety of oils and emulsified agents in water-based skin lotion formulations.

Such derivatives also are employed as lubricants in alcohol-based fuels for internal combustion engines, especially those of race cars, aircraft, and model airplane engines.

References

1. "Linear Alcohol Ethoxides," Tech. Bulletin, Texaco Chemical Co., Houston, TX (1993).
2. Formulation courtesy Ballard Medical Products, Draper, UT (1995).
3. Nishizuki, K., and Mayuzumi, F., Japan Patent #2,126,929, May 15 (1990). Note: New patent system designation in Japan's patent office in this year. The patent date is very important.
4. Nagashima, S., Japan Patent #63,159,484, July 2 (1988).
5. Szentivanyi, N., Knotek, L., and Ondrejmiska, K., Czech Patent #269,506, Aug. 31 (1990).
6. Dickinson, E., *Proc. 4th Intl. Conf. on Gums Stab.*, Vol. 4, p. 249 (1988).
7. Kroustein, M., and Eichberg, J., U.S. Patents #4,056,494, Nov. 1 (1977) and #4,126,591, Nov. 21 (1978).

8. Haurowitz, F., *Biochemistry—Lipids*, Wiley and Sons, NY, p. 289 (1955).
9. Lin, S., and Detroit, W., *Intl. Symp. Wood Pulping Chem.*, Vol. 4, p. 44 (1981).
10. Alexander, R., "Colloid Chemistry", Vol. VI, Reinhold Publ., NY, p. 332 (1946).
11. Rowland, B., U.S. Patent #2,981,630, April 25 (1961).
12. Maloney, H., U.S. Patent #2,158,987, May 16 (1939).
13. Sankyo Corp., Japan Patent #80,129,458, Oct. 6 (1980).
14. Seeman, J., and Seeman, W., German Patent #702,576, Jan. 16 (1941).
15. Necas, M., et al., Czech Patent #251,118, April 15 (1988).
16. Diery, H., et al., German Patent #2,938,623, April 9 (1981).
17. Diery, H., et al., Europatent #25, 998, April 1 (1981).
18. Linden, H., German Patent #2,738,539, March 1 (1979).
19. Herz, A. (Eastman Kodak), *Croat. Chem. Acta* **313** (1980).
20. Mathur, V., *Indian Drugs* **19**, 323 (1982).
21. Kravtsova, V., Shutova, A., and Aleksandrova, E., VINITTI, No. 917a, (1974).
22. Showe, K., Japan Patent #26,959, March 16 (1981).
23. Horst, K., and Schild, H., U.S. Patent #2,213,984, Sept. 10 (1941).
24. Kloetzer, E., Kiousteldis, J., and Schmidt, G., Europatent #339,310, Nov. 2 (1989).
25. Sawamura, K., and Hayashi, M., Japan Patent #326,767, Feb. 5 (1991).
26. Bridger, K., Fairhurst, D., and Vincent, B., *J. Colloid Interface Sci.* **68**, 190 (1979).
27. Nishizuki, K., and Mayuzumi, F., Japan Patent #2,107,327, April 19, (1990).
28. Haurowitz, F., *Biochemistry—Proteins*, John Wiley & Sons, NY, p. 151 (1955).
29. Allen, T., General Electric Corp., Schenectady, NY, private communication (1984).
30. Conia, J., and Voelel, S., *Biotechniques* **17**, 1162 (1994).
31. Molecular mass and three solution-phase amine groups make single-terminal amine group attachment both unlikely and unstable. Also, where three amine groups are able to coordinate with the spermatozoa's surface, a fourth should also coordinate, which would reduce the required field alkalinity. Either an "L" formation occurs, in which terminal and alpha amine groups just fit acidic sites on the protein sheath (leaving two amine groups free in the liquid phase), both terminal amine groups attach in chelate form leaving alpha and beta amine groups extending into the solution phase, or the reverse of this process, the two central amines dock, leaving terminal groups in solution phase. These geometries provide approximately equal theoretical dispersancy.
32. "Cell Robotics Applications," Cell Robotics, Inc., Albuquerque, NM (1994) (see equipment description in Chapter 13).
33. Photograph courtesy Dr. Jerome Conia, Cell Robotics, Albuquerque, NM. Live photo taken in 10:1 saline solution dilution of seminal fluid. Video scan image enhancement by Protouch Imaging, Albuquerque, NM.
34. Witco Organic Div. Tech. Bulletin #254, Houston, TX, March (1993).
35. As this book was going to press, over 25 trade names of such products were being marketed, both in drug stores and by mail order. All are low-to-moderate-strength aqueous solutions, some colored, some flavored, and all expensive, typically $15–35 per pint!

CHAPTER

7

Dispersant Functionality in Nonaqueous Media

A significant fraction of applications dealing with dispersed solids in liquid phases involves organic liquids. These include systems which are permanently retained in an organic liquid state or slurry (polishes and food products), systems which are but initially produced in an organic liquid state but whose function lies in a modified state (oil-based paints) and systems in which dispersants are added and retained on the dried powder to be later dispersed in organic liquids (pretreated pigments, ink pastes, color premixes).

Twice or so each decade a broad review article appears summarizing new developments in dispersants, primarily for oil-based paint applications, and ranking these with regard to specific properties of gloss, hiding power, quality of deep tone, rheology, and other factors. While such reviews serve numerous short-term purposes,[1] they largely represent consolidations of new product information. Most lack the structural information on dispersant composition, the character of the various applicable pigment surfaces, and specific differentiation from previous materials periodically altered by their manufacturers.

The basic concept of oleophilic (''oil-loving'') dispersing agents refers to a solid phase not readily wettable by an organic phase. This in turn suggests the solid surface is at least in part hydrophilic. In such systems three competitive reactions between molecules exist: (1) interaction between dispersant and the solid surface, (2) interaction between the dispersant and the organic liquid phase, and (3) interaction between the liquid phase and the solid surface. For effective dispersion of hydrophilic solids in oleophilic liquids, the relationship must be: (1) > (2) > (3).

To understand this relationship better, the various energies must be examined

Table 7.1 Heats of Wetting by Organic Liquids for Rutile

Liquid	Dipole Moment (Debye)[a]	ΔH of Wetting (erg/cm^2)
Water	1.87	550
Ethanol	1.69	397
n-Pentanol	1.60	413
n-Butanol	1.59	410
n-Butyl iodide	2.12	395
n-Butyl amine	1.15	330
n-Butyl chloride	2.05	502
n-Butyric acid	1.72	506
n-Butyl aldehyde	2.72	556
n-Nitropropane	3.60	664
n-Hexane	0.0	135
n-Heptane	0.0	144
n-Octane	0.0	150

[a] 1 Debye = 10^{-18} esu cm.

Data taken from Refs. 2 and 3.

quantitatively. The energy of attraction of hydrocarbon molecules to each other is extremely small, as evidenced by methane, CH_4, whose molecular weight is very close to that for water (16 vs. 18). It has a liquefaction temperature of −161°C compared to water's +100°C. Ethane, with a weight 66% greater than water, liquifies at −88°C. Thus the energy of attraction between hydrocarbon molecules is dominantly van der Waals and does not enable liquefaction at ambient temperatures until a molecular weight of 72 is attained (pentane). Heats of wetting for hydrocarbons on inorganic solids are only about $\frac{1}{3}$ those for water on average.

Heats of wetting for the pigment rutile titanium dioxide by a broad spectrum of organic solvents are compared in Table 7.1.

Similar information has been collected on calcium fluoride (fluorspar), fumed silica, aluminum silicates, carbon black, and Teflon. When heats of wetting are plotted against the dipole moment of the liquid, a nearly linear relationship results, the slope of which is a measure of the surface polarity.[2]

7.1 Influence of Adsorbed Moisture

As deduced from Table 7.1, water is a difficult liquid to displace from solid surfaces once these are immersed in organic liquids. To establish effective wetting between liquid phases and solid surfaces, the dispersant must overcome any adsorbed moisture layer. Should the powder bearing its aqueous film be incorporated in a polymer phase, breaking apart the agglomerates, whose structure is assisted by this film, becomes very difficult for those reasons cited.

Further, these moisture films slowly allow a breakdown in the polymer–solid boundaries, checking, and eventual pigment or filler flaking and system disintegration.

Adsorption of water films on pigments does not necessarily render them unwettable by organic polymers and solvents. It does increase one particle surface's affinity for another, which excludes organic molecules within agglomerates. Figure 7.1 contains the results of a study on titanium dioxide[4] showing the adsorption energy of benzene on various degrees of wet surface. At full surface coverage, approximately 0.5% H_2O, the interaction energy of benzene for the wet titania surface is nearly equivalent to that for pure water, 520 erg/cm^2.

In the dispersion of ultrafine, chemically precipitated silica, moisture carry-in on the silica surface is not static but moves within the agglomerate once an oleophilic solvent is added (e.g., mineral spirits). Nuclear magnetic resonance (NMR) studies, as well as heats of wetting studies, have demonstrated[5] that exposure of the silica to no more than 30% humidity carries enough water film into the system to provide full liquid-phase mobility within the agglomerate structure. This water moves from the interparticle contact zones to the interstices. Sufficient mobile water exists to saturate fully these convoluted tetrahedral sites. Mineral solvent originally present in these locales (necessary to permit full distribution of the silica particles throughout the liquid phase) is displaced because of the considerably higher heat of wetting by water compared to the alkane.

Interstitial water and surface tension explain the dominance of trimers and

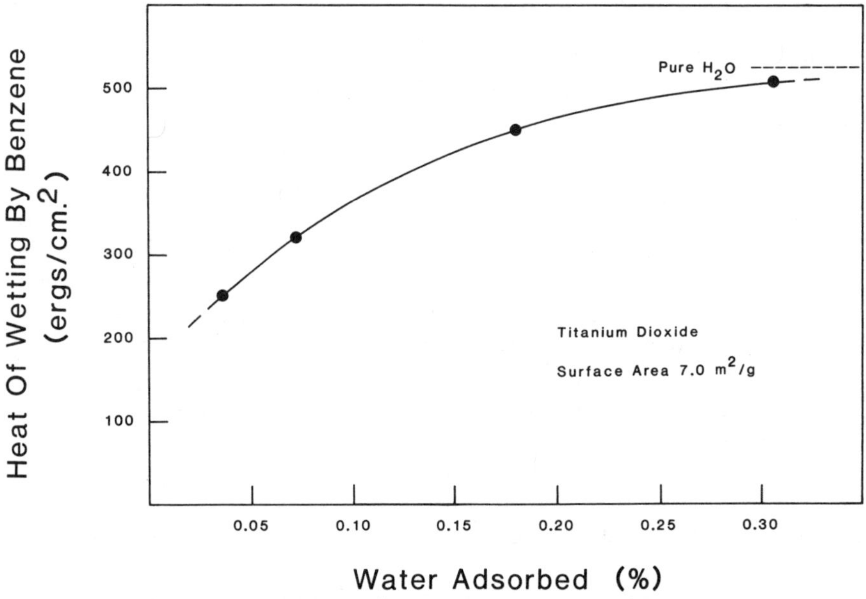

Figure 7.1. Aromatic solvent adsorption energy on wet TiO_2.

Table 7.2 Wetting Energy for Hydrocarbon[a] on Oxide Pigment Surfaces

Material	Physical Form	Energy of Adsorption (ergs/cm^2)
Copper	metal powder	60
Silver	metal powder	74
Lead	metal powder	99
Tin	metal powder	101
Iron	metal powder	108
Ferric oxide	alpha	107
Titanium dioxide	anatase	92
Stannic oxide	powder	111
Barium sulfate	precipitated	76
Graphite	oxidized	122

[a] Heats of adsorption measured with heptane.

tetramers in electron micrographs of silica and other pigment-filled polymers in thin-film examinations. Energies of wetting for metal oxides (including metals) and other inorganic pigments with alkane solvents is given in Table 7.2.[6]

Note that heat of wetting increases with the electromotive force (EMF) of the metal, indicating a high fraction of the surface is converted to oxide or hydroxide. Metallic iron's surface is essentially the same as ferric oxide's.

Pigment pretreatment is one solution to organic system dispersion that has become more popular as a mode of manufacture in the last quarter-century for materials designed for oleophilic applications. Most pigment manufacturers have found pretreatment results in greater improvement in dispersion (as measured by color strength) than that obtained by altering binder properties via solvent dilution. One common testing procedure compares viscosities in water of the untreated pigment over an applicable solids (loading) range with the treated material in solvent–polymer, for example, alcohol acrylic resin.

Several procedures for pretreatment involve water-phase addition of dispersants followed by flocculation and drying, often where the dispersant for oleophilic systems becomes the flocculent in the aqueous one. In other pretreatment modes, a dry premix is made, the dispersant added to the pigment in a high-shear mixer, usually at elevated temperature (almost by vapor-phase distribution), and the blended product aged to ensure thoroughness of surface interaction or attachment.

Three modes of attachment for inorganic surfaces and oil-phase-directed dispersants exist: charge association, dipole association, and chemical reactivity.

7.2 Solvent Role in Dispersion

In dispersancy procedures for both laboratory evaluation and commercialization, the role of solvent–solid interaction on the pigment's dispersibility in viscous oleophilic resins cannot be overrated. During brown iron oxide formulation, the selection of a solvent based on its energy of interaction with the pigment surface (as shown earlier with titanium dioxide and nitropropane) will control the quality of dispersion within the polymer, maximum tint strength, and color development, as well as general smoothness in application. It may not, however, always improve shelf-life stability of the finished formulation.

Reasons for instability can be many, most common of which are: (1) increased surface area to polymer ratio increases the interaction potential of that polymer with the solid surface and (2) where catalytic effects may be present (as with zinc oxide in PVC resins and active metal oxides with acidic polymers) shelf rheology may change adversely and those application properties related to rheology may degrade. Because of these long-term interactions, solvents with low-to-moderate interaction energies are recommended[7] for alkyd resin formulations, especially where prolonged storage is required.

In the example of rutile cited in Table 7.1, the sole practical solvent capable of replacing water is nitropropane. In two formulations of military paint, one employing a xylene–MIBK mixture (average dipole moment 1.4) and the other nitropropane (3.6), the difference in weathering resistance was dramatic, as shown by the gloss degradation with prolonged marine air exposure in Figure 7.2.[8]

Polar pigments, which have low surface adsorption energies (or surface ten-

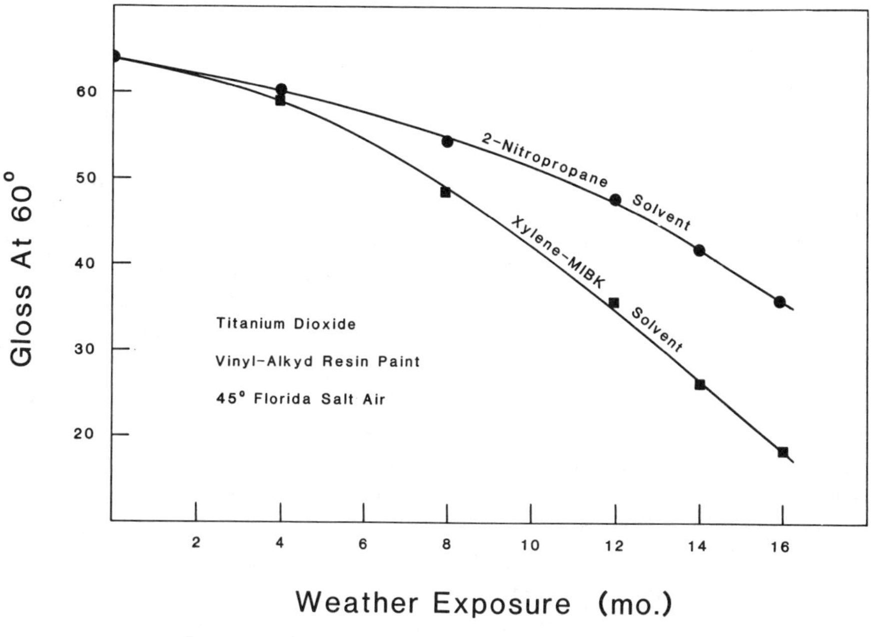

Figure 7.2. Solvent influence on weather resistance in alkyd TiO_2 paint films.

sions) for organic solvents, always are difficult to distribute and stabilize mechanically in these systems. The dispersant's role is to adsorb either through hydrogen bonding or direct reaction and alter the solvent phase image of the pigment surface.

7.3 Alcohols as Oleophilic Dispersants

Most elementary of the noncharged species capable of adsorbing on oxide–hydroxide surfaces and presenting a hydrophobic tail to an oleophilic phase are alcohols. These can be primary alcohols with the hydroxyl group at the terminus of the chain or diols where the second alcohol group is along the chain, not at the opposite end, for reasons described in detail in Chapter 5. The data in Table 7.1 suggest alcohols, especially long-chain ones, must interact with water adsorbed on polar surfaces. Heats of wetting on such surfaces vary slightly with chain length, the combined effect of slightly decreasing dipole moment and slightly increasing van der Waals forces. Alcohols average about 400 ergs/cm^2, insufficient to displace water at 550 ergs/cm^2.

Adsorbing alcohols onto bone-dry surfaces, as through vapor-phase attachment, and preventing the solid–alcohol from prolonged contact with ambient water vapor, will render the surface oleophilic. When incorporated into alkyd resins, polyolefin premelts, or other hydrophobic systems, the powder will disperse readily. Chain length is a factor in dispersion, but the eight carbon rule (oleophilic–hydrophilic balance point) is certainly a workable guide for agent design.

This is illustrated by the employment of stearyl and lauryl alcohols to waterproof cotton (cellulose) fabrics. The dipolar alcohol group readily attaches to the polyhydroxyl cellulose. When washed by conventional laundry procedures, the fabric slowly releases dirt but does not easily wet (hydrophobic). However, when dry cleaned (tetrachloroethylene solvent), the fabric wets immediately and grime is quickly flushed away.

Ground quartz with a D_{50} of 2.2 μm, to be used for dispersion in mineral spirits and incorporation into alkyd resins, has been treated by drying at 125°C and immersing in propyl alcohol. The thoroughly wet product is decanted and introduced wet into a long-chain hydrocarbon (octane or decane) and is then ready for introduction into the hydrophobic resin. This technique relies on *n*-propyl alcohol (and higher homologs) at low concentrations entering into a molecular solution with higher paraffinic hydrocarbons. Following incorporation of the pigment–solvent mixture into the resin, water penetration is prevented, except at the immediate air interface. Upon prolonged standing while exposed to the atmosphere, the silica–solvent–resin system prior to drying often develops a slight haze due to displacement of the alcohol dispersant by atmospherically diffused water vapor. Replacement of the *n*-propanol with higher homologs, decanol for example (the highest liquid alcohol), prevents this haze and increases "open bucket" life.

Attempts have been made to bond alcohols chemically to acidic oxide surfaces by the following reaction[9]:

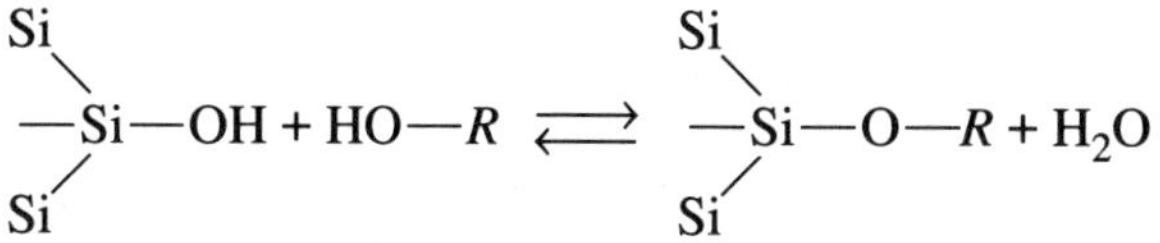

(7.1)

Primary alcohols show a greater stability towards the reverse hydrolysis than secondary ones, based on infrared spectroscopy and length of time for hydrolysis. Furfuryl alcohol, having an unsaturated ring structure adjacent to the alcohol group, is more stable yet. Attempts to stabilize long-chain alcohols on solid surfaces by additions of ammonia, which infrared spectra indicate forms an intermediate adduct, have shown promise. The products, however, are not oleophilic but hydrophilic, indicating the ammonia molecule occupies a position along the adsorbate's exterior surface (solution directed).

Similar investigations have been carried out on alumina, where a Lewis acid site is a prerequisite for alcohol molecular fixation.[10] However, even with this siting mechanism, the alcohol molecule displays a strong mobility across the alumina surface. When such adducts are heated to 400°C in vacuum prior to the alcohol reaction, the olefin form of the alcohol is produced quickly, indicating a surface-catalyzed dehydroxylation reaction to take place.

However, most such materials are thermodynamically unstable around high moisture and slowly hydrolyze back to their respective hydroxyl states.

Attempts have been made to stabilize long-chain alcohols on oxide pigments by adding trichloride and tetrachloride salts (aluminum, thorium, zirconium) as intermediaries.[11] The mechanism here appears to be an initial reaction of the alcohol with the metal halide followed by a secondary hydrolysis of the remaining halogen groups to hydroxyls which coordinate with the pigment surface. Because the metals are slightly more eletropositive than silica or titanium, the structure acquires an increase in stability. The method is an in situ formation variation on the silane technology discussed under Section 7.10. Titanium dioxide pigments have been produced by this technology and commercialized.

Short-carbon-chain alcohol dispersants function well with expandable lattice clays made oleophilic by quaternary amine treatment (bentones). Because of the high solvation energy of alcohols and electrostatic repulsion of particles (resulting from the increase in the electrical double-layer diffusion from a dipole at the interface), even short-chain alcohols, as methanol, disperse these products in paraffinic and aromatic hydrocarbon systems. The alcohol-dispersed products find use in paints as viscosity-enhancing agents (thixotropes), in high-molecular-weight hydrocarbon greases, as suspending agents in abrasive bearing oils for cutting and machining, and in caulking compounds and sealants. Alcohol loss by slow volatilization is not a problem once the oleophilic bentonite particles are thoroughly dispersed because system viscosities are high enough to impede greatly alcohol diffusion through the complex particle network.

The use of diols where each hydroxyl is on a terminal carbon might be expected to coordinate in a chelate configuration on acidic surfaces. Here a hydrocarbon loop would be presented to the liquid (oleophile) phase. While this represents one potential energy state, the single surface coordination with the hydrophilic tail in the solution phase represents an equally probable state. Unless the molecule is a complex polyglycol, hydrocarbon dispersancy is improbable.

Where a diol molecule is structured such that both hydroxyl groups are on adjacent or vicinal carbon atoms (vic), a significant increase in adsorption energy occurs via the chelation mechanism to stabilize the structure.[12] While the agent may be hydrolyzed on the surface, a ring structure remains with water molecules being introduced into the complex solid–dispersant ring structure. The resulting alcoholate-coated material, titanium dioxide, for example, with a tail of 8 to 20 carbon atoms configured in an "L" arrangement, readily disperses in mineral spirits and secondarily in alkyd resin. This results in a high-brightness pigment for gloss finishes.

7.4 Cationic and Anionic Alkyl Species

Agents in the charge category most successful in effecting dispersion of polar surface solids in oleophilic media are the anionic and cationic molecules with long hydrocarbon tails:

$$Na^{+}\,{}^{-}O{-}\underset{\underset{O}{\|}}{C}{-}(C_{12}H_{25}) \qquad \text{and} \qquad (C_{12}H_{25}){-}NH_3^{+}\ Cl^{-} \tag{7.2}$$

It has been demonstrated that a C-8 hydrocarbon represents the chain limit capable of functioning in both aqueous and nonaqueous systems. Thus longer molecular chains are more likely to adapt to the oleophilic phase. Weight becomes the commercialization tradeoff. With longer hydrocarbon tails, weight progressively increases and with it, the cost to disperse the solid.

Unlike aqueous dispersions, it is often impractical or functionally difficult to dissolve the dispersant in the liquid phase and allow it to diffuse onto each particle's surface. Some systems are far too viscous to permit efficient diffusion. Addition of the pigment as a dry powder always results in the carry-in of an adsorbed water layer, which causes the particles to agglomerate (flocculate) and diminish the effectiveness of diffusion. Also dispersant molecules may form micelles with their hydrophilic heads oriented to the interior, preventing head-to-surface contact at the solid interface. Finally, the dispersant, especially where it contains an ionic head, may have very limited solubility in the oil phase, far less than the hydrophilic surface requires.

With very fine particle size pigments, such as titanium dioxide, carbon black, and phthalocyanine blue, where diameters are 0.2 μm or finer, dispersion becomes very difficult in such viscous organic media as alkyd resin paints, polyester layup formulations, diallyl phthalate casting resins, etc. While solvent letdown to yield lower grinding binder viscosity provides better wetting and polymer coating of deagglomerated particles, practical limits exist, both economically and functionally. The finer a pigment's particle size, the greater will be the surface moisture carry-in by polar surface materials and the greater will be the shear necessary to redistribute individual particles from agglomerates to a homogeneous dispersed state.

The following general rules have been formulated for the development of new formulations where previous experience is not forthcoming. These are based on Imperical Chemical Industries' (UK) experience in paint and polymer formulations.[13]

1. Select that dispersant most adaptable to the surface chemistry of the solid and most resembling the polymer liquid phase. This takes the form of an HLB (Chapter 2) assessment, but differs in that both ends of the dispersant molecule must be evaluated independently.
2. Incorporate the tailored dispersant at the mill base stage of formulation. This may require some level of solvent dilution.
3. Employ a reduced level of resin solids in the grinding vehicle during the incorporation stage.
4. Adjust the optimum ratio of dispersant to pigment to wet out fully all pigment surfaces. This can be calculated from a theoretical basis or determined experimentally by methods shown later in this chapter.
5. Increase the pigment content in the mill base to ensure a workable mill makedown viscosity.

During mill dispersion some size reduction may occur, either accidentally or intentionally. New surface activity will increase the demand for dispersant and must be compensated to enable the full range of solid properties (e.g., mechanical and optical) to be realized in the final product. Fresh surface chemistry is much more energetic with regard to agglomeration than is old surface. Thus ensuring adequate dispersant in stages (2) and (3) cannot be overly emphasized.

One of the more unusual applications of surface-active agents to adapt a hydrophilic surface to an oleophilic medium occurs with mixed-layer clays, a reference to which has been made earlier in this chapter. Two of these clays are bentonite, a montmorillonite having a platelike silica–alumina–silica structure, and hectorite, with a lathlike or fibrous structure having silica–magnesia–silica layering. These two silicates have cations occurring on the silica interlayer to balance out the charge. Cations are largely alkali and alkaline earth ions, although nickel and other species have been found. Because of their loose association and their great number, these ions are readily exchangeable. Typically, bentonites will possess a cation exchange capacity (CEC) of 800–1000 microequivalents (0.8–1.0 milliequivalents) per gram. Alkali cations are extremely mobile. Natural bentonites and hectorites have a great affinity for water and are employed as gellants in pond liners, basement sealers, food thickeners, etc. They gel in water at a few percent and produce an impermeable barrier.

The agent $R_4N^+Cl^-$ contains a relatively strong cation capable of substituting readily into the interlayer positions of both bentonite and hectorite.[14,15] When all ion-exchange sites are replaced, the surface becomes expanded and very hydrophobic. The extent of expansion is dependent upon the chain length of one or more of the *R* groups. Gellation properties of the quaternary ion exchanged clay are retained, except they function with hydrocarbon liquids, not water.

Because of their high surface area when expanded, bentonite and hectorites have the capacity to increase viscosity in oil-based paint formulations and to gel hydrocarbons at low addition levels. Figure 7.3 is a rheological profile for bentonite and hectorite treated with dimethyl ditallow quaternary ammonium chloride (DMDTAC). The tallow component consists of mixed fatty chains from C-12 to C-18. Note that rheological minima commence just short of the exchange capacity, indicating a molecular spacing–siting conflict.

Better dispersed species have low shear viscosities below 500 cp in solvent. Effectiveness of the hydrophobic agent can be measured readily by assessing the viscosities of very dilute suspensions in toluene. Because the two minerals noted impart thixotropic character, high viscosity at low shear is accompanied by reduced viscosity at high shear, a two-state characteristic useful in lubricating bearings of rotating shafts. Natural hydrocarbon oils and greases are near-Newtonian in viscosity and are less able to fluidize upon shear.

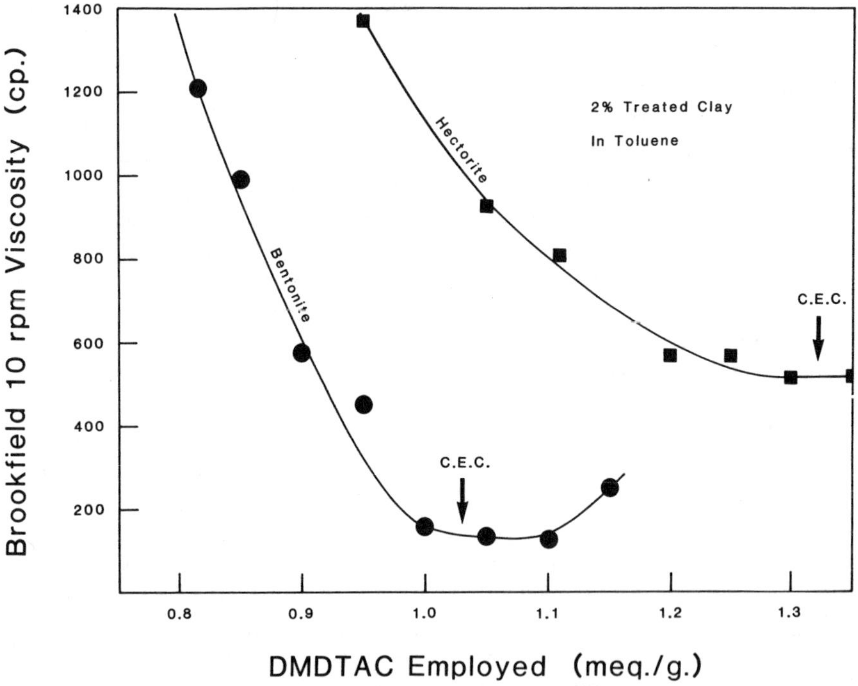

Figure 7.3. Bentonite and hectorite rheology with cationic agent.

Other cationic agents may be employed for producing these oleophilic thickeners, including dimethyl tallow benzyl ammonium chloride (DMBT) and trimethyl tallow ammonium chloride (TMT). The dimethyl ditallow and dimethyl tallow benzyl (reduced weight) quaternary ammonium cationics have been found to be among the better agents to develop oleophilicity, although trimethyl and other alkyl chain compounds have been employed as co-agents. Chains other than pure hydrocarbons, including glycols and ether linkages,[16] can be employed on the quaternary molecule to provide intermediate polarities. These find applications with alcohols, ketones, esters, etc. for thickening lacquers, polyester resins, and similar slightly polar organics. Most of these products, especially the sole hydrocarbon species, are at least partially removed with elevated temperatures, and their applications are limited to near-ambient temperatures.

Because of the high exchange capacity of the substrate and large molecular weight of the agent, treatment costs are very high. Most of the commercialized products, termed Bentones™ and Claytones™, require treatment levels of 40–50% by weight of the quaternary cationic agent, resulting in market prices from $1 to $2 per finished product pound (1990 basis). Where the silicate is very white, as with certain hectorites (California) and calcium bentonites (Texas), the oleophilic agents may be employed in paints and cosmetic products.

The most famous of all the applications for quaternary ammonium dispersants is that of fabric softening. There the charged head attaches to wet cellulose and polyester fibers, which have small residual negative charges, leaving the hydrocarbon tail to provide a slight amount of surface hydrophobicity. This hydrophobicity is interpreted as softness because it prevents water hardness compounds (calcium carbonates, sulfates, etc.) from drying onto the fabric and creating a hard, unpleasant texture. The drawbacks are few, the most notable of which is the reduced ability of bath towels to soak up skin moisture following showering or swimming. Overdosing laundry with fabric softener will actually produce towels with a waxy texture which are fully hydrophobic, wholly defeating their purpose! The same agents employed for producing the Bentone oleophile, dimethyl ditallow and dimethyl tallow benzyl quaternary ammonium chlorides, are the primary components in these products also.

Quaternary ammonium cationics have slightly acrid waxy odors. To enable consumer market acceptance of the product, the first commercial fabric softener masked this with a strong perfume. So associated with the softening activity has this masking odor become that competitors must also incorporate the same perfume to sell their products!

To slow down fabric softener release in driers and permit more uniform coverage, some manufacturers first adsorb the cationic agent onto bentonite or bentonite-coated paperlike sheets. At elevated drier temperatures, 70–100°C, the bentonite–cationic is thermally unstable and distributes the agent slower and better over the entire mixed fabric assemblage.

Cationics and anionics serve other purposes, but most are employed for the reverse process, dispersing hydrophobic surfaces in water. Coal and graphite are classic examples of materials readily dispersed with sodium fatty acid salts for grinding and classification in aqueous media. While the sodium salts are less expensive, the quaternary ammonium ones are less alkaline. Cleaners for metals capable of forming anionic complexes (aluminum, zinc) often take advantage of this alkalinity property for dispersant formulations.

For many decades sodium oleate has been employed to obtain quality dispersions of ferric oxide. In studies performed to determine the ultimate magnetic domain size (those regions where all molecules are magnetized in the same direction) in magnetite, Fe_3O_4, particles were required to be discretely separated. These particles ranged from 6 to 16 nm in size and required high shear and excellent dispersion to enable the single particle film to be laid down and studied. Because of size, particle breakage was not a concern. A heated Kady Mill employing sodium oleate as dispersant was one of few techniques able to provide the necessary product.

Bubble stability in dispersions increases with pH and cationic dispersants may be employed at near-neutrality to minimize this activity where foam is a disadvantage.

7.5 Surface Factors

The role of surface acidity or alkalinity, whether natural, as with ground marble (calcium carbonate), or induced, as with coated titanium dioxides, directs the choice of oleophilic dispersant. With ultrafine carbon blacks and titanium dioxides, where an acidic surface dominates, the attachment group of the dispersant (having an oleophilic tail) will display the following relationship: dodecylamine > amine and polyol containing polymers > fatty acids. Although van der Waals forces and steric factors may be involved in the abilities of dispersants to site properly on pigment surfaces,[17] these effects are minor compared to the direct chemical effects operative in these systems.

Both finely ground calcium carbonate (alkaline surface) and quartz (acidic surface) possess surfaces far too hydrophilic to disperse readily in alkyd resins, polyethylene, polystyrene, or similar oleophiles. The surface activity of these two materials can be determined readily, although by different means. From the surface area of calcium carbonate, the number of active sites, equivalent to the number of calcium (or carbonate) ions, can be determined and:

$$\text{Weight \% dispersant} = \frac{\text{surface area} \times \text{molecular weight} \times 100}{\text{wt. sample} \times \text{molecule area} \times \text{Avogadro's no.}} \tag{7.3}$$

7.6 Fatty Acids and Fatty Amines

The efficient treatment of solid surfaces with fatty acids or fatty amines to impart hydrophobicity requires an accurate measure of surface reactive groups. The calcite and quartz pigments cited previously have similar surface areas, carbonate 8.0 m^2/g and quartz, 7.7 m^2/g. With silica either the surface area method can be used, where one silanol group occupies approximately 0.4 nm^2, or the surface can be titrated (see Chapter 13) to determine the total population of silanol groups. Using these numbers oleophilic character can be imparted by adsorbing the appropriate number of dispersant molecules. The degree of oleophilicity can be determined by measuring viscosity in toluene or mineral oil at moderate solids. Figures 7.4 and 7.5 are rheogram families for three chain lengths of surfactants. For calcium carbonate, butyric acid (C-4), lauric acid (C-12), and stearic acid (C-18) were employed as surface-active agents. Correspondingly for silica, butylamine, laurylamine, and stearylamine, as well as octylamine, were employed.

Although subtle differences occur in the rheograms, general similarities exist between the two families which point to three important factors. First, chain length is important, but only up to about 12 carbon atoms. Second, once the surface is fully covered (all sites adsorbed), the rheology remains unchanged. Limitations to this phenomenon exist, however. With sufficiently large additions, dispersant molecules can intermesh in a second layer, tail-to-tail, and produce a

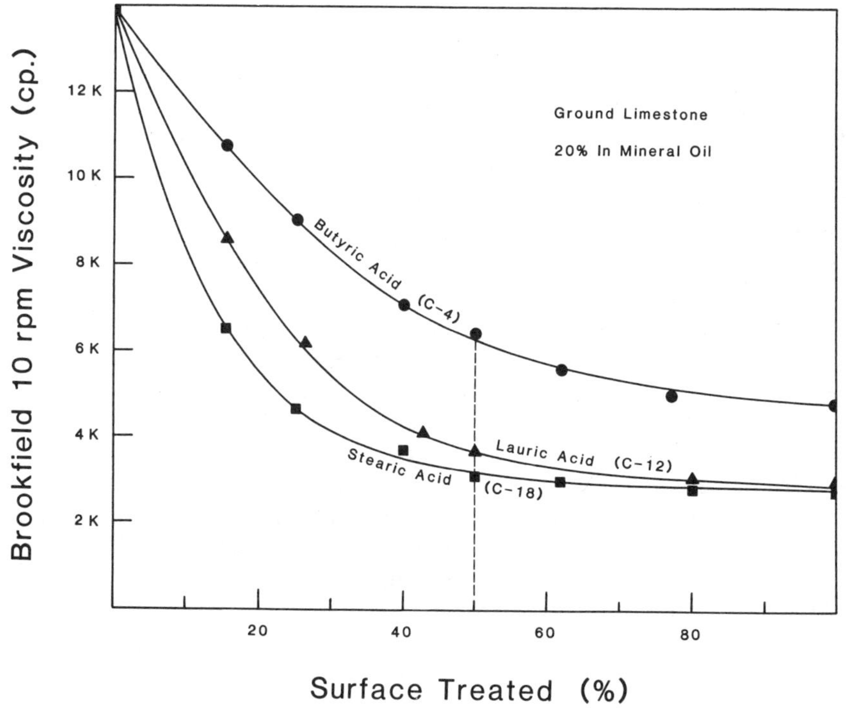

Figure 7.4. Fatty acid treated calcium carbonate rheology.

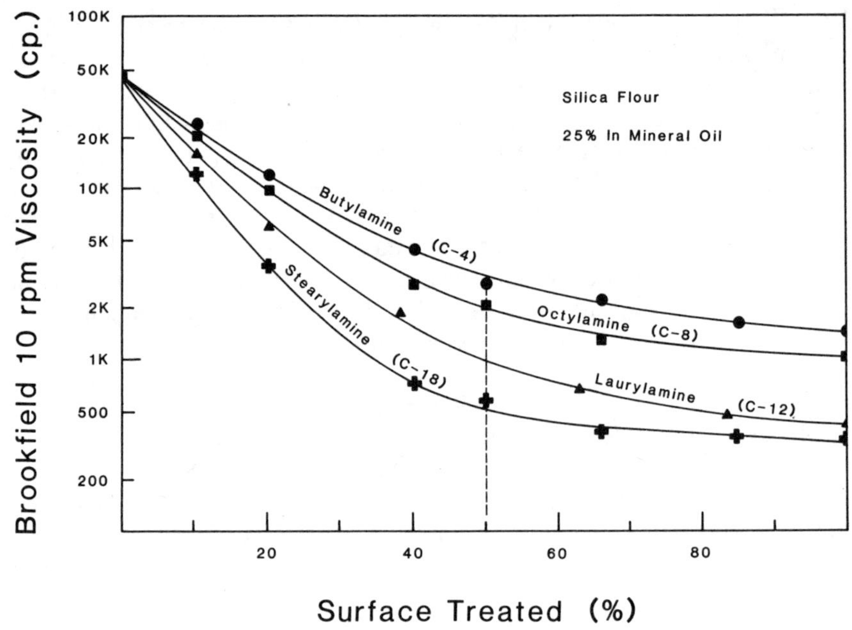

Figure 7.5. Fatty amine treated silica rheology.

slightly polar envelope, about $1\frac{1}{2}$ molecules thick. Thereupon, the viscosity rises slightly due to incipient interparticle agglomeration. Third, with the C-18 chain viscosity minima are attained at essentially one-half the surface coverage for both the acid (carbonate) and the amine (silica), reflecting the phenomenon known as "kink" formation or molecular spinning.

The cross-sectional area of the stearic acid molecule is surface dependent, but averages about 0.25 nm^2. However, by sweeping out a cone through kT energy adsorption (rotational mode), the hydrocarbon tail effectively covers, or shields, about twice that area. As a consequence, the liquid phase "sees" the surface fully covered with oleophilic tails even though only half the surface has adsorbed molecules. The phenomenon occurs also with the C-12 chains to a lesser extent but is absent at the four carbon chain length. Apparently kink formation is not as prevalent at or below 4 carbon atoms, or the conical "area" swept out is very small in comparison with the natural vibrations/precessions of the longer molecules. This phenomenon represents a true economy in dispersion because only half the amount of stearic acid, roughly corresponding to a C-11 or C-12 molecular acid, is required for full surface masking.

Although surface areas for the calcium carbonate and silica in Figures 7.4 and 7.5 are similar, viscosity and change of viscosity are disparate. Note that carbonate's rheology scale is linear while that for silica is logarithmic. Good evidence exists that commercial grades of mineral oil contain small amounts of long-chain alkyl acids resulting from sulfuric acid processing and clarification.

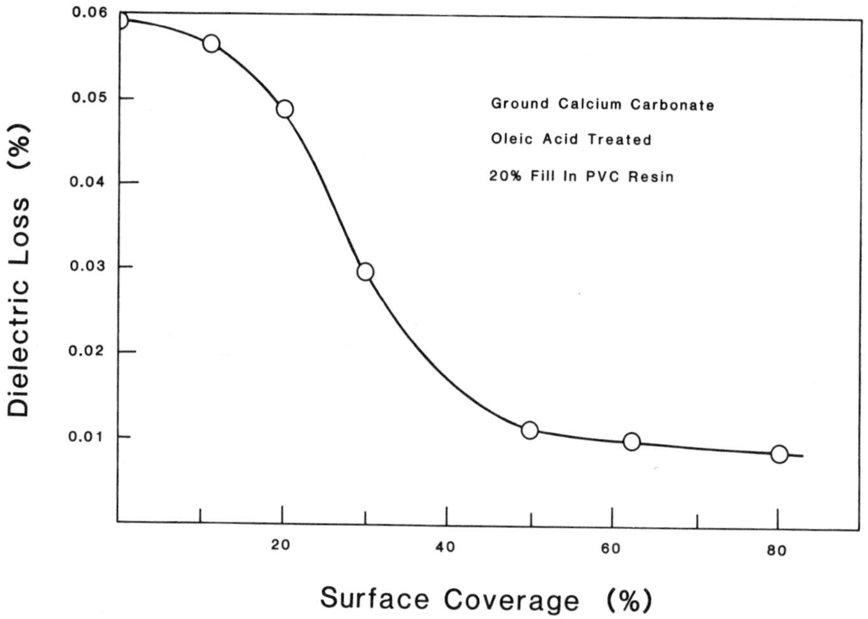

Figure 7.6. Dielectric loss in calcium carbonate filled PVC resin.

Thus the calcium carbonate test work may have contained a trace of surface modification prior to any addition of fatty acid.

When the surface is effectively covered with agent, dispersion, as measured by electrical properties, is affected dramatically. Figure 7.6 is a plot of dielectric loss of a calcium carbonate filled polyvinyl chloride extrusion resin employed in wire and cable insulation.[18] Calcium carbonate not only imparts a white color to the product, it improves the impact strength, weather resistance, and abrasion resistance and reduces cost. Typical loadings are in the 15–30% range in conjunction with other fillers. A surface modification of the carbonate with stearic acid (as shown earlier in Figure 7.4) causes a reduction in dielectric loss from 0.059% with no dispersant to 0.011% with but half of the surface covered. Increasing coverage to 100% diminishes the loss only slightly further, to 0.010%.

If loading levels of filler pigment are progressively increased, mechanical properties will eventually degrade because insufficient binder will be present to provide continuous polymer medium between particles. However, well prior to this electrical properties degrade, even with an adequately treated filler. Figure 7.7 contains both dielectric loss and breakdown potential* for a stearyl amine treated (55%) calcined kaolin filler in a phenol–formaldehyde casting resin

* Platey fillers are also employed to provide tortuousity to an electrical discharge path through the resin. The better dispersed these particles, the higher will be the voltage required to effect an electrical breakdown (arc through).

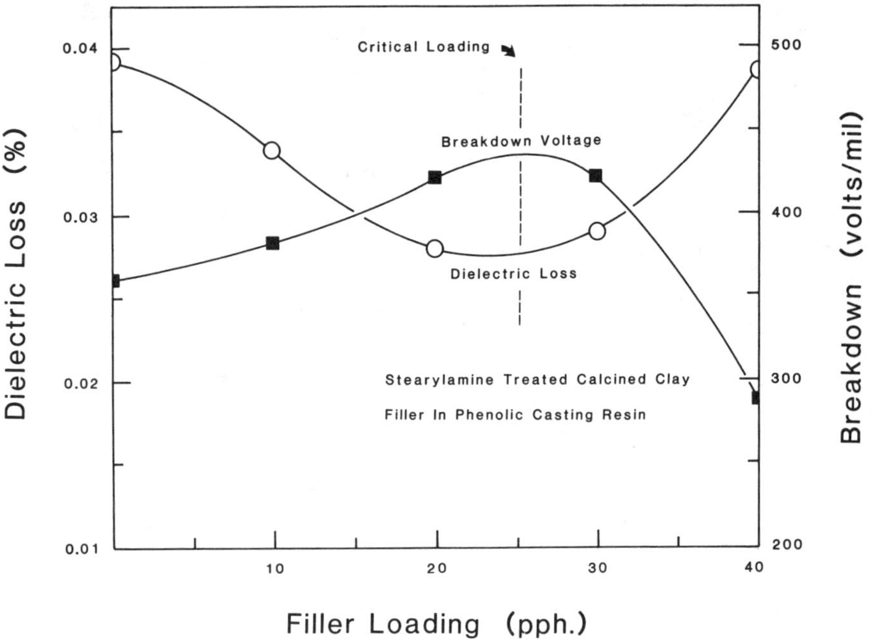

Figure 7.7. Electrical properties of kaolin-filled phenolic resin.

employed with electronic parts (chip housings, switch parts, sockets, etc.) Efficiency as a filler, based on electrical properties, maximizes at about 25 parts per hundred loading (20% by weight). Based on the differential densities of the two components, this occurs at a filler volume value of only 9.6%, which represents an average interparticle separation of 1.25 diameters. Because of the broad distribution of particle size with this filler, 5 μm down to 0.25 μm, many interparticle separations are reduced to 0.5 μm or less.

Stearyl amine and stearic acid are frequent candidates for dispersants with other oxide materials in highly hydrophobic organic systems. Both function well, provide strong adhesion to polar surfaces, and generate oleophilicity by their chain length towards the liquid phase, which they highly resemble. Alumina hydrate has long been the pigment of choice for polymer fireproofing because of its ability to release over 34% of its weight as water at polymer combustion temperatures. When polyolefin filament was introduced for upholstery and rug piling (e.g., Herculon™ fiber), regulations required it to be fireproofed. Because of the white translucency of alumina hydrate, especially the tabular form, that material was selected. However, the highly hydrophilic surface of the hydrate required an agent to render it dispersible in polyolefin melts.

Alumina hydrate for this purpose can be treated with stearyl amine in boiling petroleum spirits (125°C fraction) or toluene (111°C). Heat drives off the intermediate layer of adsorbed water and fixes the amine group directly onto the alumina hydroxyls. Once the surface is treated (50% by area), alumina hydrate immediately flocculates as a highly porous structure (terminal sediment volume > 10). After washing and drying, the predispersed product is tested by visually comparing wetting and terminal sediment volume in warm water and warm acetone (or MIBK). In water, even after 10 hours, no treated material should sink to the bottom of the container. In acetone the entire system should sink rapidly and yield a TSV value between 2.5 and 3.0. A comparison of these two parameters serves to predict on a qualitative basis predispersive effectiveness.

Interestingly, alumina hydrate so dispersed imparts not only fire retardency to polymers, it also provides increased plasticity and flexibility to the fibers extruded from it. However, because of its high loading requirement and higher density, its use elevates product costs.

Along with the hydrate, antimony oxide, Sb_2O_3, may be incorporated in flammable fibers as a flame suppressant using the same dispersant. This agent possesses a more acidic surface, and stearyl amines attach more energetically. Antimony oxide also may be predispersed using an aliphatic ester rather than amine. This dispersant apparently functions better where brominated organic compounds also are incorporated for flame suppressancy.[44]

Alumina also is dispersed for polishing and abrasive applications where it is introduced into oleophilic binders (resins, polysulfide adhesives, etc.). Treated with fatty acids, it finds a variety of other applications. With lauric and stearic acids the monolayer suggests one molecule coordinates across four oxygen atoms ($1\frac{1}{3}$ octahedra),[19,42] a much larger area than would be anticipated from the cross-sectional area of the lauric acid molecule (0.25 nm^2). The hydrocarbon tail length seems not to influence the coverage. This suggests that each octahedral face has but a single Al–OH group and that three triangularly fitted oxygen atoms (one octahedral face) expose too small an area for the organic acid molecule to be accommodated. The dispersed pigment has been employed for many other applications including polyisocyanate–propylene oxide–tetrahydrofuran terpolymers where it is incorporated for high weather-resistant marine finishes.

In train configurations, alkyl groups adsorb better than aromatic groups, which in turn are superior to ethers, based on isotherm studies of hexane, C_6H_{14} at 100%; benzene, C_6H_6 at 95%; and diethylether, C_2H_5-O-C_2H_5 at 85%.

The water-carrying minerals mentioned earlier, bentonite (smectite), hectorite, as well as attapulgite (palygorskite), have interlayer positions, not exposed easily to the surface and liquid phase, that hold water molecules by ionic coordination. If the internal ions are exchanged with hydrogen ions, stearyl amine can coordinate onto those sites,[20] displace the adsorbed water, and generate oleophilicity.

An addition of 2 to 5 percent of stearyl amine is sufficient to cover the external surface of these minerals and permit their incorporation into hydrocarbon solvents. While the external surface areas of the minerals may range from 20 to 50 M^2/g, the internal surface can increase this by a factor of ten. Once dispersed, therefore, grinding quickly diminishes their stability.[21] As new internal area approximating the original external surface is exposed, the system commences to flocculate. Sufficient dispersant to cover the entire internal and external areas is prohibitively expensive, discouraging commercialization.

While stearic acid adsorption on calcium carbonate is readily understandable based on the strong energetics of chemisorption, it is less apparent for dispersing carbon black. Here, however, the solvent role is far greater because of its lowered adsorption energy onto the solid. In a study[22] of a mildly oxidized carbon black pigment (alcoholate surface), four solvents for stearic acid were employed, cyclohexane, ethanol, carbon tetrachloride, and benzene. Based on the BET surface area of the carbon and an estimated 0.21 nm^2 area for the stearic acid molecule, an amount equivalent to one monolayer was dissolved in the solvent phase and 10% by weight carbon black introduced. At equilibrium the mole fraction distribution of stearic acid remaining in the slurry was approximately that shown in Table 7.3.

This relationship provides an indication of solvent interaction with or contribution to the solid surface. Had the carbon surface extensive oxide and hydroxyl groups, competition would arise from polar solvents, a state displayed by ethanol and cyclohexene (dipole moment = 0.55 Debye) but especially by alkali in water. The remaining two organic solvents are nonpolar.

Where ferric oxide is predispersed with stearic acid, the energetics of adsorption are sufficiently high to mask solvent effects, and little difference is displayed among the four solvents tabulated. The effective monolayer value for the specific iron oxide calculates to 0.24 nm^2 per site for stearic acid adsorption. Were the solid surface not polar to any appreciable extent, the mole fraction remaining in the solvent phase at equilibrium would be considerably greater.

Salts, especially alkali metal salts, of the fatty acids also may be employed as dispersants. With these compounds the metal ion seats upon the polar surface at negative sites (ionic or dipole). These salts are employed where color sensitivity to agent acidity is present, even though their surface affinity is less than those of equivalent acids.

Two additional classes of salts may be employed, those with divalent and trivalent cations, both of which are insoluble in water but have sparing solubility in alkane and chlorinated solvents. The first is typified by calcium stearate, $Ca(C_{18}H_{37}COO)_2$, and the second by aluminum stearate. The aluminum salt (or

Table 7.3 Stearic Acid Adsorption Efficiency on Carbon Black

Solvent	Mole % in Solution	Mole % Adsorbed
Cyclohexane	5	95
Ethanol	3	97
Carbon tetrachloride	2	98
Benzene	1	99
10% NaOH in H_2O	97	3

soap) yields the dianion, $Al(OH)(C_{18}H_{37}COO)_2$. Here a polar attachment group, the residual metal hydroxyl, may serve to anchor the hydrophilic end of dispersant molecules better to solid surfaces.

Calcium, zinc, magnesium, and aluminum stearates are used for dye pigments, such as PV Fast Red, to obtain better dispersion and color development in extruded polypropylene for decorative household products (bottles, cups, pans, handles). Although the pigment size is very small, only about 0.3% of the metal stearate is required for effective distribution of the color body into the polymer.

Aluminum stearate, aluminum palmitate, and aluminum naphthenate gained great notoriety during World War II as viscosity control agents (thixotropes) for gasoline and kerosene in the military incendiary Napalm (*na*phthenate + *palm*itate). There sufficient dispersant was added to form a complete network, part micellular and part chain bridging, to gel the fluid hydrocarbon molecules fully. Because the gel is highly thixotropic, it shears to a liquid on contact (bomb) or when extruded under high pressure (flame thrower nozzle). It acquired the field name *jellied gasoline.* Aluminum organic salts were added just prior to use because dispersion–thixotropy proceeds so rapidly. The principle has been investigated as a means of preventing broad gasoline spills from jet aircraft and tanker trucks during container rupture from crashes.

Based on the same phenomena, the hydroxy stearic acid molecule also can be used with metal salts for organic color pigments. Although many hydroxystearic acids exist, the most popular is that resulting from catalytic water addition to oleic acid, where the hydroxyl group appears approximately midway along the hydrocarbon chain. Adsorption isotherms with this agent show fewer molecules adsorb on a given polar surface than with the unsubstituted stearic acid. This may result from coordination of the hydroxyl group to the surface in an "L"-shaped configuration—half train, half tail. This is the preferred explanation based on the stretching frequency change of the hydroxyl group in the infrared spectrum. It also may be explained by a more energetic rotational activity around the hydroxyl-bearing carbon in the chain—a highly activated "kink" mechanism. By either mechanism, calcium and magnesium salts of these hybrid molecules serve as excellent dispersants for metal oxides in oleophilic systems.

The agent, 10-hydrostearic acid, has been employed as a co-dispersant for ultrafine polar surfaced solids with N-acryloamino acid esters. As such it provides exceptionally low viscosities for solvent-based spray paints, especially those employed for electrostatic coatings of automobiles, electronic cases, and metal cabinets. Similar co-dispersants with benzaldehyde–polyalcohol condensates also are reported to provide very low viscosities.[23]

Although manufactured by somewhat different procedures, the similar compounds, 12-hydroxystearate and tridecyl 12-hydroxystearate, are popular dispersants for copper phthalocyanine where the pigment is flushed into petroleum solvent.

Pigments with lowered adsorption energy for stearyl amine or stearic acid can be intermediately coated with aluminum hydroxide (much as titanium dioxide is coated), whereupon the coating becomes highly active to stearic acid (and other fatty acids). The alumino-gel must be very fine, considerably finer than the pigment particles. This usually restricts precipitation exercises to hydrolysis, as opposed to direct aluminum sulfate precipitation with alkali.

Printing inks in which ultramarine blue (a quasiceramic manufactured by fusing a mixture of sodium carbonate, sulfur, kaolinite, and wood resin) is employed as a fade-resistant pigment for printing inks and exterior blue paints, is surface modified in this manner. Ultramarine blue's surface, from high-temperature exposure, contains few active polar oxide–hydroxide sites. Once ground to an appropriate size, the pigment is slurried with dilute alum, $KAl(SO_4)_2(H_2O)_{12}$, to produce about 2–5% alumina hydrate, pigment basis. The solution is slowly alkalized with ammonia or dilute sodium carbonate and heated. Particles of aluminum hydroxide 0.01 μm or smaller form and attach to the ultramarine particle surface through differential charge association. This product is filtered, washed, and dried at low temperature. To the dry ultramarine a dilute solution of stearic acid in petroleum spirits is added with high agitation and the acid quantitatively adsorbs. The pigment remains in a dispersed state in the hydrocarbon liquid. This stearic acid treated product is vacuum dried and sold as a predispersed pigment to the ink and paint industries.

Because ultramarine blue for these applications is ground very fine, its pigment surface area is high, and stearic acid requirements to accomplish good dispersion are correspondingly high, usually about 5% by weight. This adds significantly to the pigment's cost. Where interaction (cross bonding) between pigment and polymer base is important, as with alkyd resins, oleic or linoleic acids are substituted for the stearic. A cobalt catalyst (typically naphthenate) may be added to accelerate the curing, once exposed to air.

In Section 1.3 the application of dispersing agents to milling was discussed. These applications are more commonly associated with water processing where ball and small-media milling are more efficient. However, zinc oxide is milled for use in polyester resins as special antifouling pigmentation. The grinding is performed in toluene with a small quantity ($<10\%$) of resin, then introduced into the resin phase once the product has reached final size, nominally 1.5 μm.

When ball mills are operating well, ground products will display a Gaussian distribution identical to the feed except shifted downsize, a phenomenon termed *affine transformation*. Graphical plots of these distributions as log size–probability weight percent should yield parallel curves, ideally straight lines. Where the product is skewed from the feed, as indicated by nonparallel curves, it reflects a bias on the part of the mill or the material, usually the limitation of the mill to grind. In Figure 7.8 the grinding of zinc oxide is shown with partial and with full dispersant (0.1 and 0.5% benzoic acid). Note the curve slopes on the coarser ends of the two products are nearly identical but with inadequate dispersant the fine end of the distribution curves left and approaches that of the feed; that is, fine particles are not being ground. With adequate dispersant, the feed grinds satisfactorily and the product distribution is nearly parallel to it.

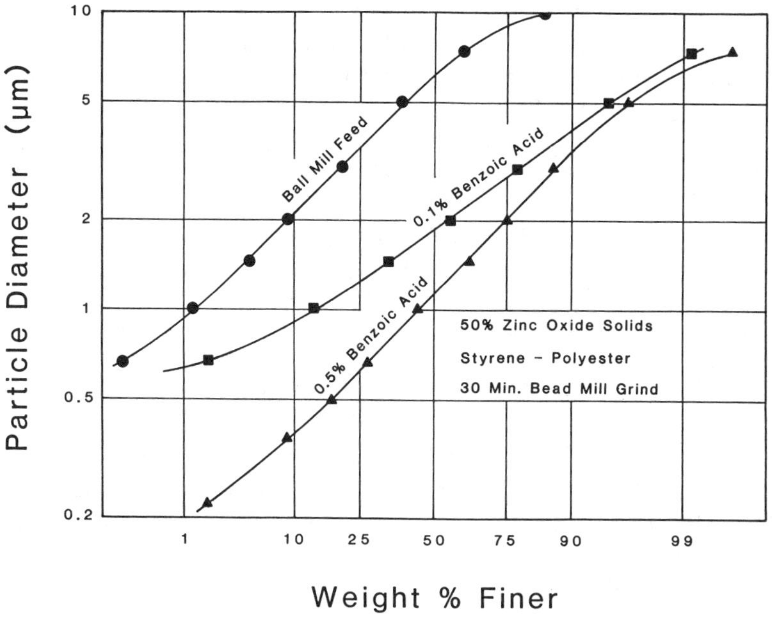

Figure 7.8. Zinc oxide ground in polyester resin with benzoic acid.

7.7 Organophosphates as Oleophilic Dispersants

In Chapter 3 water-soluble and water-dispersible organophosphates were discussed as dispersants for pigments on which alkali phosphates displayed surface instabilities. A family of organically soluble phosphates exist that adsorb onto solids by mechanisms similar to inorganic phosphates but that imbue an oleophilic character to the solid by extending an alkyl tail toward the liquid phase in place of a charged anion and cation. These have the general form:

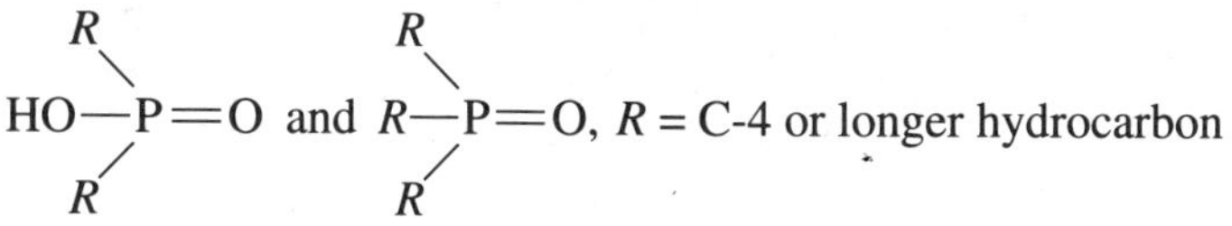

(7.4)

Popular forms of these structures are tributyl phosphate, diisooctyl phosphate, and di-2-ethyl hexyl phosphate. These are oily liquids capable of being sprayed onto polar pigment surfaces and heated to distribute the dispersant. Thereafter the product requires no further processing.

Diisoctyl phosphate has been sprayed as a dilute heptane solution onto rutile titanium dioxide at about 0.5% (pigment basis)[24] in a mulling operation after which the pigment is immediately ready for incorporation into alkyd resin paints and dioctyl phthalate casting resins.

In the tape casting of ferroelectrics for electronic capacitor use, fine (2 μm) barium titanate particles are treated with tributyl phosphate and similar species, after which the material is slurried in MIBK and ethanol for application to the mylar or other insulating surface film. Because of barium titanate's high dipole moment (its purpose for utilization), adsorbed water films are especially difficult to remove. However, it is important that these be removed for the described application as both viscosity rise and agglomeration can result. Agglomeration produces "short-circuited" ferroelectric crystals that decrease the dielectric capacitance in that region. Moist materials incorporated result in inferior products and reduced lifetimes in electrical applications. One dispersive method employed[25] involves drying barium titanate at elevated temperatures followed by direct MIBK immersion in which the tributyl phosphate is dissolved.

Although not a hydroxylated surface, calcium carbonate, especially as a superfine pigment (0.02–0.5 μm), also has been dispersed with tributyl phosphate, as well as dibutoxyphosphates in ether-substituted alcohols (Bu–O–Et–OH), for incorporation into slightly polar polymers, acrylic extruding resins, terphthallate casting resins, and cellulose esters. By improving dispersion, color development and hiding power are also improved, which diminishes pigment requirements and facilitates both casting and extrusion.[26]

Chromium dioxide, CrO_2, a black, highly ferrimagnetic material, has an unusual surface. Although green trioxide pigment's surface is essentially neutral (isoelectric pH ~7.0), the dioxide has surface hydroxyls which are slightly alkaline. On such surfaces the R_2(OH)PO form of the organophosphate family is more effective than the trialkyl form. In some formulations the R group is expanded to an alkyl-terminated polyglycol phosphate:[27]

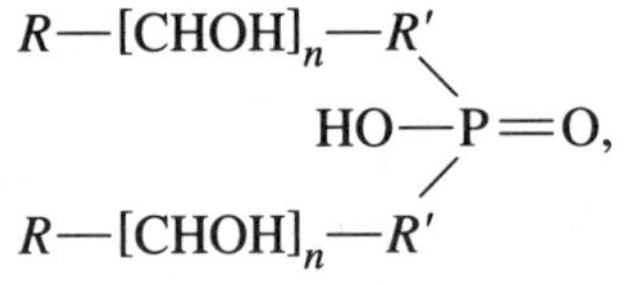

R = C-4 to C-8
R' = C-1 to C-3 (7.5)

Those additional R–OH groups on the two side chains enable improved coordination of the dispersant to the chromium dioxide surface. The common polymer employed for binding this oxide onto disks and tape is cellulose acetobutrate, a slightly polar species. In milling the oxide with dispersant into the solvent and polymer, care must be exercised not to destroy the black, cigar-shaped, particle anisometry wherein the ferrimagnetic character lies.

7.8 Titanyl Ester Dispersants

Another dispersing agent family with forms similar to the organophosphates is that of titanyl esters. Its members have as a generic formula Ti(OR)$_4$, low molecular weights, and pleasant, fruity odors due to the alcohol ester structures. Upon contact with a solid surface, one or more of the alcohol groups hydrolyze with the surface moisture on the pigment. The resulting titanol groups coordinate through hydrogen bonding with the pigment or filler surface, as follows:

$$RO{-}Ti(OR)_3 + 3H_2O \longrightarrow RO{-}Ti(OH)_3 + 3ROH \tag{7.6}$$

Because of the low toxicity of alcohol vapor (C-2 and greater), the dispersant, either in liquid form on in alcohol or petroleum spirits solution, can be sprayed onto a warm silicate, iron oxide, or other solid surface and the reaction is spontaneous. No color develops, and byproducts are easily removed without washing. Dried materials are fluffy, requiring little or no secondary pulverization.

The organotitanate dispersants were developed in the 1960s and continue to show promise in a broad variety of applications including suspending aluminum metal, flake particulate and sheet, in polyolefins. Hydrated kaolins were so treated to develop products for filling ethylene–propylene copolymer fibers to impart fireproofing (14% bound water), mechanical properties, and color potential (where the kaolinite particles carried in laked dyes as copigments in a surface nucleated form).

Although ostensibly appearing redundant, titanium dioxide itself has been treated with organotitanates. However, such a coating serves a useful purpose. Because rutile is produced by high-temperature operations, its surface is not as readily reacted as, for example, precipitated silica or calcium carbonate. Titanium dioxide can have coatings both by organotitanates or by hydrolyzing sodium titanate, Na_2TiO_3. Onto this coating is reacted sodium caprylate, $NaOOC{-}C_7H_{15}$, or short-chain carboxylic acids, followed by washing, drying, and redispersion in dioctyl phthalate for incorporation into polyester resins. Shorter-chain organic acids (having lower pK_a values) provide more adherent organotitanium complexes on the rutile surface. This in turn produces lowered viscosity in the resin. Nuclear magnetic resonance spectroscopy has shown[28] short-chain acids to produce titanyl structures with organic groups dominantly in a tail configuration. In the range C-6 to C-8 this slowly changes, and with longer chains trainlike configurations occur with an attendant increase in viscosity.

An example of unusual surface modification reminiscent of the old "C-Pigment" (calcium sulfate coated titania) is barium sulfate precipitated on titanium dioxide to produce a reduced cost pigment. Onto this coating (or copigment mixture) is hydrolyzed a triethyl octyl titanate. The product, without byproduct ethanol removal, is introduced directly into a butyl acetate–nitrocellulose lacquer. Later process modifications[45] employ a secondary dispersant, polyoxyethylene lauryl ether phosphate monoethanolamine salt, just prior to introduction into the lacquer. This dispersant qualifies as near as any to an "all-purpose agent," having literally something chemical for everything pigmentary inherent in its structure!

Where an organotitanate is structured such that three ester groups are nonhydrolyzable, the mode of attachment changes considerably.[29] While this reduces the agent's applicability to oxide materials, it becomes an advantage for dispersing calcium carbonates for oleophilic applications. By producing isopropyl titanium tri-isostearate, a single Ti–OH "leg" attaches to the carbonate surface. The large molecular weight makes surface coverage weight for ground calcite high, 2.85% for a 2 μm pigment. Due to the presence of three stable, electron-withdrawing groups on the titanate central atom, the Ti–OH remaining becomes more acidic, much in the fashion of sodium titanate, Na_2TiO_3. Stability on the carbonate surface is believed to derive from interaction of this weak acid with the weak base of the carbonate. High molecular weight obviously increases the van der Waals attraction of the agent for the solid surface.

Replacing two of the titanyl ester's groups with hydrolyzable entities (e.g., isopropyl) reduces dramatically the stability of the agent on the limestone surface, as well as its rheological improvement in hydrocarbon media.

In a similar application for producing easily dispersible magnetic iron oxide (gamma),[30] a tri(dodecyl benzyl sulfonyl) isopropyl titanate is also singularly hydrolyzed, but on a surface having a multitude of hydroxyl groups. The resulting surface-coated product is sufficiently bound to withstand severe and prolonged (15 h) ball milling in xylene–cyclohexane for application in tape-coating formulations having PVA–PVC as binder.

In this application the bridging of the titanol group into the iron hydroxide matrix on the surface is a bit more easily understood. Also gamma ferric oxide is very slightly alkaline, while the alpha form (Table 10.1) is acidic, rendering the gamma species somewhat more receptive to a weakly acidic titanol group.

7.9 Siloxane Dispersants

Siloxanes also are a family with broad applications in pigment modification for oleophilic systems. The siloxane molecular structure contains a number of the mildly polar groups, which can absorb onto oxide surfaces through dipole or water intermediary association:

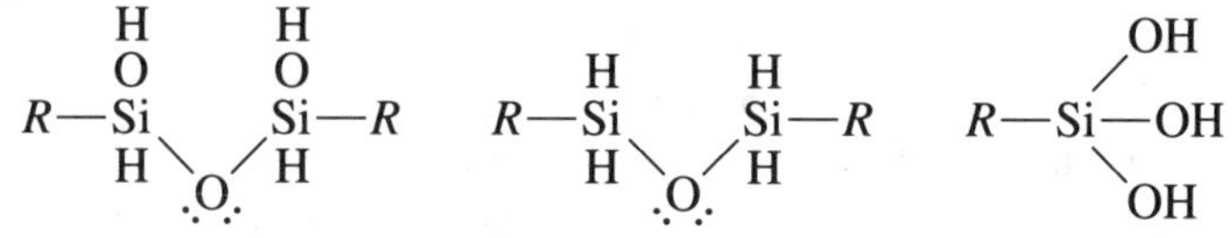

(a) (b) (c) (7.7a)

In some pigment developments the siloxane molecule contains three or more silicon atoms to increase the van der Waals force during adsorption. Triethanolamine or long-chain alkanolamines may be employed as codispersants with the siloxanes.[31] Most of these treatments, however, create a degree of particle aggregation (chemical sintering) that requires high-shear dry mixing or jet milling to break apart.

A similar pretreatment on titanium dioxide is used in the manufacture of typewriter, copy machine, and drafting correction fluids. To reduce cost, titania is often blended with talc (80–20% to 90–10%) and the entire system siloxane coated similar to the methodology just described. A typical formulation[32] is given in Table 7.4.

Binder selection is largely solvent dependent (cellulose esters with ketones, alkyds with xylene, acrylates with alcohol, cellusolves, etc.). Solvent selection is application dependent. For typewriter ribbon correction fluid, typical solvents are xylenes and toluene. For other applications, mixture of esters, ketones, and alcohols are employed. Better-grade correction fluids contain both higher pigment solids and higher binder levels. More competitive, nontrade name products replace these with higher concentrations of solvent and talc, which diminishes cost (and covering capacity).

Because the solvent must be highly volatile to enable rapid fluid drying, usually in 1–2 minutes, bottle caps that do not seat tightly cause these products to dry out rather quickly with use. By inverting the bottles in office desk storage, solvent loss causes an immediate binder seal to take effect at the cap–bottle seat that minimizes additional vapor-phase loss. Most office personnel, however, emotionally resist storing any liquid container upside down, an outlook that increases the demand considerably for these products!

Whether in paints, plastics, or office correction fluids, the degree of dispersion is reflected by the final gloss characteristics of the dried and polymerized film. Matte finishes, although often desired with correction fluids on quality stationery (high talc incorporation), are more often an indication of poor distribution and dispersion in the solvent–binder system.

Titanium dioxide bearing alumino-silica gel coatings can be made hydrophobic by coating with a siloxane where the *R* groups are amyl or hexyl (C-5 or C-6). Because the primary application falls in the lacquer family, with a slight amount of polarity to the resin, a small amount of triethanolamine is added along with the siloxane for controlled hydrophobicity. An analysis[33] of a typical pigment appears in Table 7.5.

The dried product will disperse readily in xylene for later incorporation into alkyd resins, cellulose ester resins, and similar polymer bodies. Because triethanolamine boils at about 278°C, it is not fugitive upon drying to any great extent. In similar applications organophosphate molecules (single P–OH) have been substituted for the siloxane.

Table 7.4 Correction Fluid Composition

Component	Weight % Range
Siloxane-coated rutile	35–45
Siloxane-coated talc	3–5
Organic solvent	40–50
Polymer binder	3–20

Table 7.5 Hydrophobic Predispersed Titanium Dioxide for Lacquer Use

Component	Weight %
Rutile TiO_2	~ 94.0
Al_2O_3 (gel)	0.8
SiO_2 (gel)	2.5
Bound water	1.7
Amyl siloxane	0.25
Triethanolamine	0.75
	100.0

Titanium dioxide–barium sulfate copigment has also been treated with octadecyltriethoxy silane (~1%) for employment in cellulose ester lacquers.[45] It is uncertain what mechanism is involved in the attachment of hydroxylated silane molecules to the nonhydroxylated, ionic barium sulfate. However, attachment to the hydroxylated titania appears to override any adverse copigment effect, if one exists.

Fine silica is an obvious commercial target pigment for siloxane dispersion because of the essentially perfect fit of the terminal silanol groups onto silica surfaces,[34] whether quartz, cristobalite, chalcedony, or a less crystalline precipitated species.

Ultrafine silicas produced by hydrolytic precipitation, where diameters are in the general range of 0.2 μm, may be pretreated with dimethyl siloxane. Adsorption usually proceeds in high-boiling mineral spitis (>200°C) under high localized shear, as typified by a kinetic dispersion-type mill (see Chapter 8) for an hour or more to ensure good diffusion of the agent over the entire surface. The product is well dispersed in the mineral spirits but must be vacuum dried for use. Because of high surface areas and narrow particle sizes, such predispersed products serve as excellent viscosity control and structural reinforcement agents in PVC, silicone rubber, ABS flexible polymers, and other similar polymer systems.

Many short-hydrocarbon-chain siloxanes are liquids and have the potential for direct spray-dry blend application. Such a procedure avoids the dangerous and costly petroleum solution treatment and vapor condensation techniques. The generic siloxane,

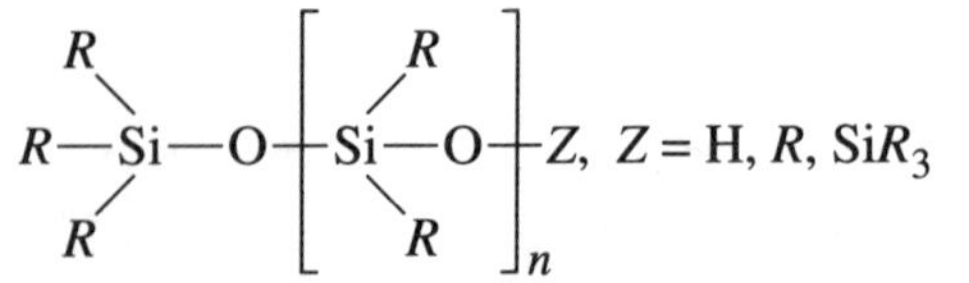

(7.7b)

has served as a treatment for rutile titanium dioxide. Dry titanium dioxide powder is agitated with a high-speed blade as a solid-in-air aerosol, the siloxane where R is the methyl group is sprayed in at 0.5% level, and the system treated further with superheated steam. The latter stage serves to partially hydrolyze the siloxane and stabilize the dispersed coating.

Methyl groups do not impart high hydrophobicity because of their short chain length. However, because the surface is fully covered and the three terminal methyl groups *R* extend into the polymer phase, the treated pigment is readily dispersed in PVC resins for pipe extrusion, wire and cable insulation, kitchen products, and plastic toys.[35]

Organosilicon compounds, especially those with methyl groups directly bonded to the terminal silicon atom, have been employed to treat a broad variety of polar surface pigments in addition to the aluminas and silicas, including titanium dioxide, zinc oxide, hematite, zirconia, barium sulfate, and calcium carbonate. The short-hydrocarbon-modified siloxane is produced primarily for economic reasons—low molecular weight. While limited testing has shown longer chains to improve both electrical properties and color development, the gain, compared to the weight addition and cost increase, is not cost effective.

7.10 Halosilane Agents for Polar Surfaces

A popular, though expensive, surface additive to render polar surfaces oleophilic is the family of organo-halogenated silanes. These are derivatives of silicon tetrachloride and bear a number of the characteristics of that chemical. The structures are:

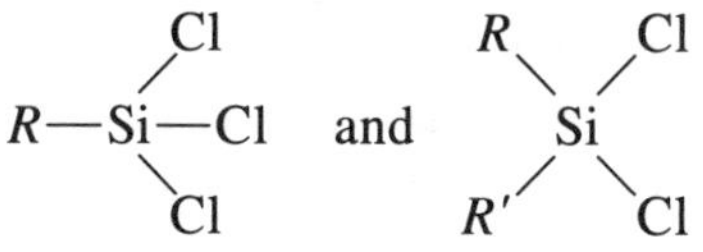

(a) (b) (7.8)

The character controlling group, *R*, may be a hydrocarbon chain as short as 2 carbons or as long as 18. In the disubstituted species (b), chain lengths are generally shorter than with the single substituted species (a).

Organochlorosilanes in theory can be attached to any polar surface. However, chemically inert species (as silicates) serve better as hosts than do alkaline pigments (as calcium carbonate, zinc oxide, and lithopone).

The basic mechanism for attachment involves both hydrolysis and polymerization reactions of the dispersant. These take place rapidly and exothermally. Once formed onto a silicate surface, their structure is difficult to define because infrared spectra interpretation is interfered with by the large population of silicon oxide and hydroxide (silanol) groups integral to the surface. However, the mechanism most defensible for the reaction is shown in Figure 7.9. Polymerization may be complete (no hydroxyls), partial, or terminated by three hydroxyls, depending upon temperature, surface moisture, and surface hydroxyl population.

True chemical bonding does not occur except with electropositive metal

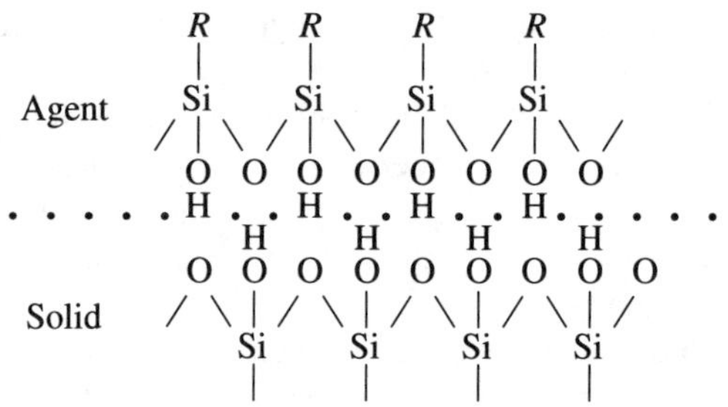

Figure 7.9. Coordination of organo-silanes on silicate surface.

oxides (iron, aluminum, etc.). Attachment of the oleophilic agent is largely through hydrogen bonds, the dispersant's own or those of the host. The *R* group may be single or double, short chain (C-2) to long (C-18), saturated or unsaturated (vinyl). Most of the chlorosilanes are liquids and may be added via vapor phase to the dry-state solid in a fluidizied bed agitator. Table 7.6 is a survey of silane-treated aluminum silicates for polyethylene wire and cable insulation and ethylene–propylene copolymer (EPC). Effectiveness was assessed by measuring viscosity at a standard loading in mineral oil. The data represent part of a much broader investigation. Level of addition was on an equimolar basis, 10 millimoles per 100 grams solid.

One of the few short-chain organosilanes that has found commercial application in dispersing polar solids in nonpolar media is that of dimethyl dichlorosilane on titanium dioxide having a 2:1 alumina gel–silica gel coating. Titania thus produced has been employed in thin films (opacifying polyethylene) where

Table 7.6 Hydrophobic Character Generated by Employment of Organosilanes

Oleogenic Agent Feedstock *R*[a]	Level (wt. %)	Mineral Oil Viscosity (cp at 10 rpm)	Hydrophobic Index[b]
Ethyl (mono)	1.64	1960	87
Vinyl (mono)	1.63	1320	92
Butyl (mono)	1.92	1720	90
Amyl (mono)	2.06	1660	95
Trimethyl	0.91	1420	91
Stearyl (mono)	3.88	960	99
Vinyl–amyl	2.00	1000	97
Amyl–stearyl	4.23	960	99
Stearyl amine	2.70	3240	83
Stearic acid	2.84	5520	31
None	0.0	>50,000	0

[a] All groups attached to trichlorosilane (if mono) or dichlorosilane (if dual alkyl groups).
[b] The hydrophobicity index is the viscosity ratio, before and after 30 min in boiling 5% alkali solution × 100.

it serves primarily an optical purpose. However, it also improves the tear strength and reduces the elongation modulus. The treated product is introduced into the polymer via heptanol prior to melting and extruding of the film.

From Table 7.6 stearyl silane is unquestionably the most effective dispersant, based both on mineral oil viscosity and the strength of molecular adhesion to the silicate surface. It is also the most expensive because of high weight requirements. Table 7.7 presents another portion of this investigation pertaining to surface moisture influence. Three silane species were employed in this investigation, at three levels of moisture, and again on an equimolar basis between species, except as noted.

Water dependence is clear from the rheology data. Trichlorosilane requires 30 millimoles of water to completely hydrolyze, or 0.54% water, while the single hydroxy species requires but 0.36%. Where very dry surfaces were incurred, the hydrolyzation–polymerization reaction is incomplete, and some surface active agent remains floating about the surface. This method of altering polar surfaces to nonpolar by organo-chlorosilane addition produces one of the most hydrophobic and stable surfaces of any additive–dispersant. It is not without its drawbacks, however. The reaction is exothermic and gives off hot hydrochloric acid vapors which are extremely corrosive and hazardous. Stainless steel equipment cannot be employed, and a chemical wet scrubber must be installed to treat the off gases.

Organosilanes function less thoroughly on alkaline surfaces, especially calcium carbonate and zinc oxide. However, other compounds are manufactured for this purpose, for example, esterified silanes, much in the fashion of the titanyl esters. These can be hydrolyzed readily on alkaline pigment surfaces through the agency of live steam. Byproducts are alcohol or similar innocuous molecules.

Table 7.7 Effect of Surface Moisture on Organosilane Dispersion

Oleogenic Agent Feedstock *R*	Level (wt. %)	Surface Moisture (wt. %)	Mineral Oil Viscosity (cp)
Trimethyl	0.90	0.09 (low)	4720
Trimethyl	0.90	0.21 (med)	3680
Trimethyl	0.90	0.33 (high)	6480
Amyl–vinyl	1.99	0.09 (low)	1760
Amyl–vinyl	1.99	0.21 (med)	1320
Amyl–vinyl	1.99	0.33 (high)	2560
Amyl–vinyl	2.51	0.33 (med)	1080
Stearyl	3.88	0.09 (low)	2560
Stearyl	3.88	0.21 (med)	1320
Stearyl	3.88	0.33 (high)	1640
Stearyl	4.85	0.33 (med)	1440
None	0.00	0.33 —	>25,000

30% solids in mineral oil.

7.11 Miscellaneous Dispersants for Oleophilic Media

The possible variations on organic molecular structures to accomplish polar surface wetting at one end and oleophilic association at the other are legion. Coupled with the possibilities of mixing dispersants to produce synergistic effects, especially on anisotactic surfaces, the examples of dispersants defy inclusion in a book this small, perhaps any book! Variations on the foregoing themes will be seen in the several examples cited in the following pages. These can provide the reader with the fundamental concepts of adaptation of molecular structures and dispersant mixtures to accomplish a select series of ends.

Unusual dispersants which have found use include the following specific examples.

1. The cyclohexyl ester of methacrylic acid has been dry blended with organic pigments showing a small amount of polarity. The distribution of dispersant is enabled not by solvent mixing but by thermally lowering the fusion point and milling the agent and pigment together, a technique similar to the titanyl ester process. The two color agents, Pigment Red and Pigment Orange, can be dispersed employing dimethyl aminoethyl methacrylate as dispersants in methyl ethyl ketone let-down acrylobutylnitrile rubber formulation.
2. Acrylic acid may be neutralized, not with alkali metals but by esterification with moderate-chain alcohols on alternate carboxyl groups to give a dispersant. Such a formulation is given in Table 7.8 to disperse titanium dioxide, various iron oxides, and calcium carbonate in Thiokol(TM)-based caulking compounds, made to color match various adobe and stucco hues on Southwestern building exteriors.

Mixed pigments are predispersed with this combination, then polymer is introduced and the viscious blend packaged in appropriate caulking tubes. Because the various iron oxides (yellow, red, and brown) have similar surface chemistry, predominantly Fe–OH groups, a master dispersed color set may be formulated and custom color blending performed for architectural applications or matching color-aged surfaces.

Oleophilic dispersancy can be better understood by comparing the simplified

Table 7.8 Caulking Dispersant for Mixed Pigments

Component	Weight %
Butyl acrylate[a]	17–18
Free acrylic acid	1–2
Thioglycolic acid	0.8–1.0
Methyl ethyl ketone	80
	100

[a] 50% esterified

molecular association in Figure 7.10 with that of Figure 4.4. For effective coating the number of free acidic groups on the molecule is small, usually less than 10, with a corresponding molecular weight in the 1000–2000 range. Note that alternate surface association again occurs because of the expandability of the acrylate chain. With slight heating, a chemical reaction will occur between the unesterified acrylic acid groups and the ferric hydroxyls on the surface. At this point the dispersant is permanently anchored on the pigment surface and has great temperature and weathering resistance when incorporated into the resin system.

Increasing the alcohol chain length for esterification from butyl to lauryl or stearyl has little effect in polyester resins because of their slight polarity. For applications in alkyd resins, the effect is greater. However, for alkyd systems, simple oleic acid treatment of the iron oxide serves as well, if not better.

Where the dispersed pigments are to be formulated in a vinyl or nitrile polymer caulk, thioglycolic acid may be substituted by azo bis cyanovaleric acid.

Iron oxides made for polyester pigmentation can also be reacted in situ with dispersant by heating the mixture described in Table 7.9 in high-boiling solvent for about an hour. This reaction produces the polyester condensate on the ferric oxide surface, much like the partially neutralized butyl acrylate in Figure 7.10.

A similar condensation pretreatment consisting of two coupled benzene rings has been employed to render copper phthalocyanine highly hydrophobic.[36] This permits its incorporation in ethylene polymers for decorative plastics and kitchen ware, as well as in formulations where toluene and xylene are primary solvents.

Where titanium dioxide pigmented alkyd resin paint is to be employed directly over fresh steel, the adhesion can be improved by employing an oleic

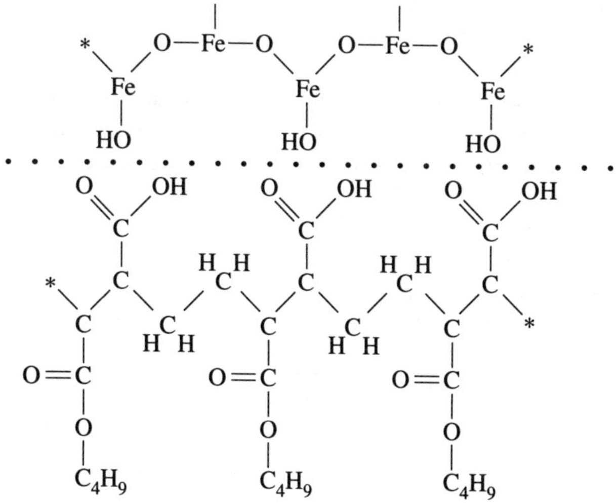

Figure 7.10. Oleophilic application of esterified acrylate dispersant.

Table 7.9 Precolor Mix for Polyester Layup Resin

Component	Weight %
Fe_2O_3 (submicrometer)	75
Glycerol (anhydrous)	10
Phthalic anhydride	15
	100

or linoleic acid treated, alumina gel coated rutile and adding a third to a half surplus acid to the formulation. On clean iron, stainless steel, and aluminum surfaces, free fatty acid coordinates with the metal surface in a tail configuration. These molecules intermesh with those on the titania and of the alkyd resin, giving a strong, weather-resistant bond.

Esterification has also been employed between tertiary alkanolamines and fatty acids of intermediate chain length in an approximate 1:3 ratio. The nitrilo triethyl hexanoate ester adsorbs strongly on titantium dioxide in hexane solvent[37,38] to give an effective dispersion in mineral spirits systems and alkyd resin paints. For its high molecular weight, only about 0.4% is required for standard commercial rutile. Alumino-gel treatments on rutile appear to disperse the pigment better than silica gel ones.

Even citrate esters have been employed to produce low-molecular-weight dispersants. Carbon black for incorporation into melamine polymers used in dinnerware has been dispersed with an ester of citric acid and butanol reacted in propylene glycol.[39] Depending upon the oxidized status of the carbon black, small additions of triphenoxy phosphate may be employed to assist in fine particle breakdown during the mill blending operation.

Glycerol esters, especially with oleic acid where the unsaturated (double bond) can react with unsaturated oleophilic polymers, are effective in dispersing alumina in organic binders. The finished coating has been designed for surfaces to be ceramically bonded (sintered) at high temperatures.[40] Alumina in this application is not especially fine, but has a broad size distribution to produce dense packing during the thermal conversion stage. As a consequence, dispersant loading is small, of the order of 0.5%.

Table 7.10 In Situ Dispersant Formation and Ball Mill Dispersion of TiO_2

Component	Weight %
Rutile (coated)	40
Alkyd resin[a]	18
Butanol	21
Xylene[b]	21
	100

[a] Soya bean modified with free linoleic acid.
[b] May be adjusted to maintain working viscosity in mill.

In a similar simple ester operation, linoleic acid and butanol are esterified in the presence of titanium dioxide (alumino-gel coated). An unusual approach is employed by utilizing the energy and localized heating of a ball or small media mill with alumina media to induce ester formation and surface coating of the pigment.[41,42] The composition is shown in Table 7.10.

Rubber pigments which require exceptionally fine particle size and polymer–pigment bonding for film and body integrity (white sidewall tires, decorative seals, condoms) may employ acrylate-aminoalcohol esterified dispersants.[43] The amino group attaches to the acidic pigment surface (organic and inorganic) while the alcohol esterifies with the acrylate. Double-bond integrity must be maintained, however, until the dispersed pigment is incorporated into the latex. This ensures chemical interaction with the unsaturation in the polymer. In these applications, the pigment is first premixed with both components of the dispersant to be esterified. Next the system is warmed sufficiently to form an ester association (*R*COOH–HO*R* union). Finally, it is transferred to the latex for final blending and milling.

Where film structures of such rubber formulations are made, these predispersed pigments commonly develop a flow pattern (marbled or pseudo-Moiré) because polymerization occurs prior to the attainment of full homogeneity. This defect often is marketed as a decorative effect in many applications.[44] Particulate matter deriving its primary property from crystalline structure or aggregate packing order often suffers property degradation when dispersed in polymer films, either through production stress (shearing, extrusion, elongation, etc.) or application stress (flexure, film stretching, inflation, etc.). Unfortunately, the better the particulate dispersion, the more probable or more extensive becomes such degradation.

Zinc sulfide for phosphorescence applications (glow-in-the-dark products) disperses readily in butadiene polymers because of the oleophilic character of sulfide surfaces. It is, however, unusually sensitive to milling or deformational stress. Particle stress from latex stretching destroys the glow effect.[45] Similarly, rubber magnets in which barium ferrite is dispersed in olefin polymers using alkyl-amines upon distortion lose a portion of their magnetic character, partly from internal particle stress and partly from particle disorientation. The better these materials' dispersion, the more detrimental the polymer stress effect.

Highly fluorinated alkyl polymers with a few hydroxyl groups or terminal polyethers are employed to disperse slightly polar surfaces in fully oleophilic systems. The basic carbon chain ranges from C-8 to C-12 and has at a minimum 90% of the hydrocarbon bonds replaced by fluorine. This produces a Teflon™-like product that is extremely hydrophobic, having aqueous surface tension values of 10–20 dyn/cm, compared to 30–50 dyn/cm for sodium lauryl sulfonate and similar anionic agents.[46] The products have very high thermal stability and are useful in dispersing moderately oleophilic solids in high-temperature liquid systems. As shown in Figure 12.12(b), they have even been employed to effect coal–oil slurries for boiler fuels.

These fluorocarbon agents also assist in lowering surface tension to minimize foaming and to increase fluidity in slurry pumping operations. Per molecular length, they are exceptionally heavy because of the substitution of fluorine for hydrogen across the chain (over 50% greater than polyethylene glycols) and expensive. They are manufactured by the 3M Corporation under the marketing code FC-*XXX*, where *X* designates various company product numbers.

References

1. Vaithyanathan, M., *Paint India* **39**, 21 (1989).
2. Zettlemoyer, A., ''Immersion Wetting of Solid Surfaces,'' from *Chemistry and Physics of Interfaces*, Amer. Chem. Soc., Washington, DC, p. 139 (1965).
3. National Bureau of Standards., Circ. #537, Washington, DC (1957).
4. Boyd, G., and Harkins, W., *J. Am. Chem. Soc.* **64**, 1190 (1942).
5. Pakhovchisin, S., Ovcharenko, F., and Mank, V., *Dokl. Akad. Nauk. SSSR* **2245**, 140 (1979).
6. Fowkes, F., ''Attractive Forces at Interfaces,'' from *Chemistry And Physics of Interfaces*, op. cit., p. 1.
7. Jertic, L., *Hem. Industr.* **35**, 79 (1981).
8. Commercial Solvents Corp., private communication (1973).
9. Azrac, R., and Angell, C., *J. Phys. Chem.* **77**, 3048 (1973).
10. Hertl, W., and Cuenca, A., *J. Chem. Phys.* **77**, 1120 (1973).
11. Nelson, W., German Patent #701,032, Dec. 5 (1940).
12. Linden, H., et al., German Patent #2,824,416, Dec. 13 (1979).
13. Cowley, C., and Schofield, J., *Polymer Paint Colour J.* **30**, 94 (1988).
14. Hauser, E., U.S. Patent #2,531,427, Nov. 28 (1950).
15. Jorden, J., U.S. Patent #2,531,440, Nov. 28 (1950).
16. *NL Rheology Handbook*, NL Industries, Hightstown, NJ (1982).
17. Von Rubinski, W., and Schieferstein, L., *Farbe Lack.* **97**, 18 (1991).
18. Conley, R., ''Processing and Property Enhancement Utilizing Modifiers and Additives in Polymers,'' Soc. Plastic Eng. Intl. Conf., Newark, NJ, p. 313 (1985).
19. De Boer, J., et al., *J. Catalysis* **1**, 1 (1962).
20. Hauser, E., U.S. Patent #2,951,087, Aug. 30 (1960).
21. Terlikovski, E., *Ukr. Khim. Zh.* **47**, 1053 (1981).
22. Kipling, J., and Wright, E., *J. Chem. Soc.* **66**, 855 (1962).
23. Taucha, E., Iwai, S., and Hayashi, H., Japan Patent #1,228,537 Sept. 12 (1989).
24. Bayer, A., German Patent #1,792,798, Aug. 11 (1977).
25. Cannon, W., U.S. Govt. Rept. Annual (R&D), Vol. 88, Paper #856,056 (1988).
26. Furusaka, K., et al., Japan Patent #16,031, Feb. 4 (1980).
27. Kovshulya, L., et al., SPSTL (SSSR), Doc. #873 (1980).
28. Aizawa, M., Nosaka, Y., and Miyama, H., *J. Colloid Interface Sci.* **139**, 324 (1990).
29. Bruins, P., U.S. Patent #4,098,758, April 12 (1974).

30. Kawasaki, K., Watanabe, H., and Seto, J., German Patent #3,038,646, April 23 (1981).
31. Koehler, K., et al., Europatent #21,262, Jan. 7 (1981).
32. Kodera, M., Japan Patent #63,256,666, April 14 (1988).
33. Koehler, K., et al., German Patent #2,946,549, May 27 (1981).
34. Tada, H., et al., Japan Patent #2,218,723, Aug. 23 (1990).
35. Braun, R., et al., German Patent #3,628,320, Feb. 25 (1988).
36. Nakamura, M., et al., Europatent #256,456 (1988).
37. Linden, H., German Patent #2,728,237, Jan. 18 (1979).
38. Linden, H., and Bornmann, H., U.S. Patent #4,165,239, Aug. 21 (1979).
39. Elmore, J., Zylla, E., and Roy, G., Europatent #361,789, April 4, (1990).
40. Tsuruta, A., and Murabayashi, M., *Yokahama KDKKKSK* **15**, 49 (1988).
41. Kansai Paint Co., Japan Patent #95,957, Aug. 3 (1981).
42. Golorko, K., et al., *Lakokras Mater. Ikh. Priment.* **3**, 56 (1982).
43. Conley, R., ''Pigment Dispersion,'' Paper #17, World Rubber Conf., ACS Rubber Div., Houston, TX, October (1983).
44. Jay, P., and Loria, D., French Patent #2,594,443, Aug. 21 (1987).
45. Glenn, J., Pioneer Balloon Co., private communication (1995).
46. Kawaguchi, K., and Ando, T., Japan Patent #62,177,070, Aug. 3 (1987).

CHAPTER

8

Mechanical Assistance in Dispersion

Dispersion, irrespective of the chemical or electrostatic alteration of solid surfaces, requires effective mechanical means to establish and maintain a state of particle separation. This must be of sufficient duration to allow agents in solution to diffuse through the liquid phase and adsorb onto fresh solid surfaces. The mechanical energy necessary to bring about complete particle separation is directly dependent upon interparticle agglomeration energy and the system viscosity into which these particles are to be suspended.

Mills capable of effecting dispersion of solids in liquid phases take a variety of forms depending primarily upon energy requirements. These fall broadly into six classes:

1. Impeller mills (agitators and blendors);
2. Ball mills (including pebble mills);
3. Stirred media mills (sand mills);
4. Vibratory mills;
5. Multiple roll mills; and
6. Ultrasonic dispersors;

Figure 8.1 is a general energy plot classification of mills[1] employed for both dispersion and comminution and their typical applications. No definitive boundaries exist between these classes, and for many applications more than one mill type may perform the necessary task. Many mills, however, have specific mechanical biases with regard to the dispersant action that may render them either preferable or impractical. Thus no firm restrictions exist for any of the mechanical devices when it comes to general utilization. Whatever the equip-

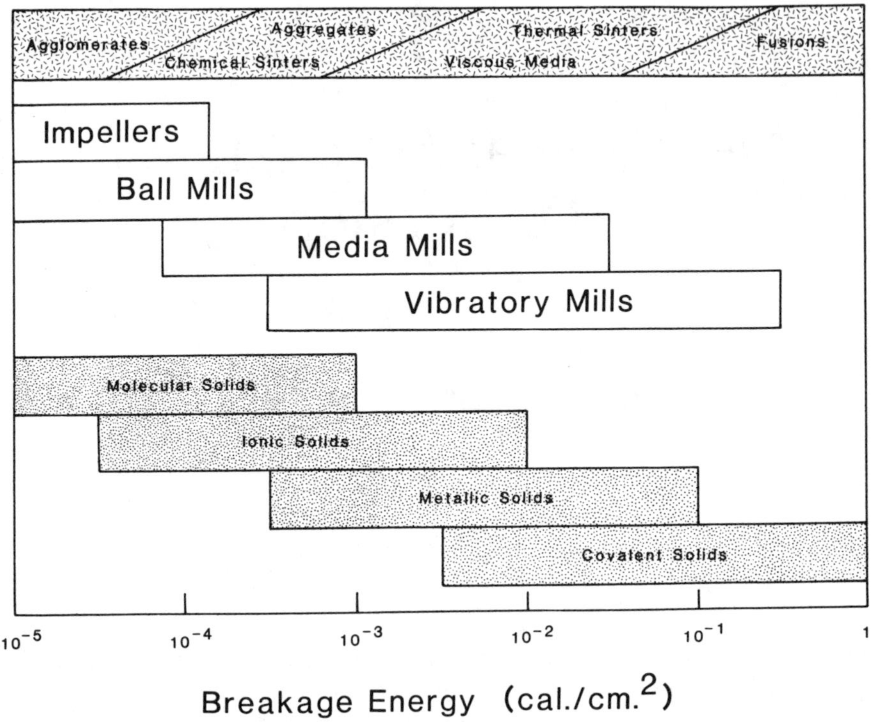

Figure 8.1. Energy diagram for dispersion mills.

ment, dispersants increase the energy yield factor in the dispersion process with these mills, whether the process energetics are as severe as comminution of hard, refractory silica or as mild as the deagglomeration of soft organic dyes.

Three primary variables dictate the equipment type employed for dispersion:

1. Viscosity of both feed and product slurries;
2. Particle size and shape of both feed and product;
3. Feed agglomerate strength.

The second variable becomes important where anisometry is a significant particle parameter in system application, as with thin platelets or long needles. Certain dispersing mills may break these into less useful fragments.

8.1 Impeller-Type Dispersors and Their Efficiency

The rate of papers published on modeling mechanics of slurry flow and mixing dynamics in containers and tanks with impeller dispersors grows steadily. Many of these are doctoral dissertations on fluid dynamics involving mathematics only justifiable and attainable in an age of desk top, high-speed computers! Frequently they contain derivations not likely to be known or useable down on the hectic floors of industrial production. In wading through these, one, perhaps two, will be found that contain insightful and useful information on improving dispersion performance or equipment efficiency.

Impeller mixers, like grinding equipment, operate at monstrous inefficiencies, optimistically no better than 0.1%. However, the energy requirements to obtain dispersion in most instances is sufficiently small that a 100% improvement in efficiency (to 0.2%) would have an almost immeasurable influence on commercial enterprises. Most equipment designers look to improve their equipment not on the basis of complex flow calculus but by minimizing foam and air inclusion, making material loading, unloading, and cleanup less laborious and reducing the cost of their equipment through better manufacturing.

A typical impeller dispersor is a mixer having a propeller-type blade inserted beneath the slurry surface, which operates at an rpm just below the point where the vortex opens the blade to air entrainment. By geometrical forming of this blade's shape and pitch, slurry turnover in the tank can be controlled with considerable precision.

All single-blade agitators create vortices because their rotational speed forces slurry material to the wall. This force is never directly normal to the shaft—the container bottom represents a rigid, unyielding resistance while the slurry surface above the blade is resistance free (except for gravity and surface tension). Slurry climbs the container wall, which, in turn, produces an air vortex around the shaft. As most mixing/dispersion takes place in the vicinity of the blade (along its surface and behind, not at the front edge), vortex formation reduces milling efficiency, as measured by horsepower hours (KWH) consumed to attain a given state. Thus most dispersion blades have shapes modified to minimize this inefficiency.

Some vortex formation is required, however. The blade, while contributing deagglomeration shear, must operate to counteract gravity and sedimenting particles in the fluid. This action in itself is a major contributor to vortex formation. The change in efficiency, as measured by power requirements, can be predicted based on several advanced computer-developed equations.[2] The usual method of reaching this evaluation, however, is that of the pilot test; that is, evaluate in a test tank and scale up by a factor of about 10 or less.

The centrifugal force vector of rotating blades always produces material axial gradients during turbulent agitation. These increase with material density and particle size. The magnitude of size discrimination can be predicted by the Kolmogorov theory,[3] which assesses the isotropic (localized) turbulence and particulate settling tendencies based on tank geometries and particle parameters.

The Cowles-type impeller, shown in Figure 8.2, is designed to minimize vortex formation. It resembles a table saw blade with teeth bent alternately up and down. Because the blade has little net axial thrust, it tends less to throw material to the wall and, in turn, minimizes deep vortices. Each of the teeth has a positive radial pitch of 20 degrees to move material out and away from the

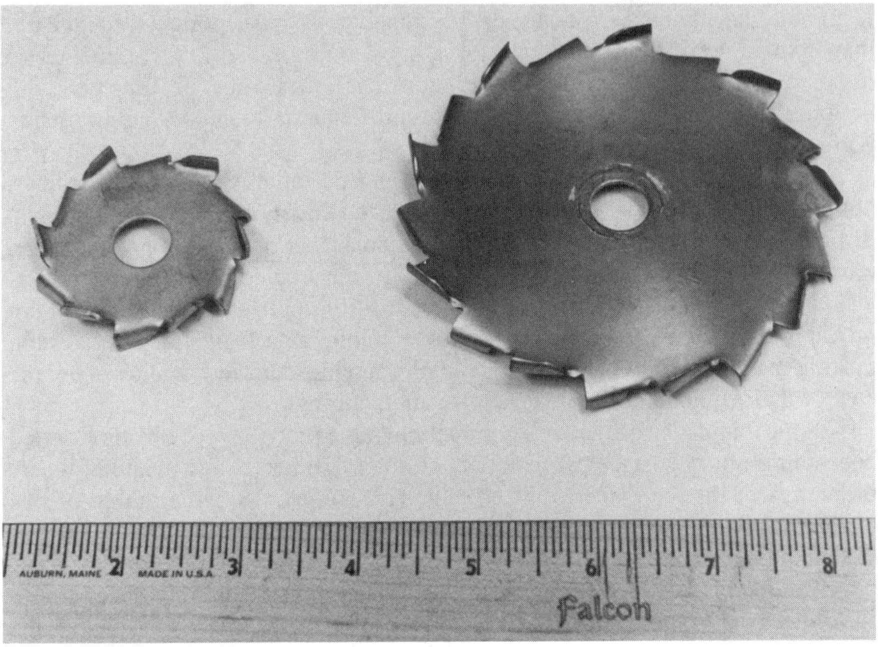

Figure 8.2. Cowles-type impeller dispersing blade.

blade after shearing takes place. This pitched tooth arrangement ensures the blade can be operated at very high speeds to induce high shear.

Some operations employ two or more Cowles blades in a stacked array along an agitator shaft in high depth-to-diameter ratio tanks. This permits higher rotational velocities without requiring large-diameter tanks. Such tanks are not without their disadvantages, however. The greater the tank depth, the greater the probability of vertical size discrimination, especially for dense particles or those with diameters above 20 μm.

Other agitator systems employ multiple blades operating on parallel shafts with blade tips either intermeshing in gear fashion or just clearing each other. Obviously, considerable amounts of shear occur within these narrow-pass regions. Typical mixers of this type are shown in Figure 8.3, with the loop style generally operating at lower speeds and the blade style at higher. For equivalent horsepower input, middle-blade dispersors commonly produce more shear and dispersancy than single-blade units. Unfortunately, they produce more internal heat, which necessitates increased cooling requirements.

Studies made with multiple-impeller mixers suggest that high-shear planes in the narrow-pass regions, which are very effective at breaking apart moderately strong agglomerates, break these at the expense of a more quiescent volume outside the impeller region.[4] Using an ideal tank height/turbine diameter ratio

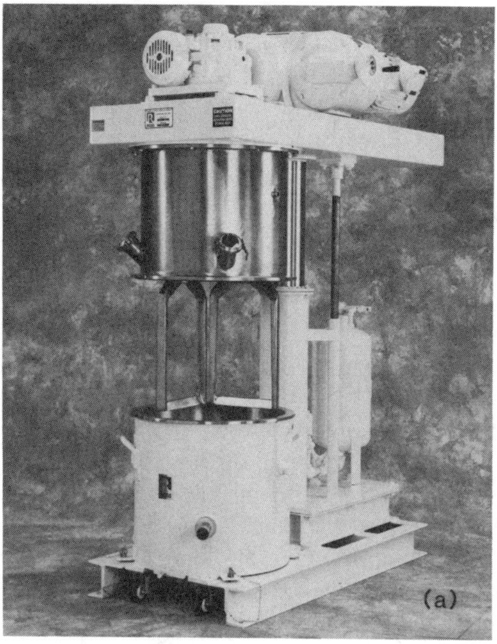

Figure 8.3. Intermeshing blade dispersor: (a) loop type.

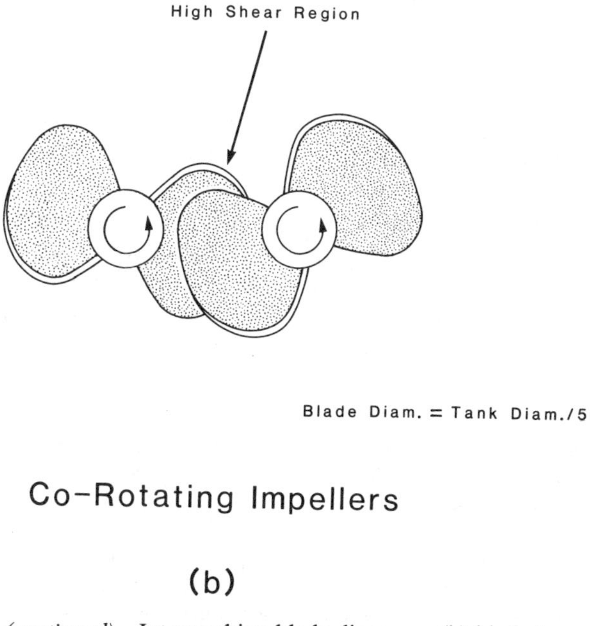

Figure 8.3. *(continued)* Intermeshing blade dispersor: (b) blade type.

of 4, a tank diameter/turbine ratio of 3, four standard baffles and a four-head Rushton-type turbine impeller, a slurry solids was produced with a density variation from top to bottom of 20% (relative) and a size discrimination of nearly 50% at +30 μm and 5% at −5 μm. These discriminatory actions decrease as slurry solids increase. However, the probability in any given time that agglomerates in high-solids slurries will diffuse or be driven into the high-shear region decreases.

As slurry solids increase, so must the tip velocity of the impeller blade. The usual rule of thumb is to increase RPM until the vortex just begins to expose the agitator head (or blade). No simple mathematical expression relates RPM to slurry solids, although some evidence exists that horsepower input increases approximately linearly with rotational speed squared,[5] in keeping with the kinetic energy equation. Other variables influencing agitator speed include particle diameter, fundamental liquid viscosity, and solids concentration. Even the "eyeball" technique of vortex exposure is not always valid because shear efficiency diminishes in an unpredictable manner as vortex volume increases.

With blade-type dispersors shear and agglomerate breakup do not occur to any great extent in the anterior region of the blade, as would be expected from an impact conceptualization of the disaggregation and dispersion process. Rather, they occur a short distance behind the blade. For hydrodynamically designed blades (i.e., typical impeller types), the inner approximately 30% of the blade radius (shaft region) remains virtually inactive during rotation, and the outer 10–15% (wall region) does little to assist the dispersion process.

Pitched-blade impellers are more efficient at breaking apart agglomerates than conventional disk-type turbines simply because they force material into the shear region faster. Because tank bottoms are fixed and rigid while tank tops consists of mobile liquid, pitched-turbine downflow is more efficient in moving particles into the shear region than is pitched turbine upflow. This becomes more apparent as the impeller velocity approaches the critical point, that velocity at which liquid is forced disproportionately away from the agitator blade, allowing the blade to rotate essentially free, that is, in air.

For efficient energy utilization with agglomerate dispersion, an agitator blade should operate in a laminar flow mode, that is, where particles and liquid possess a streamline character around the blade surface. Particles will have a Reynolds number* about 2 or less. Laminar flow occurs during the agitation range where:

$$\text{Agitator rpm} < \frac{40{,}000\eta}{D_{\text{tank}} D_{\text{gap}}\, \rho} \tag{8.1}$$

where η is the slurry viscosity in poise, D_{tank} the tank diameter in feet, D_{gap} the blade–wall gap in feet, and ρ the slurry density in lbs/ft^3.

* Reynolds number is a dimensionless parameter, the product of flow element length, velocity, and fluid density divided by fluid viscosity. The number noted is the maximum value before the system enters the transition zone between streamline and turbulent flow conditions. This transition is not sharp. However, once turbulence commences, power requirements rise dramatically and blade wear can become severe.

In this configuration power density, that is, the amount of power induced per unit slurry volume (sometimes termed specific power), can be computed from:

$$\text{Dispersing power density} = \frac{D_{\text{tank}}^2 \ \text{rpm}^2 \ \eta}{D_{\text{gap}}^2} \tag{8.2}$$

The best energy utilization of agitator tank and system, and the most efficient dispersing geometry, occurs with the arrangement shown in Figure 8.4. While these dimensions are approximations, they are the result of extensive test work in the field of materials processing.

Optimum agitation conditions have been studied[6] for a coarse, dense, experimental, solid system—silica glass sand dispersed in water. Particle size ranged from 2000 down to 100 μm (10 to 140 mesh) with solids at 50% using 0.2% sodium silicate as dispersant. Impeller speed in the study varied from 200 to 800 rpm with a tank approximately 30 cm in diameter. Optimum operating conditions

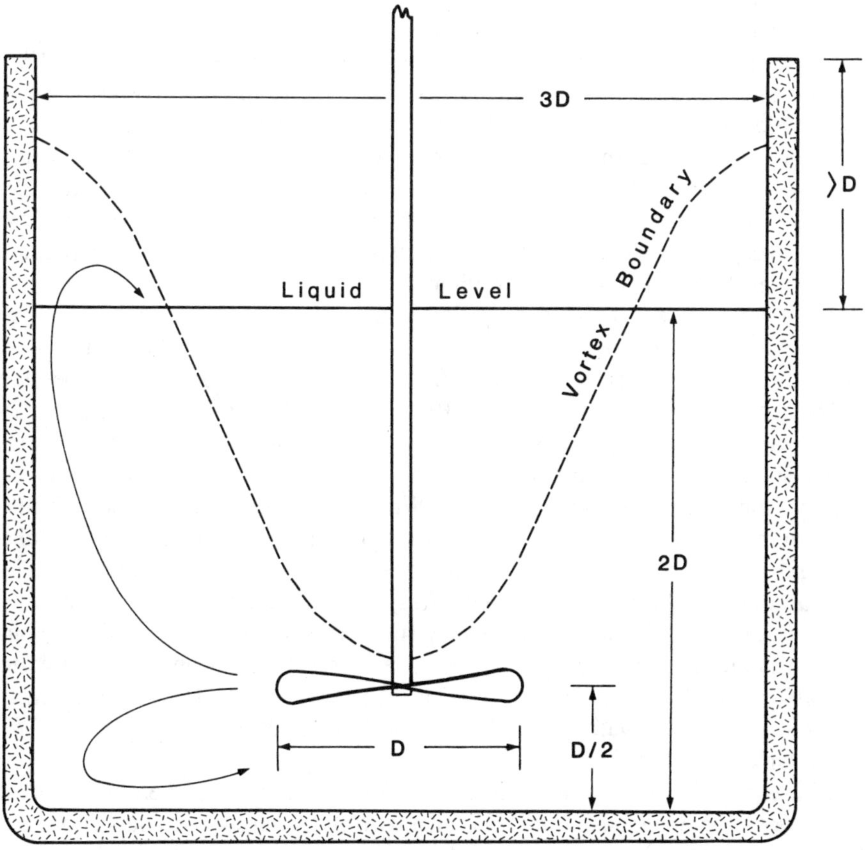

Figure 8.4. Efficient tank dispersion geometry.

were obtained with an impeller diameter/tank diameter ratio 0.35 to 0.40, impeller blade width/impeller blade diameter ratio of 0.30 to 0.35, and an impeller clearance (from bottom)/tank diameter ratio of 0.30 to 0.35. Even at optimum conditions particles with this size and density show both an axial regional differentiation (from centrifugal force) and a vertical differentiation (from gravity sedimentation).

As sketched in Figure 8.4, dispersor blades produce two toroidal patterns of agitation, one upward forming the air vortex and the other unseen under and concentric to the blade. Both redirect particles back to the blade region. First observations might suggest lowering the blade to prevent the lower toroid from developing. This invariably causes a reduction in dispersive activity and a lengthening of time to reach a set size breakdown. Conversely, raising the blade results in static activity at the tank bottom with large particles settling out and not returning to the blade region. Tanks with orthogonal basal corners, whether round or square, will produce stagnant solid in this corner around the tank wall. By rounding the tank in construction or installing a gusset in this region (conical for round tanks, pyramidal for square), large-particle isolation can be effectively eliminated.

The complete and homogeneous suspension of solid particles of a slurry in cylindrical, flat-bottom tanks requires extensive eddy currents. These need not be massive (i.e., visual), but actually may be microscopic. An investigation of such tanks employed to disperse fine particles indicates that the eddy currents required to prevent corner stagnation may be little larger than the coarsest particles within the slurry.[7] These eddies produce unusual effects. For example, a bimodal distribution (as with a two-component mixture) tends to behave in a fashion very similar to a monomodal one with an average diameter (D_{50}) equal to the average between the two components.

Most blade-type agitators are inefficient at disaggregating thermal and chemical sinters and high-energy aggregates. For these materials either a mill with higher energy should be employed or the shear blade agitator must be operated in the turbulent mode, that is, with Reynolds numbers of 100 or greater. In turbulent mixing of marginally dispersed systems (just sufficient agent to maintain suspension), flocculation occurs because impact energy exceeds repulsion energy from charge or dipole barriers. Where the dispersant is in equilibrium between surface and solution, as with polyphosphates, the kinetics of flocculation and stability of floccules formed are directly related to this equilibrium. The volume of these floccules (size and mass) will be determined by the mixing intensity, the solids content of the slurry (interparticle proximity), and the equilibrium noted. Floccules appear as increased mass particles and are thrown to the tank periphery where free dispersant diffusion is slow. Thus such artifacts may remain near the wall and produce an element of poorly dispersed slurry late in the tank discharge.

Dense particles, having a specific gravity of 5 and above, often are difficult to disperse homogeneously under laminar flow in the larger particle size ranges (20 μm and above) because the gravitational component is an appreciable fraction of the radial agitator force. Again the trajectories of the more massive particles tend to create a size discrimination between the wall and the slurry core. Although a detailed mathematical analysis of this has been made,[8] the common method of dispersion still relies on employment of turbulent flow to minimize these centrifugal and gravitational effects.

Material accumulating near the dispersor shell frequently consists of larger aggregates and material more tightly bound than average for the feedstock. These aggregates tenaciously resist breakdown unless confined to the shear region of the mill.

In studies on coarse materials under dispersion–blending conditions, major concentration profiles exist axially, the result of differential centrifugal force by particles of different mass. Minor concentration gradients exist vertically, the result of gravitational forces. Where the axially integrated velocity from an impeller blade is an order of magnitude higher than the sedimentation velocity of the coarsest particle, this effect is minimized.[9] Internal turnover from stacked impeller blade spacing of approximately one tank radius maintains sufficient turbulence to minimize the axial variations.

Tank shape affects dispersion efficiency by creating or interfering with slurry flow. In a cylindrical vessel, slurry movement near the wall is not only energy wasteful; those particulates in suspension avoid additional contact with shearing action in the blade region. Because of centrifugal classifying action, larger particles, that is, those with the greatest number of aggregates, populate this region of the container. To prvent this circular bulk inert movement, often called *synchronous rotation*, the container can be baffled slightly. Wall baffles are not massive—no more than 5% of the diameter—just large enough to redirect the slurry back to the blade for sequential deagglomeration and dispersion. This is the basis for the famed Waring Blendor, patented in the 1940s by the famous Pennsylvania bandleader (and fresh fruit juice lover), Fred Waring.

However, baffle size can be overdone. Taken to the extreme a square container might be expected to provide excellent agitation. However, just the opposite occurs, as shown in Figure 8.5. Corners serve to form stagnant cyclonic activity and isolate slurry from blade contact and dispersive breakdown.

A mineral processing plant the author visited in the 1980s had constructed such a square, concrete, in-ground dispersor to produce slurry for plant operations from a natural clay resource whose particles were 90%-325 mesh (44 μm). In the oversize fraction 5–10% were highly abrasive quartz. The dispersing cell was approximately 6 feet square by 8 feet deep, driven by a modified 3-foot-diameter bar arm blade and a 400 HP motor. Material loss was severe as a consequence of eddy currents forcing coarser particulates and aggregates away from the dispersive region. These settled to the bottom and were conveyed away to impound. When baffles were installed at midside and gussets inserted at the corners, production increased by over 20% and impounding volume diminished by the same quantity.

Baffle size must increase as slurry viscosity decreases because fluidity prevents slurry from returning to the shear region. Conventional slurry viscosities

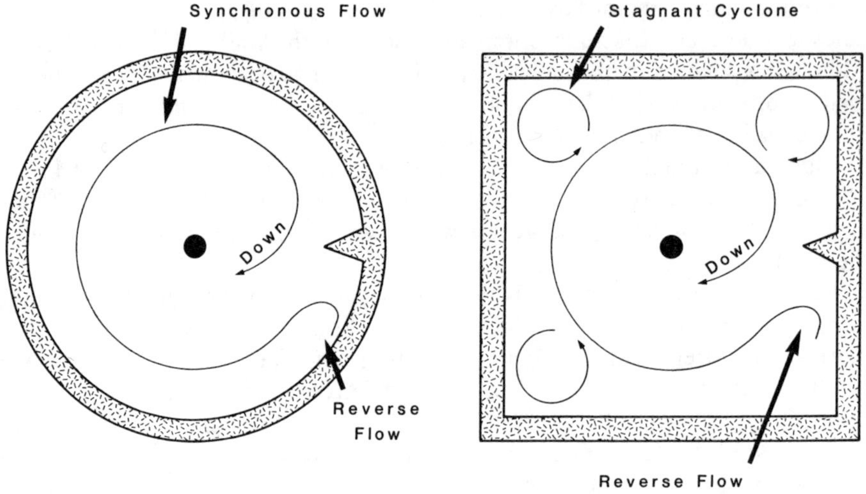

Figure 8.5. Slurry movements in round and square baffled vessels.

for chemical and physical processing usually are 2000 cp or less, and operate well with a 2–3% radius baffle. Figure 8.6 gives the effective baffle width as a function of tank diameter for three orders of viscosity variation.[10,11] High-viscosity systems are sufficiently internally turbulent from molecular kinetic baffling that baffle width can be reduced.

Baffle number is dictated in part by viscosity and in part by tank diameter. The general rule calls for baffle separation being no less than tank radius (limiting the number to six) for viscous systems and no greater than four for lower-viscosity systems. In some constructions, especially with large agitated holding tanks, vertical wall supports are designed to serve as baffles and may exceed six in number, but rarely more than eight.

Baffles introduce a static drag force on slurry movement that must be compensated by increased input power. Figure 8.6 also contains values for impeller power increase due to baffle number and size. Beyond six baffles and a $D/10$ size, power does not increase appreciably. Larger-size baffles are more susceptible to wear due to centrifugal effects where large (>50 μm) particles are in suspension.

A variation on the multiple-impeller type of dispersor occurs with the narrow-clearance, higher-shear, armature-in-shell shearing mill. This type of equipment is manufactured by several companies and includes the Kinetic Dispersion (Kady) mill and the Magni-Mixer of Premier Mill Co.[12] Here the rotor is serrated or fluted and rotates at high velocity within the shell of a fenestrated (multislitted) stator. Slurry is drawn into the stator slits, sheared by multiple bouncing contacts between rotor and stator, then pumped out as fresh slurry is drawn in.

This type of dispersor has relatively close tolerances. It functions especially well with creamy-textured materials such as cosmetics and food products, where

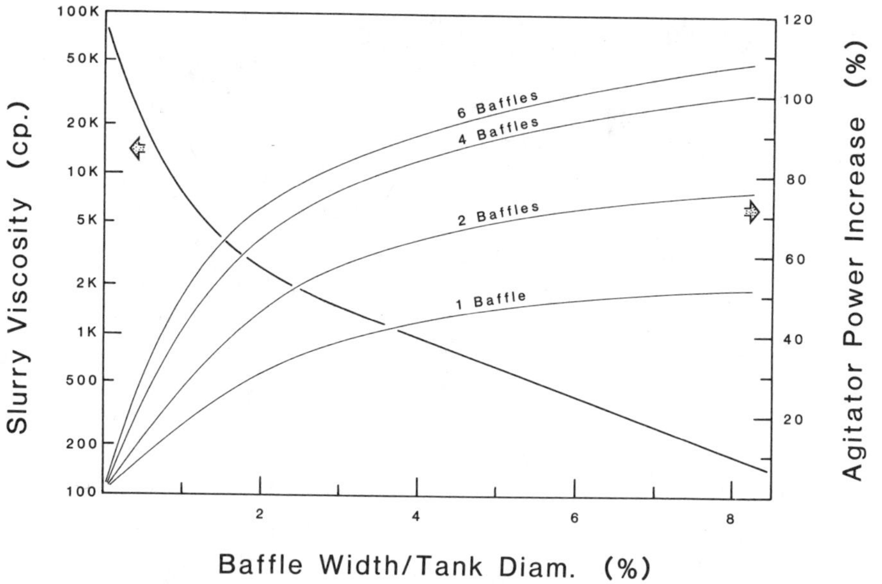

Figure 8.6. Baffle and tank diameter relationships.

product thixotropy allows the slurry material within the rotating head to display low viscosity but reform into a nonseparating cream or paste immediately upon exiting.

Armature-in-shell mills also have been used successfully to disperse very fine, laminar, and acicular magnetic particles ($<0.1\ \mu m$) in oil for magnetodynamic coupling systems,[13] for example, fluid brakes, clutches, etc. A scanning electron micrograph of such a system dispersed with a Kady mill and an appropriate dispersant (see Chapter 7) at 50% solids and aligned under applied magnetic field is shown in Figure 8.7. The mill is less successful where strongly bound or very coarse aggregates are present in the feed slurry.

Head construction for these mills is shown in the closeup photograph of Figure 8.8. A number of manufacturers produce mills with slight design variations on this type of agitator head.

Other agitator geometries include paddle wheels, turbine blades, intertwining screw impellers, and slotted discs. Manufacturers of these agitators publish general application guides for specific impeller design and most applicable rheology range for attaining good dispersion with a broad variety of powdered systems.

Some makedown systems employ two-stage dispersion where a minor percentage of "tight," or high-binding-energy, agglomerates occur. In this arrangement, the primary tank bottom may be conical with a slow moving screw or rake to force dropout agglomerates toward an exit port. Alternately a screen in the outlet line can remove coarse particles flushed off as a reject stream.

Reject material is moved via screw to a higher-energy mill, a pug mill, kneader, roll mill, etc., to break these components down. Disaggregated product

Figure 8.7. High-shear mill-dispersed magnetic particles.

is then returned to the primary dispersor along with fresh feed. This technique is especially useful with high intrinsic value materials, titanium dioxide pigments, specialty organic colors, high-tech ceramics, and electronic materials (conductive powders, metal alloys, cermets, etc.).

In two-stage milling the high-energy or secondary dispersing phase most often requires an additional increment of dispersant because increased quantities of fresh surface are produced.

8.2 Ball Mills as Dispersors

The ball mill has a long and successful history of producing dispersed slurries. Because energy can be closely regulated by varying mill diameter, rotational speed, ball size, and ball density, this device has a fairly broad range of shear

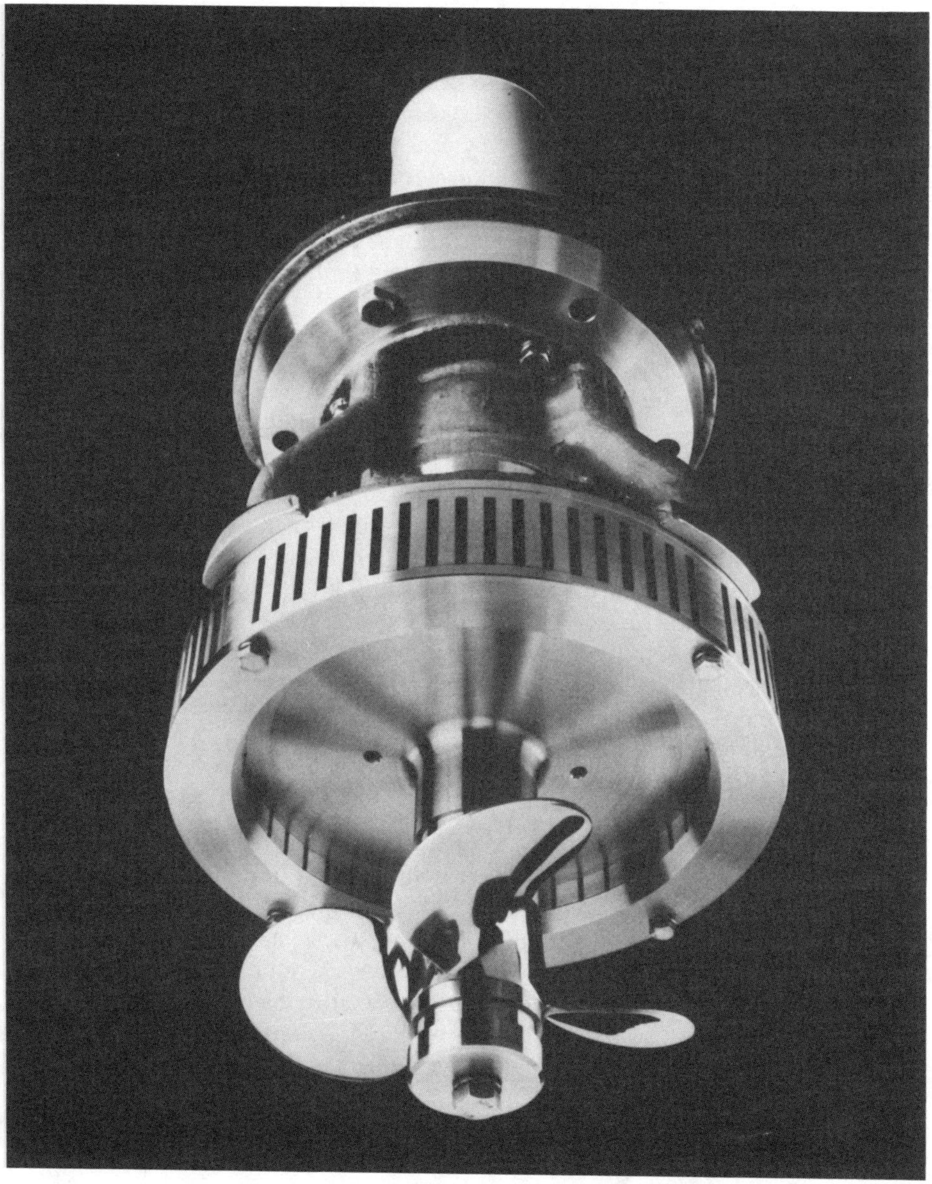

Figure 8.8. Head construction of a high-shear dispersor (courtesy Kady).

capabilities. Its shortcomings lie in the operation time to accomplish a dispersive grind, noise, and physical size. Ball mills are reasonably energy efficient as dispersors for high-viscosity systems. They are yet written into certain military specifications for such specialty paint products as camouflage and marine coatings.

A ball mill is about 25–35% volume effective for slurry processing. Thus a 50 gallon production run would require a mill of approximately 3 feet diameter and 4 feet length. The time required to mill 0.7 μm iron oxide in a solvent-thinned polyester resin having a viscosity of 10,000 cp with benzoic acid as dispersant will be minimally 8 hours. Typical power consumption with 1 inch (25 mm) steel balls will be about 16 kilowatt hours, a comparatively low operational expense.

The ball mill's chief energy inefficiency lies in rotating a massive steel shell that performs very little work, even where lifters (baffles) are installed on the shell wall.

Ball mills for dispersing color-sensitive pigments (dyes, calcium carbonate, talc, kaolin, mica, zinc oxide, etc.) avoid employing steel balls or steel linings (Ni-hard, etc.). These are replaced by corundum (alumina) or mullite (ceramic). While a mill's lining has no effect on milling time, ball density reduction from 7.8 to 3.5 g/cm^3 essentially doubles it. Decreasing ball density also increases the limiting size attainable for ultrafine particulates in viscous systems; that is, the mill achieves poorer overall dispersion. Although energy consumption is reduced (by about at third) with the ball substitution, this is of secondary importance because energy costs to disperse with ball mills are a very small component in the formulation expense.

Steel balls may be employed for dispersive milling of iron oxide formulations, deep red and deep blue colors and carbon blacks in oil-based polymers because the metal (or metal oxide) color contamination passes unnoticed. Lighter colored pigments as whites, organic yellows, and chrome greens almost always dictate corundum or ceramic media.

Media size will be dictated both by mill diameter and slurry viscosity. The work effort by mill balls is confined to breaking apart aggregates and agglomerates of micrometer-sized particles in assemblages of 100 μm or less. For ceramic or corundum media with viscous slurries (i.e., 5000–10,000 cp), a maximum $1\frac{1}{2}$ inch (38 mm) top size balls is dictated. These may be substituted by $\frac{3}{4}$–1 inch (19–25 mm) steel where color is not a consideration. Even with much lower viscosities (i.e., 3000 cp), ceramic balls smaller than 1 inch are rarely used. The smallest practical steel balls arc $\frac{3}{8}$–$\frac{1}{2}$ (10–12 mm) inch.

Balls mills have advantages in loading and unloading. Batch ball mills, which are frequently employed in dispersive operations, are constructed with a shell port through which loading and discharging takes place gravitationally. Because large void spaces exist between balls within the mill, slurry drains quickly. Such mills often have a length-to-diameter ratio of 1.5 to 0.5. A typical mill employed for dispersing applications in shown in Figure 8.9.

Mills also may be constructed with a hollow trunnion (shaft-bearing system) into which slurry is pumped, traverses the mill length, and exits through a similar hollow trunnion on the mill's opposite end. This arrangement is termed a *continuous flow mill.* It usually has a large length-to-diameter ratio and functions better if baffled because of the risk of slurry bypass or short circuiting, especially across the pool surface within the mill.

Figure 8.9. Batch-dispersion ball mill (Courtesy Paul O. Abbe).

Continuous flow mills often have a classification stage immediately following (i.e., a vibrating screen or centrifuge), enabling oversize material to be removed and fed back in a loop to the feed port.

Ball mills have further advantages in their low rate of energy input and large shell area, which reduces greatly heat buildup from motor energy. This is an important concern in dispersing temperature-sensitive materials (organic dyes, pharmaceuticals, food products, electronically conductive, dielectric and magnetic materials).

For large batch sizes, ball mills are less practical. A 1000 gallon dispersive batch would require typically a mill 8 feet (2.4 m) in diameter by 12 feet (3.7 m) in length, the maximum usually found in slurry-oriented industries. Installations for these mills are often designed with loading capabilities from a floor above but discharge from the mill floor.

More commonly ball milling for dispersing difficult organic systems is restricted to 6 ft (2 m) diameter units (by 8 ft (2.4 m) long), primarily for the simplicity of loading and discharging slurry.

Low-viscosity slurries (i.e., those less than 2000 cp viscosity) are not recommended for ball mill dispersion. High fluidity provides low resistance to ball–ball and ball–wall impact, which results in considerable ball wear. Ceramic microfragments can appear in the products as abrasive contamination. Table 8.1 shows ball mill contamination compared to other forms of mill dispersion. Note that both pin and jet milling are performed dry, that is, without dispersants.

With more viscous slurries, liquid clings to the balls and remains on the

Table 8.1 Product Contamination from Milling Dispersion

Mill Type	Media	Feed	Feed Size (μm)	Product Size (μm)	Contamination (wt. %)
Ball	alumina	limestone	19000	−44	0.11
Ball	mullite	limestone	19000	−44	0.15
Ball	alumina	limestone	6300	−44	0.05
Ball	alumina	limestone	6300	−10	0.08
Jet	—	limestone	−50	−10	0.004
Pin	st. st.	limestone	6300	−44	0.07
Ball	alumina	carbon	−174	−10	0.05
Pin	st. st.	carbon	6300	−10	0.09
Jet	—	carbon	−174	−10	0.002

Note: Mill balls were 1.5 in. (38 mm) and mill diameter was 48 in. (1.3 m), st. st. = stainless steel

surface as the balls are transported up the mill wall and cascade. Thus ball–ball impact always works upon the slurry and maintains a constant mixing action in the process. For this reason mills may be filled to 40–60% of the mill volume with liquid in place of the approximately 25% value restricted for dry grinding. To increase this liquid transport by the balls, mills may be fitted with "lifters" or "flights" on their interior walls. These project from the wall approximately $\frac{1}{2}$ ball diameter. As liquid volume percentage in the mill increases, however, dispersing efficiency decreases slightly. Balls must emerge from the liquid phase during rotation and cascade back into it for good dispersive activity.

Both ball mills and small media mills, to be described later in this chapter, have limits on product size fineness because of surface imperfections and fracture depressions in the impacting media. This limitation can be detected in the product from examination of the very fine end of the particle size distribution with log-probability graph paper, as explained in Chapter 13, or by periodic hand lens examination of the grinding/dispersing media. Badly pitted balls from a −20 μm calcium carbonate dispersion operation are shown in Figure 8.10.

8.3 Small-Media Mills

Small-media mills, often called sand mills, bead mills, or stirred ball mills, were developed expressly for dispersing ultrafine pigments in medium-viscosity slurries. These mills are armature powered, are either vertically or horizontally oriented, and function independent of gravity. They may be operated in batch or continuous modes.

Unlike ball mills, small-media mills are not centrifugal force limited in rotational velocity. Although practical limitations exist in speed, they most often have an rpm at least a factor of 10–20 times that of ball mills. Media are fre-

Figure 8.10. Grind size limitation from pitted media.

quently 6 mm or smaller, of annealed glass or plastic, and always spherical in shape. For high-energy comminution, as well as difficult dispersion, alumina media may be employed. Small-media mills also are quiet.

A typical horizontal mill is shown in Figure 8.11(a) with the disc arrangement better defined in Figure 8.11(b). In the vertically designed mills, centrifugal force throws the media toward the mill wall, then up the wall, where it cascades back into the vortex created by rotor agitation. In horizontal mills the core region is devoid of media during rotation. Mill core volume with either design, comprising approximately the first 30% of the radius, produces little shear activity, either grinding or dispersion, because media are moving too slowly in that region. Several mill designs employ a large cylindrical agitator to improve upon motor efficiency. Annular volume adjacent to the wall, 5–10% of the radius, also is a region of little work because media are slowed from wall drag and synchronous flow.

Dispersion energy varies approximately as the square of rotational speed integrated over the work radius, but optimizes where media possess maximum speed while retaining eddy currents within the mill. As with agitator blades, dispersion

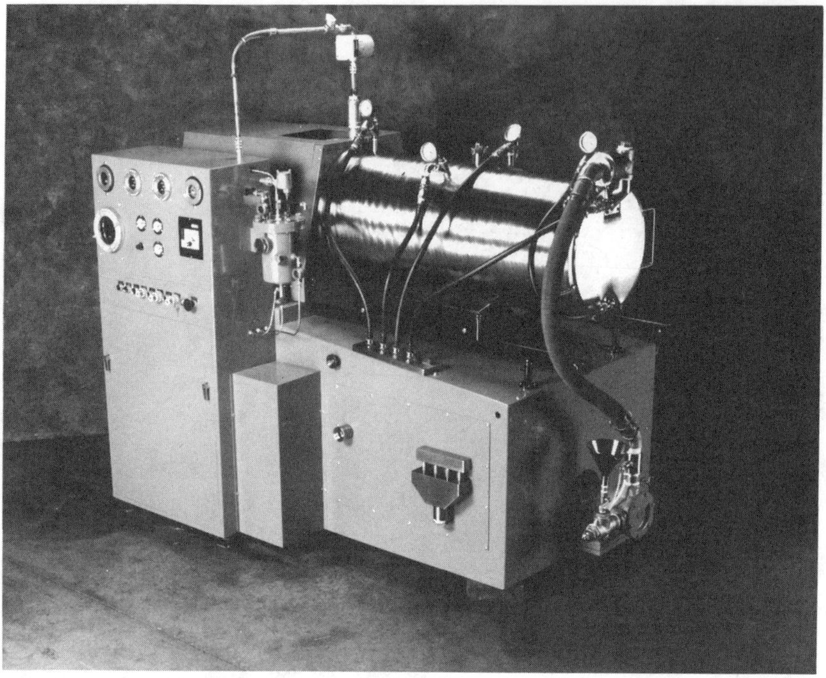

Figure 8.11. Horizontal bead mill: (a) general view. (Courtesy Premier Mill)

does not take place at the frontal contact between agitator and media, but in those regions where media are flowing counterdirectional to each other, as shown by the shaded areas in the schematic of Figure 8.12.

To effect greater efficiency, manufacturers increase the number of regions (or cells) with eddy currents through the agency of numerous agitator arms, blades, discs, or fixed pins set into the mill wall. Again, design limits exist because countercurrent flow must occur within the mill in localized cells to create the shear and impact necessary to break up agglomerates. The number of these cells is in part media size dependent and in part shell diameter dependent. They vary from about 5 per axial foot (1 per 6 cm) of mill to over a dozen. For high-population cell design, agitators consist of pins, while for low-population design, bars or discs serve better.

Although specific mill energy consumption is rarely a concern in dispersion operations, mill efficiency as reflected by heat generation can be of importance. As mill viscosity increases, progressively greater portions of the input energy are lost as heat. This heat rise may be detrimental to the solid, the dispersant, or the stability of the system, especially in compound dispersions.

With ''pure'' dispersion (no comminution), solids are formulated as high as practicable within a mill. The combination of solids, liquid, and dispersant com-

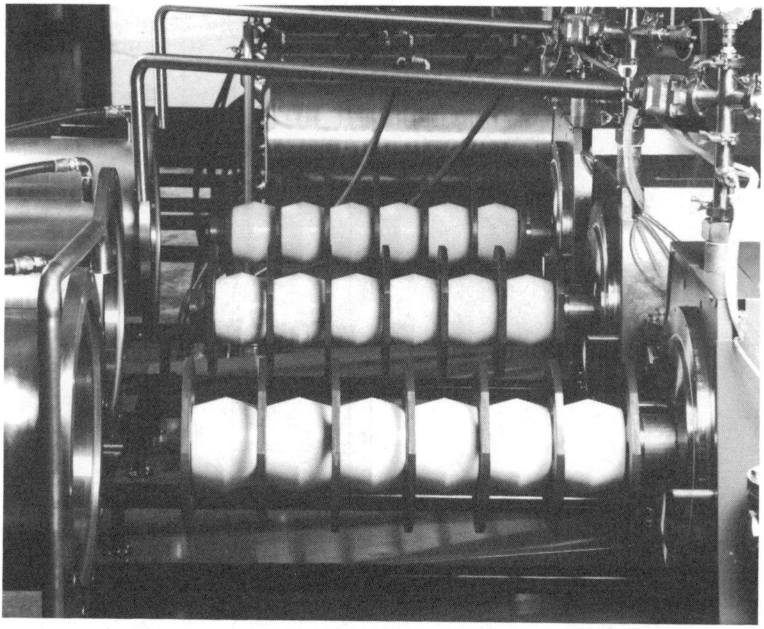

Figure 8.11. *(continued)* Horizontal bead mill: (b) disk arrangement.

monly is assembled in a separate makedown tank and pumped into the mill. Viscosity within the mill will be high initially, effecting high shear and high agglomerate breakdown efficiency. As dispersion progresses, viscosity reduction is dramatic, often falling by two to three orders of magnitude. The final, or product, viscosity, determines the initial dispersant addition, even though feed slurry viscosity may well be at the mill's limit of capability.

An unusual application for small-media, low-energy dispersion occurs with mineral rheological control agents. Creams and viscous suspensions for cosmetics or pharmaceutical applications contain an agent that when well dispersed imparts very high thixotropy to the system. This character is needed to maintain particulates in suspension for market acceptance, prolong shelf life, and to ensure the homogeneity of the contents. Such control agents have been traditionally carboxymethyl cellulose derivatives.

However, inert mineral agents, including mixtures of attapulgite (palygorskite), bentonite, and more recently, high-purity halloysite, are finding applications in this field. Attapulgite and bentonites have a slight tan color and contain a sizeable quantity of surface exchangeable ions. All three materials are relatively inert chemically toward other components and in reponse to those components. Halloysite, in contrast, is extremely white and nearly spectographically pure. Attapulgite and halloysite consist of assemblages of acicular particles bound by weak geochemical forces. Neither material shows adverse effects in

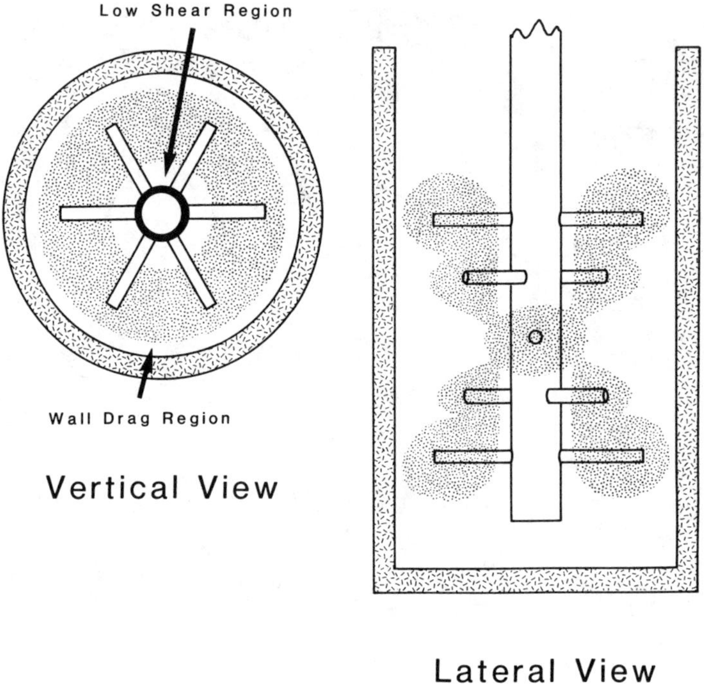

Figure 8.12. Dispersing activity within a small-media mill.

the body and may in certain instances perform as an antidiarrhetic. Surprisingly, when particles of halloysite are well dispersed then allowed to dry on the skin, the finished film has a soft, silky-smooth feeling resembling talc.

To realize fully the thixotropic characteristics of both attapulgite and halloysite, the tightly bound bundles must be defibrillated without needle breakage. Routine grinding provided by hammer and ball mills breaks the needles, both as clusters and within the individual particles themselves. Continued grinding produces essentially equidimensional particles with greatly diminished thixotropic character.

Use of a small-media dispersing mill and appropriate chemical defibrillating agents enables the needles to be individually separated and their rheological properties to be developed, as shown in Figure 8.13. This is one of only a few systems in which product viscosity actually increases with improved dispersion. Slurry enters a mill having a consistency resembling dairy cream and exits a paste having a thixotropic index of 10 or greater. Within the mill where high shear exists, viscosity is low. The time required to reform (hysteresis loop) upon exiting is very short, usually a few seconds or less.

Small-media mills serve a broad variety of functions in pigment, cosmetic, electronic, and similar industries. These include dispersion, blending, homoge-

Figure 8.13. Bound halloysite needles separated with a dispersion mill.

nizing, grinding, morphology altering, defibrillating, etc. Table 8.2 provides typical operating parameters for a variety of materials processable in these mills.

Media size and density are dictated not only by the energy requirements of the system but also by production time requirements of the commercial operation and by the mill's volumetric capacity. In some systems where a recirculation loop with external classification has been installed, increasing media density shortens residence time in the mill and improves productivity.

Milling studies performed for both dispersion and size reduction show effi-

Table 8.2 Typical Operating Conditions for Various Types of Dispersion

System Character	Media Composition	Media Size (mm)	Viscosity in Mill (cp)
Soft agglomerates	polymeric[a]	0.5–1.5	1000–4000
Hard agglomerates	glass[b]	0.5–2.0	1000–3000
Sinters	glass, alumina	1.0–3.0	2000–3000
Soft comminution	alumina	1.0–6.0	1000–2000
Hard comminution	alumina, st. st.,[c] zirconia	1.0–6.0	1000–2000

[a] Polymeric: nylon, divinyl benzene copolymer, polystyrene, polycarbonate, Teflon.
[b] Glass: annealed borosilicate glass.
[c] St. st.: stainless steel.

ciency improves with decreasing media size (in direct contrast to ball mills) optimizing at about a 10:1 media-to-feed size ratio and remaining nearly constant to about 20:1. This ratio is calculated from the coarse end of the feed material (for example, 0.4 mm beads for −325 mesh materials).

Because dispersion arises from both media–media impact and from close-pass shear during countermovement (the distribution between which is poorly known), the viscosity of the system must be overcome by increasing rotational velocity, media diameter, or media density. With larger media, the number of effective collisions per unit time fall off (media population decreases as the cube of diameter but contact area increases only as the square). However, energy per contact is proportional to the bead's mass, and hence the diameter cubed. Shear, however, increases as the square of the media diameter, being influenced by surface area. These factors must be balanced against the rheological characteristics generated within the mill, as shown in Figure 8.14.

Where high energy is required, as in hard material comminution, larger mill media size dominates (20:1 ratio). In dispersion, where deagglomeration requires low energy, the frequency of contacts dominates (10:1 ratio). The fundamental limitation, however, is dictated neither by mill media impact energy or frequency but by the ability to separate media from product and discharge the mill, once dispersion is achieved. This usually mandates media no smaller than 20–30 mesh (0.6–0.8 mm), irrespective of feed or product particle size.

Although mills have been designed where the entire charge, media and mate-

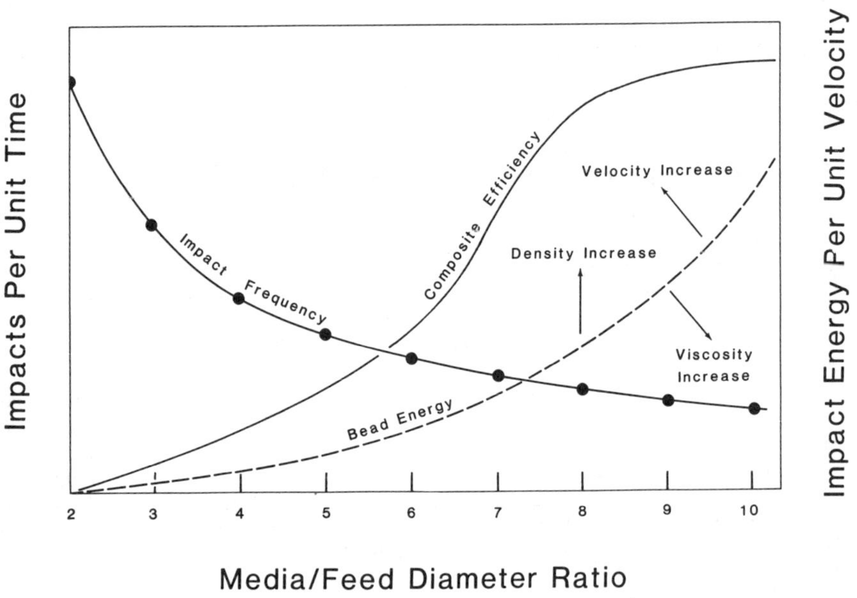

Figure 8.14. Media size factors in dispersion mill.

rial, is emptied and separated through dilution and centrifugation,[14] this has not proven satisfactory in commercial practice.

Mill RPM for the examples in Table 8.2 will be governed both by agitator geometry and shell diameter, and falls in the range 500–1500 rpm. While increased RPM increases dispersion activity, it also increases mill wear, color contamination, and heat generation. Mills are most often scaled from pilot testing to production on the basis of agitator tip velocity. Thus production equipment has the same media impact and media shear velocities as bench or pilot test mills.

Small-media mills, because their energy can be finely tuned, have certain capabilities not found in other dispersion milling operations. Just as acicular particle bundles can be defibrillated, plate assemblages can be delaminated, if interplate bonding energies are significantly lower than intraplate energies. Specific milling examples of this unique application of small-media mills will be given in the section on surface chemistry modification in Section 10.1.

Insofar as particle assemblage breakdown occurs and no dispersant demanding surface is generated, viscosity will not increase appreciably. Prolonged comminution, however, which appears to generate edge and planar face surfaces in closer proportion to their occurrence in nature (whether particles are acicular or plate-shaped) will cause dispersant demand to increase. Because viscosity change is not great in the transition zone where particle disaggregation is slowly replaced by particle fracture, torque meters or ammeters cannot adequately define the operational end point of defibrillation or delamination. Manufacturers instead employ a set milling time for set processing conditions (percent solids and dispersing agent) based on initial particle morphology assessment (usually by a dependent parameter) to establish their production parameters. For this approach to be valid, feedstock composition must remain constant.

Small-media mills are effective dispersors for ultrafine pigments in low-viscosity media, water, alcohol, glycol, etc. With expensive materials such as copper phthalocyanine pigment and ultrawhite mineral thickeners, discrete particle dispersion is an economy more than compensated by higher grinding costs associated with media mills. Figure 8.15 is an energy versus thixotropy development plot for white lathlike sepiolite. Two dispersive grinds were performed at various energy inputs, one by ball milling with 1 in. (25 mm) ceramic balls and the second by small-media milling with 0.8 mm beads. While small-media milling costs are 2–3 times greater, rheological development results in a savings 10 times this increase.

Where toxic or flammable solvents are employed in bead mills, the tanks are sealed and liquids pumped in with dispersant added or metered in through a sealed line to the small-media mill. The tank is grounded and the system generally operated with a cooling jacket[15] to minimize vapor pressures within the mill.

Atmospheric gases, notably oxygen, may be damaging to some milled systems (metals, pharmaceutical materials, sensitive chemicals). To prevent this, media mills have been constructed to operate under a vacuum.[16] These mills are

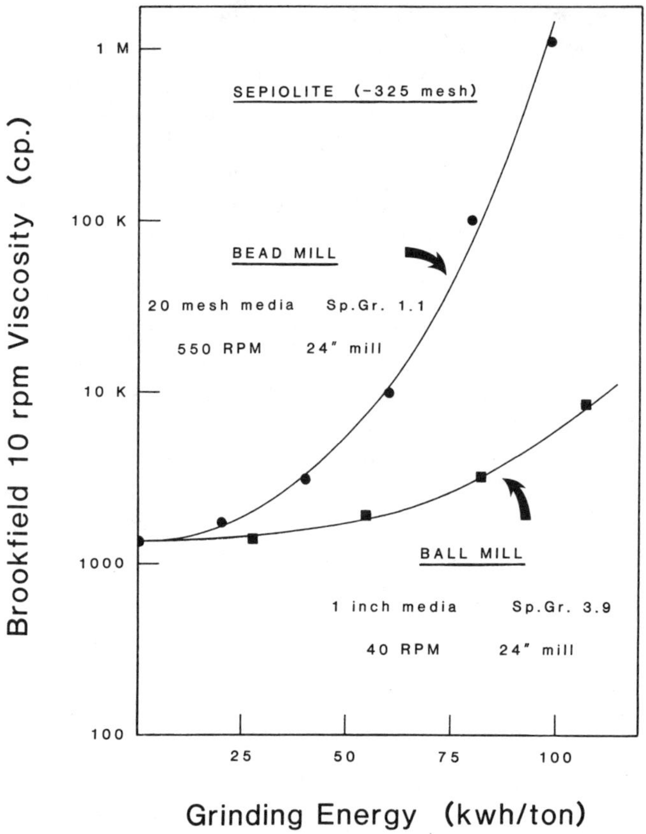

Figure 8.15. Ball versus bead milling of sepiolite mineral pigment.

also employed to prevent foam formation and gas entrainment with slurries where those characteristics are an important factor. The construction of these mills is not especially complex, but loading, dispersing, and unloading are time consuming.

8.4 Vibratory Mills (Wet-Mode Operation)

Near the upper end of energy-intensive milling are vibratory mills, These require higher-energy input than ball mills and overlap slightly in application with small-media mills. Vibratory mills are more efficient in dispersion where high-energy operations are dictated, most often with ultrafine pigments and highly viscous, nonthixotropic carrier liquids. General mill design, like that of small-media mills, falls into two classes, vertical tank and horizontal tube.

The vertical tank mill is typified by the Sweco unit shown diagrammatically in Figure 8.16. The tank compartment, filled nearly full with media, is mounted on a series of springs. The mill's motor turns an off-center weight causing the tank to dip in an inductively coupled compliance to the weight motion. Thus the tank and its media load have no net rotation. Vibration frequency is generally in the 1000–1500 reciprocal minute range and represents either the fundamental or first harmonic of the assembly's natural period of oscillation. Vibrational frequencies outside this period result in severe bearing wear and wasteful power consumption.

As the tank dips, media are forced into oscillatory motion and strike each other. Because the core of the mill has little dip, it is usually kept hollow to reduce motor load. Media vibrate in all three axial directions. Media may be spheres, cylinders, prisms, or any form not having apices (projections break off quickly to contaminate the product). Vibratory mill media vary in size from 6 up to 25 mm in diameter. They include alumina, zirconia, stainless steel, silicon carbide, tungsten carbide, mullite, and chert in a variety of sizes and shapes. A few examples are shown in the photograph of Figure 8.17.

The horizontal tube design is shown as the Palla mill in Figure 8.18, a double-tube system in which one tube counterbalances the second in mass. Vibration

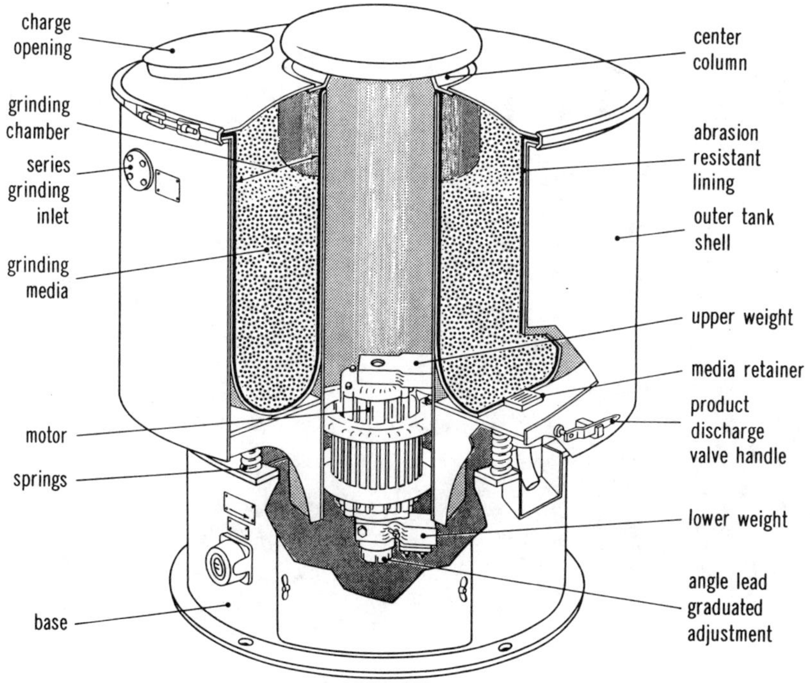

Figure 8.16. Vertical tank vibratory mill (courtesy Sweco).

Figure 8.17. Typical dispersion–comminution media for vibratory mills.

Figure 8.18. Palla horizontal vibratory mill (courtesy ABB Raymond).

arises from shaft-rotated adjustable off-center weights contained in either end of the support structure, which can be "tuned" for media weight and size. Because of the two-tank arrangement, differential media may be employed in the tubes for stage grinding. The two tubes may be operated in sequence, in parallel or independently.

For size reduction vibratory mills are often operated dry, coupled to an external air classification circuit where oversize is removed from fines and returned to the mill to improve size reduction efficiency. In the wet mode a vibratory screen or centrifugal classifier is not effective as a followup stage because of the viscous character of the slurry within the mill. Thus wet operations are frequently batch operations. Because of their high-energy input, vibratory mills are employed for dispersion where rheological demands are severe. Classic in this regard are inks with submicrometer pigments.

Figure 8.19 is a comparative dispersion plot of milling time versus agglomerate reduction as measured by Hegman grind gauge for four ultrafine pigments, carbon black, titanium dioxide, copper phthalocyanine (blue), and toluidine red.[17,18] Note the time ratio for copper phthalocyanine is great (1.5 versus 25 h), whereas that for carbon black is more modest (18 versus 65 h).

In milling and dispersing of white pigmentary cosmetics or pharmaceutical-

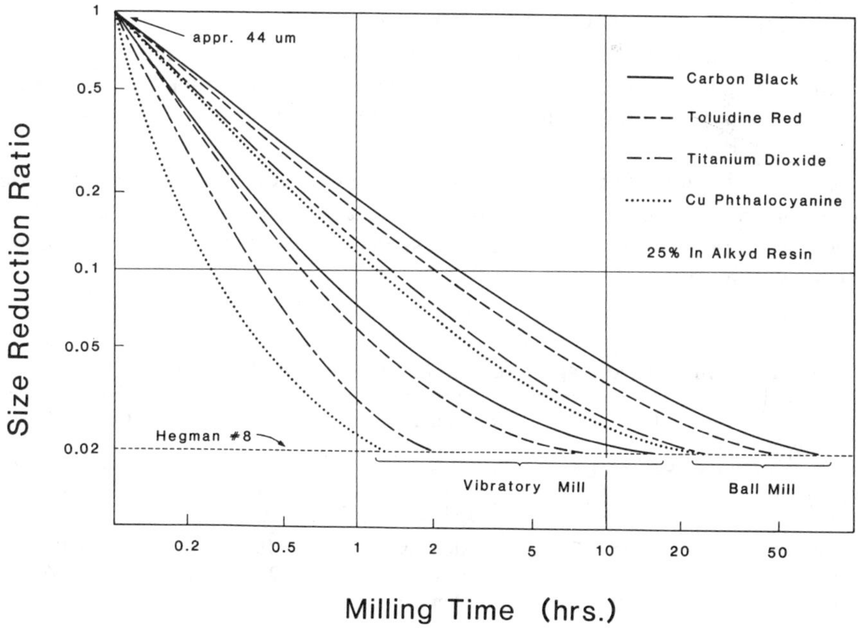

Figure 8.19. Comparative dispersive grinds of ball versus vibratory mills for fine particle inks.

grade diluent products (calcium carbonate, for example), color contamination and metal contamination resulting from overzealous milling and poor dispersion may be as important considerations as dispersing efficiency or processing time. Color contamination can influence psychological sales value, and metal contamination may cause adverse interactions with drugs, organic agents, or fragrances employed with the carbonate in the final formulated products.

Talcs, ball mill ground with metal media, often display an adverse chemical effect on perfumes added at a later stage during the manufacture of body powders. Certain manufacturers have employed the same perfume with their product for such a protracted period that its quality is associated with that odor. Slight oxidation during prolonged shelf life can adversely affect sales even though the mechanical or cosmetic properties of the product remain wholly satisfactory. As will be discussed in more detail in Chapter 10, activated surface and inadequate or poorly selected dispersants can detrimentally affect the interaction of adsorbed organic molecules.

Similar phenomena are associated with calcium phosphates, carbonates, and calcined clays ground and dispersed in toothpaste as mild abrasives. The presence of metal mill materials can alter the taste of flavor additives and make the product taste old or even slightly rancid! A new, highly effective, abrasive was tested in the 1960s for a popular brand of toothpaste because of its fineness (100% -5 μm), whiteness, purity, and compatibility with soluble fluorides. It was rejected, however, because milling contaminants altered the traditional flavors employed in the toothpaste's manufacture.

While vibratory mills with silicon carbide or stainless steel media may perform equally well or better with regard to particle size reduction and dispersion with viscous pastes, those employing alumina media have fewer problems of the types just noted.

In an evaluation of dispersive milling of acicular iron oxide into a polymeric resin for roll coating, the presence of mill contamination was an important consideration, not because of color but abrasiveness. The results of this evaluation are shown in Table 8.3.

Table 8.3 Comparative Dispersion and Properties from Different Mills

Mill Type	Media Size (mm)	Media Type	Time (h)	Contamination (%)
Agitator	—	—	3.6	0.001
Ball mill	50	Ni hard	7	0.023
Ball mill	25	Alumina	16	0.017
Media mill	2	Glass	2.1	0.002
Vibratory	10	Alumina	2.0	0.009
Kady mill	—	—	1.7	0.001

Note: 3 ft diameter ball mill, 25 gal Kady mill, 13 gal media mill and 25 gal agitator mill with 6 in. blade. Feed size, -325 mesh (44 μm) dry iron oxide at 30% solids; product size, -1 μm (100% dispersed); liquid phase, solvent-reduced cellulose acetate.

Although deagglomeration efficiencies were similar for all mills, the time to achieve a set fineness (Hegman 8) and the contamination contribution, which displayed a broad range, became the deciding parameters for mill selection.

In assessing the comparative performance of dispersion mills, the ultimate decision should include the following additional factors:

1. Manpower requirements (loading, control, discharging, cleanup);
2. Installation cost (direct and ancillary equipment);
3. Maintenance cost (media, wear parts, pumps, percent downtime);
4. Contamination contributed by the mill;
5. Noise and vibration;
6. Potential for leakage, splattering, and spills;
7. Floorspace (footprint).

8.5 Multiple-Roll Mills

Multiple-roll mills are often employed to disperse small slurry batches and high-viscosity systems. Mills are most often of the two- and three-roll configuration. Rolls have precision-machined, mirrorlike cylindrical surfaces, are adjustable for clearances down to about 25 μm or less for fine dispersions, and rotate at relatively slow speeds. Slurry is fed onto one roll by means of a troughlike distributor and is carried into the nip region with the second roll. With some mill designs the feed or takeup roller has a chemically embossed surface to aid in slurry adhesion, feeding, or transmission. A typical commercial unit is shown in Figure 8.20.[30]

For highly viscous feedstocks one or more of these rolls may be heated to reduce slurry viscosity. Surface velocity of the second roll is parallel with, but may be slightly greater than, the first to induce shear into the slurry. In three-roll mills a third roll takes material off the second in the same fashion but has a narrower nip clearance than that between the first and second roll. With either configuration the final roll has a trailing blade, sometimes heated, which removes slurry and forces it into a funnel-shaped drain.

Where single-pass operations provide insufficient shear to obtain the required dispersed state, a portion of the product is returned and blended back with subsequent feed on the first roll to increase viscosity (greater exposed surface) and shear, especially with ultrafine powder components. Three-roll mills provide the highest shear of all mills, which in general follow the order:

three-roll mill > small-media mill > ball mill > blade dispersor

Unforunately, multiple-roll mills tend to provide the slowest volumetric rate of dispersion because of their highly limited surface area and nip region.

Multiple-roll mills are frequently recommended for dispersing such ultrafine

Figure 8.20. Commercial three-roll dispersion mill (courtesy Kieth Machinery Corp.).

components as carbon black and fumed silica. These materials are milled into polymer systems first before any other solid component is introduced into the slurry. Solvent is added sparingly and then only after milling has produced too viscous a slurry for subsequent processing. Fumed silica performs better if not prewet with solvent prior to addition because solvent acts as lubricant and viscosity reductant, allowing silica agglomerates to deform between the rolls and then reform thereafter. A synergistic dispersing agent may be employed with such ultrafine powders (as glycerine and glycol with fumed silica) to separate particles slightly upon shearing and prevent their later reformation into agglomerates. Such agents must be carefully selected to prevent bridge bonding of the ultrafine particles.

8.6 Dispersion with Ultrasonic Energy

Most energy intensive of the processes for dispersion is that obtained through subjecting the slurry to ultrasonic agitation, or cavitation. Commercial ultrasonic tanks, or baths if small, have one or more transducers mounted directly on the tank wall. These transducers are vibrated at frequencies varying on the low size from about 18,000–20,000 Hz up to 75,000–100,000 for the large units.[19] Energy consumption rises proportional to sonication frequency. Special equipment has been made whose frequency lies in the megahertz range, but this is highly inefficient for mechanical dispersion.

With ultrasonic dispersors, the frequency of vibration is too low to inductively couple and translate particles, especially near-micrometer-size particles, synchronously in a liquid system. Wavelength of the oscillation is of the order of centimeters. However, ultrasonic transducers impart very high energy density into the slurry. The mechanism for deagglomeration lies in cavitation effects arising adjacent to particles or within loosely associated particle assemblages and this inability to follow the induced wavefront projected by the transducer. The ability of particles to follow the oscillating wavefront is directly related to particle size (one particle size analyzer has been developed based on this principle).

Cavitation effects around particles results in liquid boiling on a local basis.[20] This localized high energy is believed literally to blow particles apart. Evidence also has been submitted that selective vibrations destroy part of the resistant layer of adsorbed water around each particle and around agglomerates.[21] Thus, for a given material density and particle size distribution, a specific frequency (or narrow range) exists that will oscillate the components in an agglomerate to the point they break away from each other.[22] Where free dispersant is available in solution, these particles become permanently dispersed. Jewelry cleaners are based on this principle and employ strong dispersant concentrations as a consequence. Less expensive, and less efficient, home units employ 60 Hz (from line frequency) for cleaning. Better units generate either the 20 or 90 kilohertz range.

Most vulnerable locales on the surfaces where ultrasonic cavitation begins are at crystal imperfections, internal step lattice corners, and projecting points.[23] These are, coincidently, the same locales that attract other particles to form floccules or agglomerates from individual particles.

Ultrasonic energy can cause water trapped in crevices and incipient cracks to boil, propagating those cracks yet farther. Thus ultrasonic exposure can actually cause size reduction if continued overzealously.

In an ultrasonic dispersor the wave front spreads out radially from the transducer pole piece and its energy falls off inversely as the square of the distance from the transducer locale. Thus dispersion is most probable in the immediate vicinity of the transducer array. Most dispersion units have a secondary method of agitating the suspension to provide a continuous slurry movement to and around the pole piece(s).

Figure 8.21 represents part of an investigation into the rate of -2 μm particle separation from ground, agglomerated, microfine, gold-in-quartz slurry using a one-paddle agitator blade (0.6 kW) coupled with a small ultrasonic generator (0.5 kW). To assess which energy component was the prime contributor to the system, ultrasonic transduction was operated intermittently (15 minutes on, 15 minutes off).

The transducer is most effective early in the dispersion process. After an agitator blade has broken all agglomerates apart and dispersant has diffused to the surface, the addition of supplantive energy has but a small effect. Ultrasonic energy has its greatest advantage early in mechanical separation processes.

Silver halides for photographic films have been effectively dispersed[24] to enable a higher "speed" (photon capture efficiency) to be obtained than with simple mixers during the gelatin formulation process. Ultrafine particles (-5 μm) of gold and silver can be broken away from quartz and feldspar and dispersed, as can pyrite from coal, with the proper selection of dispersants (alkanolamines and sodium xanthate, respectively).[25,26] Ultrasonic dispersors have been designed to function in narrow constrictions (parallel plates) of pipe where slurry flows between two batteries of transducers. Because of the high energy density, the transit distance to accomplish dispersion is very short, usually only a few centimeters.

Large dispersor tanks have transducers located at numerous positions around the tank to distribute better the high-frequency energy produced by the oscillator

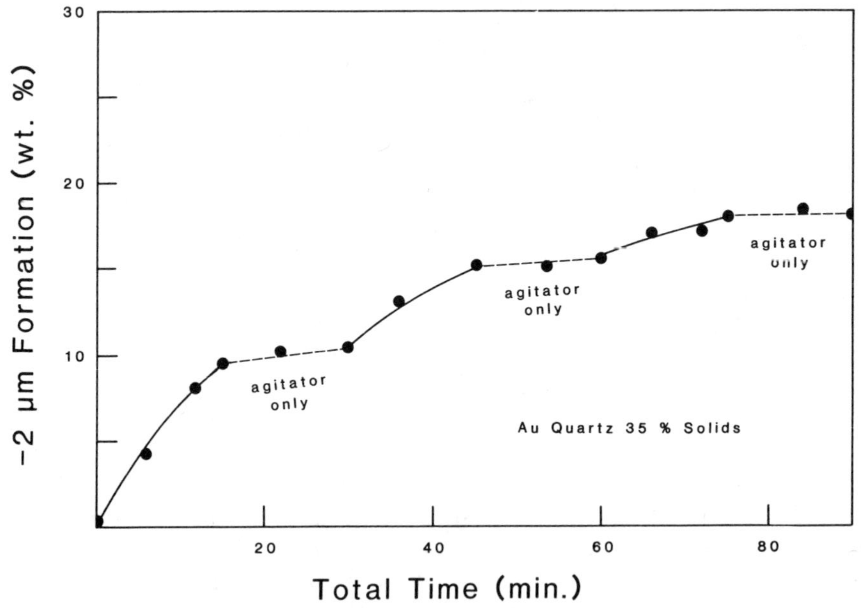

Figure 8.21. Effect of ultrasonic energy on agitation in dispersion.

circuit. Energy densities of 500 to 1000 watts per square foot are common. To couple this energy to the slurry effectively, solids must be high, 65% or greater being preferable. Also, as ultrasonic energy in the high-solids dispersor is absorbed within a very short distance (a few centimeters), tank design is critical for efficient utilization of transmitted energy.

8.7 Mechanics of Dispersed-State Comminution with Mills

While ball mills are employed effectively to disaggregate particle assemblages, more commonly they serve to effect size reduction. The same may be said for both small-media and vibratory mills. Ball mills may be operated both dry and wet. However, wet milling enables a finer grind (by an order of magnitude), generally a more efficient power utilization, and improved efficiencies in loading and discharge. Dispersants are effective aids in size reduction in all wet grinding mills.

The various function of the dispersant can be illustrated by their actions in a ball mill.

The first function of a dispersant in wet grinding lies in lowering viscosity in the mill, which permits higher solids. While grinding dry represents a 100% solids feedstock, slurry solids can be obtained at 60–65% with high fluidity, easier feed and discharge, elimination of dusting, and reduction of difficulties in many grinding operations.

By lowering viscosity, particles are maintained in the milling region, that is, that volume where cascading, agitating, or vibrating balls (or other media) impact. Lowering viscosity also prevents fine, solid particles from adhering to the media in thick layers and producing a cushioning layer during the cascading activity. Similarly, fluidized material in the mill prevents material from adhering to the mill wall, being carried around out of the milling field. Fluidization also prevents formation of a softened wall bed, which reduces the impact energy upon about 10% of the particles during the cascading, rotating, or vibrating activity.

Evidence also exists[27] that those chemical agents which effect dispersion during ball milling diffuse into incipient fractures and prevent these from "healing" in the mill. Serving as propents, they allow cracks to be held open, enabling later impacts to continue propagating these fractures throughout the solid particle. This has been the primary action proposed to explain an order of magnitude finer grinding achievable with wet milling than with dry.[28]

Third, the formation of new fractures from milling results in particles in which bonding, whether covalent or ionic, is highly strained as a consequence of stress rupture. Because atomic displacements and diffusion are involved, dry grinding processes leave new, highly activated surfaces requiring extended periods to reform and return to ground, or low-energy, states. This will be illustrated in considerably greater detail later in this book (see, for example, Table 10.2).

Wet milling does not have as pronounced an effect upon surface atomic dislocation phenomena, probably because of constant ionic exchange in progress at particle–liquid interfaces. Grinding with dispersant removes the adverse structure effect completely. It also permits higher mill loading and better energy conversion to comminution. The remaining parameters, surface potential and chemical activity, bear this out. Gas adsorption isotherms display a considerable difference between wet and dry grinding. Wet curves typically follow the pattern of Type II isotherms—materials having a near-homotactic surface. Dry grinding produces Type IV isotherms, commonly associated with energetic pores, differentiated surfaces, and catalytic activity.

Table 8.4 contains data from a commercial study for calcium carbonate pigment production, which compares dry, wet, and wet-with-dispersant grinding and the quantity of fines generated, as measured at 10 and 5 μm. This fine range is the practical limit of dry ball milling for calcium carbonate, due to material buildup on both balls and mill walls and a resulting cushioning effect.

Ball milling of 1 μm magnetite iron oxide (Fe_3O_4) particles in an aliphatic-modified, MIBK-diluted, polyester resin with and without 0.5% oleic acid as dispersant was illustrated in Figure 1.8. Both systems, milled 6 hours followed by a cast film formation, were thin sectioned to evaluate dispersion. The use of dispersant resulted in aggregates at that milling time having a size only $\frac{1}{5}$ to $\frac{1}{10}$ that produced without agent. Further, homogeneity, even of aggregates, was superior with the oleic acid. The dispersed sample, while not yet fully broken down to the desired discrete particle status, shows aggregates of about 10–20 particle diameters (1000–8000 particles) remaining, while without dispersant the aggregates contain millions of particles. Eventually, both milling techniques will reduce the system to discrete particles. However, milling with dispersant requires considerably less time and likely results in fewer broken needles.

Figure 8.22(a) is a time-dependent milling graph showing the loading/milling effect of oleic acid on the grinding effectiveness of magnetite iron oxide with time, and Figure 8.22(b) the results.

Table 8.4 Effects of Dispersant on Grinding of Calcium Carbonate

Grinding Method	Dispersant Level (%)	% Mill Solids	Weight % (−10 μm)	Weight % (−5 μm)
Feed	0	—	30.2	8.1
Dry	0	100	55.1	20.8
Wet	0	51[a]	78.2	42.0
Wet	0.1	66	82.3	61.5
Wet	0.2	66	88.9	72.4
Wet	0.3	66	90.1	73.2
Wet	0.4	66	91.6	74.9
Wet	0.5	66	91.5	75.0

Note: Dispersant is SOPA (MW 1200).
[a] Maximum solids obtainable at milling viscosity.

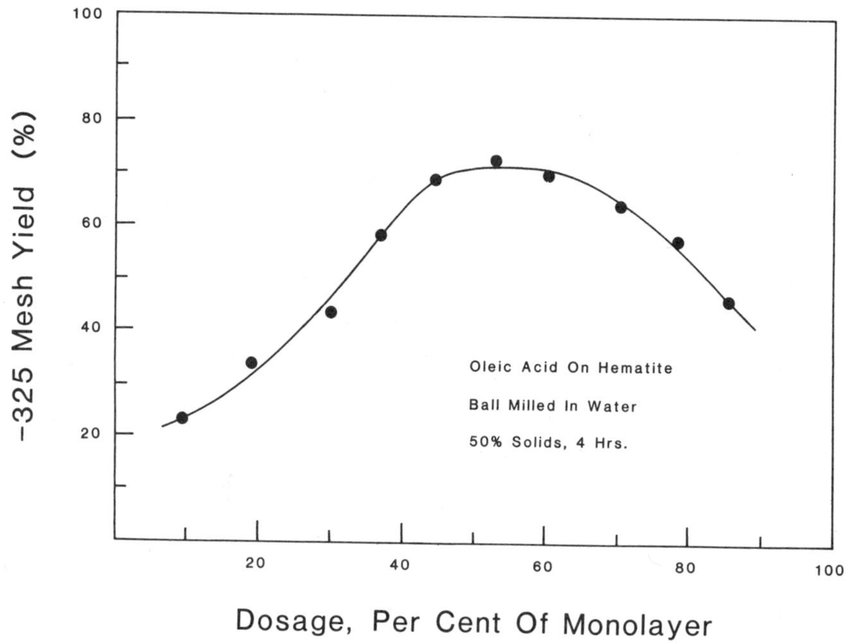

Figure 8.22. (a) Loading effect of oleic acid on iron oxide in a mill.

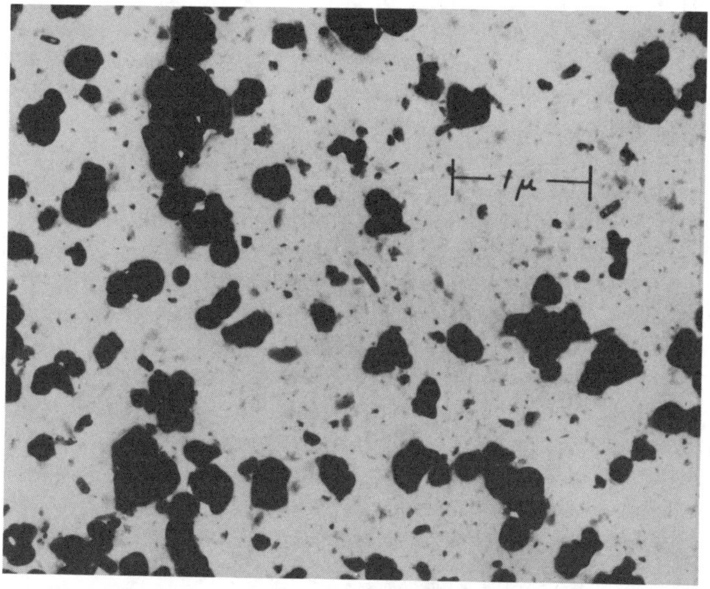

Figure 8.22. *(continued)* (b) Thin film showing the effectiveness of dispersion.

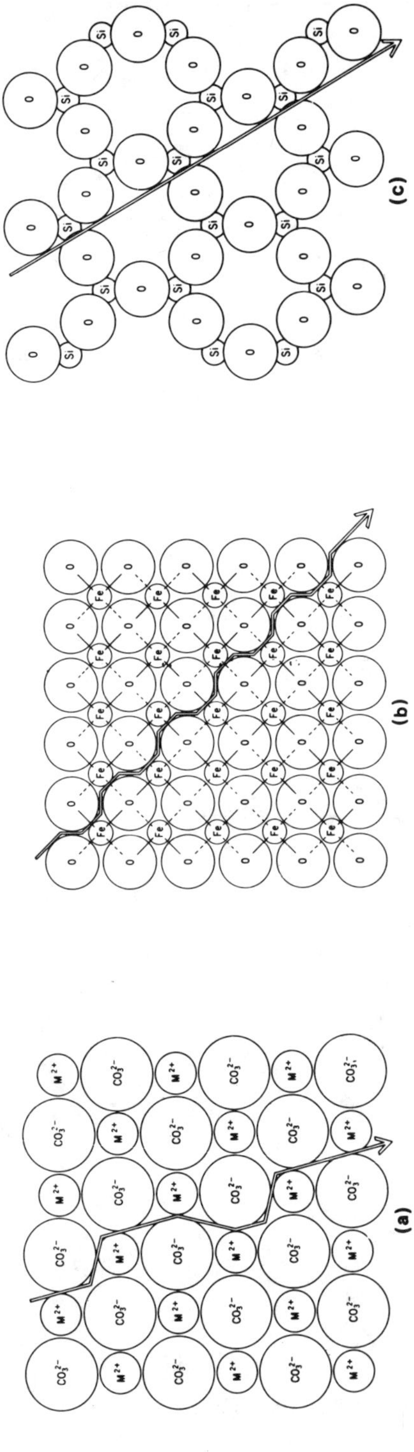

Figure 8.23. Solid-state bond rupture from grinding: (a) calcium carbonate; (b) ferric oxide; (c) silica.

Dispersing, or at least reducing dipolar flocculation of, iron oxide particles not only allows them higher mobility in the grinding phase, it produces a more fluid coating formulation and superior alignment of particles for magnetic imaging applications.

Grinding by any equipment creates massive stress centers along planes of fracture by the process of molecular rupture and atomic movement. How these surface atoms respond and for how long determines their effect on dispersants and applications in general. Figure 8.23 schematizes three classes of pigments: (a) an insoluble ionic but slightly alkaline, salt, illustrated by calcium carbonate, (b) an intermediate acid–base surface, exemplified by hematite, and (c) an acidic oxide, portrayed by quartz. Fracture planes are shown as ordered for the two oxides and random for the carbonate.

The rupture of silica (or silicate) bonds from grinding produces free radical silicon and oxygen species. Except in vacuum, sufficient moisture is always present to permit these two to form silanol groups, as illustrated in Figure 8.24.

As a consequence of grinding, the silanol group count rises and their acid strength increases, at least for a short period. This effect can be measured readily with a laboratory pH meter, where freshly dry ground quartz will show a drop in pH of 0.2 to 0.4 pH units. Silicate grinding also shows a surface acidity change, as illustrated by the 30-day aluminum silicate milling study in Figure 8.25.

The second example given, ferric oxide, similarly forms free radicals at the point of surface rupture, followed by reaction with moisture to form ferric hydroxyl units on the surface. If the surface is dominantly acidic (66% with red oxide), bond stress shifts the surface to yet more acidic conditions. These have been measured (see Figure 10.1) by surface titration with sodium dodecyl sulfonate (SDS) and dimethyl ditallow quaternary ammonium bromide (DTAB). The former measures anionic hydroxyl sites, while the latter measures cationic acidic sites. Acidity rises to about 72% in the first few hours of milling but returns to its original state in 1 month at ambient temperature.

The third example in Figure 8.23, that of limestone, has yet another mode of

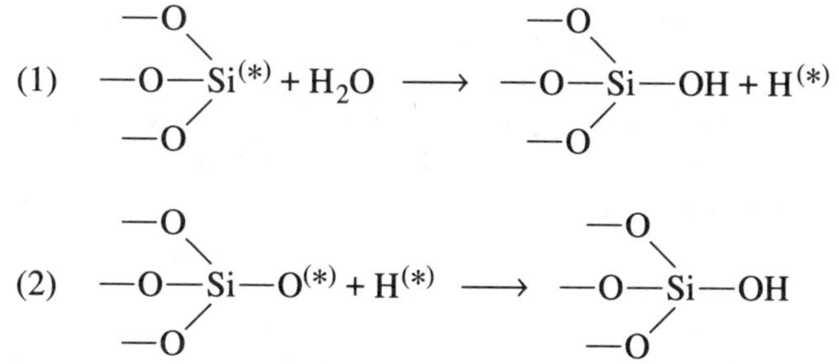

Figure 8.24. Free radical mechanism for silanol formation on silica.

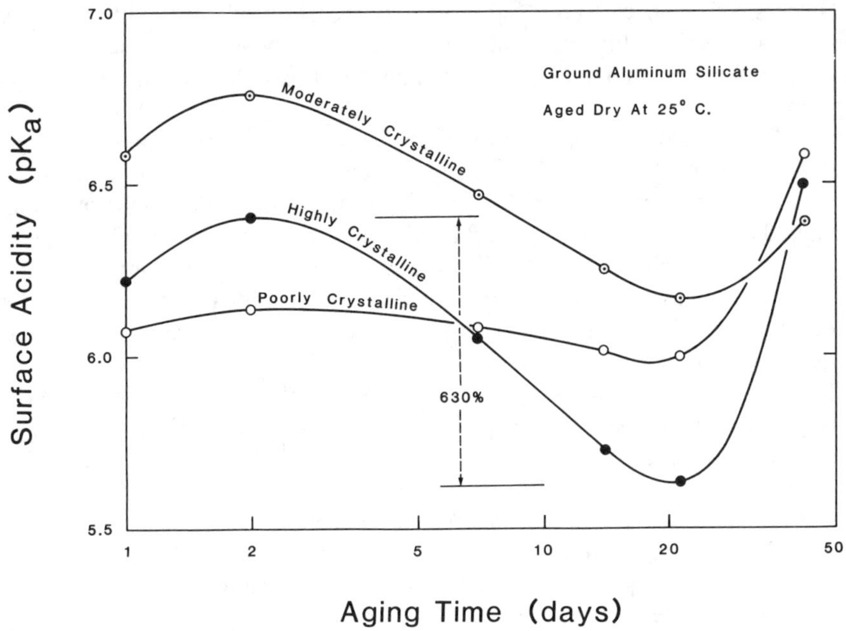

Figure 8.25. Surface acidity change of aluminum silicate following grinding.

activity. Where fracture planes isolate two carbonate ions in adjacent positions, in contrast to silica, where addition takes place, decarboxylation and structure loss can occur:

$$CO_3^{2-} + H_2O \rightarrow CO_2 + 2\ OH^- \tag{8.3}$$

While this reaction is hypothetical, studies on limestone grinding for desulfurization of flue gas[29] showed a reaction to occur that produced a rise in pH from 9.4, the equilibrium value for fresh carbonate, to 9.6–9.7 following milling, the value depending upon the intensity of grinding. This freshly ground material exhibited an elevated reactivity with sulfur dioxide (see Table 10.2). It might be postulated that all freshly formed surface carbonate ions could undergo such a reaction. However, were this in evidence, the pH would experience a much greater rise than has been observed. Adjacent anion repulsion is both a more probable motivation and more statistically likely, based on actual hydroxyl levels produced. As with silica and ferric oxide, this is a short-lived reaction, the presence of ubiquitous carbon dioxide quickly reversing the reaction, once surface reorganization takes place.

Calcium carbonate grinding into the low or submicrometer range represents a special case of dispersion that merits detail. In this commercial example true size reduction takes place, and dispersion is an associated requisite, either to continue comminution or to produce a finished product. In either instance mill rheology presents a complex situation.

The complexity is illustrated with the size reduction–dispersion rheograms in Figure 8.26. Feed to the mill is a −325 mesh ball milled calcium carbonate, whose particle size distribution averages (D_{50}) about 10.7 μm. Other information is provided directly in the diagram.

Three rheograms were collected during the course of the milling operation, at time zero, 15 min, and 30 min, the total milling time for product. Product particle size was approximately an order of magnitude finer than the feed. The rheological profile for the feed at 60% solids dictated sodium polyacrylate at 0.2% (solids basis) for a minimum in viscosity and a range of 0.15–0.25% to maintain no greater than about 1200 cp, the optimum mill viscosity for grinding.

However, 0.2% feed dispersant level represents a product dispersant demand level that is grossly low and a viscosity much too great for the mill's operation at the 30 min termination.

The optimum dispersant concentration for the nominal 1 μm product lies in the range of 0.5–0.7%, much too high for feed. Thus no condition of dispersion satisfies all operational constraints. Because viscosity is much more critical for the fine distribution (note the narrowness of the profile as milling progresses compared to that of the feed), a practical solution to the rheological dilemma lies in metering the dispersant during the course of the milling and maintaining the dispersant level along the "leading" edge of the rheological profile. This is illustrated by the intermediate rheogram.

Rheogram narrowing is the result of the high surface area of the material in the mill and a lowering of the top size, making the system less Newtonian and more thixotropic.

Viscosity control may be solved with a secondary or makedown tank connected in a recirculation mode. In this tank dry material is fed manually or automatically, based on or controlled by mill rheology. The latter may be directly controlled through a torque meter or an ammeter in the motor circuit, although the former is the more sensitive monitoring device. Feed solids may be typically $\frac{1}{3}$–$\frac{1}{2}$ of product solids, but dispersant present is predicated on full solids. The system recirculates through the makedown tank, which is typically no more than 30% of the system volume to minimize overload to the mill. An agitator in this makedown unit prevents solid powders from lumping onto the tank bottom.

This quasistatic mode of makedown with a demand feed system is very effective at maintaining relatively constant shear within mills yet maximizing efficiency in dispersion over the full grinding range.

Dispersant metering can be manual but also can be accomplished through a solution feeder controlled by the mill's torque meter.

Bead size for the operation noted shows an optimum, as seen by the grind data displayed in Figure 8.27. Grind time minimizes at about 8 mesh media (2.6 mm) because smaller media have insufficient energy upon impact to break fine particles after breaking coarse ones in the nip of the beads.

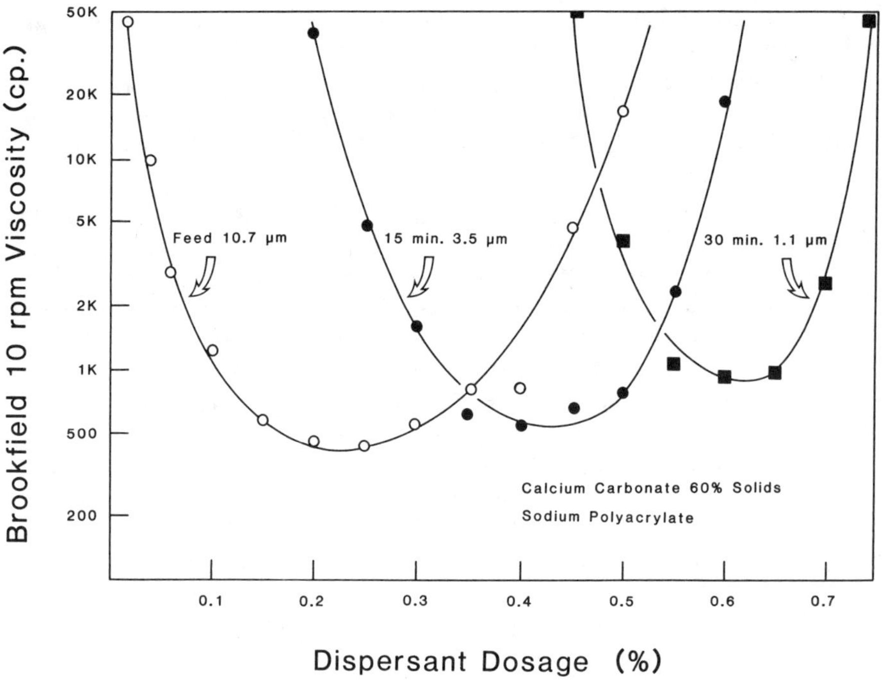

Figure 8.26. Rheology change during calcium carbonate milling.

Size limiting capacity of the media is more clearly revealed in the full particle size analyses of three of the different media grinds in Figure 8.28. The primary differences lie in the -5 μm portion, where fine media fail to grind effectively.

Not all dual-purpose milling operations (comminution plus dispersion) display a dispersant criticality similar to that shown for calcium carbonate, particularly where surface chemistry does not change appreciably during the milling exercise (alumina hydrate, organic pigments, coal, etc.). However, those materials whose isoelectric points lie outside the pH range 6 to 8 in most instances do display influential pH changes.

8.8 Effects of Temperature upon Dispersion

As slurries are most efficiently formulated at high solids levels, that milling activity required induces heat into the system during the operation that elevates slurry temperature. Temperature rise can be a valuable adjunct in attaining dispersion.

Temperature increases solution kinetics, moving dispersant molecules faster,

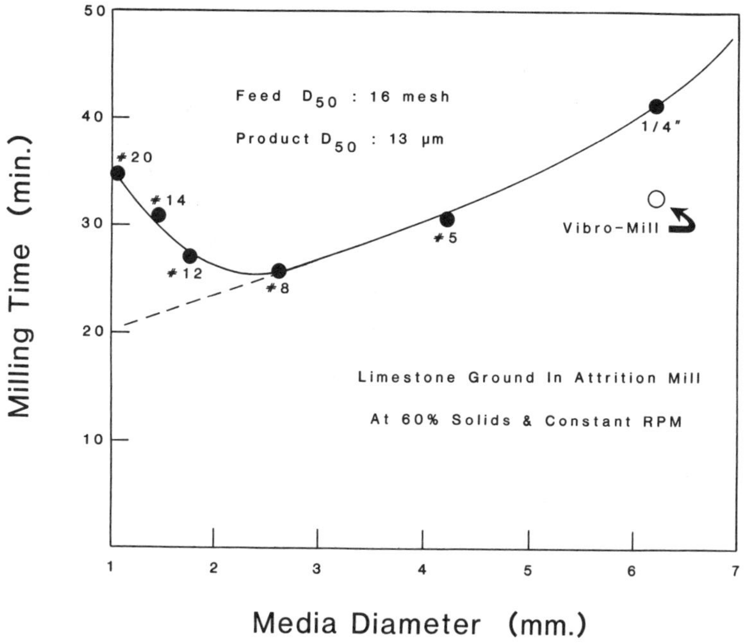

Figure 8.27. Influence of bead size on calcium carbonate milling rate.

allowing them to deform, disrupt, and diffuse into floccules and to adsorb upon naked surfaces before floccules can reform. Thus heating, in concert with efficient mechanical shearing, can significantly reduce the time to reach a given dispersed state.

However, heating frequently reduces the maximum solids attainable for a given dispersed state. This derives from increased entropy (see Chapter 2), arising from faster, more randomly moving particles, longer mean free paths of motion in suspension, and an effective increase in the particulate–solution envelope volume. A temperature increase also shifts dispersant–solution equilibrium toward the solution phase, thereby reducing surface charge and/or dipole layer thickness. These two phenomena, equilibrium rate attainment and effective particulate volume, may serve countervailing ends by influencing separate slurry characteristics. As a consequence, heating, like dispersant dosage itself, can easily be overdone, creating detrimental effects both in production and application.

Other positive aspects of temperature on dispersion include the following effects:

1. Liquid viscosity drops with elevated temperature, allowing more rapid shearing activity to take place at a set solids level.

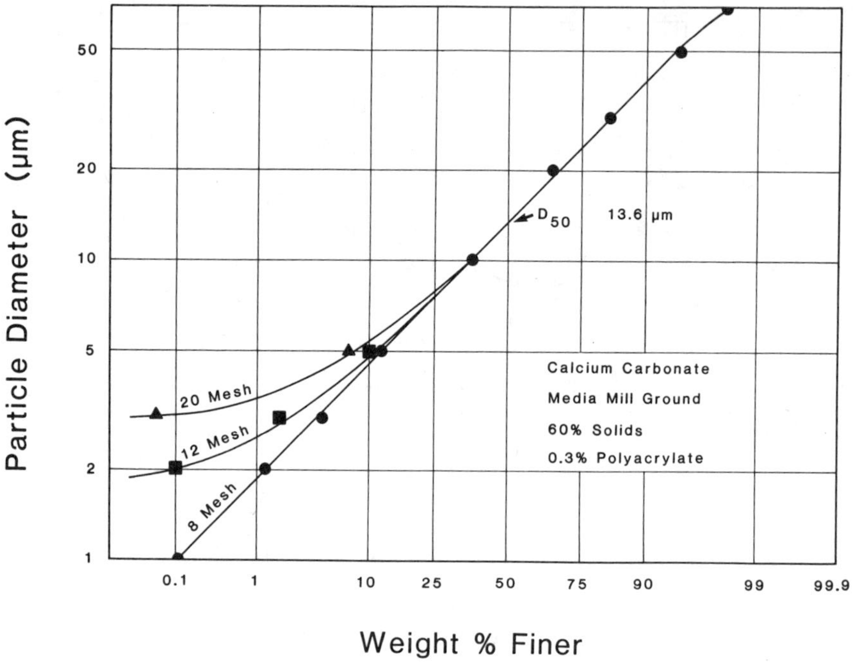

Figure 8.28. Particle size discrimination with bead size in calcium carbonate milling.

2. The increase in solution kinetics is important for large molecular dispersants, but less important for smaller ones, in attaining final surface–dispersant species configurations. It can shift train-to-tail configurations and their formation time.
3. Temperature increases *kT* spinning and can reduce the quantity of long-chain molecules required to cover effectively foreign character surfaces (oxides in oleophilic media).
4. Thermal elevation improves entropic stabilization with dipolar dispersing agents.

Negative characteristics induced by temperature rise include the following phenomena.

1. The chemical stability of the surface–dispersant bond is reduced.
2. Increased particle–particle impact energy may drive particles through the dispersion barrier, especially with dipolar agents, inducing floccule formation.
3. ''Soft'' surface coatings, such as silica on titania, are more readily dislodged, exposing the host surface character.

4. By increasing the rate (but not the quantity) of oxygen uptake from air and narrowing the activation energy gap, partial and premature oxidation of polymers may occur.
5. Polymerization may commence between double-bond species (acrylate monomers or linoleic acid).
6. Shock can be induced into the system where cold components are added to a heated, milled formulation.
7. Lowering slurry viscosity reduces the effectiveness in attaining dispersion with ultrafine components such as fumed silica and carbon black (see Section 8.5).
8. Elevated temperatures increase solvent volatility and loss.

Finally, the employment of thermal assistance may require alteration of component addition, especially where catalysts, promoters, solvent, or thermally sensitive materials are involved.

References

1. Data taken in part from: Rose, H., Decema Monograph No. 57, 2nd. European Symp. on Comminution, p. 27 (1966).
2. Einenkel, W., *Chem. Ing. Tech.* **51**, 697 (1979).
3. Shamlou, P., and Koatsako, E., *Chem. Eng. Sci.* **44**, 529 (1989).
4. Brucato, A., Magelli, F., and Nocentini, M., *Chem. Eng. Res. Dev.* **69**, 43 (1991).
5. Buurman, C., *Processtecknologie* **4**, 21 (1988).
6. Rao, K., et al., *A.I.Ch.E.J.* **34**, 1332 (1988).
7. Baldi, G., Conti, R., and Alaria, E., *Chem. Eng. Sci.* **33**, 21 (1978).
8. Wang, L., *Diss. Abstr. Int.* **52**, 487 (1991).
9. Tojo, K., and Myanami, K., *Ind. Eng. Chem. Fundam.* **21**, 214 (1982).
10. Tatterson, G., *Fluid Mixing and Gas Adsorption in Agitated Tanks*, McGraw–Hill, NY (1991).
11. Laity, E., and Traybal, H., *A.I.Ch.E.J.* **3**, 176 (1959).
12. Premier Mill Co., HM Series Tech. Bulletin, Reading, PA (1993).
13. Westinghouse, Georgia Kaolin, private communication (1967).
14. Matter Mill Co., Norcross, GA.
15. Drais Perl Mill, Drais Machine Co., York, PA.
16. Dainippon Toryo Co., Japan Patent #32,724, Feb. 22 (1982).
17. Conley, R., *J. Paint Tech.* **4**, 73 (1972).
18. Schmidthammer, J., Sweco, Inc., private communication.
19. Alexander, P., *J. Research London* **3**, 68 (1950).
20. Shoh, A., 9th Symp. Ultrasonic Ind. Assoc. (1977).
21. Esche, R., *Acoustica, A.B.* **2**, 208 (1952).

22. Heuter, T., and Bolt, R., *Sonics*, John Wiley, NY, p. 242 (1955).
23. Haul, R., Rust, H., and Lutzow, J., *Naturwiss*. **37**, 523 (1950).
24. Jones, J., and Herold, R., Res. Rept. No. 54078, AF (DOD) Proj. 33/616-2289 (1954).
25. Conley, R., NASA Prop. #11-02-4783, Washington, DC, Aug. 31 (1982).
26. Tarpley, W., U.S. Patent #4,156,593, May 29 (1979).
27. Shand, P., *J. Am. Ceram. Soc.* **44**, 21 (1961).
28. Rose, H., et al., *Silicat Tech.* **13**, 200 (1962).
29. Conley, R., and Johns, D., Rept. #76C114-1, Anlin Corp., Houston, TX (1976).
30. ''Three Roll Mills,'' Kieth Machinery Corp., Lindenhurst, NY (1990).

CHAPTER

9

Environmental Aspects of Dispersion

In the preceding eight chapters both functions and mechanical facilitation of dispersants in commercial processes and other endeavors have been discussed. During the three-quarters of a century that dispersants have been employed, industry and society have slowly gained an awareness of the many problems industrial chemicals can and do create in the general environment. This chapter will give a brief overview concerning the problems these materials can create in the environment and methods currently in use to prevent or minimize their damaging effects.

While most dispersants are employed at low concentration levels, often less than 0.5%, certain specific members of the group can impact severely upon the world at large, either at low concentration or by massive accumulations. The four areas of concern with disposed materials are:

1. Direct toxicity;
2. Modified toxicity;
3. Water and soil alteration;
4. Worker handling hazards.

All dispersants employed for food processing, cosmetics, and pharmaceutical products must pass extreme scrutiny by the U.S. Food and Drug Administration before being introduced into the public marketplace.[1] This is not true for a large segment of commercial products handled and used by the public, but not intended to be introduced into or onto the human body, paints, for example. A tacit assumption exists that such materials will have limited contact with the body and will not be ingested.

The four primary treatment methods for disposal of dispersants (often with their carrier solids) to render them innocuous are:

1. Chemical modification on-stream and flushing;
2. Chemical insolubilization and impounding;
3. Untreated, direct impounding;
4. Thermal destruction.

Materials in the first category are very few and consist largely of simple molecules, such as ammonia and sulfates. The second method is the more popular and should not be viewed as an out-of-sight, out-of-mind approach. Precipitation and burial allows materials time to develop stable morphologies, after which ground waters or material transit of the products pose no more problems than natural minerals to aquifers and surface stream bodies.

Class 3 techniques are those truly problematical. For water-soluble materials, even impounded in geochemical and polymer barriers, the risk of leakage and aquifer transfer always exists.

Thermal destruction carries a highly positive image (''nothing left''), except it is a large and expensive process, requires excellent scrubbers, and still has ash disposal requirements for systems where solid carriers accompany their dispersants. Where acute toxicity of a given compound exists (and these are rare), thermal destruction is not only justified, it may be the only practical method of preventing long-term contamination of drinking water resources.

9.1 Phosphates

Perhaps most publicized of the dispersant problems are those with the complex alkali phosphates. Most of these compounds are not toxic materials, and some are even employed in food products as preservatives and antioxidants. In ponds, lakes, and reservoirs where water is not highly dynamic (compared to rivers), phosphates present a serious problem, but not due to toxicity. Orthophosphate, most probably as the species monohydrogen phosphate, HPO_4^{2-} (sometimes called secondary phosphate) is readily assimilated into the biological chain of algae, water-dwelling plants and shoreline weeds, grasses, etc. Phosphate is a necessary component in all living biochemical schemes and is a major component in fertilizers (usually in reduced solubility form). In many of the aquatic flora it represents one of several growth-limiting stages.

Where excessive amounts of soluble phosphates appear in the water resources, plant uptake is accelerated, and growth and reproduction increase dramatically. Eventually such species as algae, which form at the air–water surface, spread so rapidly they block off oxygen interchange across the boundary, and animal life within the water die from suffocation.[2] This process is termed *eutrophication* (''high nutrition''). Polyphosphates (e.g., sodium hexametaphosphate) are nonmetabolizable and function to sequester or tie up calcium, magnesium, and ferrous iron ions in these water bodies. However, polyphosphates, irrespective of their structure, fall prey to reactions catalyzed by acidic mineral surfaces lining the edges and bottom of natural water bodies (silica, silicates, and sulfides) and, aided by decaying vegetation, convert to the secondary phosphate, which is almost immediately metabolized by plants.

Unquestionably, the greatest source of waste polyphosphates lies in dispersants employed in laundry, car wash and miscellaneous cleaner products. The second source is waste water from clay, titanium dioxide and other pigment processing and dispersions used to fill and coat paper products. Because polyphosphates are almost always in equilibrium between pigment surface and solution phase (typically 25–75%), flushed solution carries residues to the point of exit, the industrial sewer drain.

Polyphosphates employed in paint (usually potassium tripolyphosphate) are retained in the film and only very slowly are exsolved through the agency of weather elements and runoff. Most of these flushed entities make their way into the soil and are slowly metabolized by plants growing in the immediate area of the outwash.

This mechanism is represented by the reactions:

$$\begin{aligned} Na_2O(NaPO_3)_6 + 2H^+ + 5H_2O &\longrightarrow 6NaH_2PO_4 + 2Na^+ \\ 2NaH_2PO_4 + 3Ca^{2+} + 4OH^- &\longrightarrow Ca_3(PO_4)_2 + 4H_2O + 2Na^+ \end{aligned} \tag{9.1}$$

Because of their solubility, alkali polyphosphates would appear to migrate extensively via groundwater. However, because soils are composed dominantly of clays, phosphate ion is taken up quickly, held on the clay surfaces, depolymerized slowly, then reacted with migratory calcium, magnesium, and iron ions diffusing through the soil. These cations react to form insoluble products that localize the phosphate immediately. Eventually, this degraded, but localized, phosphate is metabolized by plant roots.

Studies undertaken on the rates of dispersant decomposition in clay processing impounds, where large amounts of polyphosphate are discharged, are shown in Figure 9.1. Note that at the shallow depths of 30 cm (1 ft), degeneration of polyphosphate to orthosphosphate is moderately rapid at pH 4, but is an order of magnitude slower near pH 6. The presence of atmospheric gas uptake is negated with deep burial because almost no change in decomposition occurs once burial to a depth of about 10 feet occurs (approx. 3 months). Depending upon porosity and type of accretion, soil below about 18 in. (45 cm) does not receive atmospheric gas diffusion. This occurs because of tight packing with near-log-normal size distributions. Where distributions are monosize, air diffusion can occur much deeper. In sampling old tailings piles for precious metals, where the impound material (quartz and feldspar) was near monosize and about 20 mesh, atmospheric diffusion was shown to penetrate at least 25 feet (8 m).[3] The precise role of atmospheric gases, most likely carbon dioxide directly and oxygen indirectly (by oxidizing vegetation and accompanying organics), in complex phosphate breakdown in impounds is not well known.

Where polyphosphates are discharged into sewer lines and bypass soils, the

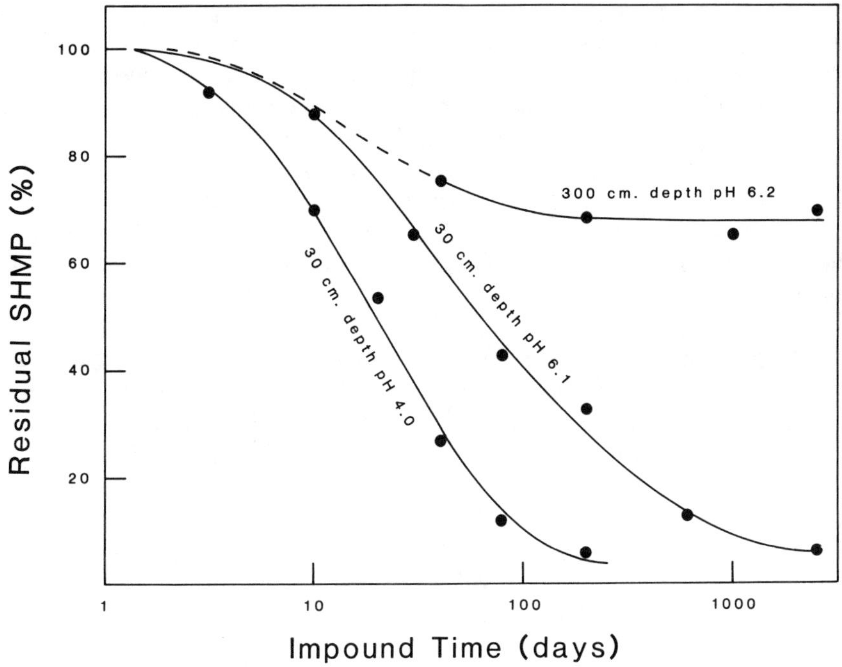

Figure 9.1. Depth and acidity influence on polyphosphate breakdown in impounds.

lifetime of the structure will be affected by the water pH, soils in suspension, hardness (divalent ion content), and turbulence.[4] In many instances sodium hexametaphosphate can travel several hundred miles (and many weeks) prior to any depolymerization or chemical breakdown and localization. Once the species reaches a terminal water resource (lake or pond), it accumulates. Southern Lake Erie, which drains the entire Cleveland, Ohio, residential and industrial basin, in the 1960s showed concentrations of phosphates approaching 0.1% (1000 ppm) in places.

Commercial operations, where polyphosphates are employed for in-plant operations and do not accompany the product into the market place (see Chapter 1), discharge these compounds at significant tonnage. In a Georgia plant where pigmentary clays are produced, waste phosphate amounted to several million pounds annually in the decade of the 1960s. At this level of discharge, some method of chemical correction must be employed. As the discharge contains waste acid streams and low-grade reject clays, the phosphate degrades quickly. Addition of limestone, and in limited instances, free lime (calcium hydroxide), neutralizes the acid and converts the polyphosphate to insoluble calcium acid phosphate.[5,6] This operation must be conducted with the assistance of impermeable impounds.

Impounds allow excess water to evaporate while the insolubilization process continues, usually several weeks. Eventually the impound fills with waste feedstock and a new facility must be constructed. The advantage to this methodology, beyond immediate elimination of soluble phosphates from the waterways, lies in forming the insoluble species at considerable depths. There it becomes localized on a permanent basis because the impound is isolated from groundwater seepage, aquifers, and rain runoff. Calcium treatment is a necessity because of the limited reactivity of this compound once deep burial occurs.

Organophosphate salt dispersants represent a special case of phosphate introduction into the environment. While the tertiary organic derivatives (tributyl phosphate, for example) have both high toxicity and extended life in the environment, the ionized salt compounds are less viable. Because alkyl or aryl groups are attached directly to the phosphorus atom, they metabolize differently in the soil and in water than inorganic phosphates, primarily due to their stability and slow interaction with polyvalent ions. In the presence of oxidizing bacteria (near the soil surface and near the water–air interface), *R*–P and *R*–O–P bonds are ruptured and the alkyl group (*R*) is slowly reduced to carbon dioxide. The phosphorous molecular core reverts to hydroxyl phosphorous groups, and the system reforms to a conventional inorganic phosphate. Only the duration differs—it is greatly prolonged.

The series of organophosphates based[7,8] on ethylenediaminetetracetic acid (EDTA) are relatively harmless and slowly oxidize in impounds, usually through complexation of trivalent cations, either added to the waste dispersant stream or leached from the soil. Many of these complex phosphates can be decomposed at low temperatures, negating the problems of alkali formation and fluxing.

9.2 Other Inorganic Dispersants

The remaining families of inorganic dispersants consist of alkali metal aluminates, borates, silicates, and, to a lesser extent, carbonates. All of these react adversely with stream life, soil, and human skin. All are strongly alkaline, quite hazardous, and corrosive, especially to zinc and aluminum metal structures.

These dispersants can be chemically inactivated by reaction with soluble calcium ion. Reactions are rapid (of the order of seconds). However, the use of ground limestone can be highly ineffective as an inactivating agent. Both silicate and aluminate form a passive and impermeable coating of calcium silicate and calcium aluminate on the surface of the fine carbonate particles that renders them inactive after a few seconds of contact. With a -325 mesh particle size of limestone the reactivity is less than 20%. Microscopic studies show the depth of penetration (or reaction) on the limestone particle is 10 μm or less. Sodium carbonate does not react with limestone at all. Alkali borate reacts slowly with the limestone but eventually forms a barrier surface similar to the silicates.

Employment of free lime (calcium hydroxide) to insolublize impounding phosphates is far more effective, but also is more costly and more hazardous. Its highly exothermic heat of reaction often boils out caustic lime solution and presents a major danger to local workers. Dry lime feeders operated in the open air, especially near impounds, tend to clog quickly by atmospheric reaction with water vapor and carbon dioxide. Dangerous manual cleanout is then required.

9.3 Organo-Acid Polymer Salts

Salts of polyacrylic and maleic acids have a low level of toxicity to humans, animals, and plant life. Their solutions are nearly neutral and display little effect on delicate life balance ecosystems. Because of their organic nature, they slowly biodegrade to sodium bicarbonate, water vapor, and carbon dioxide. Most of the species tie up calcium and magnesium in complexes to a limited extent and may alter sensitive stream balances where these ions are important.

Polyacrylates and their homologs may be impounded as slurries to enable oxidative decomposition of the structure via exposure to ultraviolet radiation, the atmosphere, and bacterial decay. Unlike polyphosphates, however, the polyacrylate family requires long periods (over a year) for photo-oxidation and bacterial destruction to decompose. Chemical breakdown is assisted by pond aeration. Also unlike polyphosphates, acidification and calcium treatment have little effect on species localization and diffusion prevention.

Although high-temperature methods ($T = 1400°C$) destroy most polyorganic acid salts, these are not cost effective because of the difficulty in separating the dispersant from the aqueous carrier fluid prior to introduction into the combustion equipment.

Recycling is at best a limited tradeoff since separation and concentration from waste water are costly, and, once complexed or sequestered with divalent ions, they are not readily regenerated. Where introduction into waterways represents a hazard because of sensitive aquatic life, polyacrylates may be adsorbed upon activated charcoal, alumina, and zeolites. The latter two can be heated to destroy the adsorbate and reused but are less effective in subsequent cycles.

Anionic salts of fatty acids are readily biodegradable, but those with aryl groups (e.g. sodium alkylnaphthalene sulfonate) are less so.[9] Besides alkalinity, a certain amount of topical irritation results from their exposure, and decay products ultimately involve naphthalene derivatives, which may be carcinogenic.

Certain sodium aryl sulfonates are actually employed to recover waste products, such as tall oil (fatty acids with small quantites of unsaponifiable components) through the process of acidification of soap skimmings.[10] Tall oil from wood pump breaks as a separate phase, while the acidified, bleached solution is dispersed, or wet out, onto the lignin components.

Many of these anionic salts do not form precipitates with polyvalent ions. They may be precipitated into impounds by various amino acids, especially PABA (para-aminobenzoic acid). The precipitate is biologically degradable and is oxidized by way of benzoic acid to carbon dioxide and water. Sodium benzoate breakdown, however, is very slow.

Normal thermal destruction creates the problems of fluxing because of the generation of sodium carbonate, a material that strongly adheres to ceramic components and many metals in thermal oxidizers. Other techniques for disposal and destruction are available from the manufacturers.[11]

Cationics disposal creates fewer problems. These agents are generally neutral to slightly acidic and break down under bacterial action to carbon dioxide, water, and ammonia. Their biodegradability has been studied at considerable length by detergent companies because of their incorporation into both laundry soaps and antistatic, fabric softeners. Except for a mild irritation of the nasal lining when exposed as dusts to applications personnel, the products show little adverse skin contact problems or biological activity in waterways and soils.

In certain instances disposal causes soil-borne clays to be accidentally treated with the cationic complex, which renders them hydrophobic. Thereafter, they may float on water surfaces. These usually drift to shore or attach themselves to aqueous plant stalks, where they slowly decay to innocuous byproducts.

Cationics may be destroyed thermally without fluxing problems because of the absence of metal cations. In their place nitrogen, or ammonia, is given off, depending upon the combustion temperature. The destruction of cationic dispersants from manufacturing is infrequent because of their high costs, high energy of attachment, and engineering care exercised to minimize wastes and production costs. Their employment in industry is limited to only a few high-cost items, such as organoclays. The major source of disposed cationics is in wastewater from laundry operations.

9.4 Simple Amino Compounds

Aminoalcohols and morpholine have several characteristics in common. First, their moderately alkalinity ($pK_b \sim 7.5$–9.5) can produce damaging effects on plants in soils and animal life in water sources. Because soil acidity derives in part from dissolved carbon dioxide or silica and silicates, the interactive products, salts of weak acids and weak bases, are readily hydrolyzed and thus are not localized or truly neutralized. Second, alkali-loving bacteria in the soil slowly convert these to nitrites, nitrates, and free nitrogen. Third, these compounds have low-to-moderate levels of toxicity to aquatic animal life, primarily because of their alkalinity.

In addition to their alkalinity, alkanolamines have low vapor pressures (10–30 mm Hg) and mildly offensive odors, which are readily apparent to personnel working with them. Inhalation on a prolonged basis represents a toxic exposure, again because of alkalinity effects upon human respiratory linings. On strongly acidic surfaces, the materials are nearly quantitatively adsorbed, but on mildly acidic surfaces, (silicates, ferric oxide, and rutile) at high dilution a small equilibrium shift does occur and solution requirements for dispersion usually exceed the acid–base neutralization value. This is reflected by the retention of butanolamine on hematite, kaolinite, rutile, and quartz pigments buried to various depths, as measured by surface vapors. Portions of an in-ground emission study appear as Figure 9.2.

Alkanolamine washout solutions should not be discharged directly into sewer systems, nor flushed into earthen impounds. The molecules diffuse slowly through soils and damage root systems. The diffusion rate is pH dependent, being faster in alkaline soils. If introduced into waterways, health problems to fish and other aquatic life can result.

Bentonite-lined impounds, unless secondarily sealed and lined with impervious membranes (as polypropylene), are less effective with these organic molecules than with polyphosphates. Further, evaporative impounds lose a sizeable fraction of the chemicals to the atmosphere because of the material's vapor pressure. Alkanolamines in the atmosphere are strong haze formers and act as nucleating agents for rainfall and water droplet formation. They also react with sulfur and nitrogen oxides to form acidic intermediates which also are droplet nucleators leading to haze formation.

Waste alkanolamines can be burned to form carbon dioxide, water vapor, and nitrogen using catalytic alumina, high-temperature hydrocarbon oxidizers. Where waste vapors are collectable by a scrubber or flue gas condenser, they may be recycled into plant process streams.

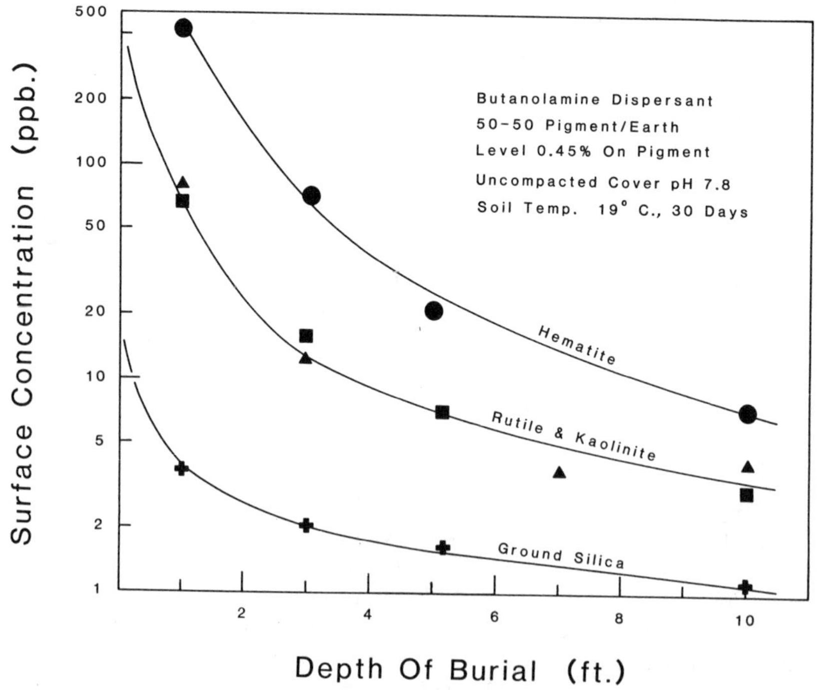

Figure 9.2. Surface diffusion of aminoalcohol from impounds.

9.5 Macromolecular Nonionic Dispersants

Many of the complex molecules based on ethylene oxide polymerization go the route of glycols—actively bio-oxidized in the soil to harmless byproducts. Many are innocuous in water bodies but may consume dissolved oxygen and deplete from the water this necessary component for animal life (chemical and biochemical oxygen demand). Where breakdown results in the formation of glycol (ethylene glycol), serious hazards to animal life result, as it can oxidize through ultraviolet radiation and other agencies to oxalic acid, a highly toxic substance to most animal life, whether living in or drinking the water.

Many polyethylene glycols are so innocuous, they are permitted in food, cosmetic, and pharmaceutical products as viscosity-control, lubrication, and thixotropic agents. They are not broken down to any great extent in the human body.

Amine-based macromolecular dispersants, however, are always problematical. Determining toxicity is not as simple as reviewing the molecular structures for toxic chemical groups. Ethylenediamine, propylenediamine, butylenediamine, and triethylene tetramine are highly toxic agents. Yet spermine (dipropylene butylene tetramine), highly similar in structure, has been shown to be a biochemical component in the body.

Toxicity testing is a prolonged process,[12] and long-term hazards, as with carcinogenic effects, may be known only after many years of testing. Thus most regulatory agencies, to maintain a cautious position, review these agents on the basis of active groups and assume hazardous activity based thereon.

9.6 Organosilanes and Organotitanates

Organosilanes, especially the halogenated species, are extremely dangerous, both directly and as derivative byproducts. As noted in the processing section on silane application, the chief hazards lie in hydrochloric acid vapors, which can be generated with great exothermic evolution.[13] Waste chemicals cannot simply be buried because of immediate reactions that take place with moist soil.

The treated products, those silicates and other oxides with surface coatings of alkyl-substituted silicic acid [R–Si–$(OH)_3$], are relatively innocuous and biodegrade to free silicic acid (H_2SiO_3) by aerobic bacteria. Their greatest problem is their hydrophobicity, which prevents quick wetting by soil components.

Organotitanates hydrolyze quickly in passive environments first to a hydrous titanium dioxide (likely the species brookite) and the substituent-derived alcohol. Most of these alcohols are innocuous and biochemically degrade rapidly. In many instances, however, they never have the opportunity to degrade biochemically—hydrolysis proceeds at such a pace that the alcohols volatilize into the atmosphere. Brookite slowly reverts during burial to a titanium dioxide form or a ferrous mixed oxide compound (leucoxeme or iron–titanium gel).

References

1. CFR Title 21, Parts 170, 200, and 600, USDHHS, Washington, DC (1994).
2. Bartsch, A., USEPA Publ. EPA-R3-72-001, Natl. Environ. Res. Center, Research Triangle Park, NC, Augst (1972).
3. Conley, R., MRT Rept. #2, Proj. #76-C-122, APM Corp. Las Vegas, NV (1976).
4. Menar, A., and Jenkins, D., USEPA Rept. R2-72-064, USDI, Washington, DC, June (1972).
5. Ferguson, J., Jenkins, D., and Stumm, W., ''Water,'' *A.I.Ch.E.* New York, NY (1970).
6. Buzzel, J., and Sawyer, C., *J. Water Pollut. Control Fed.* **39**, R-16 (1967).
7. Hatch, G., and Ralston, P., *Matls. Performance* **11**, 39 (1972).
8. Owens, J., Rept. ES-80-SS-22, Monsanto Chemical Co., St. Louis, MO, October (1980).
9. Barrows, R., and Scott, G., *Ind. Eng. Chem.* **40**, 2193 (1948).
10. Sadler, F., U.S. Patent #2,802,845, Aug. 13 (1957).
11. Rohm and Haas, Tech. Rept. #152, 154, 162, and 163, Philadelphia PA (1990–1993).
12. 34 CFR 15389 (DESI #4749), USDH, Washington, DC, Oct. 2 (1969).
13. Johannson, O., Stark, F., and Baney, R., Tech. Rept. AFML-TR-65-303, Dow Corning Corp. Midland, MI, Sept. (1965).

CHAPTER 10

Function and Character of Particulate Surfaces

All particulates possess surfaces whose chemical and physical character varies, sometimes between particles (discrimination by size) and sometimes within particles, that is, across their surface (discrimination by region). This latter characteristic is termed *heterotachy*. Studies indicate these variations occur often as patches, reflecting fundamental subsurface structural differences. In other materials crystal faces themselves are differentiatied (anisotachy) as a consequence of bond lengths, atomic packing, crystal field alterations, and similar fundamental variations.

While the surfaces of sodium chloride and diamond are isomorphous (each surface having equivalent character), those of most other solids, including barium sulfate, are varied. Phylloidal (leaflike) mineral pigments display this character with the greatest differentiation. For example, kaolinite, with the formula $Al_2Si_2O_5(OH)_4$, has its dominant broken (terminated) covalent bonds on its edge faces, those parallel to the *c* axis. Within each unit cell only a single bond is terminated, that of silica. This results in a silanol group, Si–OH, facing away from the particle body. The alumina octahedra's outwardly directed bonds are already hydroxyls, and only coordination bonding is ruptured with crystal breakage. Thus each edge chemical grouping contributes one Si–OH group and two Al–OH groups (from the single octahedrally coordinated aluminum atom.

A small amount of aluminum and ferric iron may substitute for silicon in such structures, producing a charge deficit on the surface that alters the surface chemistry and produces high-energy sites. However, the fraction of silicon atoms in a natural silicate with these substituted sites is small, being dwarfed in activity by the edge groups. Manmade alterations may, by contrast, generate large numbers of these activated sites chemically, whereupon the planar surfaces actually dominate the edge. Similarly, precipitation of hydroxide gels onto the edge faces can diminish, and in some instances completely eliminate, their silanol chemical activity.

10.1 Surface Acidity and Basicity

Titration of the *M*–OH groups of kaolinite (Boehm and Althoff methods, Chapter 13) indicates that almost exactly three-quarters are acidic and one-quarter alkaline (74–26%).[1] This corresponds well with an isoelectric point at pH 4.3–4.5. Variations in this value may result from the level of silicon substitution or slight structural disorder within the crystal. In geode kaolinites, the structure and composition have very nearly textbook perfection. This material, which has been investigated in detail for its unusual surface chemistry,[2] has an isoelectric pH of 4.2 and 16% of the total surface devoted to edge (chemically active) faces.

As referenced in Chapter 3 pertaining to polyphosphate dispersant breakdown, kaolinite's surface can become even more acidic (as measured by pK_a change) as a consequence of mechanical stress (fracture, high-temperature drying, thermal shock, hammer milling, etc.).

A survey of materials whose surface acidity/basicity has been determined by one or more of the methods cited appears in Table 10.1. Values are for untreated or chemically unmodified surfaces.

10.1.1 Hematite Surface

Hematite, red alpha ferric oxide, has 66% of the surface Fe–OH groups ionizing acidically and only 34% basically (Boehm method), in contrast to general chem-

Table 10.1 Distribution of Acid and Base Hydroxyl Groups on Pigment Surfaces

Substance	Form	% Acidic Groups	% Basic Groups
Silica	Quartz	98	2
Kaolinite	Sedimentary	74	26
Titania	Rutile	67	33
Ferric oxide	Alpha	66	34
Kaolinite	Hydrothermal	62	38
Chromium trioxide	Green	58	42
Ferric oxide	Gamma	56	44
Talc	—	55	45
Alumina hydrate	Alpha	48	52
Titania	Anatase	46	54
Stannic oxide	Meta	45	55
Lead oxide	Pptd.	5	95
Zinc oxide	Fumed	1	99

istry textbook suggestions of ferric hydroxide being wholly a base. When hematite is ground, the new surface produced is stressed, and this ratio changes. In measurements by cationic (dodecyl trimethyl quaternary ammonium bromide, DTAB) and anionic (sodium dodecyl sulfonate, SDS) agents, this ratio changes as grinding progresses. The progress is noted in Figure 10.1, where the DTAB : SDS ratio is plotted against milling and processing time for a crystalline specular hematite processed for precious metal liberation.

Prior to grinding the DTAB : SDS ratio is 1.90, corresponding to about 65.5% surface hydroxyl acidity. After 8 hours in a steel ball mill this rises to 2.51, or a surface acidity of 71.5%, representing a relative increase in acidity of about 10%. Little change occurs thereafter. Upon storing the material at ambient temperature for about 4 days, the ratio drops to about 2.2 (68.7% acidity). If heated for 16 hours in an oven at 150°C (drying cycle), the acid-to-base ratio change is more dramatic, and the DTAB : SDS ratio returns to 1.92, after which additional surface change is minimal.

The adsorption patterns for various molecules employed to modify or reduce energy of wetting in organic formulations[3] also reflect this change in surface ionization.

The most plausible explanation of this behavior is the disruption of edge unit crystal cells and an alteration of crystal field influence on the coordinated central

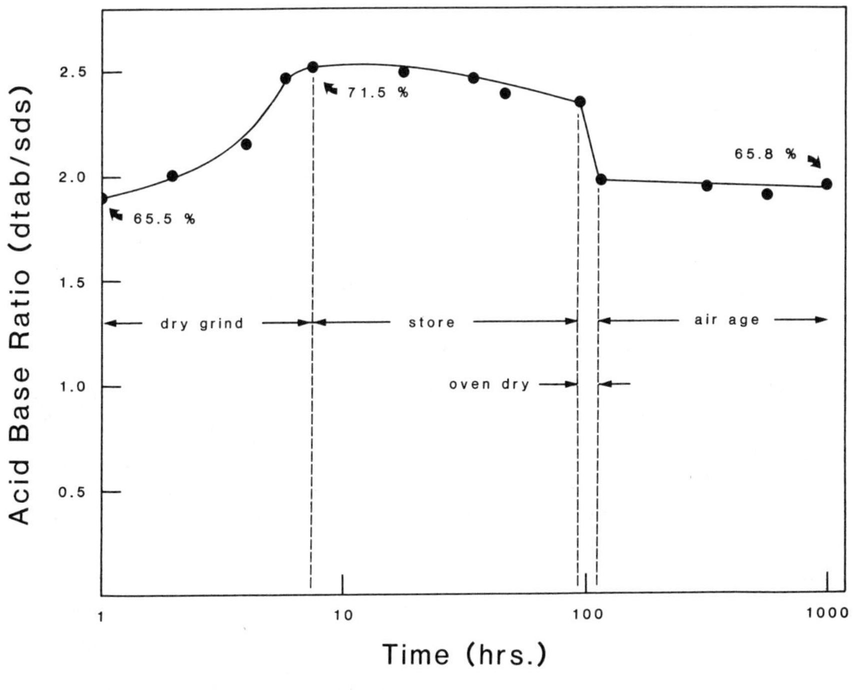

Figure 10.1. Change of surface acidity during hematite milling.

metal atom's bonding orbitals and ionization. Solids with isoelectric values above pH 7 likely become progressively more alkaline with milling. With the exception of calcium carbonate, whose isoelectric pH rises from about 9.4 to 9.65,[4] evidence to support this is sparse because few insoluble metal oxides exist in this alkaline isoelectric range.

When cyclohexane or alcohol adsorbs onto ground hematite, both form a monolayer at ambient temperatures, based on BET gas adsorption isotherms.[5] For high-surface-area iron oxide (>50 m^2/g) cyclohexane greatly exceeds methanol in molar adsorption (about 2 : 1) and will replace methanol preadsorbed on the surface. However, neither will displace adsorbed water (compare Table 7.1 for TiO_2) below the evolution temperature for water, about 125°C. These adsorption patterns confirm the dominant acidity of ferric oxide's surface.

In spite of many chemical similarities between aluminum oxide and ferric oxide, alumina behaves dramatically differently with regard to its dispersive properties. Although alumina disperses reasonably well with polyphosphates and polyacrylates, the adsorption energy of most dispersants is moderately low. Even water itself has a heat of adsorption on alumina that is approximately equal to the heat of vaporization of water[6] (9.6 kcal/mole). As a consequence, obtaining essentially bone-dry alumina for ceramics, adsorbents, and abrasives is readily accomplished. Dispersions with dipolar agents, such as AMP, which could be useful in wetcast ceramics because of their fugitivity, are not very effective for alumina. Nor does this effectiveness change appreciably upon grinding alumina. Sodium silicate serves as an effective agent but has a very short lifetime due to surface chemical reactivity unless the alumina has been calcined.

10.1.2 Calcium Carbonate Surface

One alkaline material that has been studied extensively is marble, or calcite, frequently ground for applications as a reinforcing agent in casting resins employed to produce automotive grilles, bumpers, light housings, etc. Most of these resins are modified acrylics with up to 30% calcium carbonate as filler. Calcium carbonate is employed not as much for its color as for its mechanical reinforcement properties, dominantly impact resistance.

Table 10.2 represents part of a study on the effects of ground calcium carbonates with similar size distributions produced by different means on resin characteristics, as reflected by casting viscosity. The resin contained free phthalic acid, which likely served as partial dispersant for the carbonate filler. Note that dry grinding followed by immediate incorporation into the resin generates the highest viscosity. Customer requirements for this application dictate a maximum pot viscosity of 10,000 cp.

Data in the column designated ''Reactivity'' are a reflection of high-energy surface character. The test, timing the reactivity with a standard sulfurous acid solution, suggests that aging or thermal annealing exerts a strong influence in returning the surface of dry-ground materials to a low-energy state.

While wet milling certainly provides a carbonate with the lowest resin vis-

Table 10.2 Dispersion Characteristics for Ground Marble in Molding Resin

Grinding Method	Agent (%)	D_{50} (μm)	Zeta Potential (mV)	Reactivity[a] (Acid Method)	Resin Viscosity (cp)
Dry	None	2.2	+4.8	7.3	14,600
Dry annealed 1 h	None	2.0	+4.6	12.8	12,700
Dry annealed 16 h	—	2.0	+2.7	29.4	8,900
Dry aged 30 d	None	2.0	+2.9	30.3	8,800
Wet	None	2.4	+2.3	31.1	8,300
Wet annealed 16 h	None	2.4	+2.3	34.2	8,000
Wet dispersed	0.3[b]	2.0	+2.0	36.0	8,000

[a] Reactivity is the time in seconds for 50% to react with 5% sulfur dioxide solution at 1% solids.
[b] Sodium polyacrylate dispersant, mol. wt. 2000.

cosity, the additional associated operations of filtering and drying add prohibitive cost to the material. Thus the methodology for obtaining the best material involves dry milling and extended-term aging (hopper, silo, warehoused bags, etc.).

A further examination of grinding influence upon the carbonate's surface character is derived from BET nitrogen adsorption studies shown in Figure 10.2. Freshly ground material (lower curve) is typical of Type IV gas adsorption isotherms indicating the presence of substantial amounts of fissures and high-energy surfaces (or patches).

The upper curve is representative of the common Type II isotherms, which indicate low surface energy and are employed for analyzing surface area (BET).

10.1.3 Magnesia Surface

Magnesium oxide's surface consists of essentially 100% alkaline hydroxyls. The material is insoluble in water. Magnesium metal, once exposed to the atmosphere, becomes quickly oxidized and coated with hydroxyl groups followed by partial coverage with carbonate groups. A variety of applications, including flash lamps, pyrotechnics, military flares, and laboratory chemical reactions, all employ magnesium powders, filaments, or films. Because of the metal's ease of oxidation and ignition, high heat of reaction, and position on the electromotive force scale (−2.37 volts), it is rarely ground in air or under water.

Finely ground magnesium powder may be produced by ball milling dry under a blanketing nitrogen atmosphere or wet in cold, low-vapor-pressure hydrocarbons (mineral spirits, fuel oil cuts, etc.). Wet grinding without an agent results in a terminal top size about 97% at 3 μm. However, if sodium lauryl sulfonate,

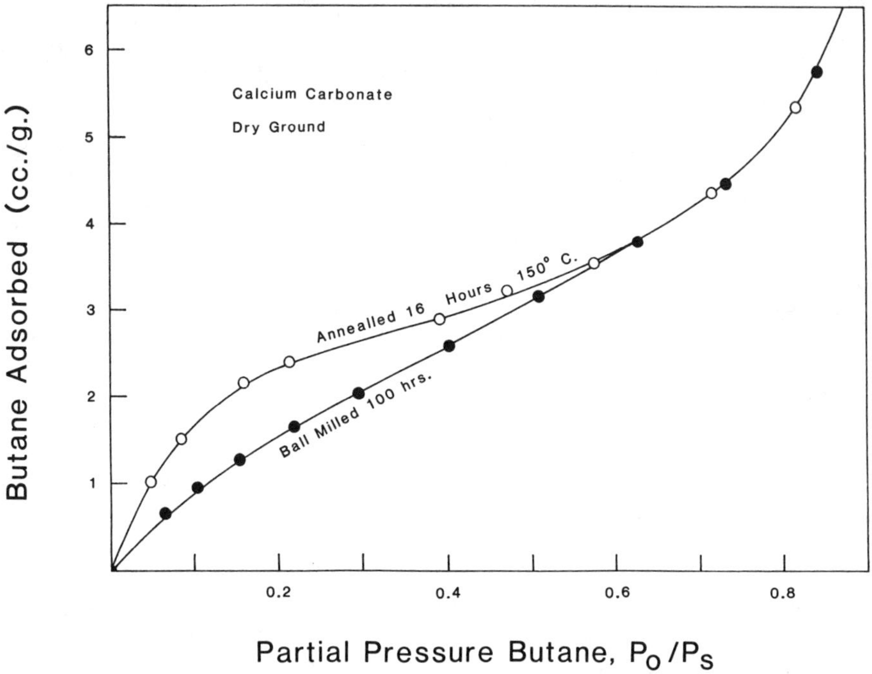

Figure 10.2. Butane isotherms on processed calcium carbonate.

about 0.1%, is dissolved in the hydrocarbon, the milling proceeds down to about 97% at 0.5 μm.[7] Ground product is often stored under such liquids to retard oxidation from atmospheric gases at ambient temperature.

10.1.4 Mixed Pigment Systems

Milling of titanium dioxide is often necessary to break apart the dimers, trimers, and other oligomers of the fundamental units, approximately 0.3 μm in size. These are formed in the thermal processes of TiO_2 manufacture (drying, fuming, etc.) and are retained in shipment to reduce dusting and atmospheric losses. The higher the percentage of discretely separated titania particles in a formulation, the greater will be its light scattering capacity. A TiO_2 trimer, for example, as a photon scatterer serves little better than a singlet. Some form of milling at the application site is a necessity to optimize optical properties.

However, milling of titanium dioxide to accomplish particle separation may induce dramatic changes in its surface chemistry, particularly its acidity and basicity. This derives not from particle fracture (fundamental particles are generally too small to fall within the milling capacity of most mills) but from mechanical abrasion and stripping of alumina- and silica-precipitated coatings from the titania surface, which changes its surface chemistry. These coatings consist of patchwork gels 30 nm and under in size. For thorough surface coverage, gels are often formed as multilayers on the titania surface.

The rutile form of TiO_2, having about $\frac{2}{3}$ of its hydroxyl groups ionizing acidically, may have an isoelectric pH of nearly 7 as a consequence of these coatings. Severe milling can alter the pigment's surface character and dispersed state. This is especially true where pH control is critical and/or where other pigments with diverse surface character are components in the formulation.

Figure 10.3 is a schematic representation of titanium dioxide pigments coating exfoliation from milling, especially sand milling.

Figure 10.4 is a shadowed electron micrograph of a heavily milled treated rutile blended with dispersed kaolinite as an extender–spacer pigment for water-based paint applications. Note the preponderance of flocculated titania particles, both homotactic (between titania particles) and heterotactic (with kaolinite). While pH elevation alone may have corrected flocculation, polymer sensitivity and other formulation characteristics may restrict it.

In compounds where titanium dioxide is a component part of an overall structure, as with barium titanates for electronic capacitor use, surface Ti–OH groups appear less acidic. Though certain crystal structure similarities to rutile exist, for most applications barium titanate is more similar to anatase in its surface chemistry. The titanates appear to have a highly stressed surface structure, one of which has the appearance of a chemical "skin."[8] This skin thickness is not related to particle size or morphology. Impurities substituted to alter the material's ferroelectric character can change this skin greatly, not in thickness but in titanol ionization. An examination of the structure in Figure 10.5 shows this more readily.

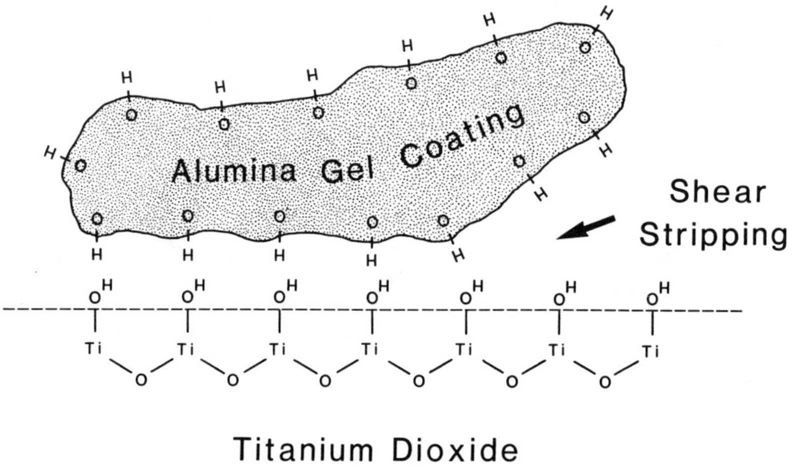

Figure 10.3. Titania coating damage from excessive milling.

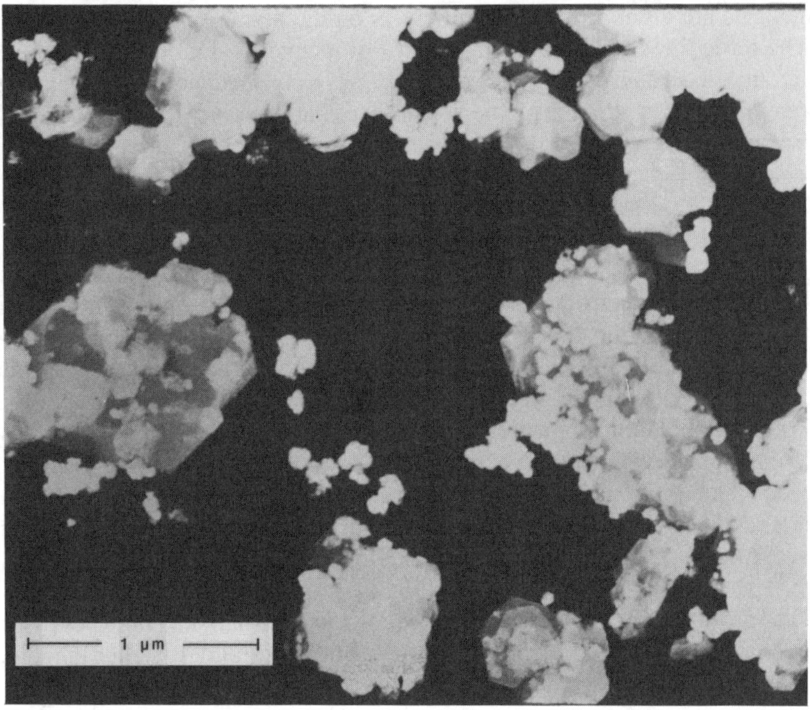

Figure 10.4. Titania–kaolinite pigment interaction.

Electropositive barium atoms feed electrons into the titania complex. These in turn shift the electron pair bond from oxygen toward its hydrogen atom, decreasing the hydrogen's ionization. Part of this activity is electron shift controlled and part is crystal field controlled. Replacement of either barium or titanium atoms with substituents will alter these electron orbital shifts accordingly.

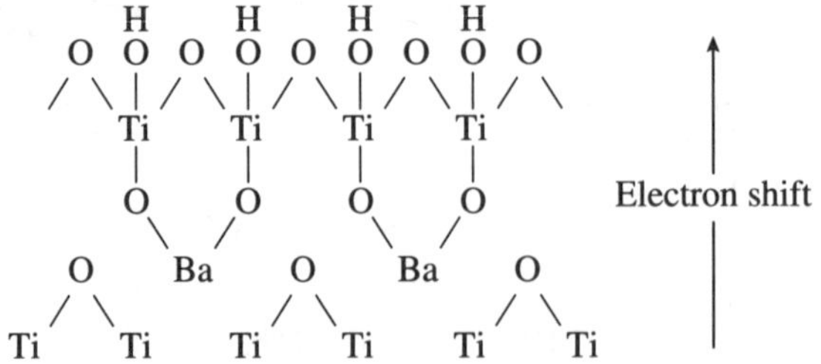

Figure 10.5. Reduced surface acidity in barium titanate.

Heterotactic flocculation is also observed with precipitated iron oxide, where kaolinite serves as both a leveling and extender pigment. Zeta potential curves for the two pigments appear in Figure 10.6, where KTPP is the dispersant of choice for the kaolinite. Iron oxide, being a relic of oxidation–precipitation reactions, has the more positive potential curve.

Two formulations were made, one at pH 6.4, the other at 8.6, of the dual-pigment system in acrylic emulsion binder. Note from the surface potential plots that at pH 6.4 kaolinite is negative but iron oxide is very slightly positive, while at pH 8.6 both are negative. The formulations were diluted down, coated onto glass, and transmission electron micrographs were taken of a small particle assemblage, as shown in Figure 10.7. Flocculation is clearly in evidence!

The formulations were highly sheared, and films were again cast on glass at full concentration. Electron micrographs were made, first by microtomed section perpendicular to the film plane (plane of applications) and the second by replication of the surface. These appear as Figures 10.8(a)–(d).

Careful inspection of the microtomed sections shows slightly superior separation and dispersion of the two pigments with the more alkaline formulation. The surface texture difference is more dramatic. The alkaline-dispersed formulation has a smoother, more reflective surface. In the acidic formulation the surface irregularity dimensions are of the order of the wavelength of light, which

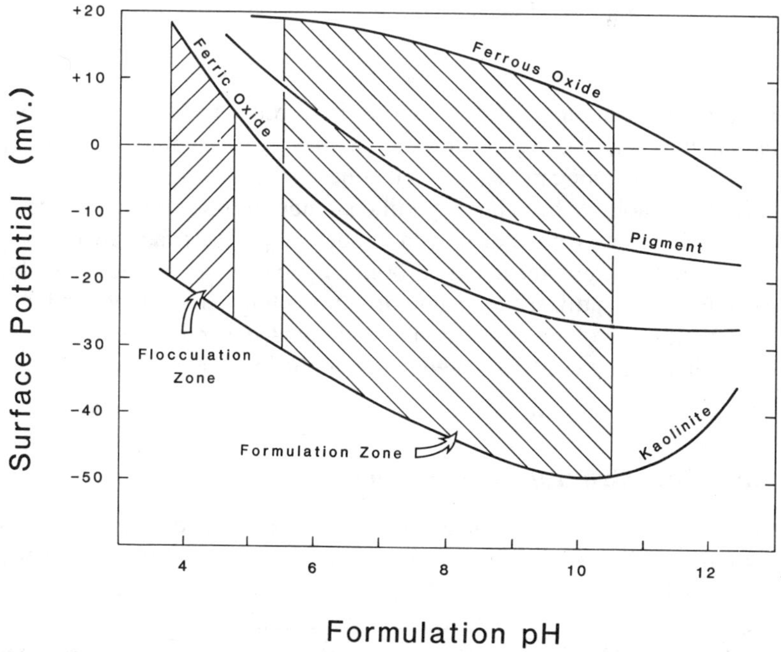

Figure 10.6. Iron oxide–kaolinite zeta potential curves.

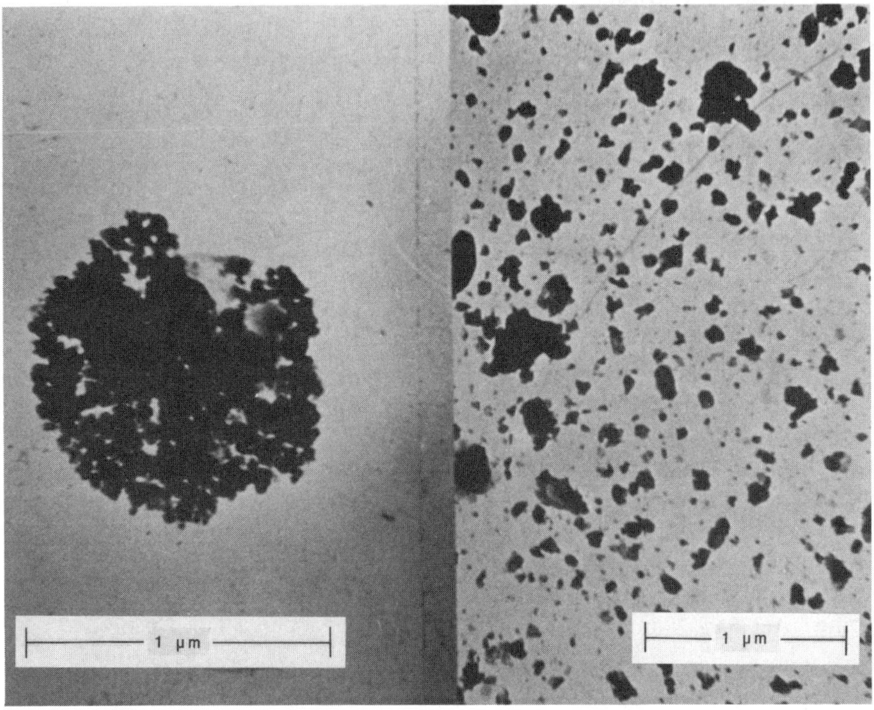

Figure 10.7. Iron oxide–kaolinite formulations, pH 6.4 and 8.6.

will induce highly random scattering during light reflection. For flat and very low-gloss applications, the pH 6.4 formulation may be acceptable. However, for semigloss to gloss applications, the alkaline makedown is unquestionably favored. Had the kaolinite been heavily milled, rendering its surface more acidic, alkaline formulation may have displayed a more irregular surface also.

In coatings with gamma ferric oxide for magnetic memory surfaces, a similar acid–base character is utilized[9] to enable firmer bonding between the binder and particles of both ferric oxide and alumina oxide (added as polishing agent), and to render particles more easily oriented under applied magnetic fields during the fluid phase.

10.1.5 Utility Ground Coal Slurries

Most solids are moderately homogeneous in subsurface structure. However, some are not and display differential properties during grinding. An excellent example is coal. Bond Work Index* figures for coarse grinding of coal lie

* The Bond Work Index is an empirically determined energy value for a variety of materials based on ball milling from a standardized feed size to a standardized product size.

(a) (b)

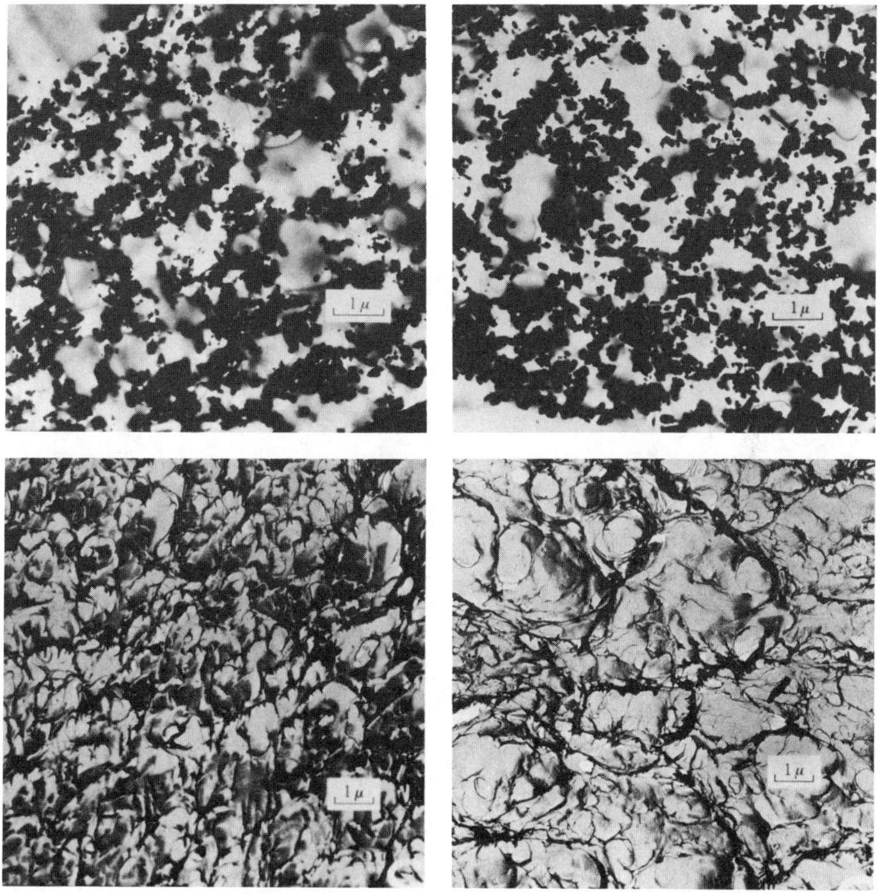

(c) (d)

Figure 10.8. Acrylic film with iron oxide–kaolinite pigments: (a) cross section of pH 6.4 formulation; (b) cross section of pH 8.6 formulation; (c) surface of pH 6.4 formulation; (d) surface of pH 8.6 formulation.

between 10 and 12. The exact value depends upon the location on the bituminous–anthracite scale (degree of softness or hardness). As coal grinding proceeds down below the sieve range where it could serve for coal–oil slurries and utility boiler applications, this value rises into the 40's and 50's. Milling energy to produce new surface rises dramatically, as do operational costs.

For efficient combustion in oil-designed burners, the coal's top size must be no greater than 7 μm.[10] Figure 10.9 is part of a Department of Energy sponsored study to grind anthracite coal in a fuel mixture of #2 diesel and heptane.[11] The feed was $-\frac{1}{4}$ in. dry coal from a crusher. Grinding was by vibratory mill at 20% coal solids using laurylamine as dispersant and $\frac{1}{2}$ in. alumina cylinders as the grinding media. A comparison grind (16 hrs) with no dispersant is shown as the

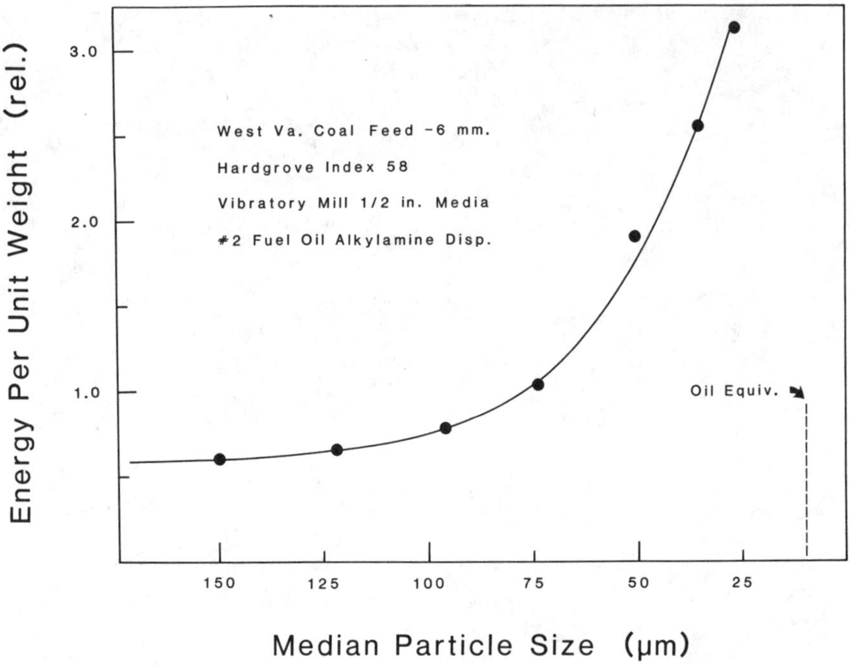

Figure 10.9. Size reduction rate of coal dispersed in an oil slurry.

micrograph of Figure 10.10. Even though coal possesses a hydrophobic surface, ground particles are highly flocculated in the oleophilic medium in the absence of dispersant.

To improve upon the exceptionally poor grind rate in the $-10\ \mu m$ region, a series of dispersants were tested for effectiveness as size reduction aids in coal–oil slurry grinding. Four of these were (1) sodium lauryl sulfonate, (2) polyfluorohexanol, (3) 2-nitropropane, and (4) a C-12 aliphatic fatty amine. Grind reduction rates in the vibratory mill improved markedly. A survey of these data shows that in all cases differences in size reduction rate occur well before the fine critical size region (7 μm) is attained. More details on this process are given in Chapter 12. Of these four recommended agents, only the aliphatic amine was adequate as a dispersant to produce fine size reduction.

Structure in coal, unlike other solids, especially with bituminous coal, progresses during grinding from a carbonaceous, rigid structure at coarse sizes (>1000 μm) down to benzopyrene tars at fine. This transition may occur as high as 200 mesh (74 μm) but no lower than 10 μm. However, with any coal the fine size cannot be attained without appropriate dispersants. Even with these the yield is very low, equivalent to a high Bond Work Index for the material.[12] Mill energy with the vibratory mill is not the limiting factor in this size reduction, though it is an economic factor, but material composition is.

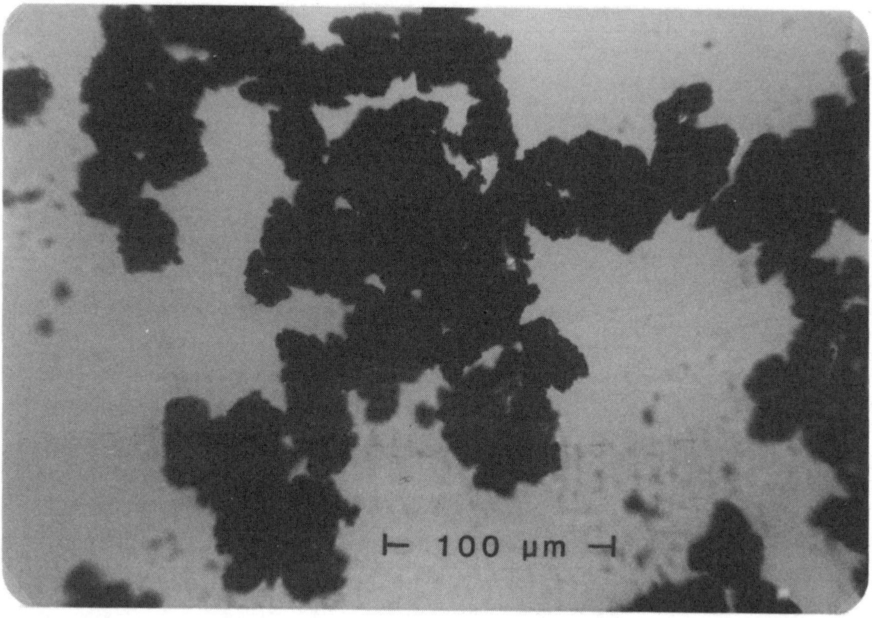

Figure 10.10. Coal ground in oil without dispersant.

10.1.6 Gold Liberation

Gold traditionally has been extracted from host rock by crushing and leaching with alkaline cyanide solution in the presence of oxygen or oxidizers. Because gold commonly deposits along veinlets in quartz and feldspar, crushing opens to solution exposure these faulted zones quickly. Grinding below 20 mesh (840 μm) is rarely necessary.

In recent years extensive gold deposits have been found in which the metal occurs as very finely divided and well-disseminated particles, almost colloids, within the crystal structure of the host rock. Gold particles may be 10 μm or smaller. Older grinding techniques fail to liberate these easily, as the metal is not associated with major rock fractures or fissures. Further, because of the metal's fineness and broad dissemination, tabling exercises (gravity enhancement) do not significantly raise ore concentration.

However, even the presence of small inclusions induces some stored stress in the host structure,[13] and prolonged grinding will eventually open these to chemical exposure. Very fine grinding of granite, however, is an expensive exercise. Thus overgrinding for low levels of gold may exceed economic constraints. Figure 10.11 is a grind-liberation plot for a 17.6 ppm (0.51 oz/ton) gold-in-feldspar ore. Milling was conducted in the wet state (cyanidation) with a laboratory 2 ft (60 cm.) × 2 ft ball mill and −18 mesh (−1 mm) feed. In the first test ore was reacted solely with cyanide solution as pH 10.6, while the second employed 0.2% TSPP in addition to the cyanide leach. Approximately 95% of

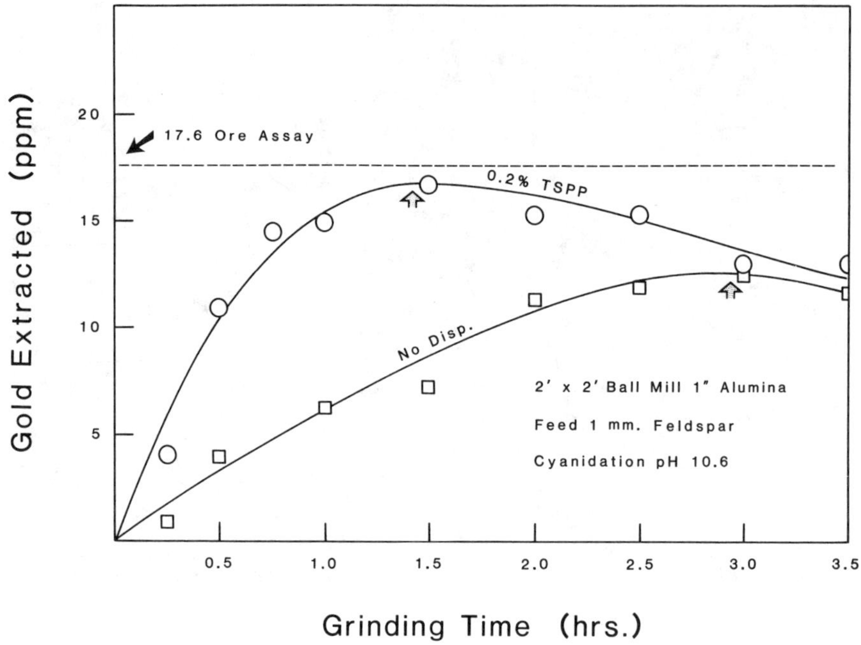

Figure 10.11. Gold liberation during cyanidation grinding of feldspar.

the gold leached in 1.5 h with TSPP (peak solution value), but only 70% leached without dispersant (peak value at 3.0 h). Both leach curves drop off slightly with overgrinding. This may result from particle welding or gold-containing solution entrapment in fines.

The combined differential of shorter grind time (50%) and increased yield (15%) more than justifies the additional costs incurred with dispersant addition.

Grind time–particle size distribution plotted in Figure 10.12 reflect the fracture sensitivity of inclusions in a host rock. Note that a ''knee'' in the particle size distribution occurs very shortly after wet milling commences. Its presence in the hammer mill feed exists but is less apparent. As milling time with TSPP increases, this knee moves progressively to higher weight percentage values of the distribution, disappearing after 4 h. The slope increase during milling, which becomes more obvious when plotted on log-probability graphs, derives from an ore hardness increase, or fracture energy increase per unit area. Particles above this size, sometimes termed the *critical mosaic*, contain frozen stress centers, usually resulting from inclusions, such as gold in the reference ore. These become initiating sites when fracture formation is induced. Thereafter, host rock becomes more difficult to fracture because fundamental bonds must be broken unaided by any energy-lowering frozen stresses. This point on the particle size distribution plot represents the mosaic size of the inclusion (not the inclusion

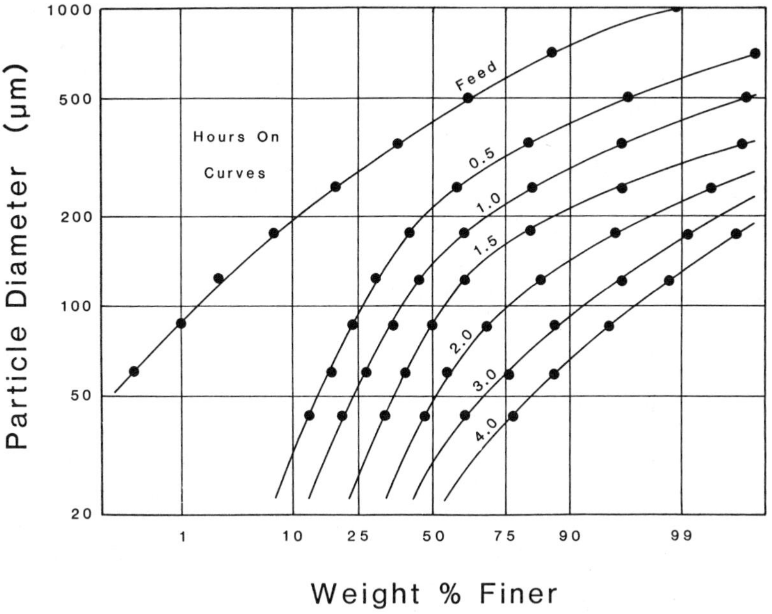

Figure 10.12. Particle size change during cyanidation grinding.

itself). For the feldspar ore of Figure 10.11, this size is about 150 μm (100 mesh).

With TSPP grinding proceeds faster and the mosaics tend to be fully exposed after about 1.5 h. Without dispersant (or milling aid) an additional 1.5 h are required. Following mosaic liberation, gold leaching is not further enhanced. This phenomenon has also been observed with a variety of other liberation studies, some of which, because they involve other interactions or because they recover the host and discard the inclusions, will be detailed in Chapter 12.

10.1.7 Pyrite Liberation from Coal

Not all processes of grinding and liberation are as economical as that shown for gold. One such process involves coal. A serious flaw regarding coal employment in utility boilers lies in the presence of metal sulfides, predominantly pyrite, FeS_2. Pyrite in fine coal not only produces high levels of corrosive sulfur dioxide in stack gases, it also produces large and rigid clinkers in the burning chamber which make pulverization and disposal difficult.

In one process developed for high-sulfur coal, ore is partially crushed, then wet ground with selective dispersant, organic amino sulfides. Sulfides are separated out, and the coal is mechanically dewatered and then introduced to the boiler feeder in the damp condition to minimize dusting hazards (fire and health).

Coal, a mixture of free carbon and aromatic tars, is hydrophobic, as is the surface of pyrite. However, coal is softer and grinds finer than pyrite. Also pyrite is approximately four times as dense as coal. Thus a well-ground coal can be separated from its pyrite impurities by density discrimination, if the coal is ground below the particle size of the pyrite, its mosaic.

In many coals this liberation size lies in the -30 to -40 μm range. Value gain from desulfurization rarely compensates for the excessive costs of grinding to that fine size, even with wetting agents added as an enhancement. Unlike gold, pyrite, once separated, has no value. Because it forms sulfuric acid upon moist oxidation, its disposal creates problems. In this instance it is truly ''fool's gold.''

10.1.8 Opaline Silica Pigments

Opaline silica, called variously chalcedony, opal, onyx, agate, and chert, can occur in a very white, highly pure mineral form. Its structure is noncrystalline and glasslike. The material has pigmentary value and is employed in two forms: hydrated and dehydrated (calcined). The surface and mechanical properties of the two forms change with thermal treatment. When ground to about -20 μm top size (98%), it serves as an excellent flatting agent for paints and plastics. Grinding to -5 μm produces a high-grade extender pigment. In either form it displays moderate abrasivity and finds use in both size ranges for cleansers, dental polishes, and metal buffing compounds.

Ground natural silicas, whether quartz, tridymite, cristobalite, or chalcedony, once severely crushed and the surface highly stressed, behave quite similarly independent of original crystal form. For example, the isoelectric point for highly crystalline, essentially stress-free quartz occurs at about pH 2.4. Less crystalline forms (as chalcedony) may have isoelectric points as high as pH 2.8. However, upon grinding followed by immediate measurement, all species drop significantly, quartz to 2.2 (more acidic) and less crystalline species to about 2.4.

Elevation of the pH to 8.0, which generates a surface potential of about -20 mV, and measurement of the total surface charge show the surface to be ionized with a charge density of 0.081 coulombs per square meter of surface area.[14] This corresponds to about 1.90 nm^2 per unit (negative) charge. Based on the molecular structure of quartz, this charge density computes to 10.5% of the surface silanol groups being ionized. At pH 10 the charge density rises to slightly above 0.8 coulombs per square meter, or about 0.20 nm^2 per unit charge. At this pH level essentially 100% of the silanols are ionized. The zeta potential at this pH is just over -30 mV, and the silicas are marginally dispersed by pH alone.

Because of silica hazards, wet grinding provides both the best method to attain fine sizes (less than 10 μm) and a dust-safe atmosphere for working. The grinding curve of Figure 10.13(a) represents uncalcined feed (D_{50} = 18 μm) and product from a ceramic sand mill where wet grinding (65% solids) was employed with both Tamol X-100 (TX-100), a long chain alkyl–aromatic polyether, and SHMP. The weight differential between the two dispersants at various particle

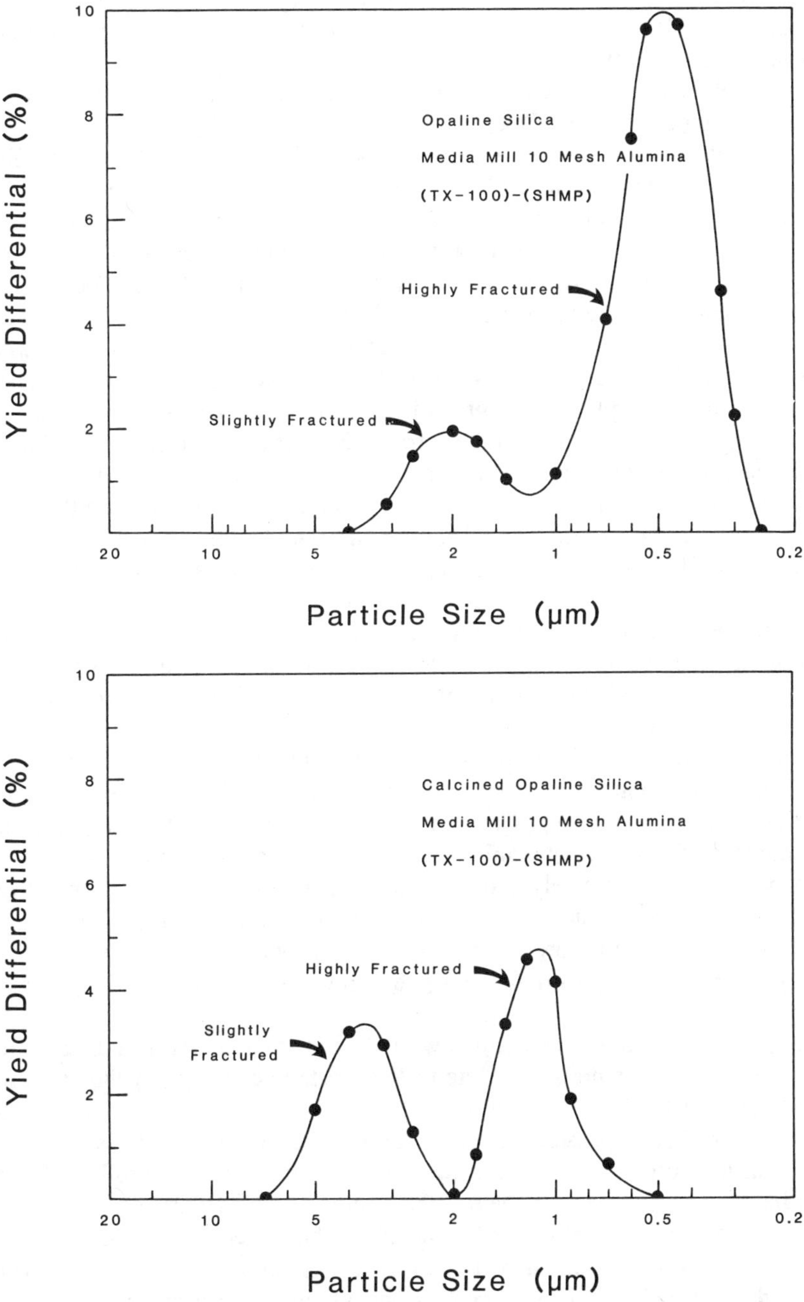

Figure 10.13. Dispersant milling of opaline silica: (a) uncalcined; (b) calcined.

sizes is plotted to show the subtle variations between the two. Note a bimodality develops between the two agents, the highly fractured portion of the feed (that below about 1 μm) showing a yield improvement of about 10% with TX-100 compared to SHMP and a minor improvement in the slightly fractured mode peaking at about 2μm.

Upon calcination, chalcedony's surface character changes significantly, as shown in Figure 10.13(b). Calcination results in a small amount of sintering (D_{max} = 7 vs. 4 μm). The superiority of TX-100 is still evident, but its effect changes, peaking at 3.8 μm (3.5%) for the slightly fractured component and 1.2 μm (4.7%) for the highly fractured. Degree of fracturing is a qualitative assessment made by particle examination with a polarizing microscope. Most interesting is the minimum differential following calcination at 2 μm, a size where maximum gain occurs with uncalcinated material.

This short milling study demonstrates that both surface and body silicon structures are altered by thermal effects. Grindability, as well as dispersant influence. varies in a more complex fashion than just rate as a consequence.

The differential yield for the highly fractured portion of the calcined pigment diminishes, which suggests a reduced acidity, perhaps due to less ordered structure in the atomic arrangement near the surface (reduced crystal field). Alternately, it may indicate calcination preferentially binds fines to coarse particles, while wet milling serves to break apart these aggregates in lieu of coarser particles.

TX-100 is a nonionic agent having a pH of 7.2, only slightly more alkaline than SHMP (pH 6.5). Its attraction for an acidic silica surface derives from multiple van der Waals interactions and is more energetic than that with SHMP, as will be shown in the next section.

Studies of noncrystalline silica[15] show heating to temperatures less than 100°C produces population densities of Si–OH groups that range from about 5 to 12 per square namometer. Their energy of reaction, related to pK_a, is about 30% higher at the lower population density than at the higher one. However, as silica becomes progressively more crystalline, its energy rises, ranging from about 320 ergs/cm^2, for adsorption of water on low-order silica, up to 800 ergs/cm^2 on quartz. Some adsorption energy variation with particle size following grinding (40 down to 5 μm) also has been observed with differential scanning calorimetry.[16]

Energy of adsorption, as measured with water, increases significantly as silica progresses from the amorphous state to the crystalline state,[17] as shown in the data on Table 10.3.

Studies of silicates possessing laminar or platelike structures often show peculiarities in adsorption behavior because of the critical size restrictions in diffusing polar molecules between the sheets of these minerals.[18,19] Hydrazine will diffuse into kaolinite to alter the structure by one molecular dimension, a process termed *intercalation*. The phenomenon, however, is best known in association with bentonites, mixed-layer clays having a silica–alumina–silica sandwich structure with alkali metals balancing out charge. Bentonites can adsorb several times

Table 10.3 Heats of Adsorption of Water Vapor on Various Silicas

Silica Species	State	$-\Delta H$ Adsorption (erg/cm^2)
Amorphous	dried gel	38
Fused quartz	annealed	140
Chalcedony	ground	430
Cristobalite	natural	531
Quartz	natural	546
Quartz	ground	729

their weight of water through intercalation and expand greatly to form gels not readily reversible. The adsorption energy for these processes, which varies with the cation substituents, is several hundred ergs per square centimeter. As a consequence the materials find broad application as pond sealers and basement liners, where they maintain impenetrability to storage and ground waters.

Calcium silicate also occurs as a laminar structure in the mineral tobermorite, Ca_2SiO_4. It has the interesting property of adsorbing water in the interlayer positions but not nitrogen, because of the high energy of polar molecule adsorption. When tobermorite is ball milled, the structure slightly disorganizes and becomes propped open. Thereafter, water penetration is quite rapid with almost an equal weight of water being tied up in intercalation.[20]

Intercalation influences dispersion, and vice versa, because of the mineral's high consumption of bulk-phase liquid. Also at least one major exposed crystal face has the same adsorptive properties as the interlayer positions. Bentonites are more readily expanded with dispersant solutions than with water alone because of the lowered energy of wetting. In some instances alkali solution alone is sufficient to prop open the structures and effect intercalation.

Acidic etching of the mineral, which replaces some of the alkali ions with hydrogen ions, develops intercalation sites with high adsorptive capacities for such complex organic molecules as tannins, humates, lignins, etc. Thus they find use when dispersed as decolorizing agents for food-grade vegetable oils, as selective catalysts for organic reactions, and as color developing coating agents for carbonless copy paper.

Acid etching also is employed for producing catalysts for the petroleum cracking (hydrocarbon breakdown) industries. It is well known that inadequate or excessive etching, which removes alumina from the structure, produces inferior material for cracking operations. At an estimated 50% alumina extraction and reprecipitation as oxide, the surface becomes highly inhomogeneous with regard to the energy of adsorption of hydrocarbon molecules. One method of assessing surface acidity, which is directly related to catalytic activity, involves gas adsorption of butyl amine (Chapter 13) and ethylene. Their equilibrium concentration between the surface and the gas phase with progressively lowered vacuum is a direct measure of the energy of adsorption.

10.1.9 Chromium Green Pigment

Chapter 2 referenced the reactivity of doped pigments as a consequence of surface acidity and basicity change. This becomes even more critical during grinding activities because of changes in surface bonding energies. The chrome green study referenced involved a color-matched pigment to be employed for state highway use (signs, toll booths, barricades, etc.), which required superior sunfastness.

The pigment, Cr_2O_3, doped with small amounts of aluminum and magnesium, required milling with solvent-letdown polymer, a mixture of methyl isobutyl ketone (MIBK) and styrene-modified polyester resin. Grinding was performed with a media mill and 1 mm ceramic media to prevent introduction of iron oxide color impurities. Chrome green processes are known to result in sinters requiring moderate energy to breakdown. During grinding fundamental particles are also fractured. Figure 10.14 is a milling time curve with the potlife time of the incorporation resin being the dependent variable. The upper curve represents chromium sesquioxide ground in solvent alone at high solids using butanolamine as dispersant, then transferred to the resin with low-shear mixing.

Clearly, dispersed-state grinding results in far fewer active sites than in situ grinding and blending, based on the striking differences in resin potlife. As noted earlier with calcium carbonate, severe grinding without dispersant shortens potlife, in this instance to the point resin setup occurred within the mill after 40

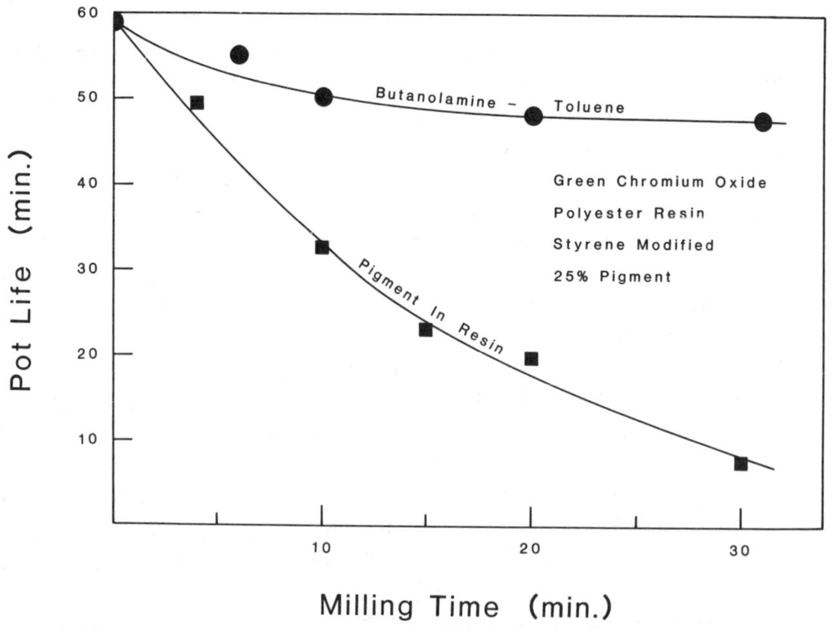

Figure 10.14. Effect of dispersant on resin milling of chrome green.

min. Separate dispersant milling of pigment may not always be practical as the dispersant may chemically interact with the pigment and alter color. While this may occur with some forms of chromium sesquioxide, the level of dispersant (butanolamine) can present yet another problem. Where dispersant equilibrium is shifted significantly to the liquid phase, the amine may react adversely with the phthalic anhydride component of the resin.

Pigments with transition metal atom cores are prone to react with electron pair donors, as aminoalcohols, and display a slow color shift due the coordination complex that forms. Copper, nickel, cobalt, and chromium compounds are highly susceptible to this hue alteration. A similar experience has been reported for Toluidine Orange in epoxy resins arising from polyamine reactions.

Readily dispersible pigments can be manufactured by simply adjusting or chemically modifying the acid–base components of a carrier resin to match those of the pigment, or vice versa. For this to be successful, an accurate determination of the relative acid–base character of the two components must be made, usually by nonaqueous titration techniques. In specialty, high-weather-resistant, marine coatings polymerized polyether polyols have been acid modified to form pre-letdown pastes of the pigments prior to incorporation into application resins.[21] The resultant formulations not only display superior weather resistance to checking and peeling; their color stability and mildew resistance also increases.

10.1.10 Milling of Zinc Oxide

Zinc oxide is an effective white pigment for multiple reasons. Its refractive index is high, 1.9–2.0; its isoelectric pH is high, 10; and it possesses excellent fungicidal characteristics in finished coatings. The oxide is manufactured in milled form as rough equants and in fumed form as needles. Zinc oxide is noted for having trapped electrons in its structure. Measurements by NMR surveys show these locate preferentially at or near the surface in the vicinity of surface cation sites and influence surface chemistry of the pigment.

Alcohols and polyols adsorb onto zinc oxide in direct proportion to the polarizability of its surface. Polarizability is directly dependent upon the method of manufacture and subsequent milling activity associated with dispersion practices. Dispersants and/or binders having polyol groups will adsorb with increasing energy as the resistivity of zinc oxide increases.[22] Zinc oxide employed as pigment filler for wire and cable insulation shows a distinctive dielectric loss at high frequencies, credited to this surface polarity.

With fatty acid dispersants, sodium salts thereof, or acidic group containing binders, this factor tends to be masked by the chemical acid–base interaction of zinc oxide's highly alkaline surface.

10.1.11 Delamination of Phylloidal Pigments

Pigments employed for producing glossy surfaces or smoothly textured films function in superior fashion if they possess a platelet morphology. Platelets lie down when well dispersed in an overlapping fashion and provide high light scatter to impinging photons. Pearlescence is created by employing mica, bismuth oxychloride, talc, and similar phylloidal materials, which orient in parallel layers in polymer formulations. The optical effect is termed *parallaxis* and *schlieren*, and causes a series of light waves to reflect off sequential but parallel layers of pigment.

Talc, mica, and kaolinite are pigments whose covalent bonding in the *c* axis direction proceeds for only a single unit cell. In the *a* and *b* axis directions bonding extends across the entire crystal. These materials are formed with sequential layers of platelets loosely associated together in an organized and hydrogen-bonded fashion. Kaolinite, for example, occurs as vermicular stacks, as shown in the micrograph of Figure 10.15. Crystal bonding values are shown for a unit cell representation in Figure 10.16. The energy per unit area necessary

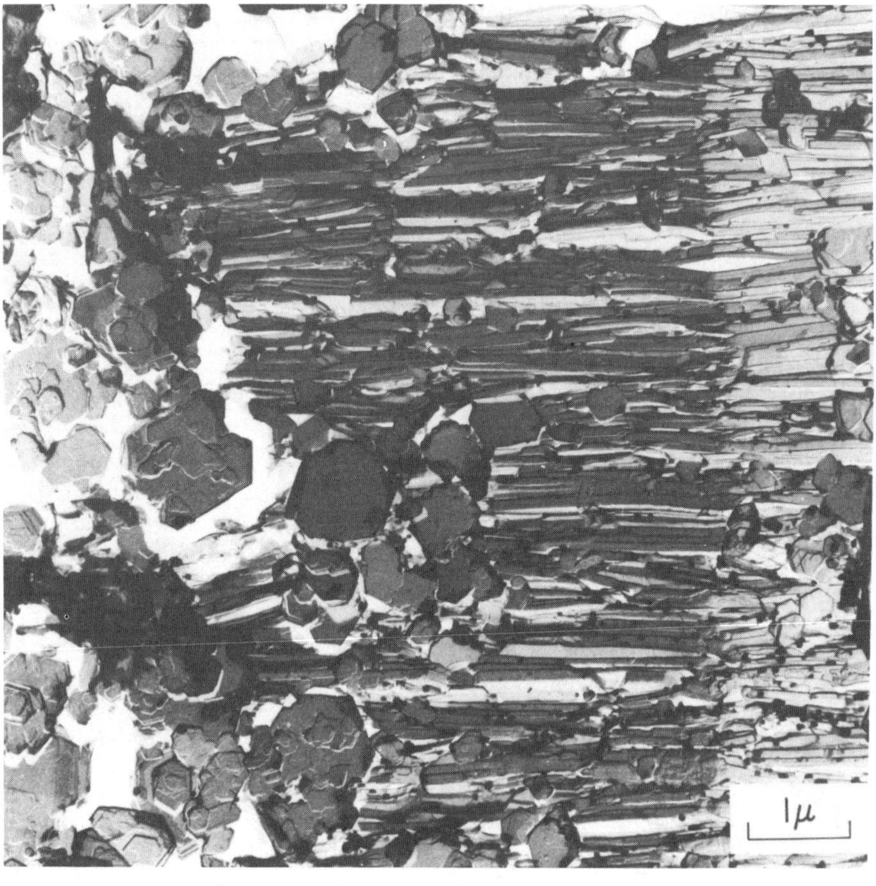

Figure 10.15. Vermicular stacks of kaolinite ore.

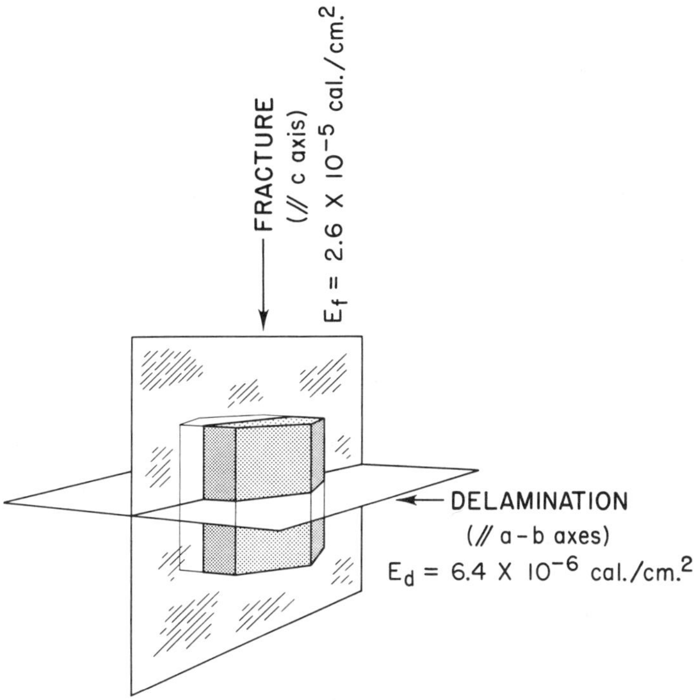

Figure 10.16. Crystal bonding (fracture) energy in kaolinite.

to delaminate (perpendicular to the *c* axis) is about one-quarter that to produce fracture (parallel to the *c* axis). Calculations of equipartition of energy indicate that random exposure of kaolinite particles to natural or manmade forces will tend to generate particles having an average aspect ratio of 6:1. Detailed investigations of sedimentary kaolins,[23] as well as the geode kaolin mentioned earlier, indicate this to be a very accurate assessment of their occurrence.

However, particle morphology higher than 6:1 provides high gloss in coatings due to leveling effects. Processes have been developed whereby 6:1 and lower-aspect-ratio particles can be delaminated to achieve much higher aspect ratios. An excellent state of dispersion, usually overdispersion, is necessary to accomplish this. A mill to provide grinding energy must be of low energy, directionally specific and shear biased in its action. If overmilling occurs, particles will be broken on a random basis, and fragments will result having essentially the original aspect ratio, only more ragged in edge texture. The most effective methodology employs the stirred bead mill (see Chapter 8). Even this mill, however, can overpower the system.

By reducing the energy through both mill rpm and media density and by employment of a chemically modified, slightly overdispersed slurry, kaolinite platelets can be defoliated to produce higher-aspect-ratio particles approaching 20:1. Portions of a recent study[24] showing the effects of two media, chemical modification and varying rotational speed (energy), appear in Figure 10.17, where 16:1 kaolinite particles are generated. Delaminated silicate produced under optimum milling and dispersion conditions is shown in Figure 10.18.

Talc, a platey (3.5:1), hydrated, magnesium silicate also was evaluated by the processes employed in Figure 10.17. This mineral can be similarly delaminated, but wet processing is generally considered to be too costly for the product quality and application. Figure 10.19 shows the delaminated talc produced with aspect ratios up to 25:1, careful control can raise this to 50:1. Talc can be partially delaminated through selective dry grinding forces in which adsorbed gases act as dispersants or interlayer lubricants. Graphite can be made to respond similarly. This method is effective in achieving very high phylloidal character because both talc and graphite have interlayer bonding energies per unit area much lower than that of kaolinite (~1 μcal/cm^2).

10.1.12 Surface Heterogeneity (Heterotactic and Anisotactic)

Three types of surface have been identified as relating to the electrical double layer and dispersive forces.[25] These are: random, isolated, and interacting

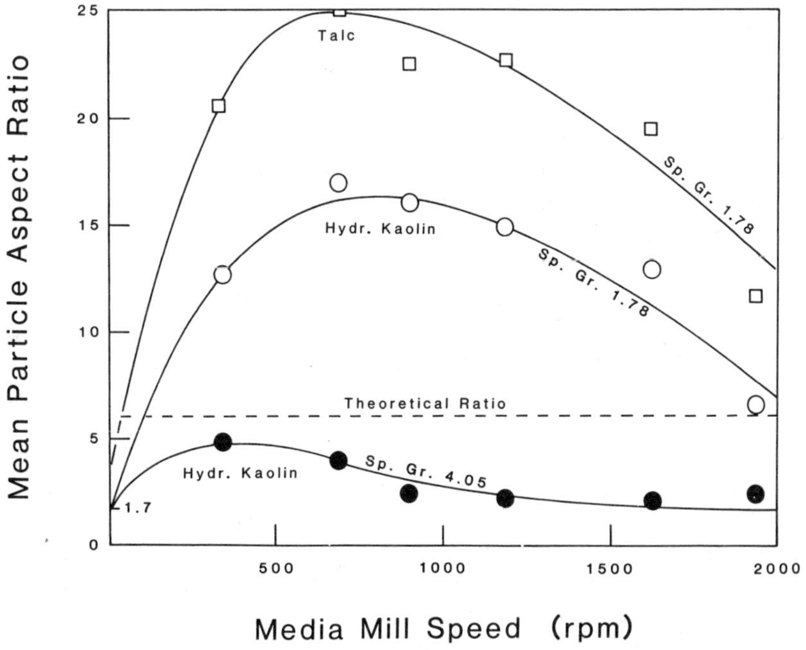

Figure 10.17. Aspect ratio dependence upon mill delamination energy.

Figure 10.18. High-aspect-ratio pigment from dispersed silicate delamination.

patches. Although most considerations of surface reactions assume a homogeneous surface, any given material may have mixtures of any of the three types. For a random heterogeneous surface the associated electrical double-layer character in that area is the same as though it were a homogeneous interface. For the patch-type heterogeneous interface, properties of the electrical double layer resemble those of the different patches.

The primary differences between isolated patches and interacting patches lie in the population or spacing of the patches. Where these are in sufficient proximity, two patches may share an influence of their double layer (inductive effect). Where the patches are sufficiently separated, little or no induced field interaction occurs. This is exemplified by materials that are crystallographically anisomorphic. Such a material is exemplified by gypsum (calcium sulfate dihydrate), which absorbs urea preferentially on certain crystal faces, allowing crystal growth to proceed predominantly in one axial direction and form highly acicular needles. It is also illustrated by phylloidal pigments as clays where primary electrical double-layer activity occurs on the alumina–silica edges, an anisotactic phenomenon, regions where but partialy coordinated Si–OH and Al–OH groups dominate.

The overall (integrated surface) point of zero charge (isoelectric point) for

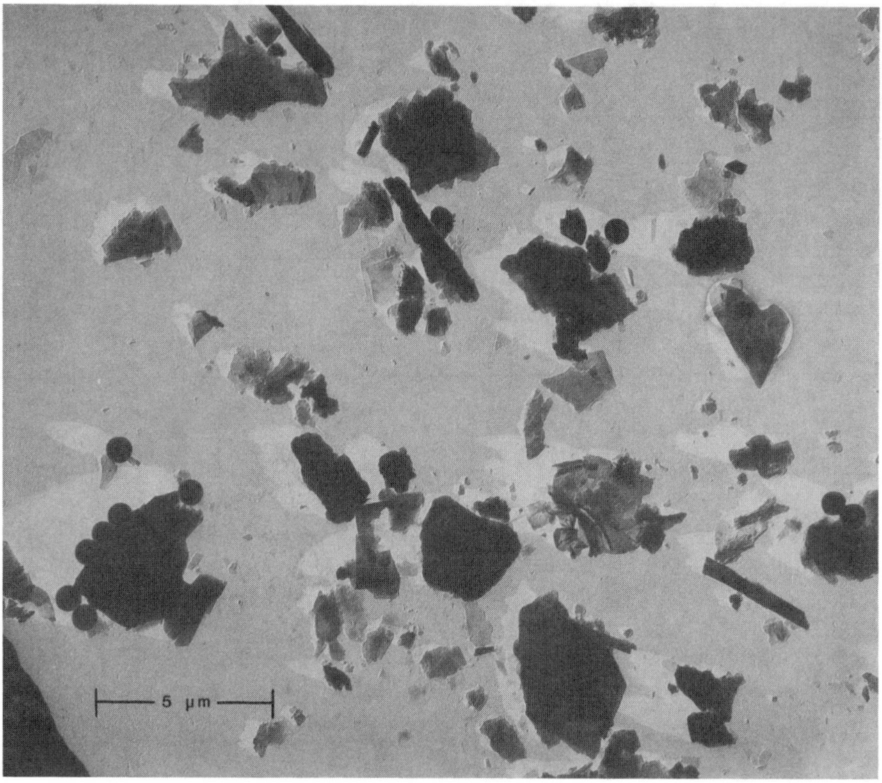

Figure 10.19. Delaminated talc produced by dispersive grinding.

any of these mixed surfaces will be a function of both the surface compositions and the solution-phase electrolyte concentration. A minimum in the electrical capacitance–potential curve reflecting the double-layer charge, which normally coincides with the zero point charge, does not coincide with mixed patch surface composition. Instead, successive, overlapping minima occur that give a broader-banded capacitance curve rather than a moderately sharp one. Were the surface truly homogeneous, by definition the two would coincide. Although it represents a complex and tedious technique, the capacitance–potential evaluation is highly useful[26] in estimating the diversity and fraction of the various patch elements on a surface.

For a patchwise surface exhibiting interacting patches, an intermediate state exists. At elevated electrolyte concentration (primarily dispersant) the patchwise state becomes apparent. At very low concentrations of solution electrolyte, the surface will appear to be a random heterogeneous surface. Thus in certain locales true charge exists at the zero-point potential, and at low levels of charge some areas remain uncharged. This is a consequence of an equilibrating dispersant on a differentiated particle.

Table 10.4 is an excellent example of distributed site energies. The measurements were computed from neon gas adsorption isotherms at different temperatures on rutile titanium dioxide.[27] The surface had not been chemically altered by secondary precipitation species, and the measurements therefore represent true structural variation across the full surface. No size differentiation was made to reflect that parameter's influence on specific energy levels.

It is interesting to compare this distribution of surface energies, which passes through a maximum in midrange, with that for activated aluminas and aluminum silicates, where the energy typically peaks at the lowest population (see Figure 5.11).

Various structural phenomena contribute to these heterogeneous surfaces: inside step structures; vacancies in the basic crystal structure (Schottky defects); interstitial atoms in charge balancing situations (Frenkel defects); equivalent charge substitutions with unequal size atoms; shear structural defects (dislocations, fractures, lattice distortions, etc.).

Because such defects create frozen stresses within particles, grinding tends to proceed such that cracks propagate through the stress locales preferentially and leave a disproportionate concentration of defects on the surface, a phenomenon associated with the gold leaching cited in Figure 10.11. Also, many defects are actually created during intensive milling and occur in higher surface populations with finer-sized particles. Thus nonuniform surfaces frequently show a skew in their occurrence with particle size, one factor contributing to size differentiating incipient flocculation.

10.1.13 Carbon Blacks

Carbon blacks are produced through incomplete oxidation of fuel gases. During their formation varying degrees of oxidation produce a variety of surfaces. How-

Table 10.4 Site Energy Distribution on Rutile Surface

Site Energy (cal/mole)	Distribution (cm^3/g)[a]	Cumulative[a] Population (%)
100	1.2	1.3
200	3.25	4.7
300	6.15	11.1
400	10.2	21.9
500	15.5	38.2
600	22.5	61.8
700	25.1 (peak value)	88.2
800	10.6	99.3
900	0.5	99.8
1000	0.1	99.9
1200	0.04	99.95

[a] Data collected by interpolation of neon gas adsorption isotherms.

ever, no specific form or chemical grouping is unique to a given surface, only the distributions of their mixtures. In fact, of all the pigmentary materials, carbon blacks represent the most heterotactic and variable of surfaces. While the basic structure of all blacks is that of a poorly crystalline or microcrystalline graphite, manufacturing can alter the material's surface from highly oleophilic to moderately hydrophilic, not through precipitation but through oxidation. Figure 10.20 includes a few of the more common surface forms which can be produced.

All of the surfaces depicted in Figure 10.20 possess some degree of polarity. Carbon black without oxidation may have terminating –CH groups, making it wholly hydrophobic. Whatever the surface, the choice of dispersants for carbon blacks may vary greatly, making these products highly variable-character pigments. The bottom species in Figure 10.20 (acidic surface) can be dispersed in water readily by way of simple alkali, short-chain amines, diamines, alkanolamines, and even ammonia. Where sodium laurate is employed (for the more oleophilic surfaces), the anionic tail portion of the agent will lie down upon unoxidized portions of the surface, allowing charge to extend into the liquid phase. Unlike many polyhydric alcohols, especially of the dispersant genre, alcoholated carbon surfaces (second example) are often moderately acidic and display similarities to the acidic surface with regard to dispersant selection. Carbon black, employed for India ink in drafting and artistic applications, may be dispersed (or redispersed when old) readily with ammonia.

Few blacks are produced with entirely oleophilic surfaces. A general guide for using these pigments in water is first to evaluate polyols as dispersants, followed by alkali addition or alkali anionics, as the system tolerates. Unfortunately, the latter tends to produce foam, which may interfere with fine black assimilation in polymers or other systems.

Carbon black can be produced with a broad range of particle size and distribution. However, the most economical of the forms have very narrow distributions from about 100 nm (0.1 μm) down to 10 nm, in which near-size particles form highly reticulated, that is, branched and networked, chains. In highly structured blacks these networks are open and cagelike, having pH values on the acidic side (~ pH 4). In low-structured blacks aggregates are more compact and ball-like, having slightly alkaline surfaces (~ pH 8). Individual particles are not held together solely by van der Waals forces but are chemically bonded through surface-active groups or direct carbon–carbon bonds. These networked structures are responsible for the difficulties in thoroughly dispersing most blacks. However, when such networks are fully dispersed, they result in a degree of system thixotropy that derives from liquid-phase immobilization, much in the fashion of food-grade gelatin.

Of all the equipment employed for dispersion (Chapter 8) of carbon blacks, three-roll mills are frequently the most efficient, followed by small-media mills with more dense media and high rpm. Ball mills and impeller-blade dispersers rarely can produce sufficient shear energy to disperse black slurries thoroughly.

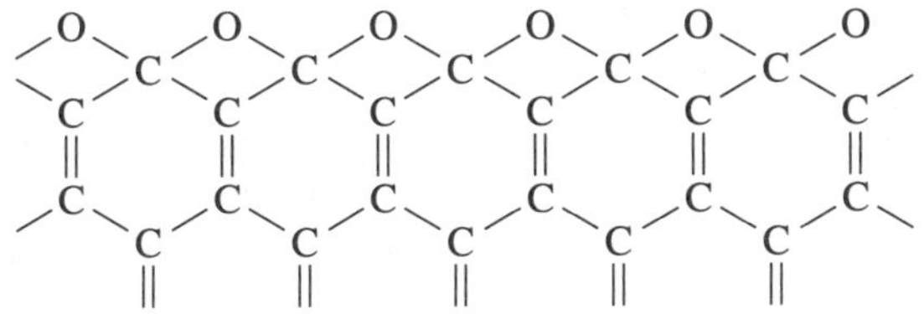

Oxide surface

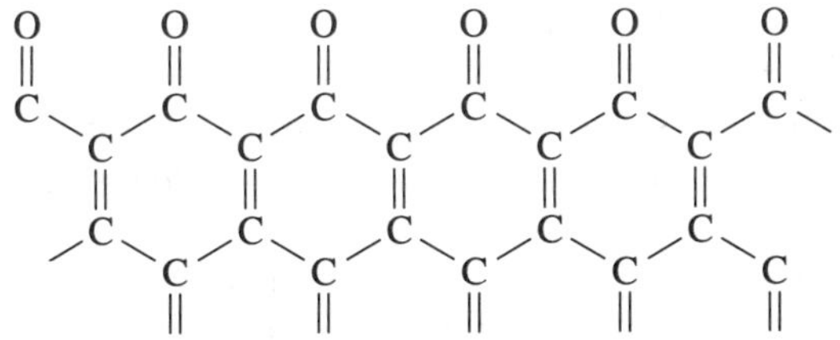

Ketone-quinone surface

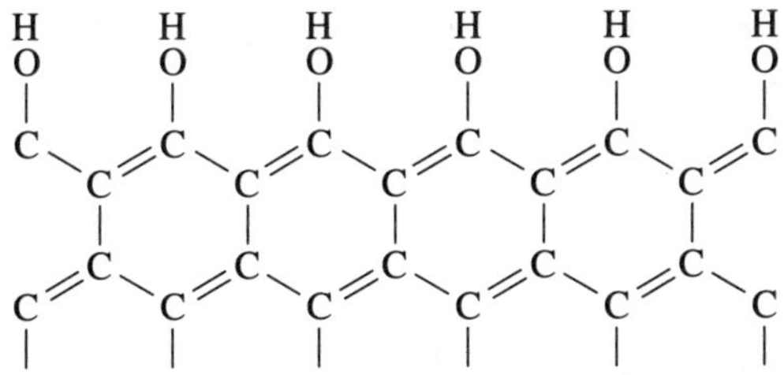

Hydroxyl surface (alcohol)

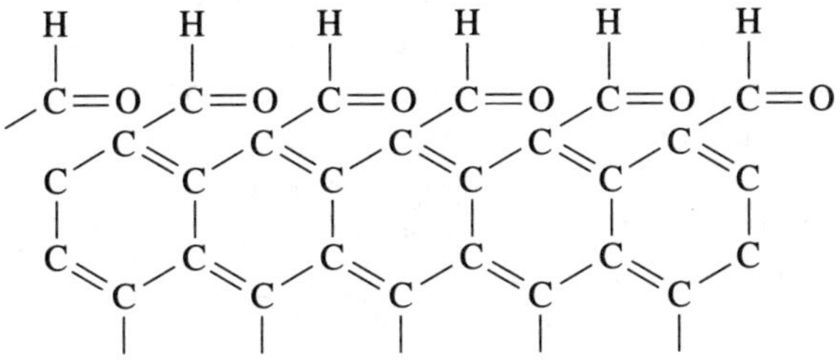

Aldehyde surface

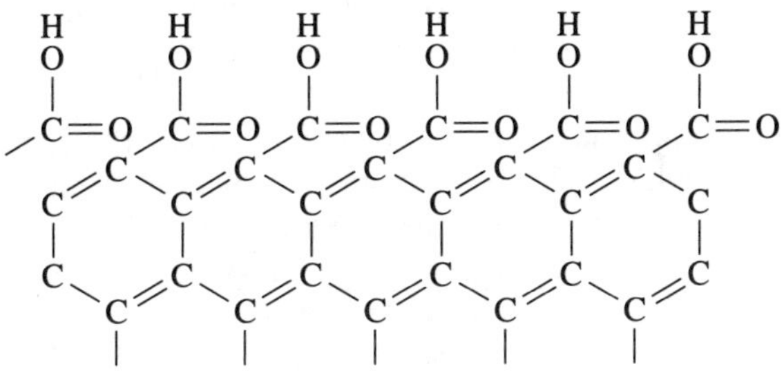

Acidic surface

Figure 10.20. Surface oxidation forms of carbon black.

10.2 Adsorption of Water

For most oxides and ionic surfaces, water vapor adsorption is a natural consequence of commercial production. The adsorbed layer ranges from monolayer to a maximum of three layers (at high relative humidity). It can produce both useful and detrimental effects.

Highly dusting powders, especially silicas and aluminas, are purposely maintained at high humidity during transfer to minimize respiratory exposures. This allows surface moisture to produce low-energy, temporary agglomerates. This same layer of adsorbed water enables high angles of repose (both advantageous and disadvantageous), ease of transfer on conveyor belts, rapid wetting in aqueous solutions, and uptake of polar and ionic dispersants.

For applications in oleophilic systems, such as paint and plastics, adsorbed water on a powder surface can have both short- and long-term detrimental effects on the finished product. If the adsorption energy of water molecules on the solid surface is high, for example, 12 kcal/mole, high shear energy will be insufficient to displace these or permit oleogenic, aliphatic amines to react directly with the silica. Instead, an intermediate complex forms of the type:

$$\diagdown\!\!\!\!-\!\mathrm{Si}\!-\!\mathrm{OH} : \mathrm{H_2O} : \mathrm{H_2N}-\underset{\mathrm{H}}{\overset{\mathrm{H}}{\mathrm{C}}}-\underset{\mathrm{H}}{\overset{\mathrm{H}}{\mathrm{C}}}-\underset{\mathrm{H}}{\overset{\mathrm{H}}{\mathrm{C}}}-\underset{\mathrm{H}}{\overset{\mathrm{H}}{\mathrm{C}}}-\underset{\mathrm{H}}{\overset{\mathrm{H}}{\mathrm{C}}}\mathrm{H} \tag{10.1}$$

where the bonding energy of the complex is limited to that between the amine and a water layer, something less than 9 kcal/mole. Were the final system to be processed at elevated temperature (molding compounds, thermosetting resins, blendor heat buildup, etc.), internal temperatures may rise to the point where this bond ruptures and particles of silicate may move about in the system and flocculate with others, also stripped of their protective dispersant.

Figure 10.21 is a schematic of a silicate surface treated in water with a fatty amine for use as a filler in an oleophilic polymer.

This phenomenon is also demonstrated by comparative isotherms of butylamine adsorbed from water and from heptane, as illustrated by the data set for aluminum silicates in Figure 10.22. In water the butylamine adsorbs through the intermediary of a water molecule, a species sufficiently large to preclude adjacent silanol adsorption.

Particles carrying an adsorbed water layer into an oleophilic resin represent islands of hydrophilic material in a hydrophobic medium. Weathering, thermal cycling, and ultraviolet radiation all serve to cause these to break away from the polymer and result in loss of structural film or casting integrity and the associated properties of color, dielectric loss, surface texture, water resistance, etc.

Methods to alleviate the water adsorption problem almost always involve expensive processing: the use of selective solvents (2-nitro-propane in Chapter 7); treatment with halosilanes or organotitanates; or high-temperature drying immediately preceding incorporation.

In addition to surface incompatibility created by a monolayer of adsorbed

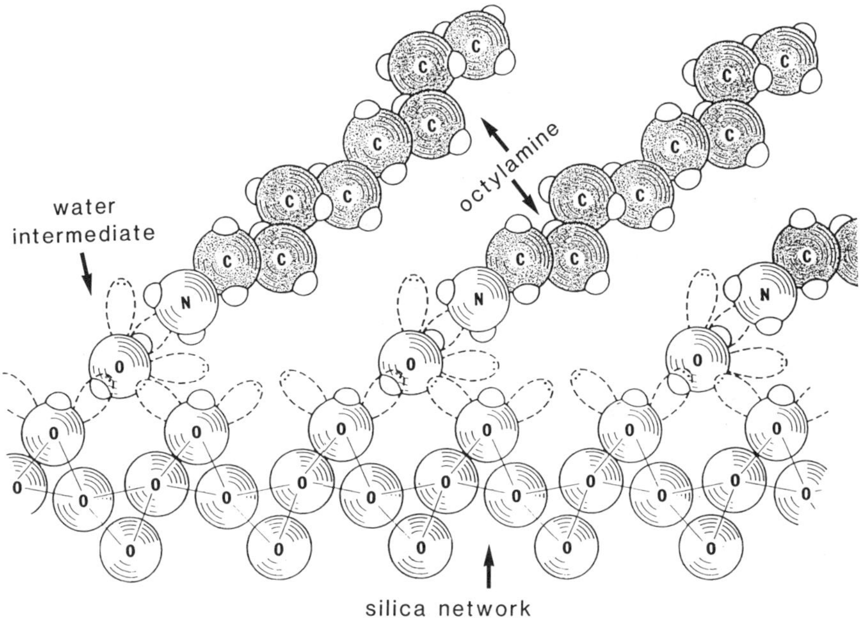

Figure 10.21. Adsorption schematic of octyl amine on silicates or silica.

water, this water provides a vehicle for unwanted chemical reactions involving ionic migrations or oxygen transfer. Lewis acids become Bronsted acids, and the limited liquid volume generates high concentrations of these, which do later damage to the filler particle or the binder itself. In paint this manifests itself in checking, hazing, gloss degradation, and flaking.[28] In casting polymers it results specifically in embrittlement, color deterioration, surface film or craze formation, and electrical property deterioration.

10.3 Cation Exchange Capacity

Charge imbalance often occurs within natural materials because of uncontrolled conditions of mineral formation in nature, and to a more limited extent in synthetic materials. Calcite ($CaCO_3$) is well known to form an almost continuous chemical series with the mineral dolomite ($CaCO_3$:$MgCO_3$), and a partial one with magnesite ($MgCO_3$), the smaller magnesium atom substituting in the crystal lattice for calcium. Small amounts of magnesium are purposely incorporated in precipitated calcium carbonate to control particle size and shape. Here no charge disparity arises because both cations are divalent. Only mechanical and optical characteristics change with composition. However, with other mineral pigments,

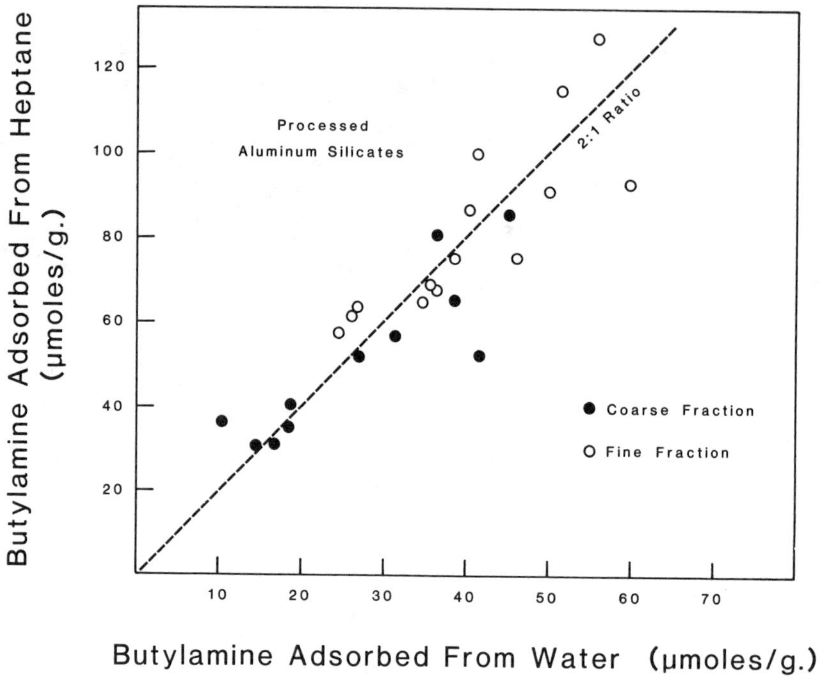

Figure 10.22. Comparative adsorption of butylamine in aqueous and nonaqueous media.

especially mixed cation species, equal size but unequal charge substitutions commonly occur.

Some examples are aluminum atoms in silicate networks, ferrous iron for ferric (magnetite), chromium for beryllium (emerald), magnesium for aluminum (spinel), etc. The quasi-Schottky defects which result may be balanced by incomplete octahedra (oxygen defect) or by adding a surface cation (or anion) near the substitution (imbalance) site. Such surface ions are highly mobile, exchangeable, and most often alkali or alkaline earth species. The population density of these surface cations in equivalents per unit mass is termed *cation exchange capacity* (CEC). In some solids, surface charge must be balanced by anions, and the terminology is *anion exchange capacity*. Because of the low concentrations involved, exchange capacity carries the units microequivalents per gram (μeq/g) or milliequivalents per 100 grams (meq/100 g).

Cation exchange capacity affects surface adsorption of polyvalent dispersant ions because the base structure is most often negatively charged. In theory, were the CEC value sufficiently great, the material would be self-dispersive. This is approximated by members of the smectite family (bentonites) with values from

600 to 1200 μeq/g, which require little dispersing agent to form thixotropic slurries.

In a given family structure (e.g., kaolinites), lattice substitutions are statistically random in nature. As a consequence CEC increases linearly with surface area insofar as crystal morphology remains constant. With decreasing crystallinity, unequal charge substituents are often higher in population density in the solid, being in part the reason for poor crystallinity. For a kaolinite having a surface area of 4 m^2/g, a typical CEC value would be 40 μeq/g, a value closely approximating the dispersant adsorption "monolayer." Calculating this value back to area per site by integrating over the full surface (if randomness in structure truly existed) yields approximately 0.60 square nanometers per exchange site. On the assumption that each of these sites represents one internal trivalent lattice substitution for a tetravalent one (silicon), simple calculations indicate the structural imbalance would be less than 0.5%, a value too low to be confirmed by chemical analysis and one that would minimally interfere with the adsorption of polymeric anion dispersants.

Some investigators have concluded that exchange capacity in the phylloidal silicates lies on the basal faces, and metal sol adsorption experiments in part substantiate this.[29] However, even with those locations population density is not adequate to account for the computed site density. Allowing each terminal (edge) aluminum atom a hydroxyl group capable of exchange yields a value that approaches, but is slightly too small for, that of the measured exchange capacity. Edge surface on some minerals accounts for only 15–20% of the total surface area, which would compact site density prohibitively.

Studies on polyester and polystyrene emulsions bearing slightly positive charges and incorporated into dilute kaolinite slurries appear to adsorb only at edge locales, as shown in the transmission photograph of Figure 10.23.[30] These colloids cannot be formulated sufficiently small to adsorb at each available exchange site. As a consequence, only a few appear to adsorb. Their charge may be sufficiently high to neutralize effectively on a heterotactic basis all available edge charge sites in their vicinity.

With ferric oxide (hematite), ferrous substitution results in a highly activated =Fe–OH surface group whose acidity exceeds considerably the values shown in the literature. However, CEC values are much smaller than with silicates, being at most several microequivalents per gram. Where the substitution is high, several percent, electron orbital overlap within the crystal unit cells induces significant magnetism. Specular hematite (also called "black diamond") is frequently so magnetic from this structural modification that it can be collected with a hand magnet. Continued substitution results in the structure approaching Fe_3O_4, magnetite. The surfaces of such structures have greatly increased acidity.

Whatever and wherever CEC sites are, they exercise a minor control over dispersant adsorption, whether ionics or nonionics, and exert an influence upon rheological properties. Within a set structural group high-exchange-capacity

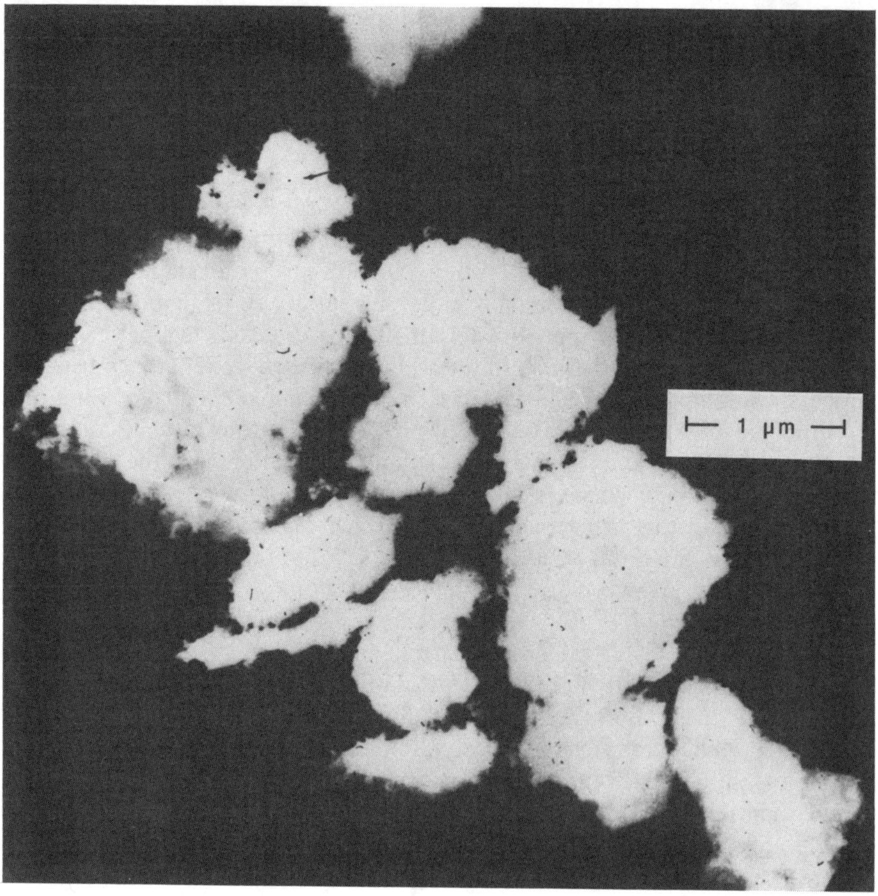

Figure 10.23. Edge charge sites attracting positive polymer colloids.

members are more likely to possess detrimental rheological problems (viscosity, dilatancy, settling, syneresis, etc.) than low-CEC members.

10.4 Particle Size, Morphology, and Surface Area

Particle size in the dispersed state exerts broad influences upon rheology and material utilization. In most commercial size distributions 70% of the powder's mass resides in the coarsest 30%, while 70% of the surface area resides in the finest 30%. Any processing operation that alters size distribution will contribute an overall effect upon the dispersive state, flow properties, and mechanical properties of the powder being processed.

Failure to provide sufficient dispersant to cover (or equilibrate) the surface adequately will result in fine particles being preferentially dispersant poor. This, in turn, effectively narrows the size distribution and shifts the rheology towards a dilatant state. Grinding, which produces a surplus of fine particles, as enhanced by a classifier or jet mill, will cause a shift towards thixotropy for two reasons. First, grinding creates a general increase in the system's surface area. Second, a surplus of fines can increase the probability of charge bridging, especially if dispersant requirements have been exceeded, between coarse particles, which in turn impedes liquid motion.

Newly created fractured area is always chemically active and will show a disproportionate dispersant demand when first incorporated into the slurry system. During drying these excessive fines are more likely to cement larger particles together in driers and impede their breakdown for dispersion at a later time during application.

Powders having and requiring large anisometries, such as needles for magnetic recording particles and platelets for pearlescence, are more susceptible to particle breakage from redisperion following thermal sintering (edge fusion) in production drying. Sintering is the result of both elevated temperatures and residual alkali dispersants. Figure 10.24 is of a hot-air-dried laminar pigment containing residual sodium carbonate and sodium hexametaphosphate resulting from inadequate washing during the filtration stage. Even after high-shear redispersion, particle aggregates exist in which the laminar character is largely lost.

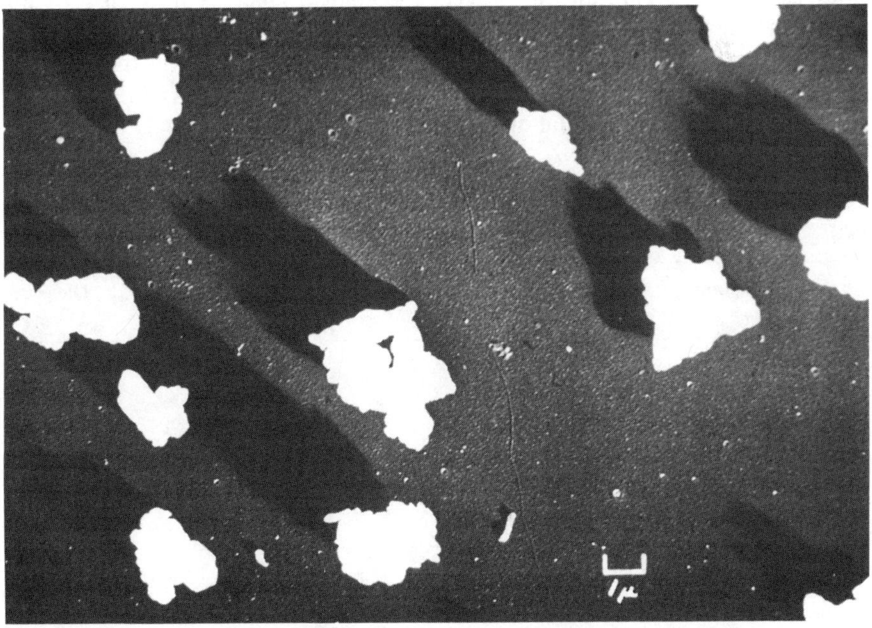

Figure 10.24. Electron micrograph of chemically sintered inorganic pigment.

To disaggregate these assemblages fully requires energy precariously close to particle breakdage. Thus an application tradeoff exists between reduced optical effectiveness and impaired rheology.

Manufacturers of calcined kaolin have in recent years amended their operations to include redispersion following the high-temperature (>1000°C) calcination stage. Cooled kiln product containing high-temperature sinters is redispersed, passed through a sand or small media mill to disaggregate the particles, then refiltered and redried. These additional operations increase the process costs but enable a grit-free pigmentary product to be produced. Residual dispersant from drying at 150°C has far less detrimental effect than that introduced prior to and carrying over in calcination!

The use of fugitive dispersants, as alkanolamines, effectively avoids the problems of high-temperature sintering. Where they can develop the required processing viscosities and where their additional cost is not a production factor, they are greatly justified. In some applications common ammonia has been employed to minimize sintering.

Certain mineral pigments may contain impurities that lie in a select range of particle size. These include, for example, micas and graphite in marble, hydrated ferro-titanium compounds and montmorillonite in kaolinites, and dolomite and chrysotile in talcs. To remove these and improve product color and other qualities, the system must be well dispersed and a selective particle size cut made to remove the detrimental impurities. Where the impurity resides in the ultrafine component of the distribution, which is most common, centrifuges must make size classifications carefully to prevent rheological changes from distribution alteration. This is illustrated in Table 10.5. The feed pigment has a distribution where some 85% of the particles lie naturally below 2 μm and the impurity dominates at 0.2 μm. An application for the finished product is as a filler-colorant in PVC wire and cable insulation.

Table 10.5 Dispersion and Classification Effects on Dielectric Properties

Centrifuge Cutpoint (μm) (D_{50})	Dispersant Level (%)	Processing Stages	Dielectric Loss (%)
none	0.00	none	0.60
0.25	0.20	f,[a] v,[b] d[c]	0.62
0.25	0.30	f, v, d	0.51
0.25	0.30	f, v, w,[d] d	0.43
0.508c	0.20	f, v, d	0.39
0.50	0.30	f, v, w, d	0.32
0.50	0.30	f, v, 2w,[e] d	0.27

[a] f, flocculate with aluminum sulfate.
[b] v, filter under vacuum.
[c] d, dry.
[d] w, wash.
[e] 2w, double wash.

Table data indicate chemical processing, including residual chemicals, dispersant, and flocculent, contribute as significantly to dielectric loss as does ultrafine clay with its associated high cation exchange capacity. Of all those unit operations practiced in powder processing industries, washing and dispersant level are those most significant in controlling the filler pigment's quality.

To effect rapid drying, pigment and other powder manufacturers often employ a dry feedback mode. This is especially true with flash driers where a 40–50% slurry does not adequately dry in the hot air chamber of the system. Here dried material is blended back into feed slurry to raise solids, which are then extruded or pumped into the drier chamber. A recycle of 25% dry material means that at process equilibrium 1% of the final product has passed through the drier at least four times. Each pass increases the likelihood of chemical sinter development. Such operations must either employ fugitive dispersants (e.g., alkanolamines, which volatilize in the drying stage) or wash their product well prior to the drying operation. Generally, the more alkaline the dispersant and the more acidic the solid material surface, the greater the probability for chemical sinter formation.

10.5 Secondary Floccule Formation and Structure

Floccule structure, both mechanical and chemical, following dispersion often has a significant bearing upon processing and, in certain cases, applications themselves. Flocculation may be a necessary component of slurry processing. Employing the Smoluchowski equation (Section 2.1), it can be shown that with moderately concentrated dispersed slurries of fine particles, the maximum number of nucleating species forms when surface charge is removed or nullified (i.e., at the isoelectric pH). With very dilute solids, the rate of floccule growth exceeds the rate of new nuclei formation. In the clarification of drinking water, for example, suspended clays, silica, and organic debris, usually well below 0.1% by weight, are precipitated. This is accomplished through the agency of aluminum hydroxide precipitation at its isoelectric pH, about 6.5, using dissolved aluminum sulfate solution followed by calcium hydroxide.

In this system, in which the population of nucleating sites (the impurities themselves) is very limited, floccule development proceeds until each particle grows very large—often approaching 1 or more millimeters in diameter. Such alumina hydrate floccules readily settle and filter, providing highly clarified potable water. Over- or underaddition of calcium hydroxide will shift the surface charge away from zero (negatively or positively), thereby greatly impeding floccule formation and growth.

As the solid concentration in slurries increases, particle proximity likewise increases, but as the cube root of the increase. These increase the number of particle–particle unions that form during incipient flocculation. Floccule size diminishes rapidly. For production methodologies in which filtration is a necessary unit operation, large floccule size is of critical importance. Approaching the isoelectric pH rapidly, as by fast acidification, almost always results in fine floccule size and poor filtration rates. To compensate for this problem, flocculation operations may be enhanced both by slower chemical addition and through addition of colorless polyvalent ions, such as Al^{3+}. Polyvalent ions induce high positive charges into negatively charged, dispersed surface envelopes (e.g., with polyphosphates or polyacrylates).

Floccule size can be increased with higher levels of polyvalent cation along with acidulant and by approaching the isoelectric pH slowly. With certain dispersants, especially polyphosphates, additional aluminum ion is required to compensate for direct reactions between the flocculent and dispersant, which are equilibrium shifted to the solution phase. As the concentration here is usually equal to or in excess of that dispersant localized on the particulate surface, the quantity of flocculant required is high, approximating the dispersant level itself. For dipolar or polypolar dispersants (as alkanolamines), where the greater portion of agent is localized on acidic surfaces and little charge compensation is required, the level of polyvalent cation required for flocculation is significantly reduced, and in some cases eliminated altogether.

Polyvalent cation induced floccules are tightly unified structures, which often grow in size with time, even during the relatively quiescent filtration process, that is, in the filter cake on the filter medium. Such floccule strengths are high, reticular in form, do not break down readily with secondary shear, and entrain solution chemicals, especially the very polyvalent salts employed for their formation. Washing on the filter requires several stages and heated water to effect solute release. Often the cake must be removed, reintroduced into hot water, highly sheared, and refiltered to produce high-resistivity pigments, especially for paint and polymer filling applications.

Where long-chain polyacrylates and polyacrylamides are employed for flocculation, reticular networks form but with large particle–particle separations. The resulting floccules are smaller, much more readily broken apart with shear, and more easily washed. In addition, the retained processing chemicals are dispersants themselves that pose few problems in many applications. For those applications requiring high resistivity, washing is more efficient because of greater floccule openness, but slower because of smaller floccule size on the filtration medium. The quantity of polyvalent cation salts required for charged anion dispersants is comparable to the level of dispersant, usually a few tenths of one percent. Levels of long-molecule flocculant are typically in the 10–50 ppm range. Retained organic flocculant, such as 50,000 molecular weight polyacrylamide, poses fewer problems where resistivity is a critical factor (as in PVC electrical insulation) than does aluminum sulfate.

Where products are to be employed in shear-cast or polymer-extruded applications, dipole-agent-dispersed, polyacrylamide flocculated pigments break apart and reform in shear-aligned patterns in the polymer. Because polymer viscosities are high, 10,000 cp or greater, reformation to reticular networks is commonly prevented, especially where polymerization proceeds quickly. Although both alkanolamines and polyacrylamides possess a dipole moment, the dielectric loss they induce is much lower than that resulting from entrainment of hydrated aluminum or other cations because of water's greater dipole moment, 1.87 Debye, and dielectric constant, 78. Also, water migration into or out of finished polymers is minimal with nonionic systems, which dramatically reduces crazing, spallation, haze formation, and fracture-induced embrittlement.

References

1. Boehm, H., *Disc. Faraday Soc.* **52**, 264 (1971).
2. Lloyd, M., Ph.D. Dissertation, Rensselaer Polytechnic Institute, Troy, NY, June (1970).
3. Foster, A., et al., *Proc. Royal Soc. London* **A136**, 3675 (1946).
4. Johns, D., and Conley, R., ''A Survey of Analytical Methods for Claus Sulfur Production,'' Proj. #76-C-114, Anlin Corp., Houston, TX, p. 8 (1976).
5. Razouk, R., Mikhail, R., and Girgis, B., ''Solid Surface and the Gas–Solid Interface,'' A.C.S., Washington, DC, p. 45 (1961).
6. Wade, W., and Hackerman, N., *J. Am. Chem. Soc.* **64**, 1196 (1960).
7. Fochtman, J., Bitten, G., and Katz, R., *Ind. Eng. Chem., Prod. R&D* **2**, 212 (1963).
8. Pulvari, C., *J. Am. Chem. Soc.* **42**, 355 (1959).
9. Fowkes, F., *Am. Chem. Soc., Div. Polymer Chem.* **29**, 245 (1988).
10. Bayles, A., ASME Paper #57-A-276 (1957); also, Sawyer, D., DOE/EPRI communication (1979).
11. Tarpley, W., Ellis, R., and Conley, R., E&MR Rept. #4, DOE Contract #C-102/78 (1979); also, MRT Rept. #DOE-79ET14237.
12. Leonard, J., and Mitchell, D., ''Coal Preparation,'' Seeley–Mudd Series, A.I.M.E., p. 1/34 (1968).
13. Conley, R., and Albus, F., ''Size Reduction and Classification of Solids—Mechanics of Fracture,'' C.F.P.A., E. Brunswick, NJ, p. 33 (1994).
14. Tadros, T., and Lyklema, J., *J. Electroanal. Chem.* **17**, 267 (1968).
15. Iler, R., ''Colloid Chemistry of Silica and Silicates,'' Cornell Univ. Press, Ithaca, NY, p. 209 (1955).
16. Whalen, J., Southwest Regional A.C.S. Meeting, Dec. (1958).
17. Wade, W., et al., ''Solid Surfaces and the Gas Solid Interface,'' A.C.S. Washington, DC, p. 36 (1961).
18. Weiss, A., *Angew. Chem.* **2**, 697 (1963).
19. Wada, K., *Am. Mineralogist* **50**, 924 (1965).
20. Brunauer, S., Kantro, D., and Copeland, L., *Can. J. Chem.* **37**, 714 (1959).
21. Kobayashi, T., et al., *Adv. Org. Coatings Sci. Tech. Series* **10**, 114 (1988).
22. Schwaab, G., et al., *Naturwissenschaften* **22**, 584 (1957).
23. Conley, R., *J. Clays and Clay Min.* **14**, 317 (1966).
24. Conley, R., 30th Ann. Meeting, Clay Minerals Soc., San Diego, CA, p. 30, Sept. (1993).
25. Koopa, L., and Van Reimsdijk, W., *J. Colloid Interface Chem.* **128**, 190 (1989).

26. Ross, S., and Olivier, J., *On Physical Adsorption*, Interscience, NY (1964).

27. Aston, J., Tykodi, R., and Steel, W., *J. Phys. Chem.* **63**, 1015 (1961).

28. Solomon, D., *Clays and Clay Min.* **16**, p. 31 (1968).

29. Schofield, R., and Samson, H., *Clay Minerals Bull.* **2**, 45 (1953).

30. Van Olphen, H., Shell Development Corp., private communication, Georgia Kaolin Company (1963).

CHAPTER

11

Formulating the Multicomponent System

In previous chapters, with but a few exceptions, single-entity factors relating to single-entity dispersants and single-entity solid surfaces have been discussed. Few commercial dispersions, however, are single-solid, single-dispersant formulations. There exist, of course, exceptions (particle size analysis, pigment manufacturer's slurries, scientific assessments, etc.), but most involve mixed pigments coupled with mixed dispersants. Many involve mixed binders, mixed solvents, or mixed polymers as well.

Upon initial inspection the task of determining what combination of dispersants will function best with a mixture of solids possessing a broad variety of surface character appears to be a complexity beyond comprehension. However, on closer assessment the system can be broken down into categories of activity that renders it more amenable to characterization and improvement.

Where multiple components are listed in published formulations, each is incorporated to serve a specific purpose. These purposes may be as sophisticated as the development of mixed optical, electrical, magnetic, and/or mechanical properties or as mundane as formulation economy. Typical formulations, as illustrated in Table 11.1 for a paint system, have components which serve useful and perhaps singular purposes. Each has been incorporated not by chance or trial and error (in most instances!), but with intelligent engineering and chemistry to accomplish a combined series of goals.

In all such formulations the two inter-relating dispersive variables, mixed pigments (or functional solids) and mixed dispersants, must function effectively together to thoroughly distribute all pigment components homogeneously throughout the slurry system. Very few formulations for commercial applications

Table 11.1 Multicomponent Formulation Assessment

Component	Weight %[a]	Use
Rutile TiO_2	20.8	primary pigment
Calcined kaolinite	13.5	extender pigment
Calcium carbonate	10.4	extender pigment
Silica	7.8	flatting pigment
Vinyl acrylic latex	14.2	binder
Ethylene glycol	3.1	viscosity builder drying retarder
Polyethylene glycol	26.0	viscosity builder drying retarder
Long chain glycol	1.0	latex disp. and emulsifier
Nonoxynol-9	0.5	titania and calc. kaolin disp. and antifoamer
Na polyoxyethylene sulfonate	2.3	calcite disp.
Amino alcohol	0.3	titania and calc. kaolinite disp.
High-mol.-wt. polyacrylate	0.1	viscosity control
Water	2 parts to 3 parts of above	solvent and diluent

[a] Formulations most often expressed as lbs/100 gal.

contain but a single dispersant simply because single dispersants are less efficient than intelligently selected combinations.

Further, the order of addition of the various components in the table is commonly a critical factor in obtaining thorough dispersion and a functional product, with dispersants and flow agents (glycols) blended first, followed in order by progressively less acidic pigments. Additional binder and viscosity building agents often are the last to be added after full dispersion of the preliminary components to ensure that interparticle and intercomponent flocculation does not occur.

An assumption often made in examining published formulations (as supplied by pigment manufacturers) is that employment of several dispersants results from assigning to each solid a specific dispersant to bring about the most advantageous particle separation. While this may have been the initiating factor in formulation research, it is rarely true for the finalized commercialized product. Were a single dispersant capable of providing the necessary total pigment separation, or even a reasonable compromise in this parameter, only one agent would be recommended. This, too, is rarely the case.

The question is posed periodically by development and production personnel why pigment manufacturers do not supply products with dispersants preattached to simplify the work of formulators. This is a fair question, as pigment manufacturers may well employ dispersants during production to facilitate their own unit operations. However, a brief survey of the industry quickly clarifies this issue. A typical pigment manufacturer may produce a given product in 5–6 particle size ranges. Coupled to size variation are manufacturing variables of morphology (equants, needles, platelets), hue and chroma variations, and drying variations (high-solids slurries, spray drying, drum drying, apron drying). Assuming for the sake of illustration that in water-based applications only 5 applicable dispersants may be available (an extremely conservative figure), the number of permutations of dispersant-accompanied powders runs into the hundreds, making manufacture, inventory, and marketing a commercial nightmare!

Second, dispersants most advantageously employed for manufacturing may not be those most appropriate for field applications. For example, sodium silicate, an efficient processing dispersant for silicates, interferes with ceramic casting employing those products. Sodium phosphate-carbonate, a common dispersant for particle size classification, produces blooming with acrylic latex paints, etc.

While certain applications may justify a dispersant accompanying a product, as alkanolamines with high-solids titania slurries and polyphosphates with kaolinite paper coating slurries, a dispersant-free powder more often enables the formulator far broader latitude in developing application properties than does predispersion.

On the following pages a few select examples of water-based, compound systems will be cited and examined in detail to illustrate the selection principles involved.

11.1 The Multiple-Solid Factor in Aqueous Systems

An excellent example of a functional multicomponent solid system is illustrated by an acrylic, water-based, tint-based formulation paint in which precipitated calcium carbonate, fine silica, and kaolinite are employed as extender pigments. The particle size for the clay (representing 60% of the extender component) was 2.1 μm (D_{50}), for the precipitated silica about 0.1 μm, and for the precipitated carbonate, nominally 0.3 μm. The system was formulated at pH 7.2 using morpholine as dispersant. Also incorporated was a nonionic corrosion inhibitor for can filling during subsequent consumer custom marketing. A color-additive-free coating having an approximately 1 mil (25 μm) film thickness was applied at 85°F, simulating hot summer usage, to an appropriately primed, impermeable, inert surface. The film is shown as the SEM of Figure 11.1. Polymeric binder has been carefully removed chemically from the dried film to reveal better extender texture, particle distribution, and orientation. In the figure platelet particle diameter appears much larger than 2 μm, due in part to instrumentation particle size analysis values for high-aspect-ratio particles being smaller than true sizes, often by 200–300% (sedimentation factor), and in part because larger particles in the distribution are more immediately apparent in SEMs (illusive factor).

The three components, of which silica is not resolved and carbonate is minimally visible in the SEM, do not form continuous or even semicontinuous films.

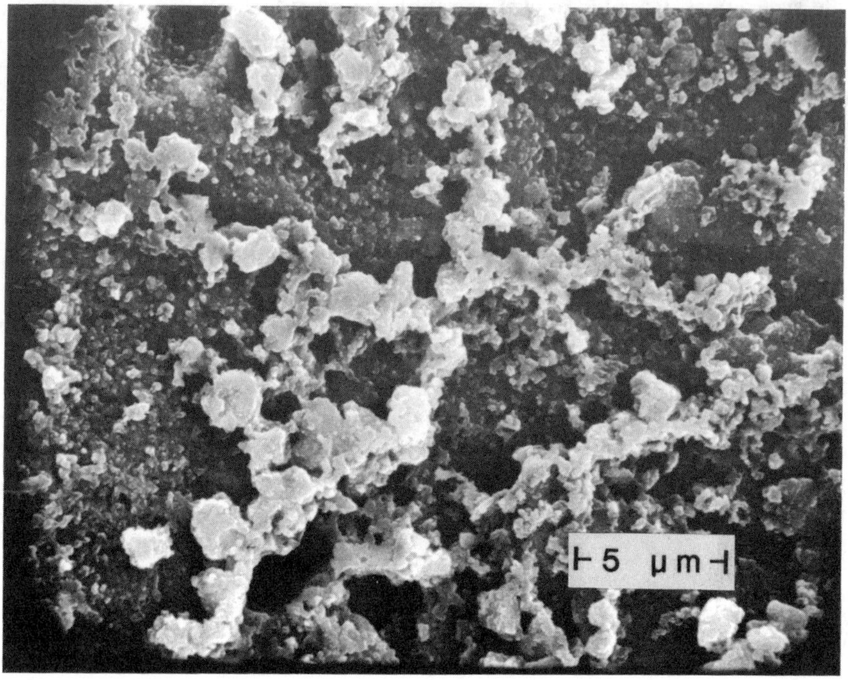

Figure 11.1. Poorly dispersed compound pigment in coating.

Rather, bridges and networks are produced. Hexagonal particles of clay can be seen forming overlapping platelets and filamentary pathways. Carbonate has flocculated around the periphery of these platelets and joined together other platelets. The optical properties of this coating (with or without subsequent tinting) display both rough surface texture (flatting effect) and exceptionally poor hiding power (covering effect) as a consequence of this "networked" structure. Film mechanical integrity suffers by inference.

This phenomenon of in-film flocculation derives from two separate activities. First, morpholine has little attachment potential for carbonate, which has an alkaline surface (see Figure 5.10). Thus at pH 7.2 carbonate particles retain some positive charge with no barrier dispersant layer. Second, while both kaolinite and silica are negative at pH 7.2 and while both attract and hold morpholine essentially quantitatively on their surfaces, volatility loss of dispersant can occur through hot weather application (morpholine T_{bp} = 126°C). The relic structure in the figure suggests dispersant level, while adequate for clay and silica during ambient container storage and secondary formulation (tinting), was most probably inadequate for warm weather application even without the carbonate. Incorporation of carbonate compounded the problem of dispersion yet further. Addition of SOPA or TSPP as nonvolatile codispersants is required for this and similar applications.

The peculiar structure of the network displayed in Figure 11.1 was predicted three decades ago based on theoretical studies of plate-equant flocculation at MIT.[1] These computer-generated studies predicted high-aspect-ratio platelet pigments mildly flocculated with equidimensional pigments would form void volume/pigment volume ratios of about 6 (16% effective volume). Visual inspection of the system in Figure 11.1 indicates this value to be a fair representation for the two-component system.

Similar phenoma are often exhibited with zinc oxide based copigment systems. ZnO has an isoelectric point near pH 9, depending upon mode of manufacture and often a mixed morphology (needles and equants). When incorporated as a primary pigment in PVC and acrylic emulsion using TKPP as dispersant in the pH range 7 to 9, where these emulsions function best, the material has yet a positive charge while most other solid oxides, especially extender pigments, are negative. In a blend with an ultrafine kaolinite extender pigment, the system aggregates, as shown in the two electron micrographs of Figure 11.2 at pH 7 and 9. This interaction results in a curding of the pigmentary system following formulation and storage (the more mundane idiom is "snotting"), requiring both high shear just prior to, and frequent mixing during, application.

This pH disparity does have beneficial effects, however. Zinc ions are slowly

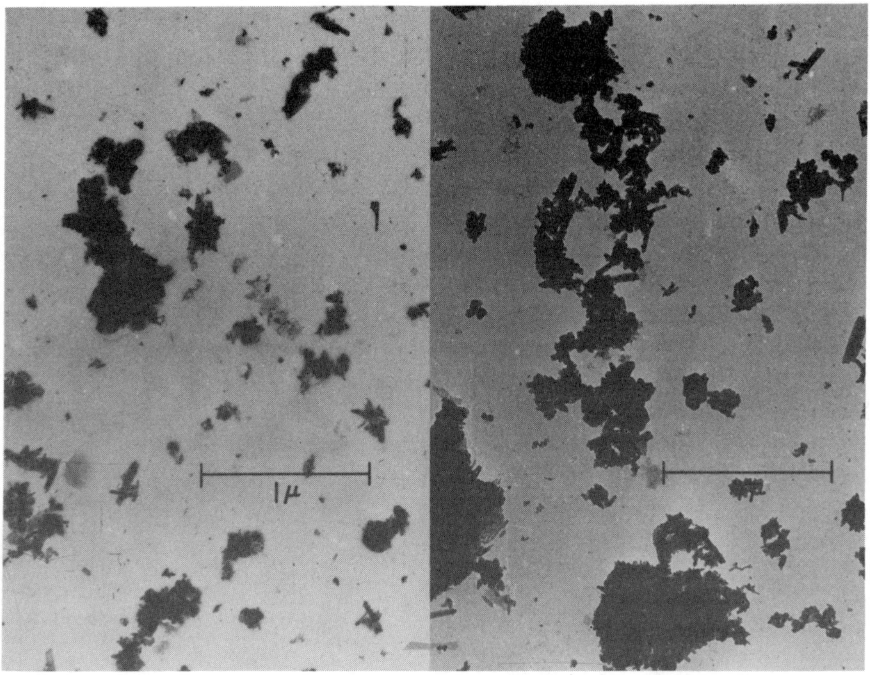

(a) (b)

Figure 11.2. Zinc oxide–kaolinite copigment in latex formulation (a) at pH 7.0; (b) at pH 9.0.

leached from the oxide and provide highly beneficial fungicidal activity (mildew prevention), especially where the coatings are employed in highly humid climates. A drawback lies in these same zinc ions being instrumental in initiating polymerization of the binder during long-term shortage. Formulation under conditions where zinc oxide has a negative surface charge is difficult. Most latex emulsions, even without zinc ions, are inherently unstable at pH values of 9.5 or above.

In Chapter 10 the comparable situation of ferric oxide and kaolinite was examined for both dispersive state and applied film quality. There a single dispersant was employed and again pH was the single detrimental factor. To effect better dispersion with nonequivalent surfaces, titanium dioxide manufacturers deliberately alter that pigment's surface to make it more readily electronegative at formulation pH values for aqueous systems.

11.2 The pH Component

A survey of the aqueous isoelectric data for a cross section of oxide/hydroxide pigments in Table 2.3 (see also Table 10.1) reveals most lie in the fairly narrow pH range from 4 to 9. The amount of alkali necessary to bridge this range and impose negative charges is small, considerably less than 0.1% based on the pigment weight in suspension. As a consequence, alkali addition, where tolerable, represents the most simplistic dispersion control to assure equipolar surfaces.

Three family choices are available to the dispersion chemist for regulating pH:

1. Strong ionic base (KOH, NR_4OH);
2. Weak ionic base (K_2CO_3, NH_4OH, Na organo-sulfonate);
3. Nonionic organic base (butanolamine, morpholine).

Table 11.2 Base Selection Parameters for pH Adjustment

pH Additive	Advantages	Disadvantages
Strong ionic base	1. small dosages 2. economy	1. bleeding 2. efflorescence 3. attacks pigments 4. handling hazards
Weak ionic base	1. moderate dosage 2. economy 3. reduces can corrosion	1. some bleeding 2. crystallizes with carbon dioxide
Organic base	1. fugitivity 2. infinitely soluble 3. surface selectivity	1. fugitivity 2. strong odor 3. surface selectivity 4. high cost

Each of these agents has specific advantages and disadvantages, some of which are listed in Table 11.2. Note that several parameters may generate either advantageous or disadvantageous properties, depending upon the solid components employed and their surface chemistry. Fugitivity, for example, may be an advantage in film drying or in producing a "chemical-free" composition (e.g., for electrical characteristics). Where the dispersing agent evaporates excessively, as with hot environmental applications (or storage), it may leave formulations dispersant poor and pigment distribution inadequate, especially in paints where underdispersancy degrades good spreadability, coverage, or other optical characteristics (as noted earlier with clay and carbonate).

The economy factor for strong ionic base employment is not a casual or insignificant parameter. The use of KOH, for example, to shift formulation pH from 6.0 to 8.5 requires less than 0.5 mg for a gallon of slurry at 30% solids by weight. Were this formulated as a paint and applied to a primed surface (typical coverage = 300 sq ft), less than 1.5 μg of alkali per square foot would exsolve and bloom, even in situations where 100% of the alkalizing agent migrates to the surface.

Surface charge control via pH agents obviously has limitations, even within the specific fields noted. Other components in the system not only may be affected by pH; they may react directly with the agents. Alkyd resins, water-dispersible acrylics, and other materials often have residual acidity resulting from their polymerization process. Gross pH shifts may alter these substances and interfere with polymerization mechanisms, film formation, or other binding and adhesion activities.

11.3 Multiple Dispersants in Aqueous Applications

By combining pH with a dispersant or by blending dispersants, several synergistic effects may be induced into an aqueous system. Employment of nonionic amines will both raise pH and form liquid–dipole barriers on acidic oxide components in the formulation.

Adding nonionics to ionics enables a dispersed state to be formed without overloading the free solvent volume with stray ions (counterions from the Stern layer). Thus a formulation showing a 0.2% KTPP plus 0.2% morpholine is not necessarily superiorly dispersed compared to 0.3% KTPP. It does, however, enable the solids level in the finished product to be increased due to improved particle mobility from increased liquid-phase mobility.

Comixtures of ionic dispersants, SHMP plus SOPA, for example, often represent variable efficiencies for different pigment components, and the final combination may be an economical tradeoff. While TSPP represents costs about $\frac{1}{3}$ those of SOPA, its rheological effectiveness on kaolinite as an extender pigment in paint is very nearly identical to that of SOPA with regard to both solids and viscosity. On calcium carbonate, however, it is far less effective. Thus a formulation embodying coextension pigments will reduce costs with such a co-dispersant system.

Because of the blooming tendencies with highly ionic dispersants, blends of polyoxyethylene sulfonates and carboxylates offer excellent alternatives to obtaining ionic dispersion. Most latex paint formulations will include mixtures of such agents with a large dosage of nonionic agents including alkanolamines.

The reader at this point may suggest to himself a simplified solution to these problems—exclusive employment of nonionics. This has special merit where ionic migration and electrical properties are salient virtues in the finished product. However, in other applications nonionics have three serious shortcomings. First, they require higher dosages than ionics. Second, their viscosity minima are usually higher than those of ionics. Finally, they can rarely produce as critically loaded solids as ionics. These shortcomings may be critical factors in systems where high solids are a necessity in application. Such applications include casting ceramics (water extraction and mold release limitations), water-based paints (critical water evaporation for nonrun, nondrip characteristics), paper coating (sheet strength deterioration with excess water), spray applications (air transit water evaporation influences surface flow characteristics), and similar uses.

Thus dispersant formulation guidelines, which are almost always economy based as a secondary (and often primary) principle, include the following considerations:

1. pH adjustment to the extent chemical considerations allow;
2. Inorganic ionic agent addition for silicates and octahedral (applicable) solids;
3. Organic ionics for nonapplicable solids (salt based, nonoctahedral, and alkaline surfaces);
4. Anionics for acidic surfaces where dispersant volatility is important or where blooming may be a problem.

A study on the permutations of three dispersants and three solids is shown in Table 11.3, where viscosity (including thixotropy) and viscosity stability are the sole functional considerations.

The influence of pH and the tradeoff between polyphosphate and polyacrylate are readily apparent. No single component in the dispersant package is sufficient in itself to obtain a working viscosity, but of the three polyacrylate provides the best rheology. Tetrasodium pyrophosphate alone produces low viscosity, low aging factor, and the lowest pH of any of the mixed systems. It is, however, pH that renders it less appropriate for the binder system. While TSPP could be combined with either sodium carbonate or AMP to elevate pH, these variables lay outside the study, which was designed around polyacrylate for long-term slurry stability.

Lowest viscosities are associated with elevated pH, due primarily to the effect of calcium carbonate in the pigment mixture. A small amount of polyacrylate is necessary to disperse this carbonate, but beyond this level (5–10% of dispersant package) the alkanolamine is adequate for both the kaolinite and the rutile. Poly-

Table 11.3 Rheological Effects in Codispersant and Copigment[a] Study

SOPA (Wt. %)	AMP (Wt. %)	Soda Ash (Wt. %)	Slurry pH	Viscosity (cp) 10 RPM	Viscosity (cp) 100 RPM	T.I.[b]	Aging Factor[c]
100	—	—	7.5	1056	339	3.12	0.99
75	—	25	8.1	592	248	2.39	1.38
75	15	10	8.0	560	236	2.36	1.33
50	50	0	8.7	440	200	2.20	1.19
50	30	20	8.9	528	248	2.13	1.21
20	40	40	9.3	592	261	2.27	1.24
15	45	40	9.4	496	238	2.08	1.33
10	90	—	9.6	200	171	1.17	1.52
6	54	40	9.5	768	352	2.18	1.04
5	95	0	9.6	216	192	1.13	1.48
3	57	40	9.6	1840	840	2.19	0.63
0	100	0	9.6	2240	992	2.26	3.57
0	0	100	9.9	5830	1870	3.11	1.71
TSPP = 1000			6.8	368	190	1.94	1.22

[a] Copigment system: titanium dioxide at 39% solids, kaolinite at 15% solids, and calcium carbonate at 6% solids; no binder.

[b] T.I., thixotropic index.

[c] Aging factor is the ratio of viscosity data immediately to that at 15 days (25°C by Brookfield). All dispersant combinations total 0.30%.

acrylate beyond the amount necessary for carbonate serves little purpose and actually degrades the system by raising viscosity and reducing pH.

These combined effects, pH to attain isopolarity and a dispersant specific to or general for all components, are the critical parameters in achieving minimal viscosity. While other conclusions could be drawn from the data in Table 11.3, more test data would be necessary to validate these.

Organic color manufacturers often simplify the work of the formulation chemist by chemically modifying their products to produce at least partially self-dispersive compounds through the grafting of an $X\text{–}R_3N^+Cl^-$ (cationic species) or $X\text{–}SO_4^-Na^+$ (anionic species) on the molecular surface. A single charge per molecule may be insufficient in some systems to disperse the color pigment fully, especially where large molecular color components are present. However, a knowledge of the surface character (anionic or cationic) will enable the necessary supplanting agent to be introduced for more thorough dispersion of the pigment(s).

In oil–water emulsion systems, color pigments may be deliberately dispersed in oil phases by eliminating surface charge, thus allowing the oil surface to wet out the pigment selectively. This is a more protracted thermodynamic process, however, than charged pigment dispersion and wet-out in aqueous phases. System mixing must be extended and thorough to prevent later "bleed" and "wash-out" of color components during application and following aqueous-phase volatilization. Pigment suppliers may reverse the above process and chemically modify slightly polar pigments with hydrocarbon tails to force these color bodies into oil phases by lowering the contact barrier energy. Flushing operations, which have been discussed earlier and which employ an intermediate-polarity solvent, also may be used to reduce polar barriers at the pigment–water–oil phase boundaries, to render the pigment and oil more compatible and to accelerate assimilation.

11.4 Multiple Solids in Nonaqueous Systems

Dispersant selection for nonaqueous systems is at once more complex and more simple than for aqueous systems. For a given solvent or polymer system, all solid components must have surfaces compatible with and chemically similar to the liquid phase. In certain applications, the polymer may react with one or more component surfaces to form a self-dispersive state (linoleic acid with bare calcium carbonate or ferric oxide). However, for multiple-component solids the probability of the polymer phase reacting dispersively with all components is remote.

In oleophilic systems, those having mineral spirits or other hydrocarbon solvents, oxide solids may be dispersed by fatty acids or fatty amines having hydrocarbon chains 12 atoms or greater in length. For a solid system composed of calcium carbonate and silica (or ferric oxide and zinc oxide), for example, both agents would be required, quantities being dictated by the relative surface areas of the two solids (and at approximately 50% surface coverage). However, prior addition of a mixture of these two dispersants in the solvent phase will result in a dispersant interaction via the acid–base mechanism:

$$C_{12}H_{25}NH_2 + C_{12}H_{25}COOH \rightarrow C_{12}H_{25}\text{-}NH_3\text{-}OOC\text{-}C_{12}H_{25} \tag{11.1}$$

While this association produces a weak acid, weak base compound of limited stability by aqueous standards (comparable to ammonium acetate), the low polarity of the medium increases its stability. This stabilization, essentially a two-molecule micelle, prevents both components from interacting with their designated solids. As a consequence, selective addition of dispersants becomes an important factor in nonaqueous formulations. In some applications solid components are more effectively utilized by individual predispersion in separate portions of the solvent, followed by combination with the polymer, binder, or other organic component comprising the system.

Where separate dispersion is not practical, the order of addition for most effective solids distribution is:

1. Solvent;
2. Dispersant with the highest energy affinity (solid *A*);
3. Solid *A*;
4. Dispersant with the next highest affinity (solid *B*);

5. Solid *B*;
6. Sequential dispersants or solids;
7. Polymer binder;
8. Ancillary agents (flow, masking and thixotropic agents, catalyst driers, etc.).

Some variations in the latter portions of the suggested order may be necessary where secondary interferences occur. In emulsion formulations with multiple pigments, where it is desired to encapsulate the pigments by the emulsified oleophilic phase, it is common practice to combine all dispersants and add the solids in order according to their dispersant demand, although this method requires more mixing shear and energy to accomplish the ultimate dispersion. Similarly, some solvent may be withheld from Stage 1 to prewet those components introduced during Stage 8. Here manufacturer's recommendations, which represent much trial and error bench testing, dictate the appropriate variations.

Where the polymer–solvent phase exhibits a higher polarity, as with polyester resins and styrene–toluene–MIBK solvents, the dispersant must have a higher tail-to-liquid-phase exposure polarity. The appropriate selection may be based either on molecular similarity (benzoic and acrylic acids, ester groups, and unsaturation) or on the basis of the hydrophilic–lipophilic balance number (HLB value).

Polyester layup resins for boat hulls, car bodies, piping, playground equipment, and similar applications may contain filler and color bodies that are dispersed with molecular configurations having polar surface affinities and resin molecular structures. Typical of agents having this characteristic are the commercial dispersants, polyoxyethylene fatty alcohol ether having an HLB of 12.5[2] and hexaglycerol dioleate at 8.5.[3]

11.5 A Few Representative Applications

In a recent recommended formulation for a fade-resistant, green, fire-retardant formulation capable of dispersion in water-based latex paints, alkyd resins, and polyester casting resins, the pigment distribution and suggested generic dispersants (without trade names) were given as shown in Table 11.4. Chromium sesquioxide is untreated and is the prime hue pigment. Titania, pretreated with aluminum hydroxide, decreases color depth. Antimony pentoxide–oxychloride is the major and catalytic fire retardant, while alumina hydrate serves both as extender and fire retardant (water loss on heating).

Both alumina hydrate and the titania have aluminum hydroxide surfaces. These attract, hold, and disperse with potassium pyrophosphate (which also lends a small amount of fire retardancy in itself) in aqueous systems. Chrome green has a near-neutral surface. As an octahedral oxide it can be dispersed with a polyphosphate, or if the ionic agent is inadvisable, sodium polyoxyethylene sulfonate. Antimony oxide–oxychloride is an acidic pigment having an isoelec-

Table 11.4 Commercial Multiple Pigment Dispersion[a]

Pigment	Pigment (Wt. %)	Dispersant, Water–Latex	Dispersant, Alkyd Resin	Dispersant, Polyester
Cr_2O_3 (nt)[b]	45	Na POE sulf.	fatty acid	polyether
TiO_2 (Al gel)	40	AMP	fatty acid	AMP
Al $(OH)_3$	10	KTPP	polyol	acid–polyether
Sb_2O_5 (OCL)[c]	5	AMP	fatty amine	AMP

[a] Total dispersant level is approximately 3%, solids basis.
[b] Not surface treated.
[c] Surface partly covered with oxychloride groups.

tric pH near 3.5. Its crystal structure exerts a poor coordinating fit with phosphate chains and would catalyze their destruction much faster than does kaolinite. The efficient agent for this component is an amino-alcohol (AMP). In addition to adsorbing strongly upon the antimony oxide surface, AMP raises the system pH to improve stability and lower viscosity.

AMP is best added first to establish isopolarity, followed by the phosphate to complete the dispersion. In the original work adjusting pH to 6.4 (low for AMP) produced higher gloss in alkyd melamine paint applications.

In nonaqueous, alkyd applications the alumina hydrate, chromium oxide, and titania can all be dispersed with fatty acids. However, this must be accomplished prior to addition of any amine to prevent interdispersant reactions as described earlier. Antimony pentoxide will not disperse well with fatty acids. It will disperse admirably with fatty amines, but these may interact with partially unreacted fatty acids and interfere with both component's dispersive states. The choice of a polyol with a hydrocarbon tail circumvents both the reactivity with fatty acids and the odor problem. Also, upon protracted storage, the slight equilibrium between surface and solvent may shift a portion of the amino group into the coordination sphere of the chromium atom in the chrome green and alter its color. Many polyols adsorb energetically on antimony pentoxide.

Polyester resin applications of this pigment system require a small amount of dispersant polarity, exempting the long hydrocarbon chain for high-solids, low-viscosity applications. The terminal alcohol group in AMP serves to disperse the antimony pentoxide readily. For the remainder of the solids, a polyol chain having a terminal –COOH group will be effective in oxide dispersion.

Yet another published formulation containing the extender pigments mica, calcium carbonate, quartz, and feldspar as a highly flatting, tinting pigment base, employs a copolymer, acrylamide–acrylic acid plus SHMP for an acrylic emulsion paint formulation. Here the acrylic acid groups react with carbonate (alkaline) and the amide groups with the remaining materials (all acidic). However, the stability and adsorption energy of the acrylamide for silicates is moderate at best. For this reason hexametaphosphate is employed for the two silicates, mica and feldspar. This formulation functions especially well where the prime color pigmentation consists of metal oxides (yellow and red iron oxides, chrome green, cobalt blues, and chromates).

Mixtures of titanium dioxide (surface treated with alumina gel) and iron oxide for exterior, oil-based, fade-resistant paints are dispersed readily by *R*–COOH agents. The sheen and rubout are influenced by the chain length, *R*, with longer chains yielding better gloss in low-acid resins and intermediate lengths in higher-acidity resins.[4]

Attempts to employ a mixture of fatty acids in the solvent phase produces neither of the above results but gives a reduced gloss with both low- and moderate-acidity alkyd resins. Because the alumina gel on titanium dioxide is a low-temperature entity and ferric oxide a high-temperature one, the former has a higher affinity for the dispersant. Also because shorter fatty acid chains are more mobile than longer ones and have essentially the same pK_a value, titanium dioxide tends to be selectively covered with the shorter-chain component of mixed fatty acids leaving ferric oxide to coat with the longer-chain component.

This surface differentiation, coupled with the smaller particle size and lower specific gravity of titania, tends toward pigment differentiation in the applied film, that is, color shifting in run patterns and on vertical surfaces compared to horizontal ones. The titanium dioxide–ferric oxide system is one where pigment separation and predispersion followed by mixing becomes an important formulation feature.

Printing of food packaging material affords another example of mixed-pigment–mixed-dispersant activity. A common film material is cellulose triacetate, having a medium-polarity chemical structure. To obtain a broad black-to-white print range, fractional tones are used to provide a gray scale, with rutile and carbon black being employed as separate but similarly dispersed pigments in a water-based ink. A commercial dispersant package contains sodium silicate (approximately 0.25%) and AMP (approximately 0.25%).[5] These two shift the pH sufficiently to create a negative charge on both pigments, good wetting, and excellent spread. AMP provides the organic anchor of the print image to the acetate surface then disappears upon hot roll drying, while silicate maintains the particles in a dispersed state until the print dries. Finally, the silicate produces a "water glass" coating on the final film for both mechanical (rub) and moisture resistance.

For high-alkalinity coating applications (ceramics, specialty granules, fused coatings, etc.), where mixtures of materials with broadly separated isoelectric points exist, a hydrolysis-stable dispersant package has been patented using a three-component mixture.[6] One application is detailed in Table 11.5.

In a novel ceramic application, a mixture of 60% titanium dioxide and 40% sepiolite (also known as meerschaum), a long, fibrous mineral, is dispersed to form a highly thixotropic system for porous, white ceramics. To obtain high green strength (structural integrity prior to firing) a polymerizable dispersant package, 0.3%, is used consisting of sodium vinyl sulfonate (telomer), sodium dihydrogen phosphite, and acrylic acid, $CH_2{=}CH{-}COOH$. The dispersant serves as an organic binder upon drying. With firing the phosphite oxidizes to

Table 11.5 High-Alkalinity Ceramic Dispersant Package

Component	Weight %
Quartz	66.0 (collective)
Zinc oxide	66.0 (collective)
Vanadium oxide	66.0 (collective)
Barium oxide	66.0 (collective)
KTPP	10.0
Ethoxylated sodium laurate	1.7
KOH (50% solution)	0.8
Water	21.5
	100.0

phosphate, bonding the alkaline oxides, and the organic components burn off, leaving a white, porous, low-density, high-strength ceramic body. It can be employed as a ceramic filter for wine and beer clarification, water purification, electrolytic cell separators, and other specialty chemical applications.

A similar dispersant–binder package has been employed with kaolinite slurries using a 65:35 blend of alkali methyl polyacrylate and sodium silicate.[7] These cast readily in gypsum molds, break free, and yield high green strength. Earlier mixed-dispersant formulations, which were formulated for spray dyring and high-density slip formation, consisted of mixtures of sodium silicate and sodium carbonate.[8] Such blends could produce ceramics subject to bloating and fissuring during the firing process.

Precipitation of alumina gels on titanium dioxide has become common practice among pigment manufacturers. However, this process has also found application on yellow (alpha) ferric hydroxides. With coating, color stability, which often shifts toward the red hue, has been improved, and redispersion, especially with fatty acids in oleophilic resins, produces superior rheology and color integrity. Surface coatings consist of various mixtures of aluminum hydroxide and magnesium, calcium or zinc vanadate precipitated via hydrolysis at pH values greater than 7. Levels of the mixed coatings are particle size dependent, but for a typical 1 μm pigment size average about 2%.

As discussed in an earlier section, satin white, a complex calcium sulfate–aluminum hydroxide pigment, has an isoelectric point slightly below pH 7 if

Table 11.6 Satin White Copigment Dispersant Package

Dispersant Component	Mole %
Methyl allyl sulfonic acid	15
Acrylic acid (MW ~2000)	15
Methyl acrylate	70
	100

manufactured properly and higher where the lime–alum stoichiometry is exceeded. Dispersing this pigment with kaolinite and titanium dioxide for paper coating formulations can take many forms, depending upon the character required in the final cellulose sheet. For high-strength, smooth-textured, white paper, a complex dispersant package, as shown in Table 11.6, has been advanced.[9]

The composition can be varied with a fairly broad latitude without altering dispersive capability, as can the molecular weight of the acrylic acid. A substitution of ethyl for methyl in the acrylate can be made, also. Based on the acidity of the general mix, the dispersant package functions considerably better on an alkaline pigment mixture (as satin white and calcium carbonate) than with acidic pigment surfaces (as rutile and kaolinite).

In paint formulations containing barite (barium sulfate), ground silica, and titanium dioxide, each component pigment can be separately modified with aluminum hydroxide or magnesium hydroxide (for superior whiteness), enabling a relatively consistent surface chemistry to be displayed. Such uniformity-of-surface generation is termed *compounded pigments* (as opposed to copigments) and is frequently a useful gambit in dispersing broadly variant materials.[10] The level of magnesium hydroxide treatment is kept as low as possible because the process results in elevated oil adsorption. Polyphosphates or amide-bearing polymers will provide a moderately uniform dispersancy in water, while oleic acid serves for alkyd resin and other nonpolar solvent formulations.

To attain the best combination of pigments and dispersants in compound formulations, a close inspection of the agents and the solids must be made to ensure proper order of addition, which will prevent dispersant interaction and ensure adequate coverage of all solid surfaces with appropriate agent(s). A thorough knowledge of the pigment manufacturer's surface or structural modifications is therefore absolutely essential to obtain optimum results in formulation and in application.

References

1. Michaels, A., Clay Minerals Soc. Meeting, Purdue Univ., West Lafayette, IN, October (1960).
2. Amerchol Corp., Edison, NJ (Promulgen agents).
3. Karlshams, USA, Columbus, OH (Caprol agents).
4. Fink, H., *Goldschmidt Inf.* **66**, 32 (1989).
5. Kawasaki, M., and Tenmaya, E., Japan Patent #1,247,470, Oct. 2 (1989).
6. Werle, R., Beurlich, H., and Kleeman, S., German Patent #3,712,206 Oct. 20 (1988).
7. Kelley, E., U.S. Patent #4,472,105, May 3 (1985).
8. German, W., *Ceramica* **24**, 36 (1978).
9. Kao Soap Company, Japan Patent #53,154, May 12 (1981).
10. Deshpende, S., *Paint India* **30**, 3 (1980).

CHAPTER

12

Special Applications of Dispersion

12.1 Material Processing Applications

In the course of manufacturing fine particulate systems, whether for pigments, cosmetics, pharmaceuticals, ceramics, polishes, or other applications, unit operations as slurries in processing are enhanced if not totally dependent upon dispersion. The effectiveness, as defined by the completeness of dispersion, often controls the properties and adherence to specifications of a company's products.

12.1.1 Wet Classification

Many powder systems, but especially pigments, are produced as broad particle size distributions. These may require grinding, desintering, separation, or other differentiating operations based on particle size. One common method of separating powders according to size is wet screening.

Most production-scale operations employ screens, or sieves, operated in a vibratory mode. The screen chamber, or tank, is vibrated by an eccentric weight rotated on a motor attached to the base of the chamber. This chamber is suspended upon a frame by means of a series of compression springs. The motor rotates at 500–1500 rpm, creating a depression or low point on the tank that follows the eccentric rotating mass. It resembles strongly the vibratory mill of Figure 8.16. Other equipment designs employ a sloping screen with agitation being bidirectional, along the axis of slope (fall line) and normal to it. Yet other units impart a vertical vibrational mode to the screen.

With the round tank design, dispersed slurry is pumped onto the center of the screen, with the oversize particles slowly being thrown centrifugally to the periphery, where they discharge through a tangential exit port. With the inclined bed design slurry enters at the top and cascades down the screen. Sub-sieve-size particles pass through the screen with the bulk of the liquid phase and are discharged through a similar but lower port or fall onto a second, finer sloping screen. Tank screens also may have a lower, finer screen.

The solids level in the oversize fraction increases dramatically, while that of the undersize diminishes, its concentration depending upon the fraction of screen oversize being removed. While dispersion quality has but a minor effect on the distribution of the screen's undersize, the oversize distribution is heavily dependent upon it. For most of these systems the viscosity must be 500 cp or lower to permit the necessary fluidity, surface transit, and screen transfer in the few seconds a given particle resides in the equipment. Slurries are of the order of 20–40% solids, depending upon particle size, narrowness of the distribution, and morphology.

Vibratory action on the screen surface is moderately violent, with slurry rising six or more inches on a tank siever. This aids in breaking apart very loosely associated floccules. However, because of the short residence time upon the screen, this activity is limited. Where secondary flocculation may occur, or where the size distribution is fairly narrow, the cam–motor unit is rotated counter to the tangential arrangement of the discharge port. This permits longer residence periods upon the screen and improved separation efficiency. By contrast, where the size distribution is broad with little tendency for loose floccules to form, the motor is operated codirectionally with the port, which throws material toward the tank rim faster and improves the transfer rate of the screen. A commercial wet-siever is shown schematically and in action in Figure 12.1.

To enable selectivity in screen drive rotation, motors are usually wound in a three-phase configuration, direction being changed by a control panel switch. With some models a high-frequency (ultrasonic) probe is attached to the center of the screen, adding another vibratory component to assist in material transfer and reduce blinding.

With inclined bed units, slurry residence is controlled by vibration frequency, bed slope, and, in some designs, a series of small weirs installed perpendicular to the fall line.

To ensure dispersion of the oversize component, as liquid phase is extracted from it, such slurries are often slightly overdispersed by 30–50%. At these dilute solids levels, the catenary viscosity profile is exceptionally shallow, often to the point where the viscosity increase is no more than 10–20% of the optimum. Oversize particles may be discarded or returned to the process stream, grinding mill, treatment, etc. Because that portion of the particulate system is larger in size, and hence has lower surface area per mass, it is underdispersed and contributes little or no processing problems when reintroduced into a closed-loop (recirculation-mode) operation.

The subsieve fraction, in contrast to the coarse, can be significantly overdispersed because (1) it was overdispersed upon entering the screen and (2) the

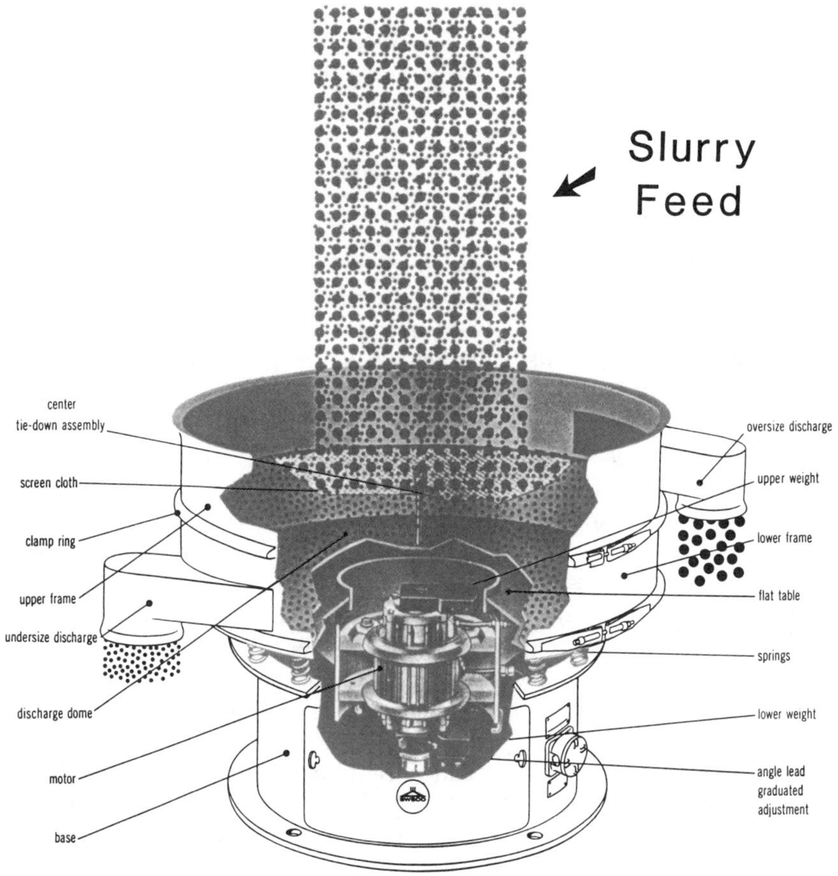

Figure 12.1. Commercial vibratory wet siever (courtesy Sweco).

solids level may have been reduced significantly. Ultimately this component must be dewatered in some fashion, which will return the slurry to a dispersant concentration closer to optimum for high-solids use. The water fraction is then returned to process makeup to utilize further the dispersant and minimize plant wastewater treatment of effluent.

12.1.2 Centrifugal Classification

Screen size classification is limited to about 325–400 mesh particles (44–37 μm). Particles above this size have relatively low dispersant demand. Where finer sizes must be classified in quantity (e.g., 10 μm and smaller), screen transfer rates are extremely slow and generally impractical. Hydrocyclones or centrifuges are most often employed for this size range. With certain high-cost products, such as pharmaceuticals, a high-frequency vibrating sieve employing Micromesh™ screens may be employed.

Centrifugal classifiers are usually horizontal, cylindrical devices several feet in diameter with a length 3 to 4 times their diameter. An axial vane and screw arrangement rotates internally. Slurry is fed into the center of the drum under pressure. Each particle is subjected to two velocity components, a radial one based on mass alone (centrifugal force) and a forward one, toward the unit's discharge port (pumping force), which is independent of mass. Depending upon flow rate through the centrifuge, rotational velocity, and drum diameter, particles have a fixed probability of being thrown against the wall before they exit the unit. The particle size at which this probability is exactly 50% is termed the *cut point* of the centrifuge. At this size one-half of the particles will appear in the coarse fraction, termed *overflow* or *solids discharge*, and one-half in the fine, termed *underflow* or *liquid discharge*. Cut points can be varied by changing rpm or pump rate. The former is preferred. A wet centrifuge (bowl type) designed for dispersed slurries is shown in schematic form in Figure 12.2.

Dispersion is extremely critical in centrifuges because, unlike the vibratory sieve, no mechanical particle–particle impact occurs to break apart loose floccules. Also some particle movement restriction occurs within the centrifuge because of increased solids near the outer wall. Efficiency in size cutting improves with dilution,[1] ideally being performed at 1% solids or below. However, economics in process time and later dewatering activity dictates higher solids, usually in the range of 20–30%. At this solids level floccules tend to drag out finer particles by occlusion. This decreases the efficiency of cutting and loses fines to the coarse fraction. As with screening, oversize materials may be returned to previous processing stages in a closed-loop operation (recirculation), made into other products or discarded as process waste.

A slow-moving screw operating inside the centrifuge pushes out oversize particles collecting along the wall through a coarse reject port. As with screening, the oversize becomes dispersant poor while the undersize becomes dispersant rich. To minimize this effect on later operation stages, the centrifuge feed is often at higher solids than later stages require to compensate for solids dilution.

Again, slight overdispersion may be employed. Unlike screening, the cumulative solids within the centrifuge are highly compacted and under great shear stress. Thus inadequate dispersion of the feed will shift the viscosity up one leg of the material's rheological profile and render the oversize "cake" highly resistant to shear and screw transfer, often to the point of dilatancy because of the narrowness of particle size and high solids. The solids level can be extremely critical, and centrifuges are often damaged by their screw drive being sheared off because of this high-solids, low-dispersant stress.

Predicting dispersant level in such operations is complex because viscosity, especially at 20–30% solids in the feed stream, is insufficiently sensitive to

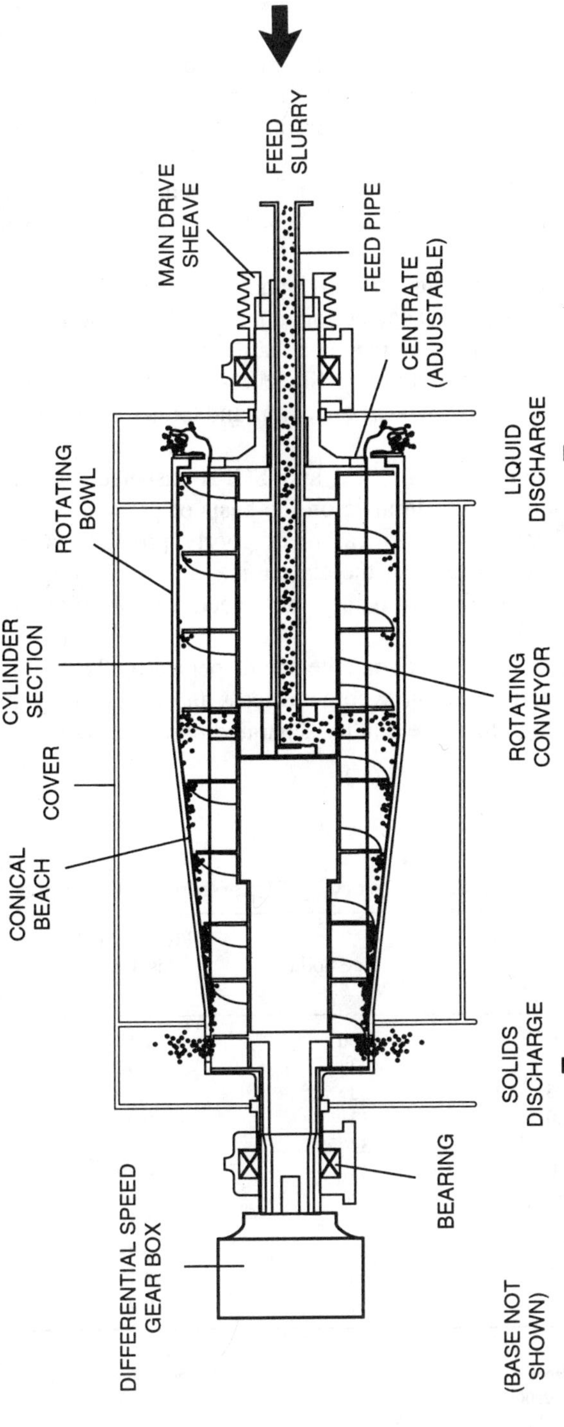

Figure 12.2. Dispersed slurry centrifuge, bowl design (courtesy Bird Machine Co.).

dispersant loading. To aid centrifuge operators with proper dispersion of feeds, centrifuges often are equipped with torque meters or motor ammeters that indicate when the wall cake is becoming difficult to move. Meter output may be directly connected via a control network to dispersant addition equipment, dry feeder, solution pump, etc. Dispersant addition also may be performed manually by an operator who constantly monitors the centrifuging performance. Table 12.1 is an example of material purification with a centrifuge as a function of dispersant level. Table 12.2 demonstrates rheological improvement where a fine size component is detrimental.

Limestone impurities are mica, quartz, and pyrite, while those of kaolinite are smectites (mixed-layer clays) and colloidal iron oxides. Limestone impurities contribute abrasivity to the product in paper coating applications. Impurities also decrease brightness (whiteness) slightly. Smectites are ultrafine-size, dark clays that in kaolins increase viscosity, degrade color, and hinder other ceramic application properties.

With calcium carbonate both TSPP and SOPA disperse all components, requiring the centrifuge to discriminate on the basis of particle size or density. By contrast, AMP disperses calcium carbonate poorly, producing weak flocs of oversize particles, while dispersing the clay, mica, and pyrite impurities. This artificial size effect and discriminatory or selective dispersion allows for improved impurity differentiation.

With clays, both kaolinite and smectite are dispersed by TSPP, SOPA, and AMP. However, smectites are on average at least an order of magnitude finer than kaolinites and, therefore, are readily separated on a real size basis through effective dispersion.

Table 12.1 Calcium Carbonate Recovery by Dispersion and Centrifugation

Dispersant Type	Dosage (Wt. %)	Centrifuge Solids (Wt. %)	Product Acid Insoluble (%)	Process Loss (%)
None	0	Feed	1.12	0
TSPP[a]	1.0	33	0.92	6.9
TSPP	1.5	33	0.94	7.7
AMP[b]	1.0	33	0.52	1.1
AMP	1.5	33	0.48	1.3
SOPA[c]	1.0	33	0.89	6.5
SOPA	1.5	33	0.94	8.6
AMP	1.0	20	0.43	1.1
AMP	1.5	20	0.38	1.4
SOPA	1.0	20	0.77	1.2
SOPA	1.5	20	0.75	1.4

[a] Tetrasodium pyrophosphate.
[b] 2-Amino 2-methyl 1-propanol.
[c] Sodium polyacrylate (MW 2500).

Table 12.2 Effect of Dispersion and Centrifugation on Rheology of Kaolinite

Dispersant Type	Dosage (Wt. %)	Cut Point (μm)	Brookfield Viscosity (cp) 10 rpm	100 rpm	Thix. Index
None	0	None	3600	950	3.8
TSPP[a]	0.2	0.25	2440	870	2.8
TSPP	0.3	0.50	1260	720	1.8
TSPP	0.3	0.75	1080	700	1.5
TSPP	0.4	0.50	1480	760	1.9
SOPA[b]	0.2	0.25	2220	880	2.5
SOPA	0.3	0.50	1120	740	1.5
SOPA	0.3	0.75	1020	710	1.4

[a] Tetrasodium pyrophosphate.
[b] Sodium polyacrylate (MW 2000).

Centrifuge feed solids = 20%.
Brookfield solids = 70.0%.

Selective impurity removal can improve a number of application properties other than color. Oleic acid surface-modified calcium carbonate (aragonite form) is employed as a filler for wire and cable insulation because of its white color, high loading capability, and excellent electrical properties. However, the trace presence of mixed-layer clays (usually 1–2%) can have adverse effects on dielectric loss, so important in high-frequency transmission and long-line carriers. Table 12.3 contains the results of a major western electric utility's investigation[2] into upgrading of aragonite.

Table 12.3 Dispersion Influence on Electrical Properties of Calcium Carbonate Filler

Centrifuge Dispersant	Level (%)	Product Wt. % Carbonate	Dielectric Loss (%)[a]	Centrifuge Loss (%)[b]
None (feed)	0	77.6	1.28	0.0
SOPA[c]	0.5	84.1	0.91	5.7
SOPA	1.0	85.2	0.90	7.8
SOPA	1.5	85.7	0.97	8.9
SOPA	2.0	86.3	0.99	10.3
AMP[d]	0.5	84.8	0.87	5.3
AMP	1.0	92.6	0.73	7.6
AMP	1.5	98.9	0.33	15.2
AMP	2.0	99.3	0.17	25.2

[a] 20% loading in PVC resin.
[b] Loss = underflow/feed.
[c] Sodium polyacrylate (MW 2500).
[d] 2-amino 2-methyl 1-propanol.

Note that both the type and level of dispersant are influential factors in improving the dielectric loss of the filled PVC insulation. This derives from the selectivity and fugitivity of AMP.

Data in all three tables reflect efficient removal of an adverse impurity by judicious selection of dispersants.

12.1.3 Wet High-Intensity Magnetic Separation

With the advent of sophisticated magnet technology has come practical process equipment for removing weakly (para) magnetic materials from nonmagnetic hosts, especially with regard to white pigment upgrading by pyritic and hematitic impurity removal. These magnets have field strengths from 25 to 50 kilogauss and consume large quantities of electrical energy. To be economical slurries must be highly dispersed and have very short residence times. Because of economic restraints, magnets operate with as high-solids slurries as electromagnetic mobility permits. An example appears in Table 12.4.

To effect particle movement of feebly magnetic materials in a slurry, the slurry viscosity must be optimally low. It requires excellent dispersion. Finely divided hematite (Fe_2O_3), goethite (a hydroxyl form approximated by FeOOH),

Table 12.4 Effect of Dispersants on Magnetic Separation Efficiency

Dispersant Employed	Dosage (Wt. %)	Slurry Solids (%)	GE Brightness[a] at 458 nm vs. MgO	Color Index[b]
none	0	25.0	87.5	12.7
TSPP[c]	0.2	25.0	88.3	12.0
TSPP	0.4	25.0	88.9	11.8
TSPP	0.6	25.0	88.7	11.7
Na silicate	0.2	24.6	87.9	12.1
Na silicate	0.4	24.6	88.3	12.0
Na silicate	0.6	24.6	88.1	12.0
AMP[d]	0.2	24.6	87.6	12.5
AMP	0.4	24.6	87.9	12.0
AMP	0.4	24.6	88.0	11.9
SOPA[e]	0.2	25.1	88.5	11.9
SOPA	0.4	25.1	89.2	11.5
SOPA	0.6	25.1	89.3	11.4
none	0	35.1	87.2	12.9
TSPP	0.2	35.1	88.1	12.3
TSPP	0.4	35.1	88.7	11.8
TSPP	0.6	35.1	88.6	11.9

[a] Brightness measured on nonmagnetic fraction.
[b] Brightness differential at 400 and 700 nm.
[c] Tetrasodium pyrophosphate.
[d] 2-Amino 2-methyl 1-propanol.
[e] Sodium polyacrylate (MW 2500).

or pyrite (FeS_2), flocculated onto a coarse calcium carbonate, kaolinite, or barite ($BaSO_4$) pigment particle, is exceptionally difficult to remove. Application of high shear and additional dispersant in a stage immediately preceding wet, high-intensity, magnetic separation enables magnetic impurities to increase in mobility and transfer across the magnetic cell to become entrapped in the mesh or fluted pole pieces. A solids compromise to effect separation and maintain production is often around 25%. Dispersant selection is based on the surface character of the two components being separated, a critical factor as shown by the table.

A number of deductions may be gleaned from the foregoing study. First, overdispersion is not universally advantageous. Second, AMP and sodium polyacrylate appear to be more beneficial at higher addition levels. Third, a lower solids level (25%) functions better than higher (35%). Finally, the dispersant effectiveness of the agents is:

$$\text{SOPA} > \text{TSPP} > \text{Na silicate} > \text{AMP}$$

This preferential activity is specific for the system under study (kaolinite) and should not be considered a broad categorization for all materials.

12.1.4 Flotation and Foam Separation

An older but still practiced form of mineral separation is flotation. This operation involves dispersing a mixed mineral system and adding a chemical agent that causes air bubbles to wet one component selectively and carry it to the surface, where it can be removed mechanically from the bulk liquid as froth.

Dispersion of the two components is an absolute necessity, as occlusion of the floated component by the host seriously impedes the separation practice. Second, while air bubbles are large enough to carry to the surface minor components and particles of the host, this results in a loss of the major component. In some instances host material is the desired product, for example, calcium carbonate with graphite or mica impurities. In others floated components are the desired products, for example, base and precious metals in granitic host rock.

With some mineral combinations components are separated mechanically in situ but have about the same size and density so that classical centrifugal or hydrocyclone techniques are ineffective. A dispersant may be available that is effective on one component but ineffective on the second, causing a major size differential to exist through flocculation versus dispersion. This process is termed *selective dispersion*. Thereafter, centrifugal separation becomes efficient. A case in point is the mixed-layer clays (dark) occurring in aragonite calcium carbonate (white) shown in Tables 12.1 and 12.3.

Dispersants employed in flotation are moderately broad ranging and include sodium silicate, short-chain quaternary amine chlorides, sodium dodecyl sulfonates, sodium alkyl thiophosphates, sodium cresylate, etc. Frothing agents are exotic and must work efficiently to attach air bubbles, yet not interfere with the

Table 12.5 Dispersant Employed for Flotation Separation of Minerals

Desired Mineral Component	Reject Component	Dispersant
Kaolinite	Quartz	Na cresylate
Calcium carbonate	Pyrite and graphite	Na xanthate
Graphite	Pyrophylite	Na silicate
Kaolinite	Bentonite	Polyglycol
Calcium carbonate	Bentonite	Butanolamine

dispersed state of the host material. Such agents include chemically modified xanthates and long-branched chain alcohols.

A select series of materials and agents are given in Table 12.5.

12.1.5 Selective Bleaching/Leaching

Mineral pigments are often naturally contaminated, primarily by iron oxides and secondarily by organics, weathered titanium products, and manganese compounds. The dispersant acts to ''open up'' contaminated surfaces and allow the activity, chemical sequestrant, reductant, oxidant, etc., to reach color bodies and/or impurities. For efficiency this requires a cooperative pH. Iron oxide color bodies require an acidic environment for leaching, yet, as was shown in Figure 2.5, acidity produces a flocculated system with zeta potentials being well above −30 mV. Unfortunately, the most effective pH for kaolinite bleaching occurs at a point where floc structure is most rigid, pH 3–4.5.[3,4] As a consequence acid leaching necessitates constant agitation to break apart floccules and agglomerates and to allow fresh chemical activant to diffuse in and leachate to diffuse out.

Only one dispersant functions well on the acidic side, orthorhombic phosphorus pentoxide, o-P_2O_5, a SHMP derivative. This is a highly effective bleach/

Table 12.6 Iron Leaching from Kaolinite as a Function of Dispersant and pH

Dispersant Type	Dosage (%)[a]	Slurry pH	Slurry Solids (%)	Brightness Gain[b] % vs. MgO
TSPP	0.3	6.5	35	0.5
TSPP	0.3	4.5	35	2.2
TSPP	0.3	3.5	35	4.1
SHMP	0.3	4.5	30	2.8
SHMP	0.3	3.5	30	4.3
o-P_2O_5	0.3	3.1	30	4.0
o-P_2O_5	0.25	3.6	30	4.7
o-P_2O_5	0.20	4.0	30	3.8

[a] Bleach dosage 0.35% solids basis with H_2SO_4 as acidulant except with o-P_2O_5 (self-acidifying).
[b] Brightness gain measured at 458 μm.

leach adjuvant, as shown in Table 12.6, where hydrated ferric oxide is removed chemically from clay. Its shortcomings of high cost, short-term stability, and attack of iron and aluminum equipment unfortunately preclude its broad commercial use.

Alkaline bleaches exist for other tinctorial contaminants, such as sodium peroxide, sodium hypochlorite, sodium peroxysulfate, and sodium peroxyborate, which function well with inorganic polyanionic dispersants. These are often employed for removal of tannins, humates, and other organic impurities in pigments and in laundry products to remove food and grass stains, greases, and other residential organic debris.

12.1.6 Mill Grinding Aids

Most grinding of ores, especially coarse crushing and very fine grinding, is performed dry because of the design of comminution equipment, such as, jaw crushers and jet mills. However, ball mills are frequently operated both in wet and dry states because of their enclosed structure. In wet size reduction, additives are often employed, termed *grinding aids*. With ball mills in use for nearly a century, a host of chemical agents have been tried, perhaps as much through wizardry and chance as through scientific efforts!

Grinding aids, while defined differently, are simply dispersants employed to

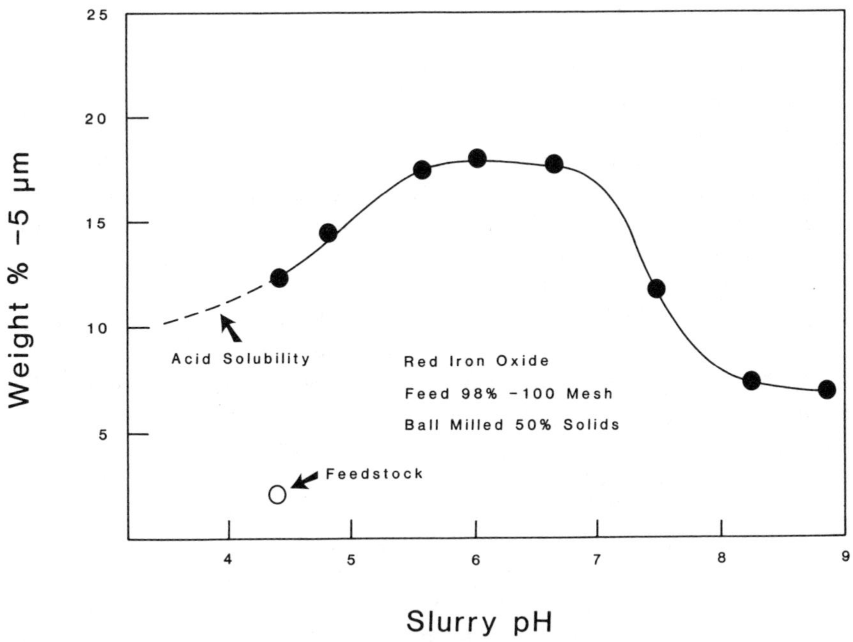

Figure 12.3. Wet milling of hematite using alkali as dispersant.

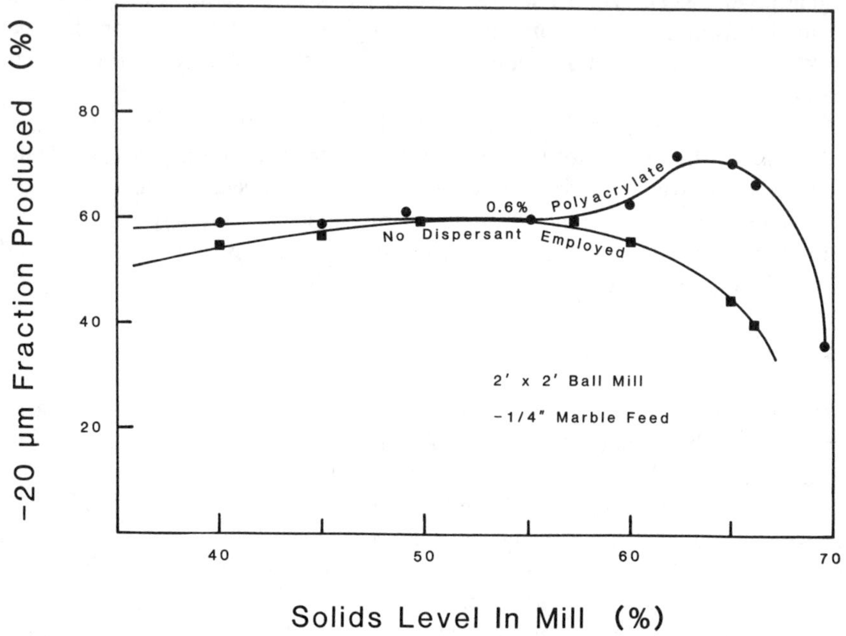

Figure 12.4. Wet milling of calcite with sodium polyacrylate dispersant.

reduce grinding energy. These affect the operations in three ways. First, the additive increases fluidity (viscosity reduction), which allows higher percentage solids in the mill and increases production throughput. Second, it reduces the tendency for material buildup on mill walls and on the balls themselves, both of which would reduce the impacting effectiveness during mill rotation. Third, a reduction in viscosity shortens mill loading and discharge time and decreases mill downtime, which also increases productivity. Such agents range from the complex organic materials to the simple alkali. In Figures 12.3 and 12.4 are two examples expressing this range, sodium hydroxide on red iron oxide grinding and sodium polyacrylate with marble (calcium carbonate).

Note the dispersing agent's effectiveness does not show up until high solids are attained, that is, those approaching critical volume.

In the 1960s a new form of mill was invented termed *a stirred ball mill* or *sand mill* (although sand is no longer employed). A detailed description of this equipment was given in Section 8.3. These mills are not centrifugally limited, and rotational speeds in the range of 1000 rpm are often employed. Dispersants, also known as grinding aids, must be employed in most instances to improve milling performance and reduce machine wear.

Those rheological constraints occurring with ball mills also apply to sand or media mills; that is, high solids improve efficiency (KWH/ton), and media must

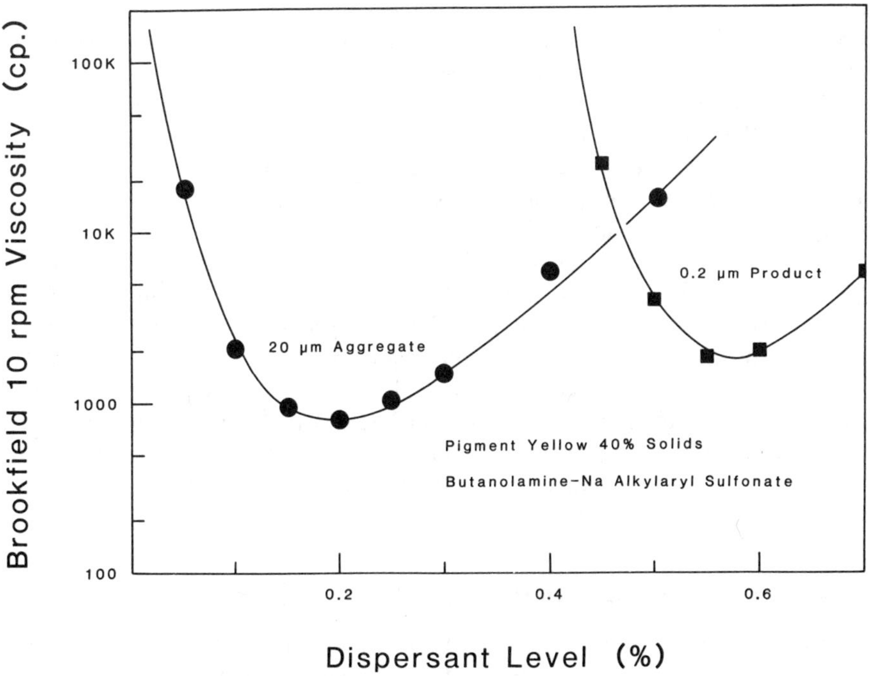

Figure 12.5. Grinding rheology of pigment yellow.

be fluid and capable of being drained from the container once grinding is complete. For this operation, good dispersants are a prerequisite.

Media mills are most often employed for very fine grinding, where particle size distributions are predominantly <5 μm. Size reduction ratios, like most mills, lie in the 10:1 range. High surface areas generated in milling can cause dramatic increases in viscosity. In very fine grinding the system can build viscosity so rapidly that mill paddles or radial arms, or their coatings, shear off.

Depending on media size, grind size, and other factors, a mill viscosity around 1200 cp is recommended for efficient size reduction (i.e., best utilization of time and electrical energy).

To prevent media mill damage from high viscosity, special dispersing techniques have been introduced. Figure 12.5 shows a typical catenary rheological profile for Pigment Yellow GB having a D_{50} of about 12 μm in aggregate form. Because of the steepness in the underdispersion leg (low agent concentrations) of this curve, the mill is highly sensitive to viscosity to the point where grinding efficiency drops dramatically or mill damage ensues.

On the right-hand portion of the figure the second curve represents the rheological profile for the product, having a nominal D_{50} of 0.2 μm. This behavior parallels to a degree that of calcium carbonate milling shown in Figure 8.26.

Optimally dispersed feed represents poorly dispersed product. Optimally dispersed product represents severely overly dispersed feed. Overheating in the mill can alter organic pigment's color. The practical solution involves dispersant metering over the grinding period, as with calcite. Ingenious efforts have been employed by mill users to circumvent this problem and have been discussed earlier. In the example selected the most effective dispersant is a mixed agent, approximately equal parts butanolamine and sodium alkylaryl sulfonate. While continuous addition of dispersant is desirable, manual control can be effective using slight overdispersion as the guide.

Dispersant demand is not directly proportional to surface area (0.2% vs. 0.65% for a tenfold size reduction). This observation in association with the gradual rise in the overdispersion leg of the rheograms (both feed and product) indicates sodium alkylaryl sulfonate on organic pigments is also in equilibrium with the water solution phase. Because of the character of the product resulting from commercial manufacture of Pigment Yellow, true comminution does not likely occur, only disaggregation. Fundamental properties are reported below 0.5 μm. The dispersant demand increase more reflects diffusion into particulate clusters rather than adsorption onto new surface.

12.1.7 Particle Size Analysis in Quality Control

Most processing operations involving products carrying particle size specifications must be constantly monitored and quality control maintained at each stage along their process stream. This will include not only finished products but also the processing stages where problems can be corrected before they reach later, more expensive stages in the operation. Manufacturers must have particle size measuring equipment in place to accomplish these ends.

It is beyond the scope of this text to go into the various pieces of particle size measuring equipment, and adequate texts exist on this topic.[5] Of the five or six basic methods, light scattering, optical blockage, volume displacement, sedimentation velocity, and fractionated stream flow, all require the ultimate in dispersion. Also, all operate at exceptionally high dilutions, often at less than 0.001% solids. The highest percentage solids (about 0.5%) occurs with X-ray adsorption sedimentation practices.

The prime concern in particle size analysis is to maintain discrete particles, which enable any measurement being made, whether in a collective array ("field scanning") or as a stream of sequential individual particles ("stream scanning"), to be uniquely associated with that individual particle or group of particles. To effect this, great care is exercised in both slurry preparation and slurry processing.

At these extreme dilutions, viscodynamic effects essentially cease,[6,7] and the concept of overdispersion disappears. Often the dispersant dosage is 5–10 times optimum for high-solids viscosity control. This ensures that no solution effects will alter the discrete state of the individual particles and that kinetic particle collisions will not produce floccules of any size, not even dimers.

To obtain sufficient shear to break apart flocculated or associated assemblages occurring with dry powders, slurries may be formulated for optimum viscosity, or even slightly overdosed and agitated under high shear for extended periods, typically 20–30 minutes. The slurry is then diluted under shear as additional dispersant is introduced. An alternate procedure is to remove by pipet a small aliquot of concentrated slurry and introduce this into a shearing agitator filled with dispersant solution. Shearing at low solids is entirely inadequate to reduce particle systems to discrete particles.

Typically, a laboratory blendor is employed, and a solids level 5–10% below the critical level is formulated and agitated. This is diluted further by one of the previously described procedures. The diluted phase is placed in an ultrasonic bath for 1–3 minutes to ensure a thorough breakdown of any remaining assemblages in the system. Some instruments have self-contained ultrasonic dispersors and bypass the agitator phase completely. A pair of interesting studies of residence time in an ultrasonic bath and the attendant particle size measured are shown in Figure 12.6.[44]

Some studies suggest that protracted exposure to ultrasonic baths may in itself reduce particle size, and be, a form of comminution. This may occur, as shown for calcium carbonate in the foregoing figure, where samples from a hammer mill are analyzed immediately and after 6 months aging. In the fresh material

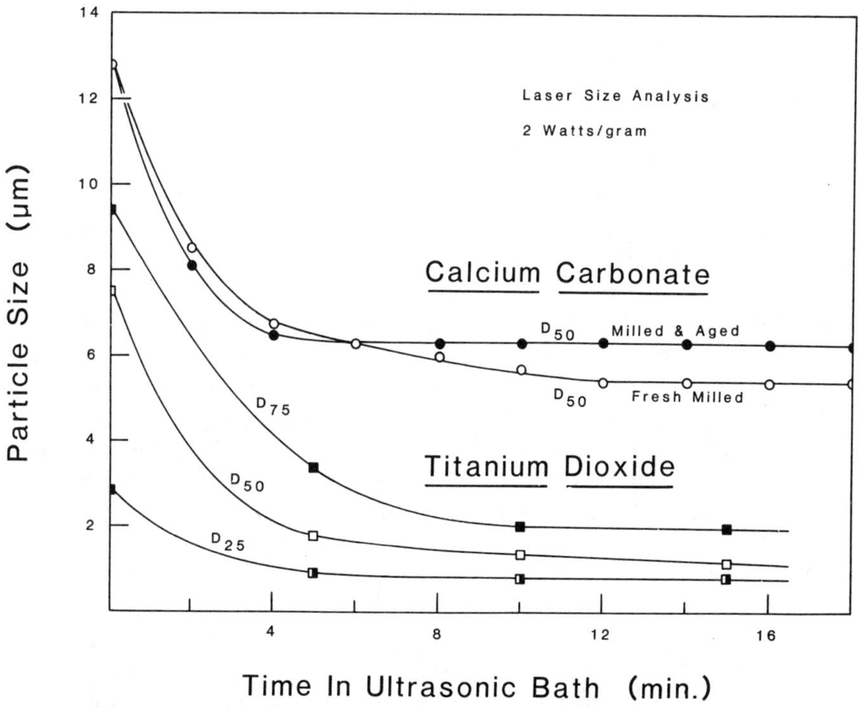

Figure 12.6. Particle size versus ultrasonic dispersion time.

some particles possess residual fractures running through the major portion of the particles. Upon intense cavitation, gas bubbles may develop in these incipient fractures that with time propagate the fracture though the solid, flake off small fragments, and render the median size (D_{50}) slightly finer. Breakage in the study appears to continue up to about 12 minutes of cavitation time, after which median size remains constant.

With titanium dioxide ultrasonic energy is insufficient to break down primary particles. The data show a constant size at three levels of assessment, D_{75}, D_{50}, and D_{25}, to be reached in about 9 minutes. Most studies which suggest "ultrasonic grinding" indicate preferential generation of very fine components rather than breakdown of a broad range of sizes.

A second dispersion problem unique to particle size is the formation and dispersion of air bubbles in the system. In optical scattering, light blockage, and volume displacement (e.g., Microtrac Laser, Hiac and Coulter methods), an air bubble is sensed as a particle and is counted as such. Most air bubbles generated in these analytical makedowns are large and bias the distribution towards the coarse end. Often they can be detected through distribution analysis employing log probability plots (see Chapter 13).

Those dispersants employed for particle size analysis can be more exotic and expensive because their quantities are minute. They must be nonfoaming, nonfugitive, noninteractive, and stable. Often the pH of the systems is elevated with alkali to ensure the maximum negative values for zeta potential, if charge dispersants are employed. Sodium hexametaphosphate, sodium silicate, sodium carbonate, and sodium polyacrylate are frequent candidates for analytical use. In the Coulter method of size analysis (electrosensing zone technique), the employment of electrolytes actually enhances the analysis by making the slurry more electrically conductive. This lowers electrical noise and makes resolution of the finer particles more accurate.

Dispersing organophilic powders is not especially complex, and analyses can be performed with Hiac, Sedigraph, and laser scattering size equipment. The dispersant is most often sodium dodecyl sulfonate (SDS), dodecyl trimethyl ammonium bromide (DTAB), oleic acid, or fatty amines. Makedown procedures are identical except for exercising caution against air inclusion, which is more severe with nonaqueous solvents. Some dispersants contain a few percent of antifoaming agent, for example octyl alcohol.

Reference has been made periodically in this book to the influence of dispersed particle size distribution and morphology on mechanical, optical, rheological, and electrical properties of the applied slurry in finished form. It is most important to understand just what a given dispersed size actually means.

All instruments reference their measurements to a theoretical sphere and a diameter derived from that sphere. The instrument infers the sample material's behavior corresponds to that of a set distribution of spheres. Few materials, however, enter the instrument in a form even remotely resembling spheres! For example, a laser size analyzer may interpret a 10 μm long by 1 μm square needle of gamma ferric oxide as a 4 μm sphere.

Different instruments assess particle dimensions by different physical phenomena. As a consequence, a sedimentation equivalent sphere (with Sedigraph, Brookhaven, Horiba, etc. instruments) is not the same entity as an electrozone equivalent sphere (Coulter, Elzone), a laser diffraction sphere (Microtrac, Malvern, Celas, etc.), or an optical blockage equivalent sphere (Hiac, Climet, PMS, etc.), none of which will resemble a microscope sphere. It is the confusion of these "equivalent spheres" under the best of dispersion conditions that creates difficulties in establishing powder mixture systems capable of accomplishing a desired application purpose.

Anisotropic particles yield different sizes depending upon the method of measurement, as shown in Table 12.7. Attempts to predict optical properties from such data may result in wholly erroneous conclusions. Of all the particle size assessment methods, the most accurate, as well as the most tedious and time consuming, is the direct measurement approach with optical or electron microscopy, where the sample has been shadowed by metallic sputtering.

Particle packing by true dispersed spheres will not provide the same physical assemblages as the same distribution of "equivalent spheres" with anisomorphic particles. As a crude relationship, packing density of a film composed of an instrumental log-normal distribution of particles, as determined by laser scattering instrumentation, decreases in the following order:

spheres > equants > platelets > needles

Variation in packing density can range by a factor of 3, or even more where particle anisometry is great. Blending of fine particles of one morphology with coarse particles of a second to produce a quasi-log-normal distribution can yield highly unpredictable packing density results. This is especially true with ceramics where firing density becomes an important quality factor in the product (breakage strength, voltage breakdown resistance, moisture adsorption, etc.). Knowledge of the method by which a given size instrument interprets diameter is important in applying the results to accomplish a given purpose and product property. As a consequence of varying instrumental interpretations, instrument

Table 12.7 Equivalent Sphere Interpretation by Various Measurement Techniques

Instrumental Method	True Sphere Diam. 10 μm (μm)	5:1 Hexagonal Platelet, 20 μm (μm)
Electrozone volume	10	10
Sedimentation	10	7
Optical blockage	10	12
Laser scattering	10	9.3
Microscopy	10	18
Micromesh sieving	10	13

purchases should not be made solely on the bases of cost, speed, simplicity, or size range.

Most commercial sizing instruments function well as quality control tools. If the sample is run repeatedly, the results should be essentially equal. For production control this is often sufficient to meet production and quality goals. If it is known, for example, that a certain distribution, as measured by a specific instrument, produces certain desirable characteristics (packing, flow, opacity, etc.), then that instrument can function satisfactorily, even if it provides erroneous size data. Irrespective of the misinterpretation of true size, quality control technicians must ensure that a consistent and thorough method of dispersion, with both agent and agitator, is employed for sample preparation.

In a standard method of preparing samples for laser size analysis, isopropyl alcohol is often recommended for dispersing the particles in an ultrasonic bath. With silicates and many other powders, considerable oversize interpretation results because floccules, rather than discrete particles are being scanned.

Severe problems in relating size to properties occur within distributions where morphology varies with size. This occurs with certain kaolinites (higher-aspect-ratio platelets at finer sizes), with disaggregated needles (finer particles include fragments from fracture), and in multicomponent mixtures (e.g., fly ash). In such situations, microscopy, for all its shortcomings, may hold the definitive answer to true size.

Similarly, distributions in which density changes with size will yield an erroneous size distribution where size is determined by sedimentation procedures (Sedigraph).

12.2 Dispersing Metal Powders

Applications for dispersed metal powders fall into six broad categories. These include:

1. Paint (decorative and anticorrosive);
2. Decorative ceramics;
3. Electrical and electronic uses;
4. Mechanical soldering;
5. Catalysis; and
6. Ceramic/cermet premixes;

Decorative paint applications involve primarily aluminum and ''bronze'' alloy powders having flake morphologies. Decorative ceramics applications include artistic embellishments, designs, and rim bands on dinnerware and ceramic tile (dominantly gold or platinum). Flake morphology also is preferred here. Decorative uses also include jewelry (cloisonne).

Electronic uses are dominated by the printed circuit board industry where conductive elements may be printed directly[8,9] onto polymeric substrates. Such uses require highly conductive powders, commonly copper or silver, their alloys, and mixtures.

Dispersion of metals is a special case of surface affinity, and systems must be designed specifically for the metals employed and the liquid phases into which they are distributed. In all applications of dispersing metal powders in organic liquids, the dominant factors to be considered are the dielectric constant of the solvent phase and the electron affinity of the metals. With more electropositive metals the surface is populated almost exclusively with tightly bound oxides and hydroxides. With noble metals only a small fraction of the surface is occupied by such groups. The ability to attach a dispersant molecule onto these surfaces necessitates having accurate information on the metal surface characteristics. These may change with aging.[10]

Noble metals are characterized as chemically inert and assumed to have non-compounded surfaces. While this may hold true for gold and platinum in some instances, it is not true for all others. Silver particles, for example, which have aged even short periods in air, show portions of their surface to be altered to oxide and sulfide and sometimes halide by reaction with atmospheric ozone, hydrogen sulfide, and chlorine. The average person is painfully aware of this surface chemistry when prized silverware is removed from storage having heavy coatings of dull black sulfide tarnish!

Metals above hydrogen in the electromotive series are assumed to have oxides, hydroxides, carbonates, mixtures of these, or even halides on their surfaces. Making these assumptions, thermodynamic modeling of the parent metals gives characteristics that agree with surface wetting characteristics and the reaction of surface-active agents.[11] The character is that of a strained oxide, hydroxide, or carbonate, indicating the substructure differs significantly from that of true oxides and carbonates. This strain, or covalent bond stretching, alters the ionization of the hydroxyl groups sufficiently that their pK_a or pK_b values change, often by nearly an order of magnitude. A change of hydroxyl character may alter the selectivity of dispersant (anionic vs. cationic, amino alcohol vs. charged species, etc.).

Zinc and copper, although greatly separated in the electromotive series (-0.76 vs. $+0.34$ V), can have almost identical surfaces, even though zinc exhibits a hexagonal close-packed crystal structure and copper is cubic close packed.[43] Each is in contact with twelve other metal atoms in their crystallographic structure. In addition, they differ in atomic size by less than 4%, accounting for the nearly continuous series of brasses made from their mixtures. The surface of both metals contains hydroxy-carbonates, as shown in Figure 12.7.

As zinc oxide's isoelectric point occurs above pH 9, at all practicable aqueous pH values the surface is ionized positively:

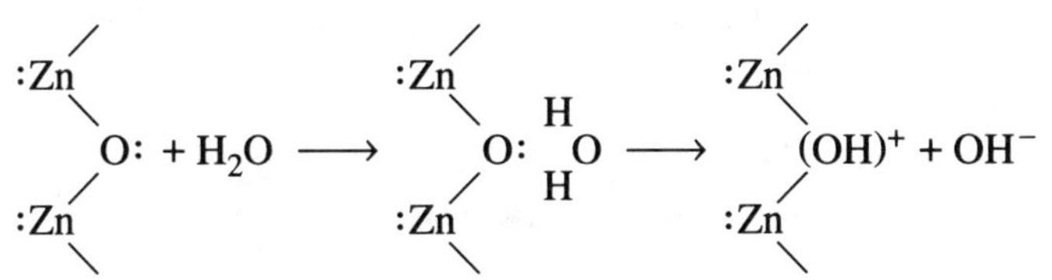

(12.1)

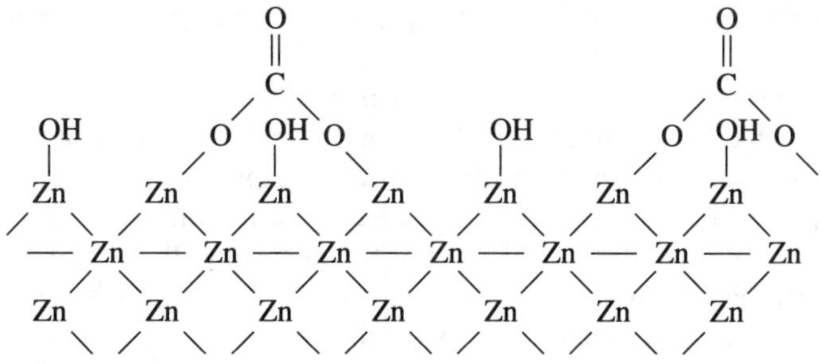

(A) Zinc–hexagonal close packed structure, 001 crystal plane

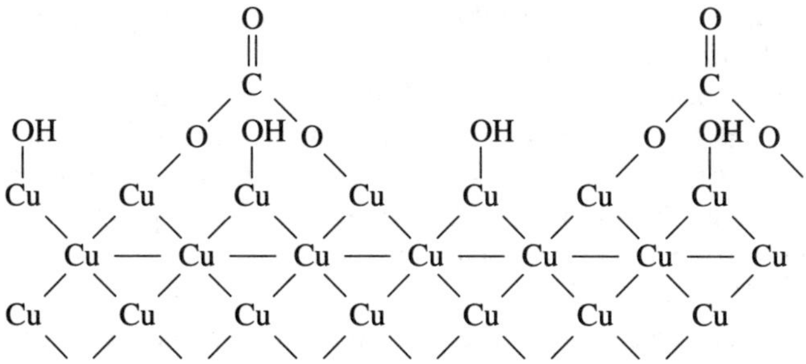

(B) Copper–cubic close packed structure, 111 crystal plane

Figure 12.7. Similarity of zinc and copper metal surfaces.

Selection of a functional dispersant must compensate for this alkalinity. At pH values only slightly acidic (5.5–6.5), employment of ammonium polyacrylate will allow an interaction to take place similar to that shown earlier for alumina and aluminum silicates (see Chapter 4). For very short-term applications, orthorhombic phosphorous pentoxide (H Calgon) will serve similarly. However, o-P_2O_5 is an aggressive dispersant that will quickly enter into a fixed chemical, rather than dispersive, reaction with the oxide to form zinc orthophosphate:

$$x\ Zn(OH)_2 + (HPO_3)_xH_2O \rightarrow x\ ZnHPO_4 + (x + 1)\ H_2O \qquad (12.2)$$

In the slightly alkaline range advantage can be taken of zinc's ability to form sp^3 coordination hybrids (ammine complexes) with ammonia and water soluble amines. Thus a long-chain alkanolamine, H_2N–R–OH, where R represents at least 4 carbon atoms, can enter the coordination sphere of oxidized surface zinc atoms and extend their hydrophilic tails into the aqueous phase, as depicted in Figure 12.8.

In recent years zinc powders have been employed as anticorrosives for paints designed to protect steel. Their function somewhat parallels magnesium anodes

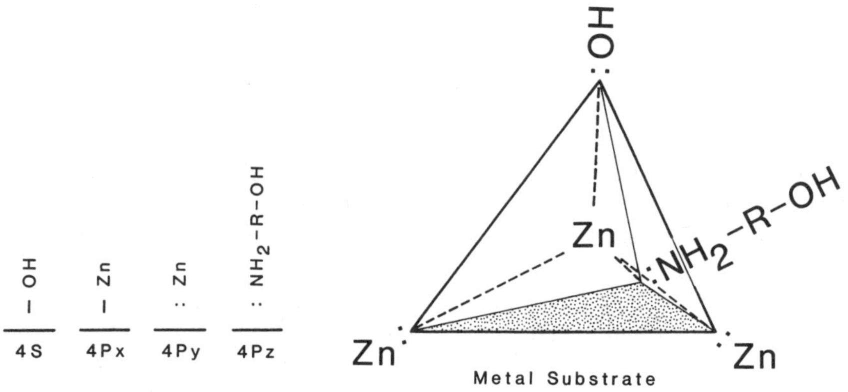

Figure 12.8. Alkanolamine coordination onto zinc metal for dispersion.

in water heaters—they become sacrificial toward diffusing water and oxygen, carbon dioxide, and other acidic atmospheric gases (SO_x and NO_x).

Powdered zinc functions admirably in this role. Its disadvantage lies in its density, 7.1 g/cm^3, which causes severe formulation caking upon storage. Caking can be reduced through modification of particle size, morphology, and binder viscosity. Because of their high settling velocity (see Table 1.1), zinc powders must have very fine particle sizes (which also aids in anticorrosive activities), preferably around 3 μm or finer. As the metal is generated through high-temperature reduction-fuming processes, powders form either as fine needles or as near-equants. The acicular form has greatly retarded settling compared to the equant form.

Zinc's high chemical reactivity biases its formulations to alkyd resin bases in preference to water bases. As these resins possess residual acidity from manufacturing, they form readily a surface reaction product (usually via linoleic acid) with the metal (via the oxide–carbonate surface) and become self-dispersive. In the absence of an inherent acidic agency, oleic, stearic, benzoic, succinic, caprylic, or other long-to-medium aliphatic or aromatic acid will serve to provide dispersancy. Zinc metal carries an adsorbed water film, which is displaced by these organic dispersant reaction products.

The metal may be predispersed in dispersant–solvent solution (mineral spirits, toluene, VMP naphtha, etc.) and added to the binder, or the dissolved dispersant may be added to the binder plus metal. The predispersed form usually provides a more thorough distribution of metal particles throughout the resin. During application the dispersed zinc particles preferentially settle to the metal–coating interface and provide a highly chemical-resistant layer toward oxidation and acidification.

Fine metallic zinc ranges from dark gray to black and, as such, is extremely difficult to pigment with other than titanium dioxide. As such zinc primers and anticorrosion paints are generally formulated to a more practical gray, or darker pigmented color for sale.

Aluminum, though higher on the electromotive force scale than zinc (-1.66 vs. -0.76 V), has a highly passive surface oxide. That is, the oxide–hydroxide surface layer is nearly an exact match in size for the parent metal surface. This fit blocks diffusion of gases through it to the base metal. Aluminum occurs in a flake morphology which, when thoroughly dispersed, allows it to settle and form overlapping particles that create an effective barrier to other diffusional gases. Its anticorrosive activity is more mechanical than chemical in nature. This pigment characteristic is termed *leafing*. The metal also is made with a less phylloidal structure in which particles distribute randomly throughout the polymer structure.

Unlike zinc, aluminum's density is but 2.7 g/cm^3, little different from conventional oxide pigments. Some forms are made with porosity, which reduces the density even more. As a consequence, little caking or differential settling occurs when these are formulated in low-viscosity slurries, especially in alkyd resin systems.

Aluminum flakes with large diameters, when thoroughly dispersed, provide a high metallic sheen to a film, paint, or ink, one which has a polished, mirrorlike (specular) finish. Small-diameter, less anisometric, and less critically dispersed aluminum particles tend to orient entropically through a film or casting, giving a random reflectance, nonspecular, pattern for a quasi-pearlescent appearance.

Because aluminum's surface is dominantly populated by =Al–OH groups, which are amphiprotic (about half alkaline and half acidic), a weak bonding attraction results with amino groups. Where dispersibility in oleophilic media is required, pretreatment with medium-length fatty acid or C-4 or greater organotitanate complexes [e.g., $(BuO)_4Ti$] is satisfactory to maintain good dispersion and restrain dilatant caking. Although dispersant can be added directly to the formulation, superior results are obtained when treatment occurs either in a presolvent phase or prior to use (e.g., by manufacturer). This is especially true for the organotitanate dispersants, which are sensitive to atmospheric moisture degradation.

In water-borne applications aluminum particles must be oxidized more thoroughly (passivated or inhibited) to prevent reaction in the liquid phase because of their high surface areas. In these applications, in which ethylene and propylene glycol ethers are common binders, the dispersant will be a polyethoxylated, polysiloxane nonionic, or anionic molecule with low alkalinity. Aluminum's reactivity, even with passivated surfaces, increased dramatically with pH values below 5 and above 8 due to surface hydroxide–oxide solubility. Where increased alkalinity is mandated because of polymer requirements, primary organoamines (butanolamine, monoethanolamine, or AMP) will usually suffice. More alkaline tertiary amines have sufficient basicity to attack aluminum flakes during prolonged storage to degrade both application appearance and dispersion. Halogenated solvents, though fairly uncommon in polymer applications, *must* be avoided because of their severe and almost explosive reactivity with finely divided aluminum particles.

Because much of the character of the aluminum pigments derives from their high anisometric morphology, great care in dispersion must be exercised to prevent high-shear mixer blades or media from breaking particles. Particle breakage also results in new metal surface, which becomes highly reactive with the aqueous solvent and agents therein.

Copper, having both a hydroxyl–carbonate surface and a capacity similar to zinc to form amine coordinating complexes, can also enter into a dispersive–coordinated state. Copper, unlike zinc, has no tendency to react with its liquid phase, assuming ozone and nitrogen oxides, common in the atmosphere, are not dissolved components. Copper also has a disadvantage with alkanolamine dispersants—the amine group complexes the metal and slowly develops a bluish hue.

The bond established between amine linkage dispersants and metal surfaces can be much stronger with copper and zinc, compared with aluminum silicates and titanium dioxide, because coordination bonding can be realized. As a consequence, the dispersant's fugitivity requires a higher temperature or longer time for the agent to release, once the dispersed metal powder has formed a film.

Copper powders dispersed in formulations quickly acquire a pale green reacted surface or hydroxyl–carbonate groups (termed *patina*) if atmospheric gases penetrate the applied film, or a bluish hue if reactions take place with acrylics, primary amines, or other polymerizing species. All remaining metal powders that are electropositive (above hydrogen in the electromotive series) have hydroxyl, oxide, and/or carbonate compounds on their surface. It is this oxidized layer that can most often be utilized in developing a highly stabilized dispersed state.

Some metal powder/particle slurries are inherently unstable, due to reactivity with dissolved atmospheric gases, reactivity with the liquid of suspension, or low surface charge per unit mass. A case in point is the formation of metal sols.

Metal sols, especially those of the very noble metals, can be produced by an electrical discharge (shorting) of metal electrodes in chemical solutions. The particle size of such sols is usually in the range 10–100 nm (0.01–0.1 μm). Often they approach a few atomic and molecular dimensions. Many exhibit color due in part to selective light scattering and in part due to photonic interactions with the valence electrons on the sol surface. The sol no longer carries the metallic characteristics of sheen and conductivity possessed by the basic metal. In suspension, sols are a limited self-dispersive state, electro-oxidized particles having a high specific surface charge and usually surrounded with a chemical species. Typical is gold in hydrochloric acid, where each gold sphere contains a halogenated and partially complexed surface ($AuCl_4^-$). Other metal sols have been produced via strong alkali, where the surface species is assumed to be $M(OH)_x^{(x-n)-}$ and n is the oxidation valence of the surface metal atoms.

Gold sols can be produced chemically by rapid reduction of the acid, $HAuCl_4$, which produces intensely colored species whose color is sol size dependent. Brilliant red sols useful in glass and ceramics form in the range of 10–20 nm. They form as monodisperse systems at dilute concentrations of the acid and as triplets or larger oligomers at more concentrated levels. The sol carries on average about one layer of water, on the exterior of which lies the ionic dispersing species. Dispersion requires a high level of charge because of the great surface area per unit mass. When these dispersed sols are mixed under high shear with ceramic components, especially silica and silicates, the exchange capacity of the solid binds the metal sol in place, even after drying. There it can be fired to produce unusual color or conductive effects in the finished ceramic article.

Silver sols created by reduction of ammoniacal silver nitrate in the presence of silicates result in gray products consisting of black sols on a white support surface. Excess ammonia in solution serves as mutual dispersant and sol stabilizer. Electron micrographs of the dried products indicate a sol size range from 50 nm down to less than 10 nm. However, upon firing the silver-support structures, the metal diffuses or dissolves into the silicate structure, giving once more a white product. The metal, in whatever form it takes, remains at the surface because it serves as a powerful bacteriostatic agent in treating contaminated water supplies.

Electron micrographs show no sol structure after firing, but lattice constants obtained from X-ray diffracton exhibit a small but distinct unit cell expansion.

Sols may be considered, in the absence of additional agencies, to be self-dispersive states highly sensitive to electrolyte composition and concentration, the latter because of close proximity between individual sol particles, which occurs even with minute percentages of metal in suspension. Slowly dissolving atmospheric gases and ultraviolet light can cause sol accretion, followed by aggregation and ultimately by precipitation.

Metal sols have been prepared commercially for decorative applications but not for paints, ceramics, electronics, or catalysis. Their instability and extremely low concentration restricts their use.

Stable metal sols must lie well below 1 μm in size to enable their permanent dispersion and suspension in water. These have been discussed previously in another section of this book. The most stable sols are those of the noble metals, Cu, Ag, Au, and the platinum group. All can be formed by an anodic reaction where the anode is composed of the parent metal.

By increasing the conductivity of the electrolyte with an inert salt (e.g., sodium sulfate) or alkali and by applying a strong electric current (by high voltage), the current density rises on the anode to the point where atomic clusters of metal (several thousand atoms per cluster) are blown off the electrode surface and carry a portion of the electrolyte on their surface for ionic stabilization. For this reason, alkali solutions are highly favored as sol promoters.

Studies on gold, silver, and platinum sols indicate their surfaces have a small oxide layer (two to three atoms thick),[12] in spite of their nobility. The oxide layer may not completely envelop the sol particle, and much of the surface may consist of metallic patches. However, this oxide plus an alkali guest at high pH is the basis for high acquired charge in suspension.

While metal sol particles repel one another greatly during their early life, a transition occurs during aging, which causes the surface charge to diminish. It has been postulated that sols eventually are attracted back to the anode (negative colloid to positive surface) where some charge dissipates, some adsorbed alkali is lost and the particles slowly lose their repulsive power. Eventually, the system will precipitate, with metal powder collecting at the bottom of the container in larger aggregates or on the walls as a thin film. This is a specialized form of electrodynamic flocculation.

Metal sols are especially attracted to silicates with positive zeta potentials. These may be employed as catalytic carriers or other systems of introducing sol-sized particles without the need for the sol-sustaining electrolyte. Chrysotile and certain forms of talc are effective for this technique. Electron micrographs show that metal sols, such as gold, are attracted to impurity cation sites, for example, Mn^{2+}, on the brucite, or magnesium hydroxide, layer and cover nearly 30% of the silicate mineral surface. Also these metal sols remain fixed following drying.

Solder powders (tin and lead alloys) are employed to attach functional elements (resistors, capacitors, semiconductor chips, and other components) onto circuit pathways as part of circuit board construction, a technology termed *surface mounting* (SMT). Solderable powders are also employed as dispersed pastes for plumbing applications and for bonding metal surfaces together mechanically.

The general compositional range for tin–lead solder powders lies between 40% Sn–60% Pb and 63% Sn–37% Pb, the eutectic. Alloys also may include a fraction of antimony for strength and silver for conductivity and hardness. Powders are typically 100–500 mesh (150–30 μm) in size, narrowly classified, and have fresh surfaces because aging degrades their solderability dramatically due to aerobic oxidation. Solder powders finer than this (e.g., 30 μm) must be introduced into nonaqueous liquids quickly after production to minimize this surface oxidation. Because both tin and lead are very slightly electropositive (EMF = -0.11 V), they respond to treatment with oleic and stearic acids. For certain applications cinnamic acid has been employed. Generally, longer carbon chains and less saturation provide better dispersion. A very practical vehicle is cyclohexanol because of its broad compatibility with many organic systems. Electrophoretic mobility monitoring studies on this and other complex alcohols[13] show increased chain length and saturation provide superior dispersion in the liquid.

Dispersed powders, either directly or in reducible form, are added to inert liquid carriers through a dispersion process and localized for applications in petroleum catalysis and specialty organic reaction promotion. The metals employed are most often transition elements (Cr, Ni, V, Co) and platinum subgroup metals (Pt, Rh, Ir, Pd). Both metal particle size (surface area) and morphology are important for catalytic applications.

Because these metals are easily oxidized (transition group) or have virtually no oxide surface (platinum group), they are more difficult to precipitate onto inert carrier solids. Where the metal powders can be made externally, they may be dispersed in a nonaqueous liquid using nonionic dispersing agents or anionic agents with a short–medium tail, typically 3–5 carbon atoms. Such dispersions do not possess great thermodynamic stability and can be driven onto silicate surfaces by slight underdispersion in the nonaqueous phase.

Alkali metal dispersions represent the most difficult of all metal systems to manufacture and formulate. Their uses are limited, being employed in highly specialized applications of organic catalysis. These metals, predominantly sodium and potassium, are often formulated in alkane liquids. Alkali metals form the most alkaline oxides of all the metals in the periodic table, react violently with water, and as such do not have a measurable isoelectric point (>14). Size reduction must take place under these liquids because of the high reactivity of the metal with atmospheric gases and water vapor. Potassium, for example, when sufficiently fine will actually take fire spontaneously in air from the energy of these reactions.

Both sodium and potassium are extremely soft and can be comminuted with low energy employing specially designed mills (cutter type). However, even the cleanest alkali metal surfaces react rapidly with the very slight levels of dissolved oxygen and water in the alkane liquid. To prevent this, a dispersant is added that reacts preferentially with the metal's surface and that has an oleophilic tail sufficiently long to minimize water molecule diffusion to the surface.

The metal is mechanically cleaned under solvent, flushed to remove corroded surface material, and transferred to fresh solvent in the mill. The second stage of solvent contains sufficient oleic, stearic, or other fatty acid derivative to cover the new surface produced from the milling activity fully.[14] As alkali metals have very low densities (about 1 g/cm^3), fine particles readily suspend in the liquid phase. Stable dispersions can be obtained with particles as coarse as 50 μm. These have half-lives (i.e., the time for 50% of the base metal to corrode to oxide and carbonate) of many months. Finer particles have longer stability, in contrast to expected reactivity.

Alkali metal dispersions formulated in this manner are not readily dispersable in partially polar solvents such as acetone, MIBK, ether, and esters (amyl acetate). For these systems the dispersant molecule must have ether side groups or secondary or tertiary amine linkages. Polyols, molecules with periodic hydroxyl groups, are unsatisfactory as these react slowly with the metal to form alcoholates. Similarly, primary amines are also slowly reactive.

Military pyrotechnics (flares, incendiaries, etc.) have been manufactured using somewhat shorter-chain fatty acid dispersants for alkali metals in ester liquids. Such formulations are highly unstable, have moderately short half-lives, and must be kept in sealed containers or casings. They are, however, extremely explosive when ignited, leaving a highly corrosive residue over their area of "application," a characteristic greatly admired by military forces.

For certain specialty catalytic applications on organic reaction promotion, finely dispersed alkali metal slurries in nonaqueous liquids have been developed.[15] The metal is first melted (usually below 100°C) and introduced into the organic liquid with high-shear agitation under an inert blanket of nitrogen or helium gas. To this mixture is added either a fatty acid or the metal soap of the fatty acid. The level of addition is about 1–2%, based on the metal content. Continued high-shear agitation causes the liquid metal to break up into small spheres, much like micelle formation, with the surface having a soap structure formed or adsorbed upon it. This system is remarkably stable, having a shelf life of many months. It is, however, highly flammable and extremely dangerous if spilled.

Dispersions of metal powders in aqueous suspension require the metals to lie below hydrogen in the electromotive series or possess a protective oxide layer on their surface. Bulk metals, which ordinarily oxidize from the 20 or so ppm of dissolved oxygen commonly in water, can be protected by coating with a slurry of a 2–15% aluminum bronze powder dispersed with alkyl sulfates, tannins, or ethylene oxide derivatives. Most aluminum bronzes are common alloy compounds in which aluminum has replaced all or part of the tin in the common 90% copper, 10% tin alloy. A typical aluminum bronze that serves for this function is Al (7%), Pb (4%), Zn (4%), Cu (85%). The most effective dispersants for this alloy are ethylene oxide derivatives, alkyl sulfates (where the alkyl group is 12 or more carbon atoms long), and solubilized tannins.[16]

Bronze particles form as platelets. These overlap onto a corrosion-prone metal (e.g., cast or extruded aluminum) surface. The dispersant serves to bond the metal powder surface, as oxide–hydroxide, to surface hydroxyl groups of the aluminum atoms in the bulk metal. Interparticle overlap plus organic chains form a nearly impervious layer to prevent, or significantly attenuate, atmospheric oxygen and water vapor from diffusing through the polymer binder, under the coating and onto the base metal.

Slurries for catalysis carrier applications onto which are to be precipitated active agents are dilute and the most desirable dispersants are fugitive to prevent fluxing and sintering of the alumina, silica, or aluminum silicate bases. Metal powder dispersants for the transition metals and platinum subgroup metals may be amino alcohols or short-chain diamines. Once the metal is adsorbed onto a solid carrier surface through precipitation (pH) or flocculation, the dispersant is removed through elevated-temperature firing. Oxides of active metals (where applicable) also may be precipitated in a similar manner and chemically reduced later to a metallic state.

A number of powdered metals are employed for chemical reactions, tracers, and as catalysts in small-scale experiments. Relatively heavy metals, silver, bismuth, mercury, and copper, if below 5 μm in diameter, can be dispersed effectively in paraffinic oils having a low-to-moderate viscosity. The metal is treated either prior or in situ with a mono-sulfonic acid salt of any metal below hydrogen in the electromotive series, for example, copper benzene sulfonate.[17] This criterion ensures water insolubility and high van der Waals force when the molecule adsorbs onto a metal surface. Salts of 1,3-benzene disulfonic acid also function but are less effective perhaps because both active metal groups attached to the benzene ring are unlikely to coincide in spacing with the surface atoms on the metal.

Appropriate levels of dispersant for metal powders can be estimated by multiplying the surface area of the metal in square meters per gram by 3 micromoles. This averages around 0.1–0.2 wt. % of agent. Because these powders have high densities, their specific surface areas are frequently low to moderate in value.

The cationic "head" of the dispersant is highly ionized because of the strong electronegativity exhibited by sulfonic acid. This cation associates with the hydroxyl (or sulfide) groups on the metal and leaves the oleophilic benzene tail extending into the solution phase.

A similar, but somewhat more complex, approach involves heating the metal powder under a slightly unsaturated paraffinic oil (nominal C-14 hydrocarbon) in the absence of air (CO_2 or N_2 blanket) to near-boiling, 225–250°C, with intense agitation. The metal powder appears to act as a low-grade catalyst to promote bonding of a hydrocarbon grouping directly to the metal surface:

metal + hydrocarbon + heat ⟶
R = C-7 to C-14

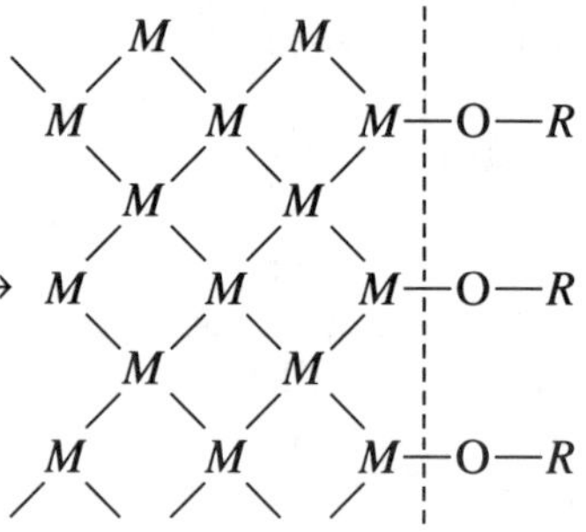

(12.3)

Shorter-chain R groups, according to infrared spectra, derive from symmetrical cracking of the parent chain (nominal half-length). This method might be expected to result in an oxidation-to-acid reaction of the oil and acid-to-oxide reaction with the surface. However, infrared data of the treated and washed product indicate a high percentage of metal–oxygen–hydrocarbon bonds. Also heating the same metal powder in hydrocarbon solvents bearing oleic acid produces a material that is considerably poorer in shear and thermal-resistant quality.

Dispersed metal powder drops out readily in the low-viscosity oil while hot but remains dispersed and difficult to separate when the oil cools and viscosity climbs, usually 1 to 2 orders or magnitude.

For more active metals, iron, zinc, tin, etc., a similar procedure may be employed using aluminum naphthenate with a high molecular weight (about $1–3 \times 10^6$), or a nonionogenic agent with a preponderance of ether linkages as:

$$H(CH_2)_x\text{-}C_6H_4\text{-}O\text{-}(CH_2CH_2)_y(C_2H_4CHO)_zH \quad (12.4)$$

A total molecular weight of 1000–2000 is most effective. For the structure where $x = 8$, $y = 16$, and $z = 8$[18] an application system results that upon heating, as with soldering, fluxes to provide superior electrical conductivity. These dispersing agents can be applied to the metal powders from toluene solution. With either treatment or dispersing agent the metal powder surface must be completely covered to obtain maximum dispersable solids in the organic carrier liquid.

The oldest method of treating metals has been updated slightly to produce highly electrically conductive powders for circuit boards, especially for production line repair. The metal powder is flake morphology silver, about $-100\ \mu m$ in size with a narrow distribution. The powder system is mixed as a slurry in hexane or heptane with agitation. Linoleic or linolinic acid is added at about 2%, solids basis. No silver salt is produced, only an association complex of the carboxyl group onto the metal surface.

The system is adjusted to about 85% solids paste and applied directly as a metallographic ink, automatedly through a computer-guided stylus or print screen, or manually with a Leroy-style pen. If longer application time is desired, hexane can be replaced by a higher homolog hydrocarbon solvent. To prevent print line spread, the system is vacuum dried immediately after printing. A trace of sodium or potassium oleate in the paste will allow the system to bond onto ceramic circuit boards when fired.[19]

A representative formulation employing this dispersant approach and using both a metal powder and an inclusion compound is given in Table 12.8. The inclusion compound serves to provide conductivity in the void spaces between metal powder particles.

The viscous formulation can be applied as an ink via silk screen or letter press, and the board or base stock then heated to 150°C for about 30 minutes. Heating may be shortened to about 10 minutes if the temperature is elevated to 180°C. Should the substrate be thermally sensitive, the temperature may be reduced to 130°C, but the heating period will require an hour or more to complete burn in. Printed lines and pads formed by this technique will have a volume resistivity of from 1 to 2 milliohm centimeters.

A similar aqueous system has been developed that avoids the flammability hazard of hydrocarbon solvents and vacuum extraction methods. In this tech-

Table 12.8 Electrically Conductive Powder Coating Formulation (Metallographic Ink)

Component	Weight %
Copper powder ($-10\ \mu m$)	77.0
Acrylic resin binder	11.5
Butylated melamine binder	7.7
Potassium oleate dispersant	1.5
Copper oleate dispersant	1.5
Hexadecyl mercaptan cyclopentadiene[a]	0.8
	100.0

[a] Inclusion compound, Ref. 20.

nique, powdered silver or copper metal is formulated into a paste by first treating with a phosphate ester having the general formula (R(OZ) O, where R is an alkyl group from 6 to 24 carbon atoms and Z a similar grouping but with 2 to 4 carbon atoms. Aluminum powder, once cleared of adsorbed water, may also be treated by stearic acid and dispersed using dodecyl phosphate, triethoxyphosphate, or dodecyl polyether (six grouping).[21] The paste can produce excellent print images for decorative or lithographic applications but is too poorly conductive for electrical applications. It may be used in water systems and especially in latex emulsion formulations.

Most organophosphate esters function better with longer chain lengths. In tests made by the investigators cited above, the following relationship held for dispersibility:

stearyl = dodecyl > octyl > hexyl > butyl > ethyl

In the past decade electrically conductive powders have been employed for filling rubber in the manufacture of pressure-sensitive switches. Metal powder is thoroughly dispersed in appropriate solvent, then introduced into the polymer at subcritical loading. Whereas printed circuit board conductive powders function best with a log-normal size distribution, rubber switches require a more narrow distribution. This is illustrated by the membrane cross-sectional sketch of Figure 12.9. The presence of finer particles to fill interstices would interfere with the "on–off" functionality.

Sufficient polymer surrounds the particles without fully enveloping them at rest (zero pressure on the left) to maintain a minimal contact status in which electrical passage is essentially nil. Upon applying finger pressure (on the right), particles in the immediate vicinity are forced together, and resistance drops many orders of magnitude. Obviously, flocculated metal powders would defeat the

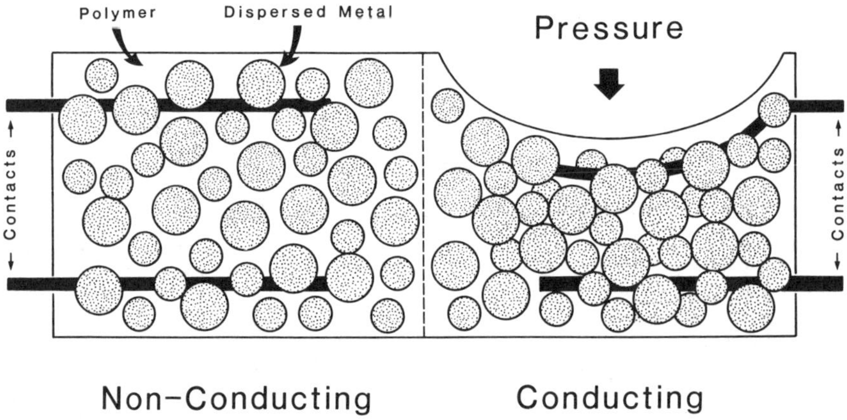

Figure 12.9. Schematic of rubber switch metal powder dispersion.

functionality in such devices. An increase in current is amplified by solid-state devices, and the particle–polymer mixture becomes an economical airless contact switch. It is employed in calculators, computers, automobile control panels, and in explosion-proof environments where make–break contact sparks could create serious problems.

Metals employed for these switches include copper, nickel, zinc, and alloys thereof. Silver is usually too expensive for the devices. The polymer may be silicone rubber or a nitrilo copolymer with acrylobutadiene styrene. Ethylene–propylene copolymers do not function as well, and many other polymers have insufficient elasticity for tactile use in the device.

Metal particles must be well dispersed—floccules, as noted, will conduct current at rest and exhibit only small resistance changes with pressure. Manufacturers took a clue in developing this process from kitchen cleaners for copper pots and silverware that contain citric acid as the primary cleaning ingredient. The agents which function best for dispersing the particles, usually in toluene or styrene, are tricarboxylic acids with a hydroxyl group appended on the molecule. The chain length must be greater than that for the citric acid molecule, however, to maintain the hydrophobic character around the metal particles.

Typically each grouping has about 10 carbon atoms. Dispersion can take place in the solvent phase directly with fairly high agitation. A small amount of polymer binder is then incorporated into the slurry and mixed, and the system is allowed to stand to enable dispersant to diffuse properly and particles to coat.

One of the more effective dispersing agents is 3-hydroxy,1,3,4-tetradecane tricarboxylic acid. The level is moderately low because of the coarseness of the metal powders, frequently having a mean diameter around 5–10 μm. Once these powders are fully dispersed, the entire slurry is flushed into bulk polymer, milled in with an appropriate mixer, heated to drive off excess solvent, and cast into the appropriate molds where the polymer cures. Volatilization of solvent causes the polymer to withdraw from high-capillarity regions, an important factor in the switch's construction. In some applications conductive elements are cast as hemispheres and attached to a rubber membrane of similar polymer composition but without metal powder.

Quality control tests of these tactile switches have shown they can function for well over a million contacts. One of their great attributes is the elimination of arcing. They are also quite inexpensive to manufacture.

Surface-coated aluminum powders are most problematical for obtaining good dispersions in hydrocarbon vehicles because of adsorbed water films inherent on their surface. Studies on alumina surfaces have shown the water bonding energy to be as high as 19 kcal/mole,[22] requiring temperatures of nearly 300°C to remove it. This exceeds normal processing practicality for commercial metal powder treatment! As a consequence, many dispersed aluminum powders have an intermediate water linkage between metal surface and dispersant that can become a source of failure in certain applications (see Chapter 7).

Using the hydrophilic–lyophilic balance method (HLB) described in Chapter 2, surface inert (Al, Zn) and electrovalently inert (Ag, Cu) metal powders may be dispersed using a surfactant with an HLB between 7.5 and 9.5. This is the basis for dispersing aluminum powder for paint applications in mineral spirits using the commercial surfactant Polysurf 210G™ at about 1.8% level (solids basis).[23] The metal powder slurry is a manageable fluid and pourable at 40% metal solids.

Active metals also respond favorably to treatment by anionic agents, where optical characteristics rather than electrical ones are of primary interest. Dispersions of aluminum powder in the 100 μm down to 1 μm range, used for decorative paints in marine environments, can be produced effectively using stearyl dimethyl–benzene ammonium chloride. They display enhanced weather resistance and reduced fouling in salt water.[24]

Where the metallic pigment has so extensively oxidized from storage, aging, or chemical atmosphere that its surface is at least three layers deep with oxide–hydroxide, an anionic dispersant will function better than on bare metal or a single oxide layer. In the study cited, sodium dodecyl sulfonate (SDS) provided better dispersion, but the formulation's salt water resistance degraded. Because other metal oxide pigments may be employed in the formulations for pigmentation, including iron oxides, chromium oxide, titanium dioxide, chromates, etc., SDS use may represent a superior compromise.

Finally, conductive high-temperature ceramic materials in which distributions of metal particles, alumina, silica, and other oxides are fired (termed *cermets*) may be formulated initially as slurries to obtain uniformity in composition and in particle size distribution. A common dispersant functional for all species is most desirable. Alkali metal polyanion agents are to be avoided as far as possible because of their propensity to flux the cermets prematurely in the firing or development process, and to provide leakage paths undesired in the products. For some of these applications amino alcohols or long-chain polyethylene oxides may serve, depending upon metal chemistry and the need for fugitivity and porosity prevention.

12.2.1. Particle Size and Shape

The particle shape of metal powders may affect secondary characteristics following application from the dispersed phase. Decorative paints, for example, take advantage of laminar morphology of aluminum and bronze powders to produce overlapping, parallel planes of metal flakes in the dried paint film. Were the particles even in a slightly flocculated state, this would not occur. Voluminous structures of the "house of cards" organization would form that would interfere with leveling and gloss development and that would provide poor coverage for the substrate. Metal flake paints are often used as protective coatings to prevent oxidation of a metal substrate from diffusion of oxygen and sulfur oxides through the applied film. Flocculated particles would reduce this capacity greatly.

To disperse aluminum or bronze metal particles in a hydrocarbon phase requires, as noted previously, an acidic alkyl molecule having eight or more carbon atoms or an aromatic ring in the tail. Frequent candidates are oleic acid, caprylic acid, benzoic acid, salicylic acid, toluic acid, cresylic acid, and their substituted derivatives. These can be reacted first in aqueous phase via their sodium salts or in nonaqueous phase via a hydrocarbon solvent. The latter requires a longer period for adsorption but provides a superior pigment.

When formulated as a paint, such systems not only level in superior fashion, the application viscosity is lower, and the bonding between unsaturated dispersants, for example, oleic acid, and the alkyd binder component produces a tight film, highly resistant to oxygen and water penetration. The coating may also be used to prime coat wooden decorative lumber where knots cause slow migration of oils and resins through, and discoloration of, applied water-based emulsion paints. Dispersed metal, oil-based paints are superior to water-based primers for this purpose.

Metals produced through vapor condensation or chemical precipitation processes occur most often as log-normally distributed spheres or equants. These generally have poor coverage capacity and hiding if employed as paint pigments because insufficient particles occur within an applied film to provide 100% blockage of photons. Metals possessing this distribution are, however, highly effective for applications where electrical conductivity is a major requirement.

Such applications must overcome the basic problem of oxidative coatings on the particles, a problem that increases with fineness of size. The solder pastes described earlier consist of fine, nearly spherical balls of tin–lead alloy (60–40 or 63–37 composition). These are produced either via shot towers or spinning discs with molten metal broken up by surface tension in inert atmospheres.

Copper, gold, and silver powders are most often generated by chemical precipitation from dissolved salts and form equants following chemical and thermal reduction. Solution concentration, chemical additives, temperatures, and agitation control the size range.

The two types of products are represented in the photomicrographs of Figures 12.10(a) and (b). Because of oxidation and difficulties in handling, particle sizes for electropositive metals are most commonly above 25 μm (500 mesh). Noble metals may be finer.

Such particle assemblages can be dispersed readily with abietic or sylvic acid, $C_{20}H_{30}O_2$, a complex naphthenic derivative from pine rosin. This substance serves not only to react with the metal particle's surface and thus function as a highly effective dispersant for alcohol based systems, it is also a highly effective, thermally activated reductant. Upon temperature rise (bulk heating, soldering, laser scanning, etc.), it will reduce the molecular oxidized film on the particles and convert this to metal. With solder pastes particles fuse to form continuous metal. With copper and silver pastes the particles consolidate. With either composition, electrical conductivity of the heated aggregate rises.

For circuit board printing inks or pastes dispersion and fluidity are critical. The dispersed powdered metal must flow through a silk screen (usually stainless steel) yet not spread appreciably upon the board after application. With the advent of newer electronic technologies, high-density circuits and microproces-

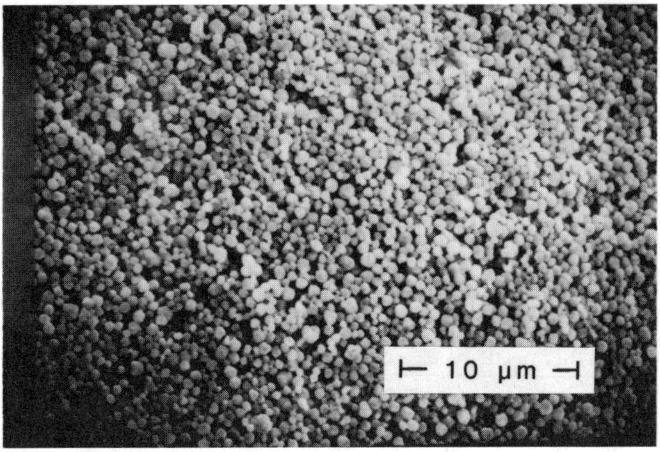

(a)

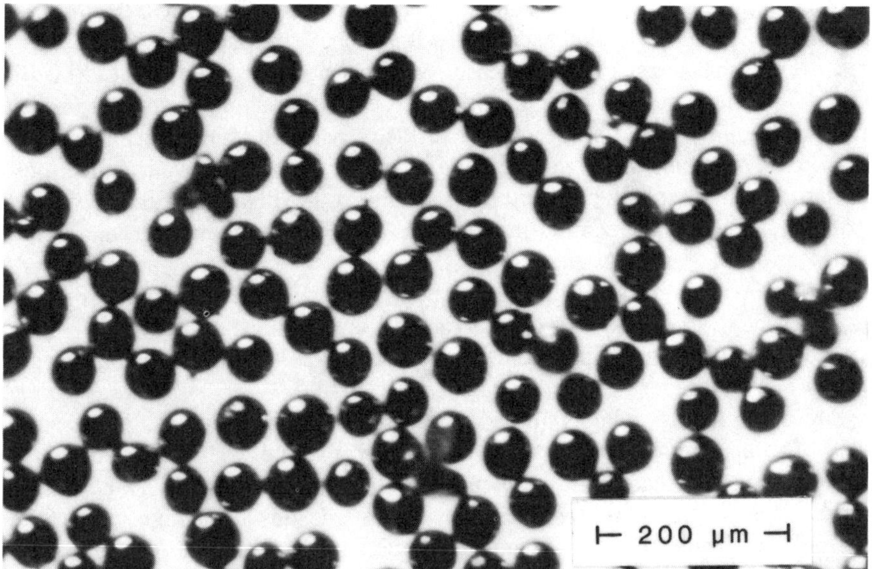

(b)

Figure 12.10. Metal powders employed in the electronics industries: (a) copper alloy powder (courtesy of Metz); (b) solder powder (courtesy Indium Corp. America).

sor chips with decreased pin spacings, this print spread characteristic has become highly critical. Printing images often must be controlled to within ±1 mil (25 μm). With dispersed tin–lead alloy powders secondary flow can be controlled through (1) narrow particle size distributions and (2) volatile solvents that evaporate quickly upon board contact.

Tin forms a surface oxide but not a carbonate. Lead forms both. However, their oxidation potential, −0.1 V, is sufficiently low that even mild reducing agents can remove the interfering coatings. As a consequence, simple reductive dispersants can be formulated from a variety of low-molecular-weight acids—benzoic, cresylic, crotonic (t-butenoic), and the like. For short-chain acids the system can be formulated using alcohol-base solvents. With longer-chain unsaturated acids (i.e., oleic and linoleic), formulations will require paraffinic solvents such as heptane, VMP, mineral spirits, etc.

Where the powder never fuses, as with copper and silver prints, narrow size distributions afford only minimal particle–particle contacts and electrical conductivity of the printed assemblage will be poor. These materials require near-log-normal distributions to maximize particle–particle contact. A dispersant/reductant as described may be employed to remove surface oxides. Under pressure, chemical reduction has been shown actually to produce interparticle chemical bonding, which stabilizes the assemblage. A resin having a lower surface tension for the metal powder than abietic acid is added to bond the particles in their close configuration and prevent surface dusting and film disintegration. Such resins are termed *binders* and must interfere only minimally with true particle–particle contact and electron transfer during finished circuit operation. Materials frequently used are solvent-loaded epoxies and polyesters.

Figure 12.11 shows the particle size distribution of a commercial copper powder (represented by Figure 12.10) that is not log normal in distribution and that is relatively dilatant in flow. Also included in the figure are coarser and finer products produced for specialty applications. By blending these in appropriate fractions, a close fit to a log-normal distribution was produced. When these were formulated as inks, printed on a fiberglass circuit board along with a small percentage of fine solder powder, and heat cured, electrical conductivities were obtained well over an order of magnitude higher than with the commercial powder alone (0.41 vs. 0.01 microohm cm).[42]

For this blend to be practical, a well dispersed, though thixotropic state, must be effected to provide a homogeneous distribution in the final print image and the maximum packing density.

A photomicrograph of the circuit board following curing but prior to drilling appears in Figure 12.12. Rippled topography in the conductive elements derives from nonuniformity and particles 100 μm in size. The narrowest lines are approximately 25 mils (625 μm) wide. In other test work, lines 2 mils (50 μm) wide could be printed with high regularity, screen resolution being the limiting factor.

Selective distributions of mixed alloys have been formulated to produce dis-

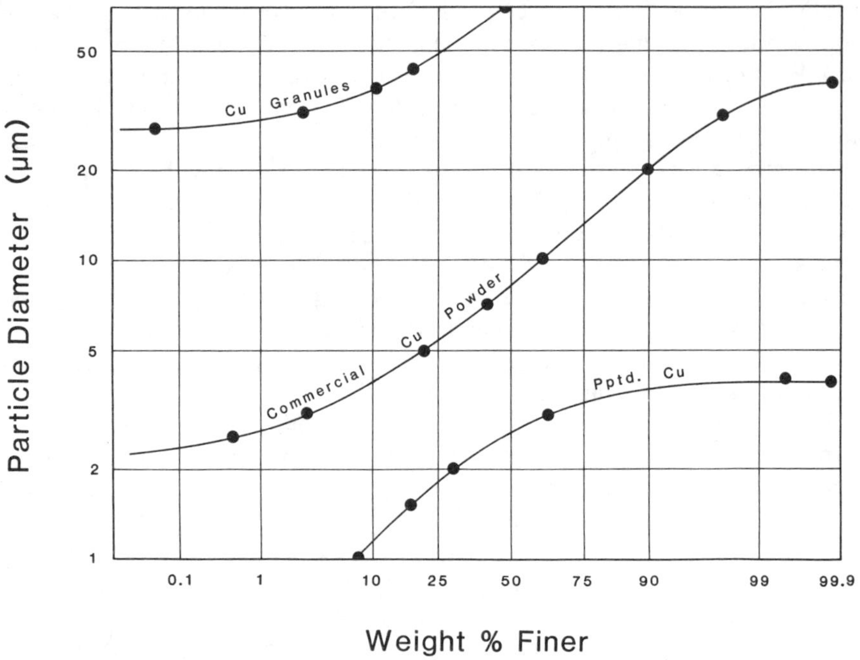

Figure 12.11. Size distributions of copper metal powders.

persions possessing maximum interparticle contacts and electrical conductivities specifically for these purposes.[25]

12.3 Two-Phase Fuel Systems

In the 1960s U.S. power generation facilities were frequently designed to be fueled by oil, usually #2 or #6 diesel. While BTU costs were greater with oil, the peripheral problems with coal were avoided: fly ash collection, grinding, dusting, haulage, electrostatic precipitator costs, burner fouling, sulfur dioxide emissions, fine coal oxidation in storage and in plant, etc. However, when the Arab oil crisis occurred in the early 1970s, these utility operations were seriously jeopardized.

The U.S. Department of Energy almost immediately funded a major technological effort to find appropriate means of dispersing fine coal in hydrocarbon fuel oils. Serious consideration was actually given to fuel jet aircraft with such dispersions, with little consideration given to the mechanical problems inherent in such a system.

The research program was predicated on developing technologies which would avoid utilities retrofitting their boilers with new and expensively rede-

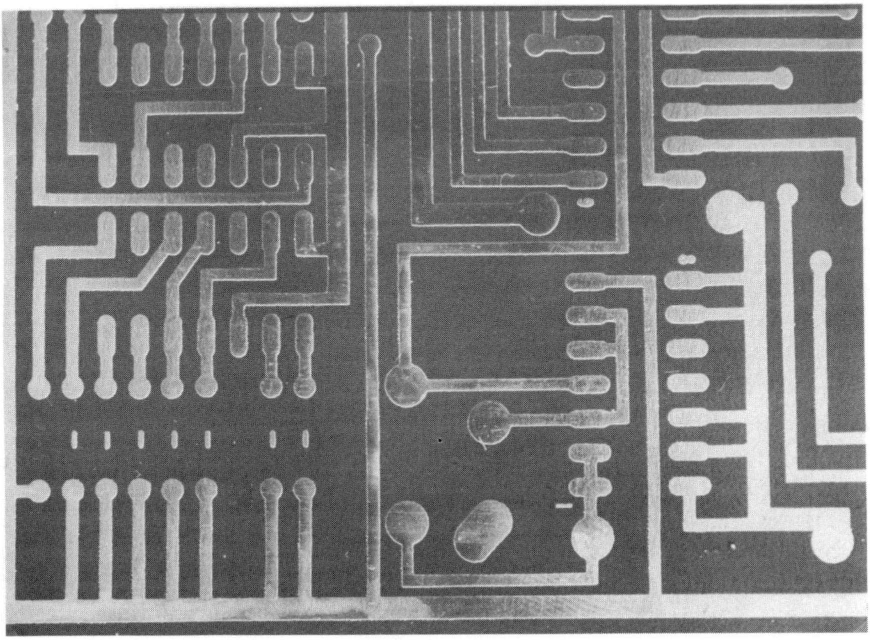

Figure 12.12. Metal powder printed circuit board.

signed firing equipment. Studies at that time showed coal particles ground to 99% −7 μm would have essentially the same burn rate as diesel fuel aerosols currently in use. The object of the investigation, therefore, was to develop effective measures for grinding coal to that size and dispersing it in oil at 25–35% solids. Over a billion dollars was spent on that exercise, including the design of two test pilot plants and two utility demonstration sites, one in Florida and a second in Massachusetts.

Problems immediately arose due to the fundamental structure of coal and its comminution. Coal has a complex composition of pure carbon bound in aromatic and aliphatic tars, coupled with organic sulfides, inorganic sulfides (pyrite), clays, and silica. The latter three end up predominantly as ash and clinkers in conventional coal-fired utility operations. However, in the very fine particle state, these impurities pass out through the exhaust ducts, where they are captured and removed via electrostatic precipitators and scrubbers. Coarse grinding of coal, with or without dispersants, can be performed readily as the mineral is poorly bonded and highly fractured from geological forces. However, as finer sizes are approach, coal's structure becomes less elemental carbon and more tar in character. This criticality in size is broadly variant, depending on coal type (anthracitic versus bituminitic) and fixed carbon ratios.

The developed technology employed a feed, crushed to a nominal −16 mesh, which was secondarily ground by a ball or vibratory mill in a viscosity-reduced

#2 fuel, or kerosene cut, employing a variety of hydrocarbon chain dispersants. This feedstock was to be ground to the desired size, and as the solids were relatively low (25–35%), viscosity buildup was not expected to occur. In practice the viscosity actually diminished because fuel oil, essentially Newtonian in character, heats up from the energy input, thereby "thinning out." In theory grinding to the fine state should be enhanced and require reduced energy in the dispersed state.

However, problems arose in attaining the desired size reduction by either mill economically. Elaborate methods of processing were employed including wet classification of the fines at elevated temperatures, which resulted in further viscosity reduction. Most grind studies resulted in milled products coarser than 10 μm, irrespective of milling time. During these investigations a vast amount of dispersant research was undertaken to obtain higher solids, more selective classification, improved energy utilization in size reduction, and pumpability. Much of the work was trial and error because the surface character of freshly ground coals at ultrafine sizes was largely unknown.

Figures 12.13(a)–(d) contain a portion of the results of that investigative effort.[26] The figures are all dual-illumination immersion photomicrographs of a high-grade bituminous coal ground in #2 diesel oil cut approximately 25% with heptane. All were size reduced in the same mill, at the same solids and for the

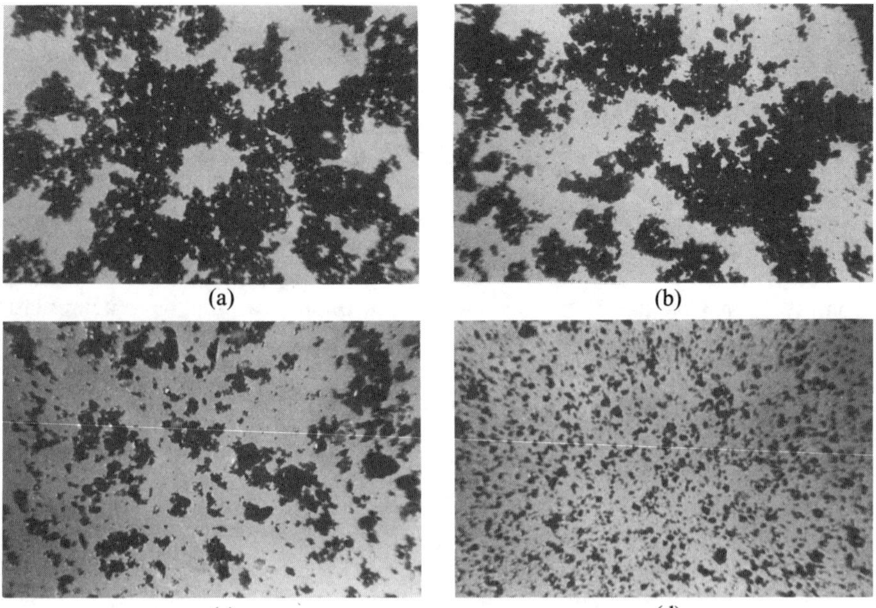

(a) (b) (c) (d)

Figure 12.13. Dispersed coal grinds in oil (500 μm fields): (a) alkyl sulfonate dispersant; (b) fluorinated hydrocarbon dispersant; (c) 2-nitropropane dispersant; (d) fatty amine dispersant.

same milling period. The dispersant level was 0.5%. Note the first two dispersants (a) and (b) show fair size reduction but poor dispersion. In (c) the size reduction and dispersion both improve, while in (d) both are excellent, with coal particles being predominantly below 7 μm. The field width in all four photomicrographs is identical, approximately 500 μm.

In performing particle size analysis of the ground coal specimens, toluene was found to be a superior dispersing liquid to heptane, suggesting that the coal surface contained more aromatic (benzene-based) components than aliphatic (alkane based).

Even #2 diesel is too viscous at room temperature with 25% fine coal to drain readily from a vibratory mill. Two solutions exist for this problem. The first involves dilution with low-boiling cheap hydrocarbons (C-6 to C-12 cut), which increases the fuel's cost and fire hazard. The second requires heating the system in the vibratory mill to reduce its viscosity, either during grinding or during the draining phase. The latter technique imposes difficulties on recirculation milling because cycloning or centrifuging coal–oil slurries to remove coarse particles is exceptionally difficult due to small density differences between solid and liquid (typically 0.15 sp. gr. units). In addition, dispersed coal ground to −10 μm immersed in hot hydrocarbon causes tar leaching into the solvent, which increases viscosity, making pumping more difficult and burner fouling more likely. Batch milling is, as a consequence, more practical.

Vibratory milling with dispersant imposes a significant increase in processing costs and can be justified only with high crude oil prices or where coal is local and very cheap.

The foregoing example is unusual in that grinding was undertaken in an oil system. Most applications of nonaqueous dispersions involve paint formulations where grinding is a term more accurately defined as wetting and dispersing, solid materials having already been ground by their manufacturer. These operations are sometimes termed *dispersive grinds* and indicate a mechanical effort to introduce dispersant to solid surfaces while in an oil fluid, rather than solid particle fracture itself.

Supplemental to the coal–oil slurry investigations, coal–water dispersive grinding has been studied intensely[27] to reduce dusting and its associated health and explosion hazards. The process also has interest in producing a system that burns more efficiently, has less fly ash, and entrains more sulfur and nitrogen gases, preventing them from entering the atmosphere.

Because coal's surface is hydrophobic, a dispersant must be employed to reduce mill viscosity and to maintain separation of particles (see Section 12.1.4) during size reduction. Either an anionic or cationic agent should serve this purpose. However, anionic materials (e.g., sodium stearate and sodium lauryl sulfonate) cause severe foaming, which in turn requires addition of combustible antifoamers (as octanol).

Coal–water slurries are not ground as fine as those of coal and oil. Typical sizes are 50–25 μm, and typically milling is performed in ball mills because of their low operating and maintenance costs as well as their enclosed structures.

As coal is ground using the cationic agent ditallow-dimethyl ammonium chloride, the dispersant adsorbs more efficiently as grinding progresses, suggesting the appearance of a more tarry surface composition. In addition, dispersant demand increased disproportionately to surface area. This increase is noted for several days following termination of the grind and maintenance of the coal as an agitated slurry. A steady increase in viscosity is noted with standing time, which may suggest fracture continuance following comminution.

The naphthalene formaldehyde sodium salt has also been employed as a milling aid. It also reduces mill foaming characteristics but can produce only about 62–65% solids, depending on the coal's position along the bituminous–anthracite scale. Static coal dispersion stability, the avoidance of dilatant settling and hard cake formation, correlates well with flow characteristics. At 50% solids and a top size near 25 μm, the 100 rpm/10 rpm viscosity index is about 0.75 with a yield stress in the range of 1 to 5 pascal seconds.

During quiescent standing, this coal slurry builds structure as shown by minimal syneresis (breaking to form a clear supernatant liquid), less than 5% by volume. The structure requires some time to form but breaks down rapidly into fluid slurry with agitator shear mixing or pumping.

Higher coal loadings have been developed by employing multibranched, high-molecular-weight, nonionic agents. These appear to generate more Newtonian flow characteristics. Alkylene oxides increase the ability to raise solids as their molecular weight per "active" hydrogen atom in the structure increases.[28] Where molecular weights are in the range 3000–6000, coal solids approaching 70% can be formulated without dilatant caking occurring.

As this book goes to press, industrial demand for coal–oil and coal–water slurries has diminished because of lowered oil costs. However, cement kiln operations and other producers of high-temperature, low-cost products continue to show interest.

12.4 Thixotropic Dispersions

Commercial thickeners are common components in a variety of personal products, foods, shampoos, cosmetics, pharmaceutical preparations, paints, etc. Two classes of thickeners are used, organic and inorganic. The former are often carboxycellulose or castor oil derivatives and the latter highly acicular or lathlike inert minerals.

Organic derivatives have the virtues of ready dispersability, transparency, low density (high molecular volume per unit mass employed), and nontoxicity. Disadvantages include degradation from secondary chemicals, bacterial and fungal decay, and cost. Inorganic thickeners are truly chemically inert—they are neither attacked by nor do they attack secondary components in the formulations.

To be effective acicular particles must be highly dispersed. In this configuration, they form an inorganic "brushpile" that interferes with liquid mobility, particle settling, and solvent-phase evaporation. Most common of the inorganic dispersants are palygorskite (formerly known as attapulgite from the Georgia town where it was discovered) and high-purity halloysite. Palygorskite is processed dry because dispersion followed by drying renders the particles highly difficult to redisperse. Most palygorskite contains a high percentage of smectites (formerly known as bentonites from the town of Ft. Benton, Montana where they were discovered) and are off-white in color.

Halloysite (named after a personage and region in Belgium) rarely occurs either as a white or a pure mineral. However, small quality deposits of it have been developed in Utah. An electron micrograph of such a dispersed mineral thickener is shown in Figure 12.14.

Dispersion of these minerals can be effected by any of the popular clay agents, SHMP, TSPP, SOPA, alkanolamines, etc. When fully dispersed they represent a paradox in dispersion science—the more thorough their dispersion, the more thixotropic their viscosity. In fact, thixotropic index is one of the better methods of assessing their degree of dispersion. At low shear, viscosities range from 20,000 to 100,000 cp.

Particles of these mineral thickeners have anisometries of 10 to 20 and particle lengths of 1 μm or under. Severe energy is required to disperse them, and care must be taken to prevent particle breakage. Thixotropy is essentially proportional to the surface area and the anisometric index of such materials.

Incorporation of the materials in a formulation is accompanied by an appropriate dispersant during the initial makedown. Additions made following wetting out of other system components often result in failure to achieve full thixotropy in and dispersion of the needles. Formulation pH is maintained as close to neu-

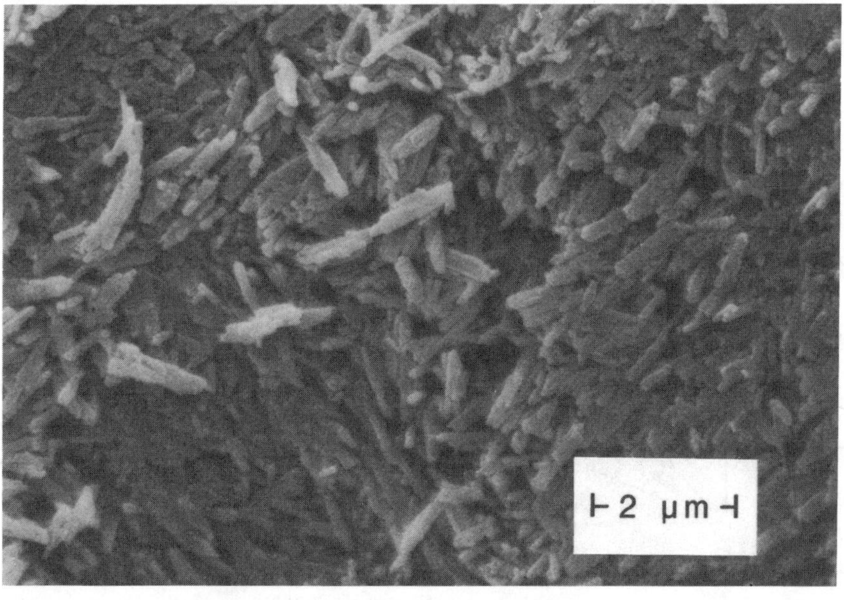

Figure 12.14. Pure halloysite particles (courtesy Utah Clay Tech.).

trality as the product justifies and polyvalent ions (Al^{3+}, Cr^{3+}, Fe^{3+}) are avoided wherever possible. This prevents the dispersant activity from being diverted by interactions with other components in the formulation. By exercising these controls, the system will remain viscous, settling will be prevented, and the system will retain its homogeneity and thixotropy on the shelf for protracted periods. The adverse effect of polyvalent cations, demonstrated by aluminum sulfate, is shown in Figure 12.15.

Developing the full dispersancy and thixotropy is not a simple task with these materials. Milling must be sufficiently energetic to disaggregate the needles from their geochemical formation networks, which possess varying degrees of bonding integrity, but be of insufficient energy to break particles into shorter structures. For these reasons ball, hammer, and jet mills are not effective size reduction devices for materials with this morphology.

Bead mills and ultrasonic disaggregators may be employed, but energy input must be carefully regulated. Because of the elevated surface areas, typically 20 M^2/g or greater, dispersive dosages often reach 1–2% with pH controlled to ~7. Under these conditions, both palygorskite and halloysite have highly negative zeta potentials (−40 to −50 mV). Interestingly, their zeta potential becomes more negative during grinding until a monodisperse system results. This may indicate the production of new surface or chemical alteration of old surface by dispersant adsorption. Milled halloysite thickening of an organic polymer system appears in Figure 12.16.

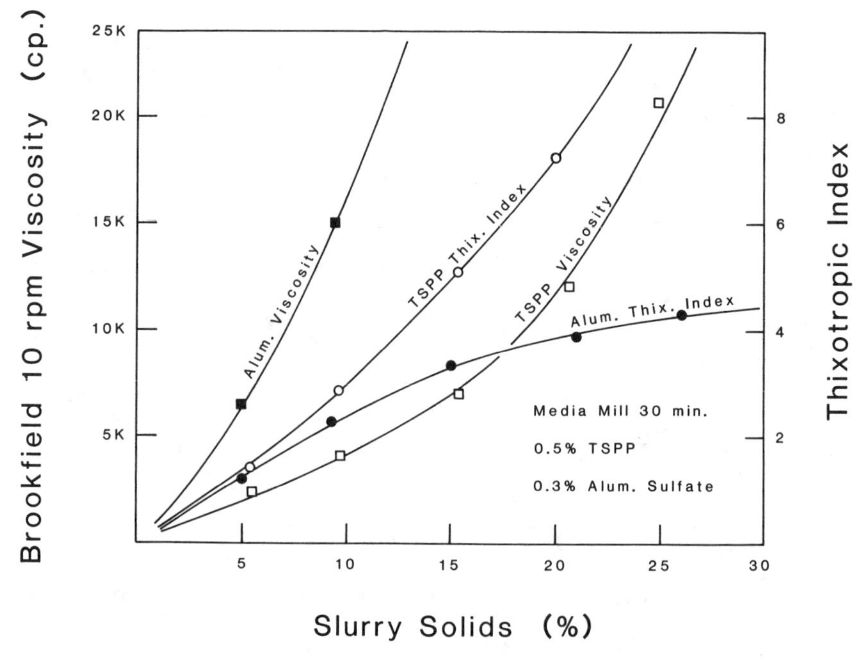

Figure 12.15. Polyvalent cation effect on halloysite rheology.

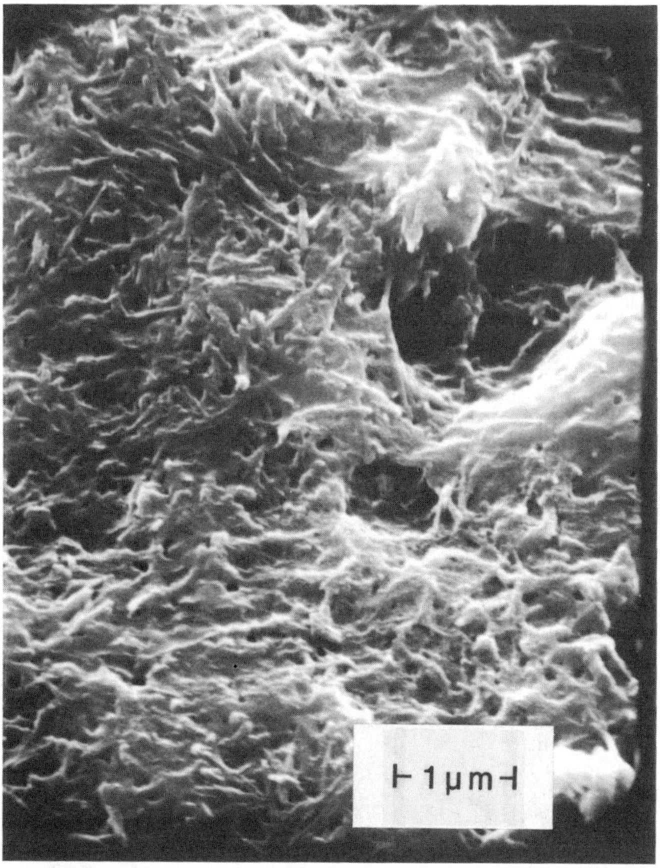

Figure 12.16. Polymer thickening with 20% dispersed halloysite.

Such acicular systems also may be treated with primary and quaternary fatty amines and long hydrocarbon chain silanes to produce extremely hydrophobic viscosity builders for oleophilic systems, for example, paints, oil-based cosmetics, and topical medicinals. Because their surface area is large, dispersant or treating agent demand can reach the several percent range and result in costly treatment operations. As a consequence these materials find their way mostly into high-end products where component costs are not signficant factors.

12.5 Liberation Milling

Liberation is the term employed in metal ore recovery that applies to grinding host rocks fine enough to break away (liberate) inclusions of metalliferous minerals. Both overgrinding and undergrinding result in poor yields. Some processes are size limited, for example, flotation, because very fine particles of host rock carry over excessively with product mineral. In others grinding limitation is simply a matter of economics—grinding becomes progressively more expensive as size diminishes and a point may be reached where increase in processing costs wholly offsets increase in metal yield.

Size reduction tends to produce gaussian distributions of fragments. Mineral inclusions also tend to form gaussian size patterns, the result of crystallization mechanics. Thus, both often closely approximate log normality in their distributions (see Chapter 2). However, the two functions are not colinear. That is, coarse host rock particles are not associated with coarse ore mineral particles, nor fine host particles with fine. As a consequence two independent gaussian functions are operating which equally influence recovery.

In Chapter 10 (Figure 10.11) a mining example was given of gold removal (or liberation) from host rock. The greater portion of liberation efforts are concerned with extraction of valuable metal components at low levels of occurrence from worthless rock. Extraction may be chemical, as with gold via cyanide solution, or mechanical, that is, by density differential. With density discrimination devices, ore may be concentrated readily by a factor of 100 or greater.

Several types of gravity separators are employed in the mining industry for liberation. Most of these are operated wet and many with wetting agents to facilitate separation. The simplest is the gravity table (Wilfley type). Ground ore particles are slowly washed down vertically across a horizontally grooved table. The table slopes both perpendicular to the grooves and parallel with them, that is at 45 degrees to the direction of particle introduction. Finer particles are flushed down first, normal to the grooves, because of their lower mass, followed in turn by progressively larger particles. The table is shaken in the horizontal direction at a low frequency (several times per second) along with a hydraulic action. Particles along each groove, tending to be monosize, move laterally with the higher-mass particles, those carrying ore inclusions, advancing over those of lower mass. This produces a discrimination based solely on density because all sizes are lumped together in the final stage. Two and sometimes three displacement fractions result, a waste rock, or ganque, a concentrate, or upgraded ore, and intermediate grade termed *midlings*. The efficiency of the wetting agent to disperse and separate these components will influence the percentage of middlings versus concentrate.

In another design, the Humphrey Spiral, dispersed particles are washed down a helical slide. More dense particles tend to collect at the exterior of the spiral because of centrifugal force discrimination between higher- and lower-density particles. The streams are geometrically split apart at the bottom.

A third density discrimination device is the variable fluid density discriminator, the Magsteam Separator™, which uses centrifugal force in a variable-density liquid to separate the phases.

Chemical methods of liberation almost always involve selective chemical leaching. After grinding for a predetermined period, the granular product is chemically leached with an appropriate, but discriminatory, agent which effectively and rapidly dissolves the ore but leaves the host untouched.

The difficulty with all such liberation devices lies not in their efficiency but in determining the fineness of grind of the feed to obtain effective separation of ore from host.

As noted in Chapter 10, inclusions (but not solid solutions) in a host rock represent stored stress regions because of differences in density, chemical composition, and thermal coefficient of expansion. Fractures passing preferentially through these contact zones do not necessarily separate them completely but assure that new fracture surface contains ore mineral at its boundary. As grinding progresses, ore particles are progressively separated from their host but are also progressively ground smaller.

On occasion liberation serves to remove small amounts of detrimental material from the host, which represents the commercial product. An excellent example is that of high-quality glass sand. Beach sand, either from current or ancient seashores, contains ferrovanadium impurities that do not melt during the glass fusion process. If incorporated in the melt, they become highly stressed microcrystalline centers inducing fracture-prone centers in the glass bottles destined for food and other household products. Should these be pressurized (as soda bottles), an inadvertent impact at or near one of these stress sites can cause sudden bottle rupture with the potential for personal injury.

These impurities, if liberated, can be removed both chemically, by a selective dispersant, or mechanically, by high-intensity magnetic separation. However, grinding, with or without selective agent, is critical with regard to size. This is illustrated by the photomicrograph, Figure 12.17, showing host rock (quartz)

Figure 12.17. Surface impurity exposure after 5% of milling cycle.

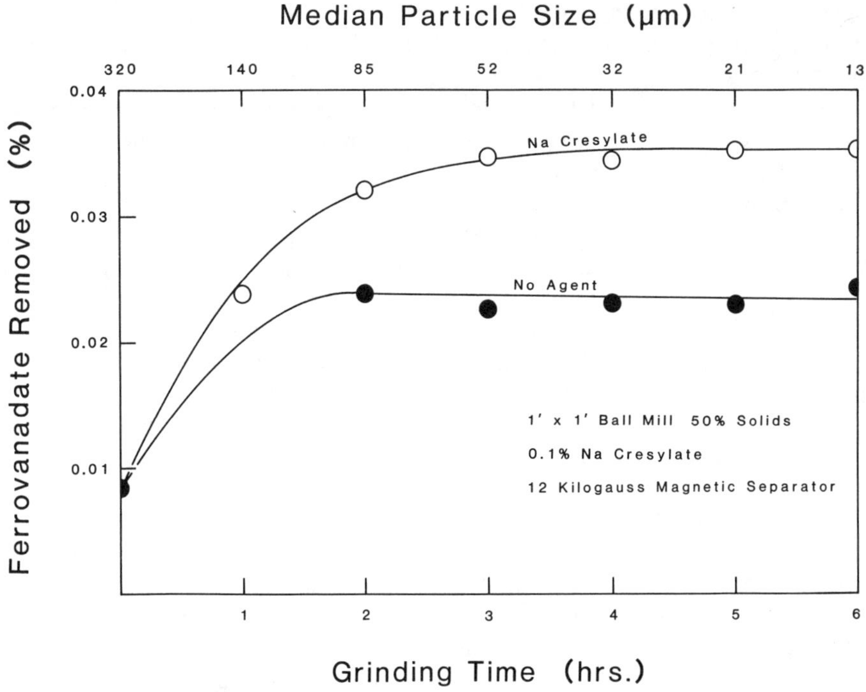

Figure 12.18. Leaching of ferrovanadate from dispersed quartz ore.

ground but 5% of the grind cycle in a ball mill.[45] Note that metalliferous ore particles (dark streaks and patches) appear on new surfaces of particles after this short grinding period. For leach-type liberation, the grind period will be very short. This is shown by the grind time—extraction curve in Figure 12.18. By scaling the test mill to the production mill size, grinding periods can be defined readily. Also, as the curve reveals, slight overgrinding does not impair liberation efficiency.

With mechanical separation the grind period becomes more critical. Extended grinding diminishes the amount of host material adhering to the metalliferous ore and increases its density and probability of discrimination in any density-sensitive separator. Overgrinding, however, reduces the size of both components, and the momentum difference now becomes small on an absolute scale, with the result that discrimination efficiency falls off.

Ore size distribution and grind size distribution derive from altogether different phenomena (crystallization versus comminution). Thus the dependent variable, separation efficiency, becomes a yield surface rather than a curve, and the maximum will most likely be a broad, mesalike area on this surface.[29] For batch processing, the solution is relatively simple—screen off fractions at various milling times with various dispersing agents and measure the liberation efficiency at each element of screen size. However, for continuous milling

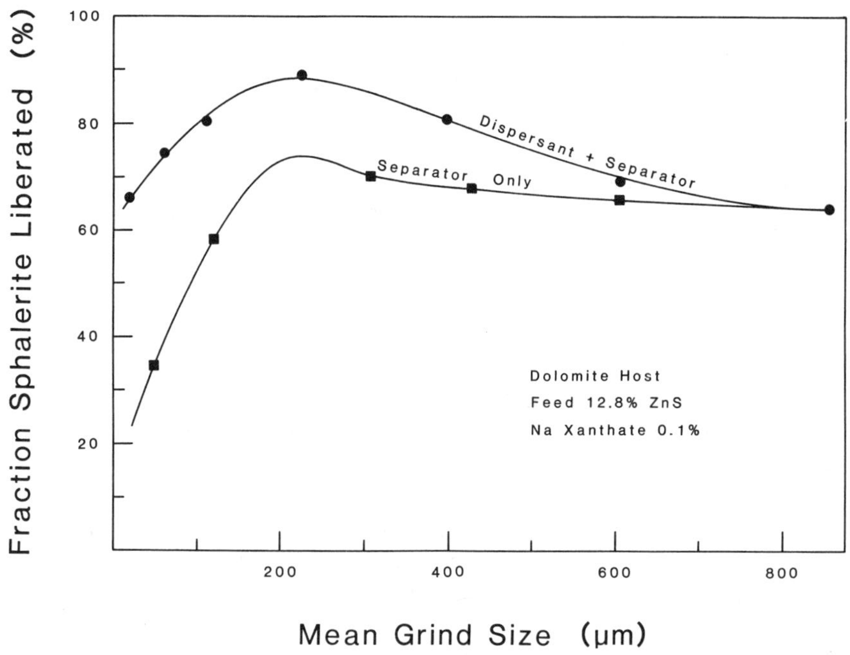

Figure 12.19. Section of sphalerite–dolomite liberation surface.

(closed circuit), where the mill output has a discriminator (screen, centrifuge, cyclone, etc.), the solution is more difficult.

Figure 12.19 is a section through such a yield surface for sphalerite (zinc sulfide) in a dolomite (calcium–magnesium carbonate) ore host, with and without dispersant, processed with a variable fluid density separator. Sphalerite's density (4.0 g/cm^3) is sufficiently greater than dolomite (2.9 g/cm^3) that particles can consist of $\frac{1}{3}$ sphalerite, $\frac{2}{3}$ dolomite and yet experience excellent separation (>90%). By employing a selective dispersant that wets out the inclusion species preferentially, the carrier mineral grinds faster without the ore mineral being equivalently size reduced. Obviously, these are differentiating probability functions, and some ore does become ground and appears with the zinc product. However, the technique offers the possibility of employing a reduced energy system in a ball mill (reduced ball size, reduced rpm, or increased mill solids) to improve both process economy and yield.

In the figure note that liberation attained at optimum milling with sodium xanthate represents an approximate 18% gain on an absolute basis and 25% on a relative basis over a nondispersed procedure. Also, both the plateau height and width (mesa on the work surface) are substantially greater with dispersant than without. The minimal grind time for the operation becomes the critical time because finer size reduction (<100 mesh) not only incurs increased costs, it decreases liberation.

This technique of arriving at liberation size for density discrimination has been investigated mathematically,[30] independent of the interplay of dispersants for this separation, sphalerite in dolomite. The complete surface is highly complex and must be solved with computer function-surface software.

Critical grind size, or critical grind time, will therefore measure the integrated size spectrum at which the liberated mineral is enhanced in density yet remain sufficiently coarse to permit mass (or momentum) discrimination by any of the devices described.

12.6 Magnetic Fluids

Magnetic fluids are composed of low-viscosity, Newtonian oils in which are dispersed extremely fine (usually less than 0.1 μm) anisometric magnetic particles. Most popular of the solids for this purpose is maghemite, gamma ferric oxide.

The primary function of the fluid is to change from low viscosity to high viscosity and back again quickly through application of an exterior electromagnetic field. Hence the magnetic force of attraction due to this field must greatly exceed the electrostatic or barrier force of dispersion. By varying the applied field intensity, a full range of intermediate viscosities can be induced. One of the more useful applications is that of the fluid coupling clutch.

Clutch designs take several configurations, one of which has two conventional cup- or blade-type fans facing each other immersed in an oil containing a 20% or greater concentration of well-dispersed anisometric magnetic particles—a magnetic fluid. A second configuration has a concentric rotor and slave stator with radial paddles passing in close proximity during rotation. This design is illustrated by the three sketches of Figure 12.20.

A magnetic field is applied in a radial or cylindrical mode. With no applied field and no rotation the particles possess random orientation and produce a thixotropic fluid in which the magnetic particles do not settle appreciably (left configuration).

As the drive shaft commences rotation under zero applied field, needles of the oxide, having an anisometry of about 5–10, align with the direction of flow, that is, in concentrically circular orbits around the inside of the chamber. This is the path of least resistance and the state of minimum viscosity. The system becomes shear thinning and highly fluid (center configuration).

When the magnetic fluid is applied (right configuration), needles align either parallel with the rotational axis or radial to it, depending on coil winding geometry. This alignment, perpendicular to direction of flow, increases the rigidity of the fluid. Also particles tend to be immobilized by the field, which adds yet another component of flow resistance. The magnetic field, originally shear thinning, transforms to a dilatant system. Because the formation–relaxation time of the needle assemblage is short, removal of the field causes the liquid to return almost instantly to low viscosity.

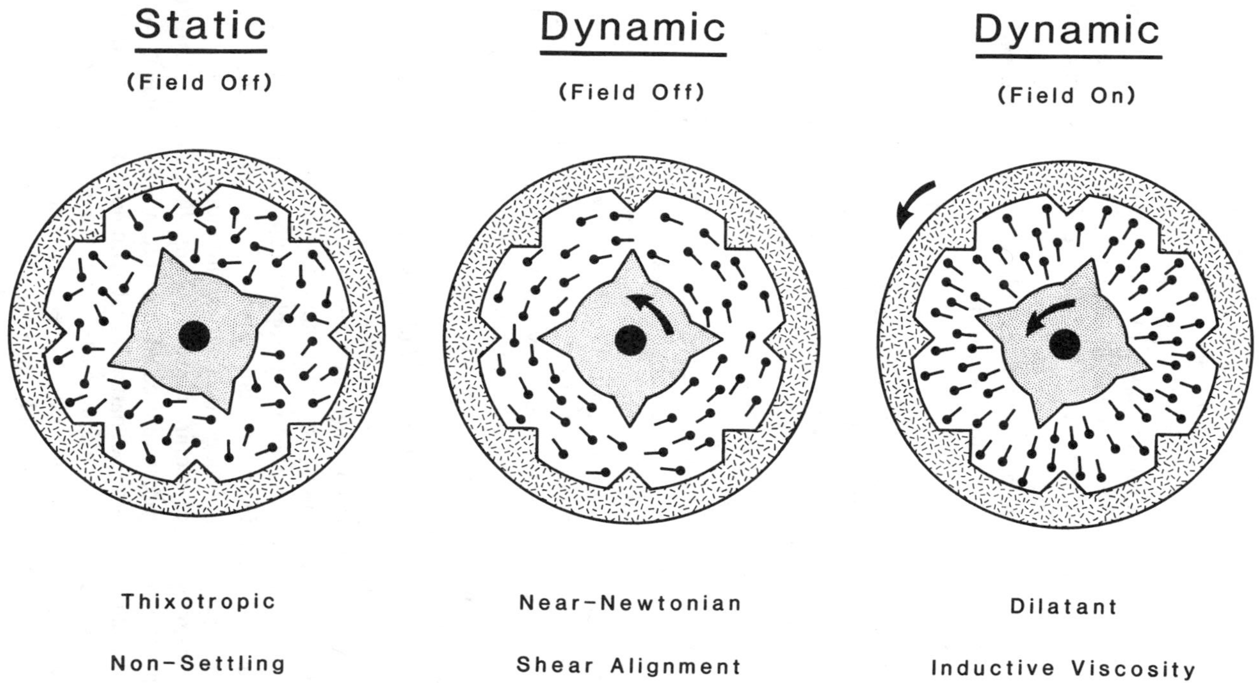

Figure 12.20. Dispersed magnetic fluid clutch system.

Coupling efficiency between the two fans or rotor and slave stator is but a few percent at zero applied field but approaches 100% with maximum applied field. Unlike the classical fluid clutch automotive transmission, coupling efficiency is not fan speed dependent but is controllable at any rotational velocity. The field intensity necessary to accomplish a full dynamic range of viscosity change is small, usually in the range 100–500 oersteds.

Magnetic particles employed for this application must be quite small for two reasons. First, larger particles of iron oxide, having a density greater than 5 g/cm^3 in an oil with a density of about 0.8 g/cm^3 would settle out upon standing and, worse, be selectively thrown to the chamber wall by centrifugal forces when operating. Second, magnetic particles must not form a rigid bridge, only a state of increased order in the fluid. However, if the particles are too small, the ferrimagnetism effect will diminish almost entirely due to magnetic domain physics.

Oxide particles can be made both as needles and as platelets for this application.[31] They are commonly heated with a fatty acid having a chain length greater than 10 carbon atoms. This can be as simple as lauric acid or as complex as N-(1,2 dicarboxyl ethyl)-N-stearyl sulfosuccinamic acid.[32] The dispersant level is high (several percent) because of the high surface area of the solid, which falls in the range 100–200 M^2/g.

Magnetic oxide particles have a relatively passive surface because most methods of manufacture involve elevated temperatures. Thus the method of forming a dispersed surface must also involve a slightly elevated temperature. Several of the methods employ an organic carrier liquid,[33] for example, kerosene, heated to 100°C or higher under a nitrogen or inert gas blanket. Once the dispersant is adsorbed or reacts with the oxide, the slurry drops dramatically in viscosity. The solid is ''dewatered'' by passing it through a high-intensity magnetic field and successively washing the wet powder with heptane solvent, followed by low temperature or vacuum drying.

When prepared by these methods, a fluid results that upon centrifuging at 4000*g* displays no significant phase separation.

Employment of oleic acid as dispersant results in a series of adverse internal oxidation–reduction reactions. Some oleic acid may be stripped from the ferric oxide surface, the material may become pyrophoric upon removing the oil carrier during production, or the dispersant may be broken down chemically,[34] resulting in partial flocculation during magnetic fluid formulation.

Working magnetic fluids require unusually high shear in their formulation. Standard impeller mixers are rarely capable of providing sufficient energy to break apart the partially chemical, partially magnetic flocs of ferric oxide. Common methodology involves multiple passes in a concentric rotor–stator mill (Kinetic Dispersion type) or with a small-media mill employing fine steel shot.

More sophisticated magnetic fluids have been formulated around barium ferrite and rare earth compounds (e.g., europium, samarium, gadolinium). The dispersant technique is essentially the same except more temperature-resistant organosilicone acid molecules are employed. Silicone oils have been employed to increase operating temperatures and oxidation resistance. Advanced systems have been produced that involve magnetic oxide–dispersant–carrier liquid curing in an electron beam to effect localized heating.[35] Such equipment has been employed for space applications and in high-risk environments, such as nuclear reactors, high-pressure boilers, and steam turbines.

12.7 Miscellaneous Systems

The number of unusual and minor systems in which dispersants fulfill an important, or even critical, role are legion. A few will be described here to illustrate general principles of materials selection and special application.

Paraffin is a mixture of aliphatic hydrocarbons averaging around twenty carbon atoms in the chain. Because of its inertness, purity, clarity, and low melting temperature, it has been employed for a broad variety of applications in the food, cosmetics, pharmaceutical, and other similar industries.

Paraffin has a hydrophobicity of 100%. It dissolves readily in heptane and other low-molecular-weight hydrocarbon solvents. In this mode slurries have been employed for waterproofing, lubrication, ski base coating, and antisqueak applications. Unfortunately, such systems are highly flammable and expose the user to the hazards of breathing excessive hydrocarbon fumes.

Paraffin can be emuslified or dispersed in water for these applications by employment of intermediate-to-long-chain, ethoxylated alcohols and similar chain length ethoxylated fatty amines. If the latter is employed, 1–2% of stearic acid dissolved (by melting) in the paraffin prior to dispersion creates a better surface for dispersant attachment.

The process involves high shear in the aqueous system, which fluidizes or melts the paraffin. The dispersant then adsorbs on the molten globules from the aqueous phase. These globules cool into microspheres without aggregating or refusing. Stability increases with fineness, which in turn requires more dispersant. However, dispersions 10 μm or finer can be obtained that have excellent shelf stability and are useful in cloth treatment for waterproofing. This operation is usually the final stage of the cloth processing (sizing, dyeing, bonding, etc.) because it renders fiber surfaces impervious to later chemicals.

Because of the compatibility of the ethoxylated agents with the carboxylated and hydroxyl structures basic to the cellulose chain in cotton and the protein linkages in wool, paraffin globules are strongly localized in the fiber bundles. Often the finished cloth is passed over a heated roll to ''set'' the paraffin onto these fibers. Once attached, the paraffin rarely can be washed out. As such it serves to protect dye colors and other processing agents.

Anionic agents for dispersing paraffin present other problems. The critical micelle concentration (CMC) for sodium oleate is about 0.5 to 1.5 millimolar. CMC is that concentration at which the dispersant changes from randomly oriented molecules in solution to an ordered sphere whose alkali ions point outward on the surface and whose hydrocarbon tails point inward to form a wholly oleophilic interior. For sodium abietate the CMC range is much broader, 0.5 to 18 millimolar. These concentrations occur in the pH range 6 to 8. This is also the range for effective adsorption onto paraffin globules to render a water dispersion. The internal competitiveness of the dispersant for itself (micelle formation), as a consequence, renders paraffin dispersions ineffective.

However, by raising the pH to about 12, the anionic agents have a preference for paraffin over micelle formation.[36] This severe alkalinity may be somewhat restrictive for certain fabric applications. Also, elevated pH renders cellulose fibers less attractive for negatively charged paraffinic particles.

An extension of the paraffin processing lies in forming dispersions of low-molecular-weight polyethylene (PE) and polytetrafluoroethylene (PTFE) polymers in water for similar applications as well as surface treating pigments for use in oleophilic resins (PVC, polyethylene, polypropylene, polystyrene, and their copolymers). Fatty acids and fatty alcohols are superior to their sodium salts. Adsorption isotherms have been constructed[37] to follow the adsorption mechanism and enable better pigments and superior products to be compounded. These show the "fitting area" for PTFE to be about twice the molecule's cross-sectional area (about 0.26 square nanometers). For polyethylene emulsions the absorption area is somewhat smaller. Adsorption of stearic and oleic acids is believed to proceed at methyl and ethyl side groups on the major spine of the chains, as well as at unsaturated bonds on the surface.

Fireproofing agents may be attached to cloth fibers by a similar, but less exotic, technique. The two most popular agents are organophosphates and organo bromides, the former for clothing worn in contact with the body and the latter more for fabrics having potentially high fire exposures, astronaut suits, automotive upholstery, kitchen towels and hot pads, and furnace, hot water, and fuel line insulation.

Typical compounds are hexabromo- and decabromo-diphenylethers. These are also hydrophobic compounds but readily disperse with polyethylene glycol and polyethylene glycol–silicone copolymer.[38] A finely powdered agent, generally -25 μm, is dispersed under high shear at about 10% solids with 0.2% of the dispersant. The finished system has the appearance of an emulsion. If slightly less than the critical level of dispersant is employed (0.10–0.15%), the system has only a short stability period, generally hours, but attaches more strongly to cottons and polyester fabrics. Unfortunately, unlike paraffin waterproofing, fireproofing compounds are slowly washed out with strong alkali detergents. Studies on children's clothes have shown that 10 washings will often halve the effective agent level on the fabric.

One final cellulose fiber adsorption phenomenon lies in attaching plant protein as an adhesive for sizing or for strengthening of paper in high-tear-resistance applications (bags, wrappings, shipping containers, etc.). In years past animal protein in the form of a glue (horse or fish variety) was added. This technique proved offensive to workers because of odors. Also the supply of animal protein was quickly exceeded by the demands from the paper and box industries. Current practice is to employ soybean hydrolyzed protein[39] for such applications. Because of amino acid linkages in this material, pH elevation is a simple method of dispersing highly sheared homogenized soya protein. The dispersion is added directly to cellulose fibers in the makedown tanks of the paper plant, pH is then lowered to an acceptable level for manufacture, and the slurry cast onto felt or wire screen for paper and board manufacture.

As this pH drops, protein flocculates onto the cellulose fibers causing interbonding. Upon drying the product has strongly bound fibers and possesses very high strength.

Boron carbide is cast in appropriate forms, usually for grinding or other applications involving the substance's hardness. The material can be made as a fine powder, rammed into molds under high pressure, then fired under vacuum to produce the finished element (bits, granules, drills, rods, etc.). By selecting an appropriate particle size for high-density packing (log-normal distribution), the particles are dispersed in toluene with short-chain amines (propyl, butyl). Longer chains with periodic amine groups separated by three to four carbon atoms[40] can also be used. The slurry is then cast into a mold, rammed under pressure, the solvent evaporated, and the finished system fired with dispersant in place. Some of the agent carbonizes, but most volatilizes. The fired boron carbide readily forms strong interparticle bonds, and the material's density can reach 98+% of the theoretical value.

Mineral wool is usually a form of fiberglass made from fused kaolins, aluminas, and silicas with alkali metals added to reduce the melting temperature. The material has excellent heat insulation and fireproofing characteristics for pipe wrapping. It can be cast into blocks, boards, and other forms, but only if dispersed. Dispersion also reduces skin exposure by workers who find the material very irritating. The fiber is chopped (short fiber) and dispersed with cationic agents. Typical of these are the imidazolines. Also effective because of the slightly acidic surface of mineral wool are amine oxides, alkylamino carboxylic acids, and their ethoxylated derivatives.[41] Solids in the 65% range can be obtained that pour as thick slurries into molds, the water drawn off, and the blocks fired to the point of interparticle fusion. Dispersing agents volatilize. The use of anionics, especially sodium salts, will reduce the melting temperature and narrow the applications, especially in high-pressure steam boilers and turbines.

References

1. Perry, R., and Chilton, C., *Chemical Engineer's Handbook*, 5th ed., McGraw-Hill, NY, Sect. 19, p. 89 (1973).
2. Conley, R., and Johns, D., Utah Power & Light—Deldessa Rept. #167-3, (1984).
3. Conley, R., Golding, H., and Taranto, M., *Ind. Eng. Chem. Proc. Design & Dev.* **3**, 183 (1964).
4. Conley, R., and Lloyd, M., *Ind. Eng. Chem. Proc. Design & Dev.* **9**, 595 (1970).
5. Allen, T., *Particle Size Measurement*, 5th ed., Chapman Hall, London (1995).

6. In X-ray sedimentation coarse particles must gravitate through a medium of finer particles and some hindered settling has been measured. However, except with high-anisometry particles (platelets and needles), the error is not serious and it usually ignored.

7. Conley, R., *Powder Tech.* **3**, 102 (1969).
8. Firestone, L., *Printed Circuit Fabr.* **11**, 28 (1988).
9. Stout, G., *Surface Mount Tech.* **5**, 30 (1991).
10. Satch, N., Bandow, S., and Kimura, K., *J. Colloid Interface Sci.* **131**, 161 (1989).
11. Wu, C., Sun, W., and Sun, X., *Huadong Hu Xuey Xueb.* **16**, 717 (1960).
12. Voet, A., *Trans. Faraday Soc.* **31**, 1488 (1935).
13. Vedovenko, N., Fedushinskaya, L., and Yaremko, Z., *Ukr. Khim. Zh.* **55**, 28 (1989).

14. Alkali metals are cast commercially under paraffin to shield the active surface from atmospheric gases, which react rapidly and combustibly. However, paraffin does not attach itself to the metal but serves only in a blanketing role. Fatty acids react with the alkali metal surface and display a limited amount of protection against corrosion when the metal is cool.

15. Hensley, V., U.S. Patent #2,487,333 and 2,487,334, Nov. 8 (1949).
16. Singer, M., Dutch Patent #77,872, April 15 (1955).
17. Bird, J., U.S. Patent #2,021,885, Nov. 26 (1935).
18. Lumina, M., et al., *Kolloidn Zh.* **40**, 1191 (1978).
19. Tyran, L., Europatent #21,439, Jan. 7 (1981).
20. Eguchi, K., et al., Japan Patent #6,389,577, April 20 (1988).
21. Ishijima, S., Kiritani, T., and Hatashi, Y., German Patent #3,020,073, Dec. 4 (1980).
22. Conley, R., and Althoff, A., *J. Colloid Interface Chem.* **37**, 186 (1971).
23. Kondo, K., Ozawa, S., and Saito, Y., Japan Patent #273,872, Mar. 13 (1990).
24. Ishida, Y., and Murakami, S., Japan Patent #241,370, Feb. 9 (1990).
25. Metz EMD bulletin #81-EM, ''Copper #120 Flake,'' So. Plainfield, NJ (1989).
26. Tarpley, W., and Conley, R., MRT Rept. #7, DOE Contract # DE-AC02-79ER10466 (1979).
27. Glenn, R., and Grace, R., Bituminous Coal Research, Inc., R & D Rept. No. 5 (1963).
28. Naka, A., et al., *J.A.O.C.S.* **65**, 1194 (1988).
29. Andrews, J., and Mika, T., Proc. 11th Intl. Mineral Proc. Congr., Cagliari, Italy, P. 59 (1975).
30. King, R., and Schneider, C., Soc. Mining Eng., Reprint #92-207, Annual meeting, Phoenix, AZ (1992).
31. Conley, R., U.S. Patent #3,115,470, Dec. 24 (1963).
32. Nippon Seiko, Japan Patent #118,496, Feb. 25 (1980).
33. Conley, R., *J. Am. Ceram. Soc.* **50**, 124 (1967).
34. Lesnikorich, A., et al., *J. Magnetism & Magn. Matls.* **85**, 14 (1990).
35. Glover, R., Svail, R., and Szczepanski, T., Europatent #418,807 and 418,808, Mar. 27 (1991).
36. Derzhanski, A., and Zheliaskova, A., *Zh. Prikl. Khem.* **5**, 2105 (1982).
37. Dolzhikova, V., Goryunov, Y., and Summ, B., *Kolloidn Zh.* **44**, 560 (1982).

38. Kaneyasu, T., and Kochi, H., Japan Patent #56,898, May 23 (1978).

39. Eberl, J., U.S. Patent #2,484,878, Oct. 18 (1949).

40. Williams, P., and Huang, Y., *Ceram. Trans.* **12**, 461 (1990).

41. Cederquist, N., et al., German Patent #2,605,633, Sept. 2 (1976).

42. ''Evaluation of A.S. Printed Wiring Boards,'' Office of Naval Avionics, D.O.D., Rept. #149-87, Dec. 7 (1987).

43. Azaroff, L., *Introduction to Solids*, McGraw-Hill, New York (1960).

44. Allen, T., ''Particle Size Measurement,'' CFPA Seminar, Chicago, IL, June (1993).

45. Murray, H., and Conley, R., ''Mineral and Chemical Composition of PGS Glass Sands,'' Covington & Burling (MRT Rept. #1), Washington, DC, p. 16a (1976).

CHAPTER

13

Instrumentation and Measurements

13.1 Particulate Surface Measurements

With the exception of direct viscosity measurements, no other parameters so significantly relate to dispersion as the physical and chemical characteristics of the particulate phase surface. The intimate association, sometimes tenuous, sometimes dynamic, of the dispersant molecule with the solid surface affects not only solids level, viscosity, and liquid compatibility; it also affects a multitude of other properties related to both manufacturing and application.

Until the development of sophisticated scientific instrumentation and its introduction into laboratories in the 1960s and 1970s, surface properties were at best guessed at and at worst ignored. This disregard explains in part the poor quality of old paint (pigment settling, poor coverage, short shelf life), rancidity of aging cosmetics, polymeric floor covering embrittlement, shortened tire life, and other adverse characteristics of dispersed particulates in liquid media.

Those solid characteristics having significant influence upon dispersion and rheology are:

1. Surface acidity and basicity;
2. Structural stress from milling or heating;
3. Surface hydrophobicity/hydrophilicity;
4. Energy of water adsorption;
5. Surface area (particle size and distribution);
6. Aging changes, physical and chemical; and
7. Chemical processing alteration (deliberate and inadvertent).

Several of these properties are interrelated or arise from common sources. Virtually no commercial powder, whether pigment, ceramic, metal, pharmaceutical, or cosmetic, at the present time is produced without manufacturer's attention being given to the greater portion of these characteristics prior to reaching the customer's plant.

13.2 Surface Acidity and Basicity Measurements

The importance of surface acidity/basicity in dispersing pigments in polymers was probably recognized as far back as Cro-Magnon cave painters, but in a less-than-scientific manner. Cave walls at Lascaux (southern France) and Altamira (northern Spain) bear timeless testimony to calcite, iron, and manganese oxides and carbon black (soot) being ground and wet out in stearic acid–containing animal fat. Early humans ground their pigments until the viscosity appeared spreadable by the animal hair brushes they fashioned crudely from the same beasts of the field that contributed their stearic acid dispersant.

Calcium carbonate, ferric oxide, and basic lead carbonate always have enjoyed excellent reputations as stable pigments in linseed oil formulations because of the reactivity of their alkaline surfaces with the acidic components in the vehicle (oil phase). Not, however, until the advent of water-based paint formulations did pigment surface chemistry become a salient, if not definitive, consideration.

Various methods have been proposed to measure a solid's acid–base character, only five of which will be considered here because their employment for many materials and for many years has endowed them with credibility. Some methods require relatively simple laboratory equipment. Some require highly sophisticated instrumentation. Some measure parameters only semiquantitatively, while others are highly precise. The methods are:

1. Nonaqueous acid–base titration;
2. Anionic/cationic adsorption and zeta potential;
3. Aqueous acid–base titration and pK determination;
4. Adsorption of Hammett indicators; and
5. Selective gas adsorption isotherms.

13.2.1 Nonaqueous Acid–Base Titration

One of the early approaches and one which has found a high level of acceptance over the years because of its correlation with other dispersion and pigment interaction properties is that developed by Boehm[1] using semimicro Zeisel analyses. In this surface assessment technique a particulate is heated just sufficiently to remove all adsorbed water then immersed in a nonaqueous liquid, typically toluene, MIBK, anhydrous alcohol, or ether, and titrated by nonaqueous solutions of acids (phthalic, benzoic, phenyl sulfonic, toluic acids, salicylic anhydride, 8-

hydroxyquinoline, sodium dihydrogen phosphate, etc.) and nonaqueous bases (morpholine, butylamine, pyridine, and sodium hydroxide).

For acidic group assessment, the powder may be treated with diazomethane in dry ether, which converts *M*–O–H groups to methoxy groups. These are then analyzed by organic microanalysis. Because no sharp endpoint occurs in weak-acid–weak-base titrations in nonaqueous media, acid groups may be determined by analyzing the Langmuir isotherm of the base on the solid (e.g., NaOH in methanol).

Weak acids, as phosphoric, may be adsorbed onto the surface via alcohol or ether to assess the alkali hydroxyls. Most of these titrations, whether acid or base determinant, do not yield horizontal plateau-type isotherms but relatively low-slope, flat lines. The isotherm must be extrapolated carefully to determine the population of acid or alkaline surface hydroxyl groups.

A variation on this methodology titrates the solid in benzene, toluene, or heptane with butylamine, employing paradimethylamino azobenzene as a colorimetric indicator for acid site assessment.[2] Alternate assessments employ quinoline (benzopyridine) in aromatic solvents and by vapor-phase adsorption (see discussion to follow) at 315°C.[3] Nonaqueous titration techniques often yield slightly higher acid numbers, especially on multivalent and multiple coordination solids, than do aqueous titrations. This is attributed[1] to the measurement of Lewis acid sites compared to Brønsted sites. These may not always correspond once an anhydrous solid is immersed in water, due to partial surface hydrolysis.

Isotherm assessments are especially aggravated by surfaces possessing multiple acidity levels, as shown by anatase titanium dioxide,[4] the more acidic having a pK of 2.9[5] and the weaker, 12.7. Inflection points may be used to correct such isotherms, which then become much more horizontal.

Titrations may employ an appropriate colorimetric endpoint indicator, much in the fashion of old aqueous acid–base titrations, for a specific range of acidity or basicity. These endpoints are not sharp, however, and require both forward and back titration with inflection-point averaging to arrive at any general measurement. Knowledge of the weight and mean particle size (or surface area) of the powder in suspension allows a calculation of both the acid–base population (microequivalents/gram) and its density (microequivalents/cm^2) on the surface.

To complicate the methodology further, surfaces may change with time, and methods employed a week or so following initial evaluations on fresh or stressed material may yield divergent results. Variations in pK_a of one full unit are not unusual under these circumstances.

Variations on the titration methodology have involved the use of radio frequency sensors and instrumentation for dielectric detection. This method actually assesses Lewis acids and Lewis bases. However, the ratio for a given surface frequently corresponds closely to aqueous pH values with regard to zeta potential, flocculation, interparticle interactions, and other, similar, properties.

Ion exchange analysis (by a strong acid or base) or neutralization techniques are, unfortunately, not applicable for many of these oxides because they may display some solubility, especially aluminum hydroxide and oxides. As discussed in Chapter 3 on phosphate dispersants, hydrolyzed phosphoric acid can enter into permanent, insoluble complexes on hydrated alumina surfaces, which renders the data useless.

A conversion of surface *M*–O–H groups to esters via the addition of anhydrous alcohol and refluxing is an alternative for alkaline surface measurement. In some instances, depending on the reaction mode, methanol will react with all hydroxyls, negating the technique for surface character analysis.

In the Althoff[6] variation of this method, a standard buret containing acidic or alkaline standard solution is employed to titrate the solvent-based slurry under an inert gas purge until a set color develops from the indicator, or electronic change results from the instrumentation. This methodology prevents the introduction of carbon dioxide into the slurry, which can alter some analyses. Because the slurry is partially flocculated, either at initiation or the endpoint, the titrant reaction is highly diffusion controlled and requires excellent agitation, another justification for nitrogen purging.

The colorimetric indicator method works best with white or very light pigments (clays, titanium dioxide, talc, silica, etc.) and less well with highly colored pigments (iron oxides, chromium oxide, carbon black, etc.) because of the requirement of observing an indicator's color change. With highly colored solids, the particulate must be allowed to settle after each increment of titer and the supernatant viewed. Slurry solids are kept as low as analytical accuracy permits, often less than 10%, depending upon the quantity of acid or base groups present on the solid surface.

Sufficient particulate must be in suspension, however, to provide active groups for the indicator to detect. Also, the titrating acid or base must be stronger than the species being titrated.

The method measures predominantly Lewis acid and base sites. However, insofar as the surface is isomorphous (having the same *M*–O–H groups, though differentially ionized), this relates reasonably accurately to Brønsted acids and bases where the powder is exposed to an aqueous phase.

13.2.2 Anionic/Cationic Adsorption and Zeta Potential

The detection of specific surface character by the influences of solid surface groups and surfactants on zeta potential was developed early by colloid chemists. However, the technique became a highly accurate tool as a result of Parfitt's work[7] with titanium dioxide pigments.

The analysis is conducted in aqueous solution and requires a zeta potential measuring instrument (not an electrophoretic mobility tester). This consists most often of a cell having two electrodes separated by about 10 cm, across which is applied one hundred or more volts. Particle electrophoretic velocity is measured with a calibrated microscope. A dilute solids suspension of particulate, $< 0.1\%$, is titrated with a dilute solution of sodium dodecyl sulfonate (SDS). Each incremental addition of SDS provides very large anions, which react at (replace) alkaline surface hydroxyl groups. The surface potential gradually decreases, that

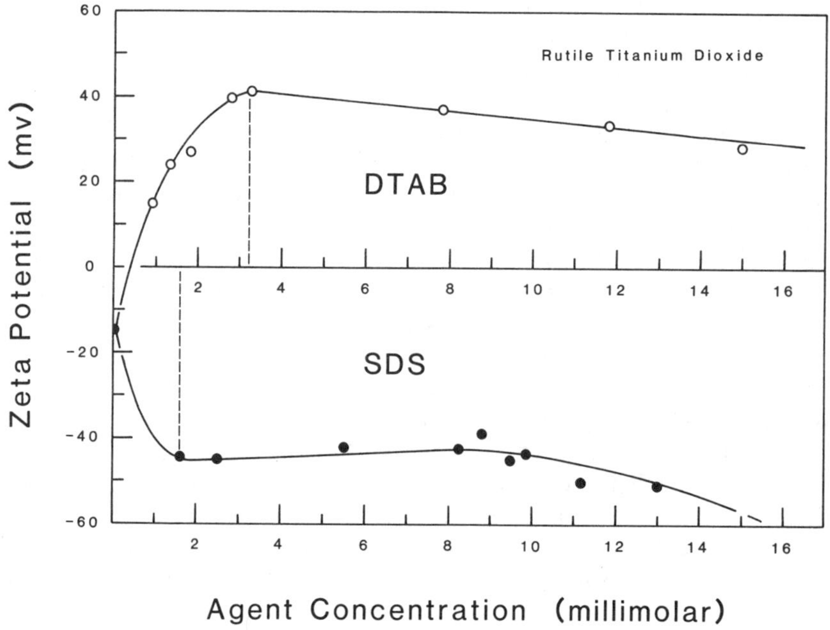

Figure 13.1 Cationic–anionic adsorption profile on titanium dioxide.

is, becomes progressively more negative. A plot of SDS concentration versus zeta potential displays a quantitative adsorption profile following a Langmuir pattern.

Once alkaline sites on the surface are fully reacted, the curve rapidly flattens, as shown in the lower portion of Figure 13.1 for rutile. The surface potential would theoretically remain constant (horizontal plateau) with further additions of SDS. However, a few solution molecules of SDS invert, intermeshing their hydrocarbon tails with those established on the pigment surface to form a partial second layer. This contributes a small amount of anionic charge to the particulate surface, reducing slightly the potential (more negative).

The theoretical apex formed by straight line extensions of the rising portion of the isotherm and that of the plateau should represent the anionic exchange capacity, or alkalinity number of the solid. In practice, that point on the concentration axis at which the maximum occurs is an excellent approximation. Usually it can be determined with a relative error of $\pm 3\%$.

For acidic hydroxyls, an identical process is followed except dimethyl ditallow quaternary ammonium bromide (DTAB) replaces SDS. These additions cause the large quaternary cation to replace the hydrogen ion sites. The zeta potential now becomes progressively more positive during the titration. Again the curve attains a maximum, and then inverts slightly due to tail intertwining. The value of concentration corresponding to the maximum represents the acidic hydroxyl surface. DTAB titration data for titanium dioxide are given in the upper portion of Figure 13.1, based on Ref. 7.

The two curves are not exactly mirror images, even considering the difference in acid–base surface population. While the cationic and anionic molecules have essentially the same length, their weight and cross-sectional area differ, and the diffusional conditions are dissimilar. In almost every instance where this technique has been applied, the SDS isotherm is sharper, as might be expected.

Certain flaws lie in the SDS/DTAB method; for example, sites must be sufficiently ionic to exchange with the large ions in the titrant. If not, the sum of acid and base sites may not equal the total number of *M*–O–H surface units. However, the ratio of acid to base adsorption (in moles) will represent fairly accurately the relative distribution (or percentage) of acidic and basic sites. The actual titration curve for the titanium dioxide species in Figure 13.1 shows an acid-to-base ratio of almost exactly 2:1 for site distribution. By comparison, this acid–base distribution as determined by the Boehm method is 67%–33% (see Table 10.1), which is in excellent agreement.

Second, for passive surfaces (high-temperature corundum, calcined kaolin, gamma ferric oxide) high solution pressures of SDS and DTAB can force molecules down onto the surface at high energy (but not necessarily dissociated)

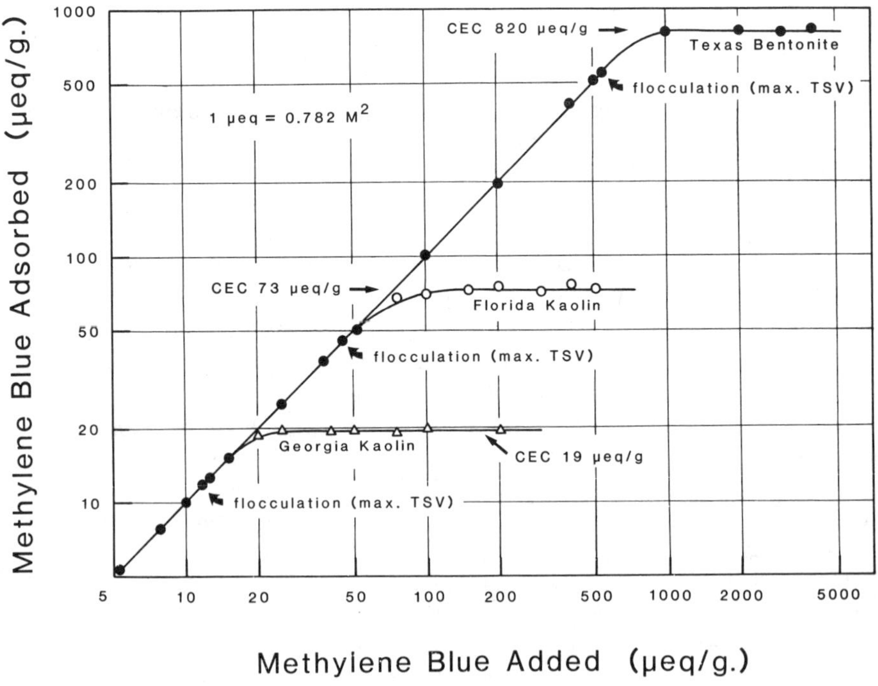

Figure 13.2 Charge-area adsorption dependency of methylene blue (taken in part from Ref. 8).

sites including corners, fracture openings, etc. Under these conditions, the technique reverts to a poor surface area tool. Similar phenomena occur with large charged dye molecules, such as methylene blue.[8] Such curves can provide both cation exchange capacity as well as surface area approximations, as illustrated by three silicates in Figure 13.2. This technique, however, is unable to discriminate between true cations and surface acid–base sites.

In addition to determining the acid–base ratio, the SDS/DTAB technique serves to evaluate total *active* hydroxyl groups on the surface. As noted earlier, some justifiable concern exists about what fraction of the surface hydroxyls remain truly neutral and do not ionize by either mechanism. This must be assessed by an independent method that evaluates the total number of hydroxyls on a separate sample, for example, gas adsorption, surface area, theoretical calculation, or other measurement.

All inorganic oxides with hydroxyl surface groups will respond to acids or alkalis if these become sufficiently strong.

13.2.3 Direct Aqueous Acid Titration

This method was advanced by Althoff et al.[6] and is a variation on the Boehm method. The titration takes place in water, through which a nitrogen purge is employed to eliminate CO_2.

A small sample is suspended in water, usually about 2% solids, agitated with a magnetic stir bar in a closed beaker having only a small, single gas exhaust port. Nitrogen is bubbled slowly with a glass frit through the slurry until the pH is constant, as read by a standard pH electrode and meter. The equipment is also useful for performing solution conductivity evaluations. The titration equipment is shown in Figure 13.3. Nitrogen must be bubbled through the slurry constantly to prevent carbon dioxide diffusion into the system and interference through the formation of carbonic acid. Once all the carbon dioxide and oxygen have been expelled by gas purging, nitrogen flow can be reduced to minimize the fluctuation effect on the pH and conductivity readings. Flow, however, should not drop below 50 cm^3 per minute (0.05 lpm). The method of calculation is demonstrated in Figure 13.4 for titration curves of silica and kaolinite. Only kaolinite is calculated in detail for graphic clarity.

The general mathematics for the titration are:

$$pK_a = pH_x + \log(B_{eq}/B_x) \tag{13.1}$$

The value for B_x is commonly chosen well up the lead-in slope to obtain the most accurate information. If B_x is chosen at $B_{eq}/2$, the equation then simplifies to:

$$pK_a = pH_x + 0.301 \tag{13.2}$$

This method has the advantage of determining the surface acidity both as the number of acid groups and as the acid ionization strength, pK_a.

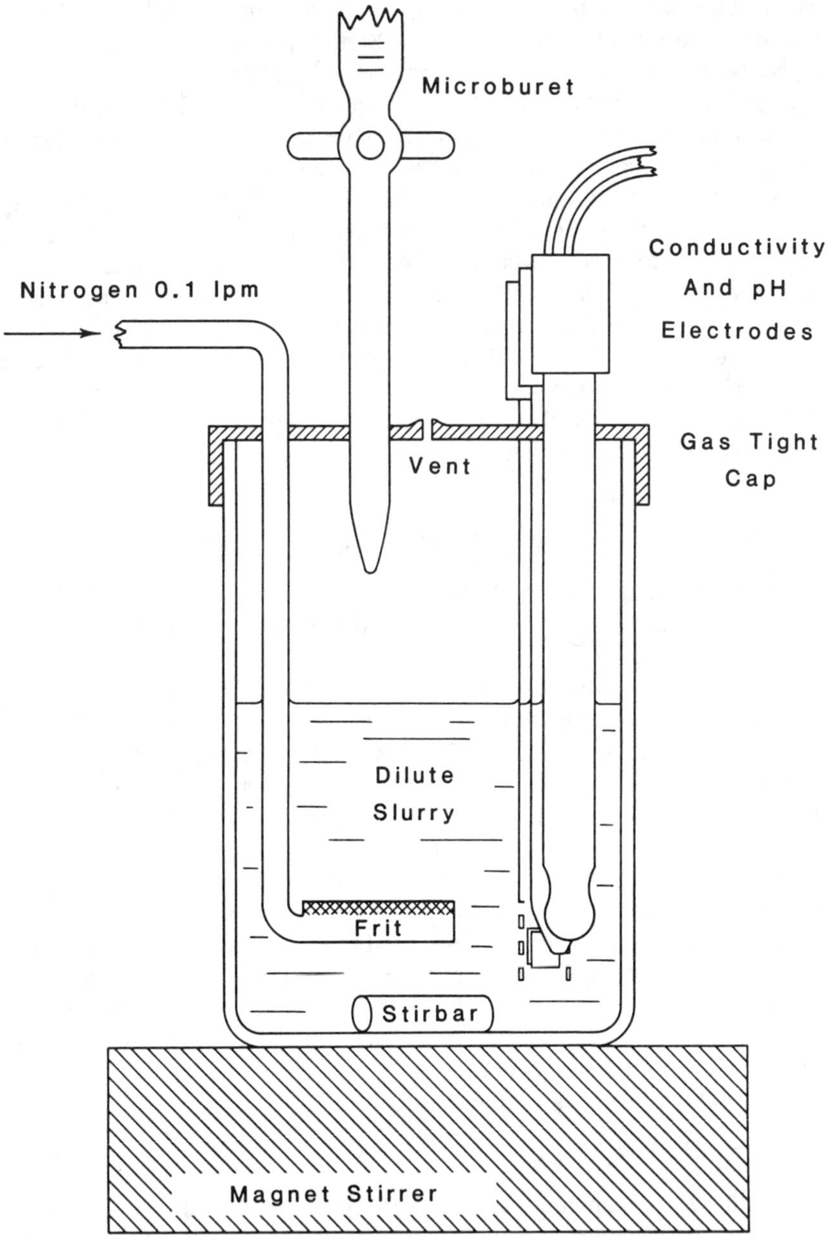

Figure 13.3 Aqueous titration equipment for acidity measurement.

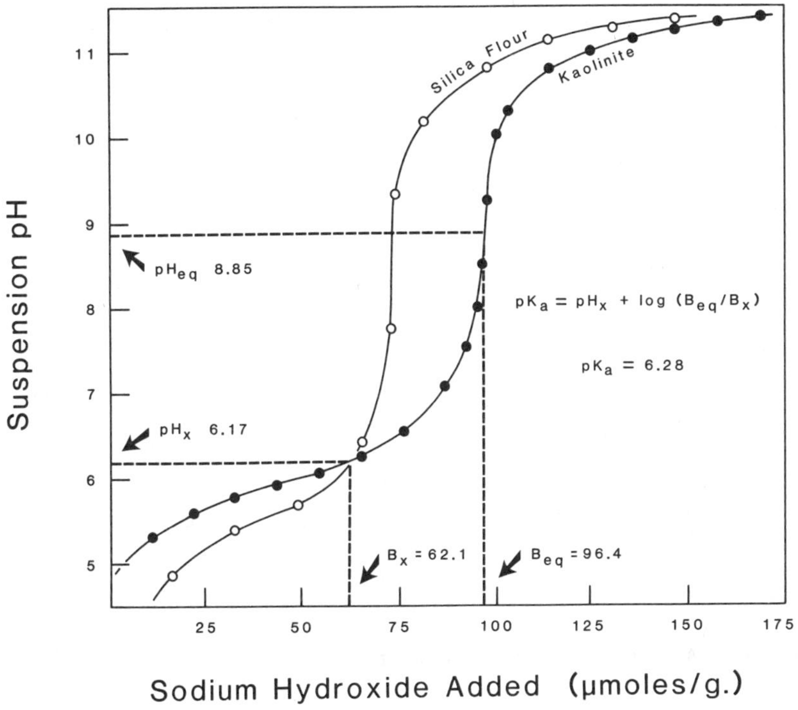

Figure 13.4 Typical aqueous titration curves.

Where hydrolysis is less than 1% and $pK_a < 10$ a check method involves the equilibrium equation:

$$pK_a = 2\ pH_{eq} - pK_w - \log B_{eq} \tag{13.3}$$

where $pK_w = 14.0$ at 25°C. Equations (13.1) and (13.3) are good counterchecks, although the former is generally more accurate. Serious discrepancies occurring between results should cause a reexamination of the titration data.

A similar technique is employed for alkaline surface groups where the slurry is titrated by strong acid. The calculations are also similar. As a check on accuracy, the two ionization constants, pK_a and pK_b (acid and basic dissociations), are related at 25°C by:

$$pK_a + pK_b = 14.0 \tag{13.4}$$

This method almost always yields a summation greater than 14, indicating some amphiprotic *M*–O–H groups are responding to both acid and base titrants. The result is illustrated by the plot in Figure 13.5. The total measured number of ambiguous (amphiprotic) hydroxyl terminal groups decreases with the spread of the isoelectric point from pH 7. With alumina hydrate, the overlap is as much

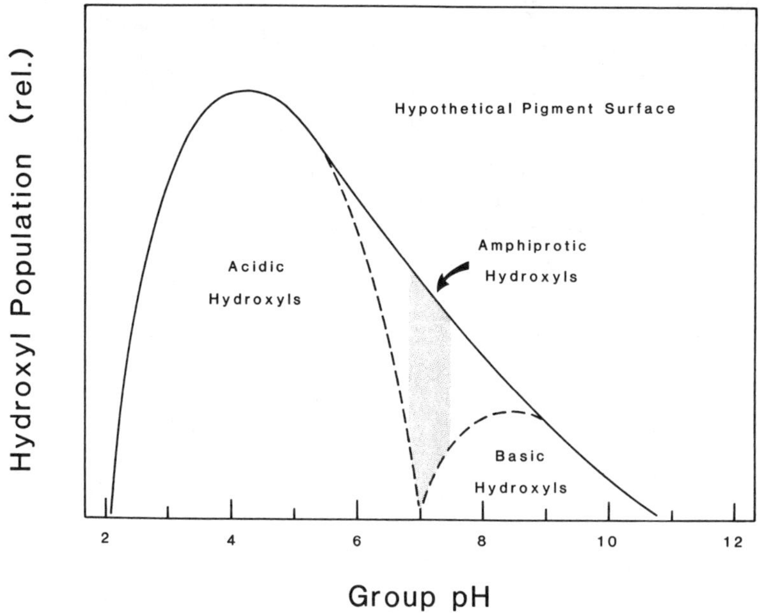

Figure 13.5 Amphiprotic hydroxyl titration response from heterotactic surface.

as 30%. With kaolinite, silica, and zinc oxide, this overlap reduces to less than 10%. Again the relative ratio of the acid and basic hydroxyl group values is sufficiently accurate to calculate percentage acid–base distributions.

13.2.4 Hammett Indicators

Use of Hammett indicators was first advocated by Paul Emmett* to assess the catalytic activity (Lewis acid strength) of silicas and silicates. It was adapted for pigments by Solomon[9] to assess the same parameter, acidity, with regard to organic acid strengths in oil-based paints and similar polymer systems.

Hammett indicators are organic compounds with large electron withdrawal groupings or derivatives capable of responding to Lewis acid strength by color

* Paul Emmett, a physical chemist at Columbia University in New York City from the 1930s through the 1960s, developed a world-renowned expertise in catalysis, the methods whereby catalysts function, show molecular specificity, are poisoned, etc. In 1937 he, graduate student Stephen Brunauer, and fellow academician Edward Teller developed the famous BET adsorption methodology during lunch at the University faculty dining room. Having no paper on hand, the mathematics were derived and written in ink on the white linen tablecloth by Teller, who promptly removed the dishes, rolled up the cloth and handed it to Brunauer! Emmett would go on to win accolades in both catalysis and clay chemistry, Teller would develop the hydrogen bomb, and Brunauer would become a world-renowned colloid chemist and cement authority.

alteration. As an example, triphenyl methanol turns bright yellow on exposure to nonaqueous log acidity (Hammett number) of −6.6. The color change for these agents is not sharp and is strictly monochromatic. Rather, an agent indicates an acid strength range. However, sufficient indicators are available to assess a specific pK_a to about ±0.5 equivalent units. The ability to assess base strength is not as practical. The actual concentration of acidic groups can be estimated (color strength) by the method.

Hammett indicator assessment has been employed primarily to measure the extent organic reactions, including oxidation–reduction, take place based on their Lewis character. Hammett acidity does not readily translate into Brønsted acidity because of the leveling effect created by water when aqueous slurries are formed.

13.2.5 Selective Gas Adsorption

One of the more sophisticated methods of assessing surface hydroxyl character is through selective gas adsorption. It has not developed as a prominent method historically because of the requirement of highly sophisticated, and expensive, gas adsorption equipment.

Any liquid having sufficient vapor pressure at readily workable temperatures (0–25°C) may be employed to adsorb onto a particulate surface. For measuring acidic groups, the gases ammonia and methyl, ethyl, propyl, or butyl amines may be employed. The technique usually compares surface area, as determined by BET nitrogen[10] or krypton,[11] with an adsorption isotherm by the reactive gas.[12,13] The gas in most instances must not have a BET profile but must exhibit a true Langmuir isotherm until the saturation pressure is closely approached.

Determination of moles active adsorbate per surface area provides a useful measurement of the distribution and size, or location, of active groups. A determination of surface area then provides the total number of acid groups, as well as the anisotactic character. Those gases noted in Table 13.1 have been employed for surface investigation. Others have been used for highly specialized purposes.

Water vapor adsorption on most surfaces follows a typical BET pattern rather than a true Langmuir monolayer one. Calculation of the monolayer does not represent the total particulate surface, however, but the fraction of surface possessing polar groups, that is, *M*–O–H group number independent of their acidity

Table 13.1 Surface Character Assessment by Selective Gas Adsorption

Gas or Family	Parameter Measured
Ammonia, alkyl amines	Surface acid groups
Sulfur dioxide	Surface base groups
Water vapor	Surface polar groups
Carbon dioxide	Crack/fracture area
Ethane/ethylene	Catalytic activity

or basicity. Water vapor adsorption can be conducted at 25°C ($P \sim 20$ mm) or even at 0°C ($P \sim 4$ mm) using good quality cryostat and gas adsorption apparatus. Water vapor data on many surfaces appear nearly quantitative for the total oxide–hydroxide groupings.

For total *M*–O–H groups methanol also may be used. Methanol has a higher vapor pressure than water (33 mm @ 0°C and 140 mm @ 25°C) and can be measured very accurately with an ambient temperature cryostat.[14] Surface hydroxyl values by methanol and water differ by less than 10%, but methanol data are faster and more accurately obtained with the instrumentation. Methanol data are also within 5% of the sums of acid and basic groups determined by nonaqueous solution or selective gas adsorption.

Because of the design of BET equipment, gas adsorption methodology possesses very high precision in obtaining molecular adsorption data. The interpretation of these data, however, is somewhat more subjective! A typical example of the methodology is shown by the time dependent butylamine–nitrogen plot of Figure 13.6 for a silicate pigment (acid population and surface area). Here two types of mineral morphology were investigated, blocky particles with an aspect ratio averaging about 3:1, and platey particles, averaging about 6:1. Based on the slopes of the two data sets, it can be concluded that blocky particles have about 50% more of their surface area dedicated to acidic groups than platey ones. A geometric evaluation of these slope values leads to the conclusion that the dominant acidity derives from edge facets.

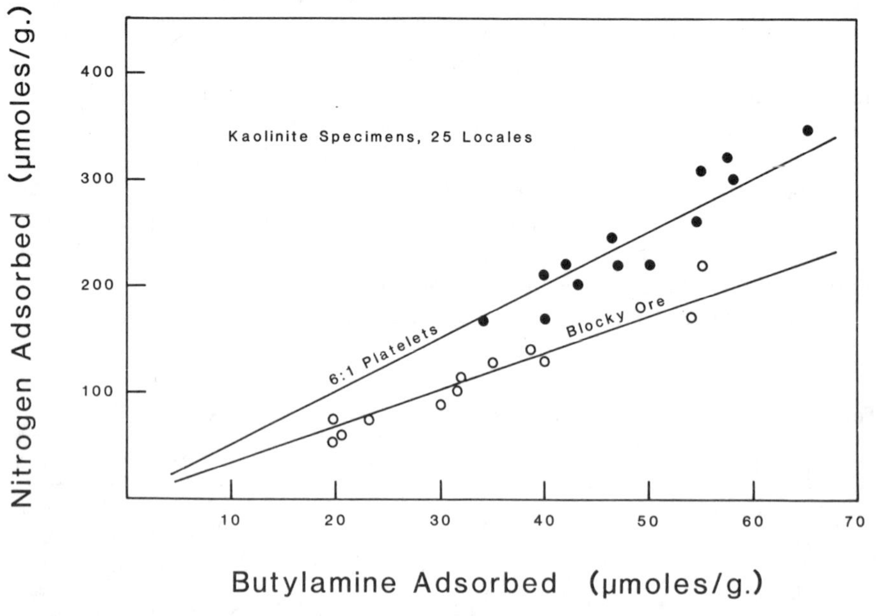

Figure 13.6 Butylamine and nitrogen adsorption on aluminum silicate.

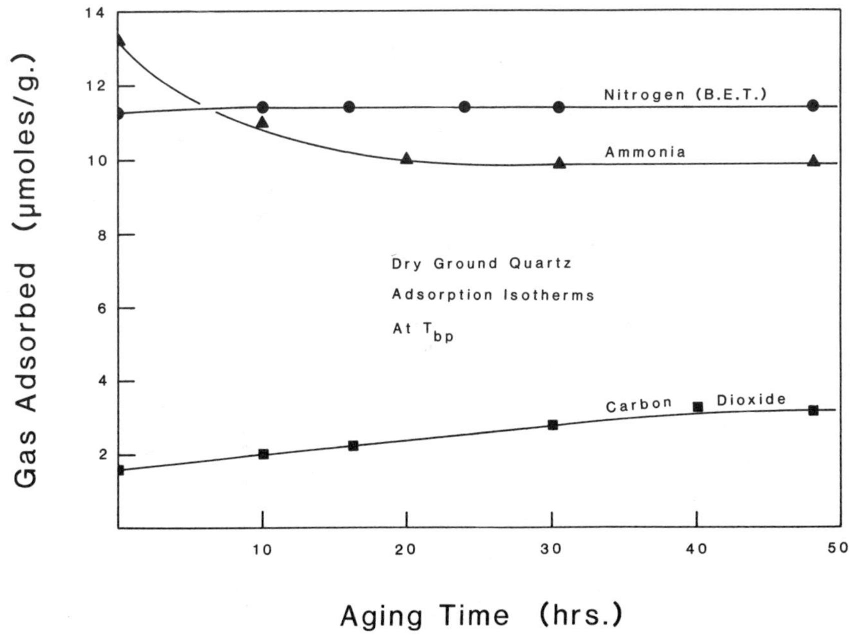

Figure 13.7 Time-dependent adsorption of carbon dioxide and ammonia onto freshly ground quartz.

The carbon dioxide adsorption plot (for measuring freshly produced crack area) for silica is shown in Figure 13.7. Such data are useful in understanding time dependent surface chemistry changes that can adversely affect rheology and applications. Ammonia adsorption, which is directly dependent upon the number of surface acid groups (or strength), decreases with time, while carbon dioxide, which is crack area dependent, increases. Note nitrogen BET area remains essentially unchanged over the same period.

In basicity evaluations, the evidence holds that chemisorption occurs on many surfaces. For example, SO_2 onto calcium carbonate is more a surface area tool than an alkali assessment tool. This results from all surface carbonate molecules being sufficiently alkaline to form a sulfite molecule at each carbonate site. On titanium dioxide or ferric oxide, such is not the case. Sulfur dioxide is a good adsorber on hydroxylated surfaces because of its high dipole moment. Note the nearly 1:1 correspondence in Figure 13.8 between sulfur dioxide and nitrogen, calculated as BET area, for calcite but great disparities in Figure 13.9 for kaolinite.

With calcite sulfur dioxide adsorption is higher for freshly milled material than aged. The average adsorption differential is about 5% but increases slightly with finer-ground materials to about 10%. Sulfur dioxide molecules are both

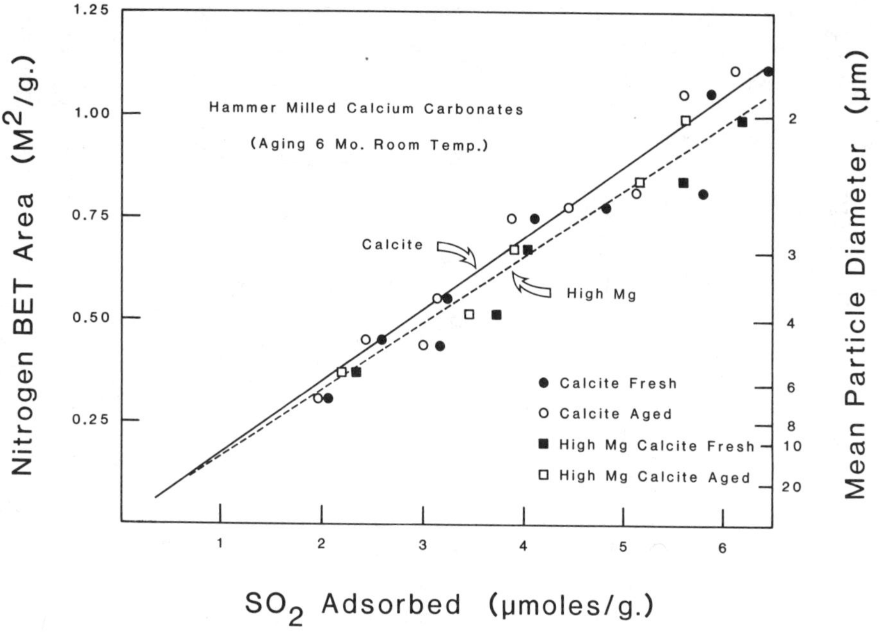

Figure 13.8 Sulfur dioxide and nitrogen adsorption isotherms on calcite.

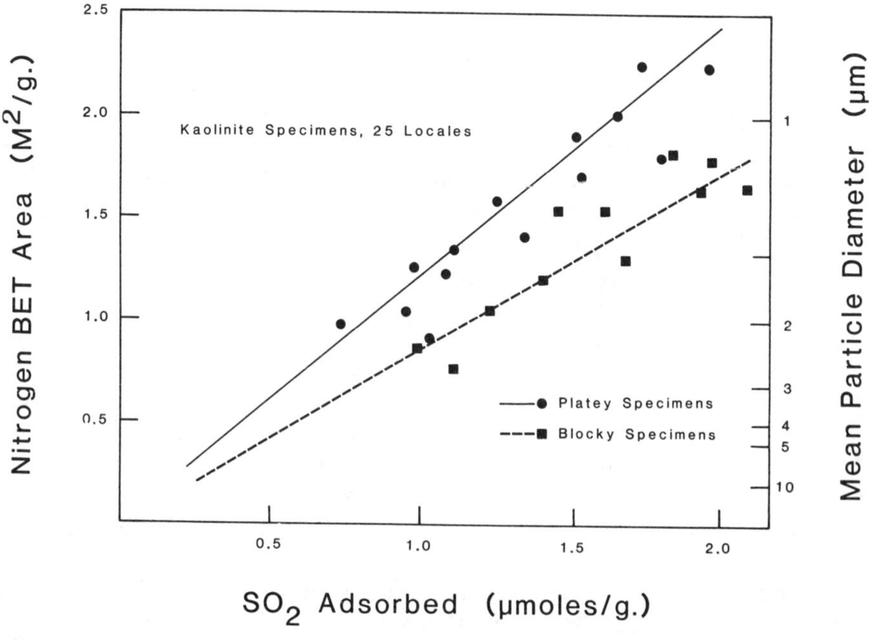

Figure 13.9 Sulfur dioxide and nitrogen adsorption isotherms on kaolinite.

planar and polar (1.6 Debye) and may diffuse at least part way into incipient cracks remaining from milling stress rupture.

High-magnesium-containing calcites appear to adsorb greater quantities of SO_2 per unit mass by about 8%. This may derive in part from greater crack formation due to embrittling of the crystal structure by magnesium substitution or be due to greater numbers of cations per unit surface area because of magnesium's smaller ionic radius.

Irrespective of adsorption source, both carbonates demonstrate crack closure and healing during the aging period, likely from interaction with atmospheric water. Such differences in surface chemistry, which affect dispersant demand and activity, reflect significantly on rheological characteristics in applications.

Sulfur dioxide, as noted earlier, chemisorbs onto calcite via formation of anhydrous calcium sulfite. This species hydrates rapidly upon exposure to atmospheric water vapor:

$$CaCO_3 + SO_2 + 2\ H_2O \rightarrow CaSO_3 \cdot (H_2O) + CO_2 \qquad (13.5)$$

Sulfur dioxide has problems in oxidation–reduction reactions on some surfaces due to the strained molecular adsorption configuration where it does not chemisorb.

Note in Figure 13.9 that the ratio of sulfur dioxide adsorption on kaolinite to surface area is also nearly constant but much less than with calcium carbonate. The adsorption profile follows a two-stage Langmuir adsorption isotherm, at least up to about $P_{sat}/2$, rather than a typical BET one. The initial components in the isotherms are the ones computed for the figure. Based on SEM data already discussed, about 15–18% of the pigment's area derives from edge facets. Sulfur dioxide adsorption values suggest that the bulk (~90%) of this first-phase adsorption, though not all, can be allocated to edge area, one molecule per one edge molecular unit.

The second isotherm component corresponds to about half that of a BET value for these specimens. Thus SO_2 adsorption likely occurs first at $-Al{=}(OH)_2$ groups, those especially active members along edge facets, followed by the more passive basal face units. This is proposed, based on alumina hydrate groups appearing alkaline (compared with silanols), to Lewis acid sulfur dioxide molecules. Further, this is borne out by the blocky specimens, which display a substantially higher edge area fraction and which adsorb nearly 40% more SO_2 per unit area than the platey specimens. This technique carries the potential to determine average particle aspect ratio without resorting to tedious shadowed electron micrographic measurements.

An organic molecule, acetyl acetone (2,4-pentanedione), also has been employed to assess alkyl *M*–O–H groups in pigment surfaces at room temperature as a gas (T_{bp} = 139°C). The values obtained are usually 10–20% lower than by wet chemical methods. Actyl acetone has several problems as an alkali-indicating adsorbate, including its ability to complex many metals by entering into *d* orbital chelation.

This section has dealt exclusively with the acid–base properties of solids in liquid suspension. Where polymers, especially alkyd resins and partially polymerized acrylics, have intentional acidity, this information is most often supplied by or available from the manufacturer. For the formulator it is equal in importance to solid surface characteristics. Polymer residual acidity (or basicity) may react with the solid or may compete with the dispersant for surface sites. It can be determined by titration, either in water by the emulsification method employing a pH meter or in alcohol or alcohol–water employing colorimetric indicators. The general technique can be found in any text on organic compound identification and characterization. The specific methodologies will not be taken up here inasmuch as polymer acidity is a common parameter whose specification is usually available from suppliers.

13.3 Viscometry and Rheometry

13.3.1 Static (Bench-Type) Units

Two tools necessary in every dispersion laboratory are viscometers and rheometers. While both measure viscosity, the latter is defined as a more automated, usually autorecording, viscometer with a rather broad viscosity range. Because viscosity is resistance to flow (or a measure of internal fluid friction), useful methods of measurement and units thereof are many. Tables and, more recently, a slide-rule calculator,[15] have been developed to interrelate this jungle[16] of units and numbers. Table 13.2 is a comparison chart of the various viscosity units for Newtonian fluids.

The centistoke* is a standard viscosity unit. Centistokes times density (sp. gr.) yields centipoise. Centipoise is an absolute metric unit (0.01 gram per centimeter per second). Other units are instrumentational or relative, geared to specific dynamic phenomena.

The broad gamut of designs and complexities of viscometers and rhcometers in the marketplace go well beyond the limits of this book. Included are selected examples and general instrument descriptions, as well as functions, for the benefit of technicians not well acquainted with the field of viscometry.

The initial question is not what instrumentation to employ but what information will be derived by a viscometer to assess dispersion. Ability to flow may not be fully related to discreteness of particulate dispersion. Processes where flow is of paramount importance (spray paints, pumping, surface coating) may

---

* Named in honor of Sir George Stokes, English physicist and engineer who discovered the constant velocity principle of sedimenting (falling) solids in water in 1886. Stokes designed equipment on the third floor of his university building with large, water-filled tubes extending down through the intermediate floors all the way to the basement. Stokes would release a weighed, measured ball, race down the three flights of stairs, and time its descent. For his efforts (scientific and athletic) in sedimentation and liquid flow he was knighted. English journals never explain why Stokes did not station a student on the ground floor with a watch and call from the window when he released the ball!

Table 13.2 Viscometer Comparison Chart

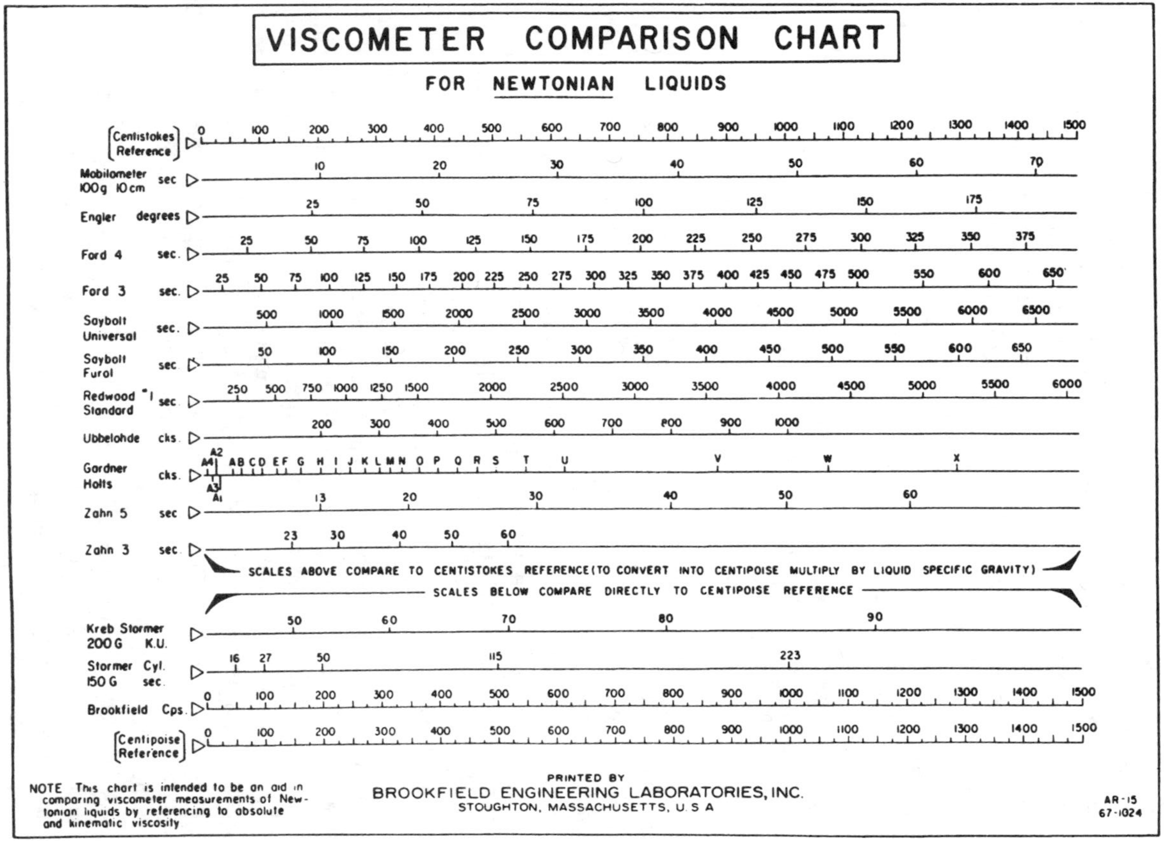
VISCOMETER COMPARISON CHART
FOR NEWTONIAN LIQUIDS
Centistokes Reference 0 100 200 300 400 500 600 700 800 900 1000 1100 1200 1300 1400 1500
Mobilometer 100g 10cm sec 10 20 30 40 50 60 70
Engler degrees 25 50 75 100 125 150 175
Ford 4 sec. 25 50 75 100 125 150 175 200 225 250 275 300 325 350 375
Ford 3 sec. 25 50 75 100 125 150 175 200 225 250 275 300 325 350 375 400 425 450 475 500 550 600 650
Saybolt Universal sec. 500 1000 1500 2000 2500 3000 3500 4000 4500 5000 5500 6000 6500
Saybolt Furol sec 50 100 150 200 250 300 350 400 450 500 550 600 650
Redwood #1 Standard sec 250 500 750 1000 1250 1500 2000 2500 3000 3500 4000 4500 5000 5500 6000
Ubbelohde cks. 200 300 400 500 600 700 800 900 1000
Gardner Holts cks. A4 A2 A3 A1 A B C D E F G H I J K L M N O P Q R S T U V W X
Zahn 5 sec 13 20 30 40 50 60
Zahn 3 sec. 23 30 40 50 60
SCALES ABOVE COMPARE TO CENTISTOKES REFERENCE (TO CONVERT INTO CENTIPOISE MULTIPLY BY LIQUID SPECIFIC GRAVITY)
SCALES BELOW COMPARE DIRECTLY TO CENTIPOISE REFERENCE
Kreb Stormer 200G K.U. 50 60 70 80 90
Stormer Cyl. 150G sec. 16 27 50 115 223
Brookfield Cps. 0 100 200 300 400 500 600 700 800 900 1000 1100 1200 1300 1400 1500
Centipoise Reference 0 100 200 300 400 500 600 700 800 900 1000 1100 1200 1300 1400 1500
NOTE This chart is intended to be an aid in comparing viscometer measurements of Newtonian liquids by referencing to absolute and kinematic viscosity
PRINTED BY
BROOKFIELD ENGINEERING LABORATORIES, INC.
STOUGHTON, MASSACHUSETTS, U.S.A.
AR-15
67-1024

sacrifice perfect discreteness for application character. Thus tradeoffs may exist between job (or function) effectiveness and finished product quality.

Second, viscosity often reflects variability or inconsistency in a formulation, either as a function of time or between batches. Such parameters as pigment settling, dispersant breakdown (or loss), binder resin polymerization, and aging changes to both the solid (from grinding) and the dispersant (chemical degradation) are sensitive factors for viscosity. While a change in viscosity does not immediately signal a change in a specific parameter, it is a strong indicator that problems are developing within the system. Good laboratory or quality control viscometers will have several virtues:

1. Portability;
2. Simplicity in operation;
3. Reasonable durability;
4. High sensitivity to viscosity change;
5. Broad range capability;
6. Accuracy and precision; and
7. Immunity to secondary parameters.

Viscometers vary in complexity from simple handheld or benchtype units for production control to online units whose output is directly coupled with a system control operation (dispersant feed, grinding time, agitation time or intensity, component addition, etc.).

Because viscosity is a kinetic property (as opposed to charge, surface area, and particle size, which are static properties), and insofar as the ratio of shear stress to shear rate is constant, viscosity will be proportional to flow time through a fixed orifice for a fixed volume. Thus the most elementary viscometer is but a cup of known volume with a small hole in its bottom.

Such simple devices are highly useful on production lines and include the Zahn and Ford cups.[17] The Zahn cup resembles a small soup ladle with a precision hole at the center of its base. The cup, with a long handle attached, is dipped into the slurry or dispersed formulation, removed quickly, and time to empty measured by a watch. Slurry running down from the cup's exterior contributes a negligible error as it drains more quickly than the cup's contents. These "instruments" are to slurry formulators what the Hegman Gage is to milling operators—simple devices kept in a drawer to perform periodic quality control on the production line.

A Ford cup is similar to a Zahn but is often employed as a benchtop instrument. It consists of a cylinder fitted to a cone with a precision drain channel (rather than a simple hole) and may be used with an extension handle for dipping or as a tabletop unit with legs attached, as shown in Figure 13.10. All such cups have a specific hole diameter for a specific viscosity range and a fill line on the container's inside surface (top edge for the Zahn cup). The time for liquid to drain from the cup (to the point where stream surface tension breaks) is measured. The cup technique is simple, easy to measure (watch second hand), inexpensive, and involves small, portable equipment. Zahn and Ford cups are made

Figure 13.10 Ford drain cup for production viscosity measurement (courtesy Paul N. Gardner Co.).

in at least five different sizes (having different hole diameters) for a range of viscosities from about 10 to over 1500 centistokes. A 1200 centipoise calcium carbonate slurry, formulated for small-media milling, will drain from a #5 Zahn cup in about 30 sec.

The formula for calculating viscosity from drain time is:

$$\text{Viscosity (cp)} = \frac{\text{sp. gr.} \times AT - B}{T} \tag{13.6}$$

where T is the drain time (sec), $A = 27.27$ (#5 cup), and $B = 540$ (#5 cup).

Numerous assumptions are made in analysis by time (Newtonian character, humidity, metal wetting, to name a few), but the method is far superior to racing up and down three flights of stairs for each measurement! Several manufacturers of viscosity-by-drain-time cups exist, each basing their equipment on the same principle. Zahn cup viscosity data are included in the conversion chart in Table 13.2.

A major improvement over the cup is the bench-type viscometer, more automated, more accurate, having a much broader range, but requiring greater sample volume. Several of these exist (manufacturers are listed in Appendix B), but the best known is the Brookfield shown in Figure 13.11. Viscometers employ a precision drive shaft with a shielded chamber connected to the slurry-immersed disc (spindle) by a small, sensitive, coiled spring. The drive shaft rotates at constant velocity to provide constant shear at the disc interface with the slurry. Drag force created by the slurry on the disc retards this portion of the shaft, and the degree of retardation is monitored by a needle attached to the spring-coupled wheel. If the drag force is zero, the immersed disc will rotate with no retardation (spring tension = zero). The instrumentation is based on the concept that torque necessary to rotate a disc at constant speed is directly proportional to the slurry's viscosity.[18] The degree this slurry-coupled wheel lags behind the directly coupled wheel is directly proportional to drag force. Each disc has a calibrated

Figure 13.11 Bench-type viscometer (courtesy Brookfield).

constant, and the viscometer has several fixed operating speeds (e.g., 10, 20, 50, 100 rpm).

Viscosity is calculated by

$$\text{Viscosity (cp)} = K \times \text{rpm} \tag{13.7}$$

where K is the disc or spindle constant. The rate of shear in the dispersed slurry is provided by the differential, change of velocity divided by change of separation distance (fixed plane to shear plane). Its units are in reciprocal seconds.

For very low-viscosity measurements, instruments employ a very large-area disc (large diameter) or even a cylinder (to increase drag area), while high-viscosity measurements utilize a low-diameter–low-area disc or small cylinder.

The rheometer is a more sophisticated instrumental version of the benchtop viscometer and usually is programmable for a constant rate of shear increase. The output is linked to a computer where the various rheological data are displayed by way of appropriate software. Because rheological applications vary over such a great range, many rheometers have mechanical discs, drums, and sensors that can be removed and replaced to provide broad coverage. One unit is shown in Figure 13.12.

For rheological profile determination, a slurry sample of 500 to 1000 cm^3 with a starting concentration of dispersant is measured. To the slurry is then added an additional aliquot of agent. Where the dispersant added is on a 100%

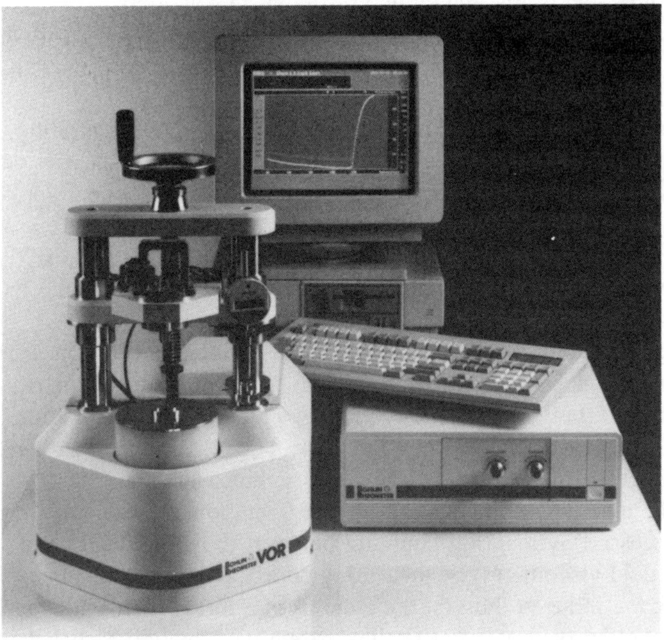

Figure 13.12 Continuously variable-shear rheometer (courtesy Bohlin Instrument).

basis (solid powder or pure liquid), any volumetric increase is trivial. If dispersant is added as a dilute solution, the solids level decreases slightly. This alters the interpretation, but usually only at critical solids levels. The slurry is brought to a standard temperature, usually 25°C, prior to measurement and viscosity is determined in a constant-temperature bath or insulated flask.

A similar technique utilizes the increase of solids level by dry powder addition. Usually a fixed dispersant level is employed, and the system must be removed with each powder increment and reagitated to obtain dry powder assimilation. This proceeds until the solids level rises sufficiently high that aliquots of dispersant must also be added. Such a method results in a two-parameter experiment, which cannot be graphed readily except to interpret maximum pumpable solids.

Many bench-type viscometers have multiple (and some programmably variable) speeds. Certain slurries (usually high-surface-area solids and/or underdispersed ones) exhibit a thixotropic behavior with a short-term memory. Thus viscosity is measured at lowest shear (rpm) first, followed by progressively increasing shear. Reversing this process will yield a time-dependent hysteresis, useful in itself to assess slurry structure stability characteristics.

Certain instruments employ a constantly increasing rpm to reach the instrument's maximum shear capability followed by an equally decreasing speed. Drag is plotted directly upon a graph to show thixotropy (the ratio of drag force to shear decreases with shear) and dilatancy (the drag force to shear ratio increases with shear).[19] This method also yields a measure of interparticle structure stability, as indicated by the magnitude of the hysteresis loop.

More sophisticated laboratory-style viscometers employ sensitive pressure transducers, which replace the spring–needle arrangement, and an electronic digital readout.

Viscometers may be adapted with a plate and cone configuration, as shown in Figure 13.13, to enable measurement at extremely low shear rates, below 1 reciprocal second (to 0.001 sec^{-1}), as encountered in low-shear applications of slurries[20,21,22] (drainage flow, certain brushing applications, etc.). With such instrumentation, the motor shaft is connected to a cone having a shallow slope, usually less than 5 degrees. The cone's center nearly touches the fixed planar plate.

The instrumentation geometry is predicated on the fact that shear rate is tangential velocity dependent. In moving from the shaft center to the rim, although rpm is constant, tangential velocity increases directly with the radius. However, due to the conical shape of the slurry, shear rate decreases with radius, also linearly, because the film thickness increases linearly. Thus, at any element of area on the disc, shear is constant for a set rotational speed. Integrated torque may be assessed by a spring coupling between the motor and cone (similar to the Brookfield viscometer) by measuring cone torque with the plate fixed, or by measuring the induced drag on the plate from a direct motor-attached cone.

Viscosity is measured, as with the rotating disc method earlier described, by the resistance to flow offered by the thin disc of liquid. The mathematics for this

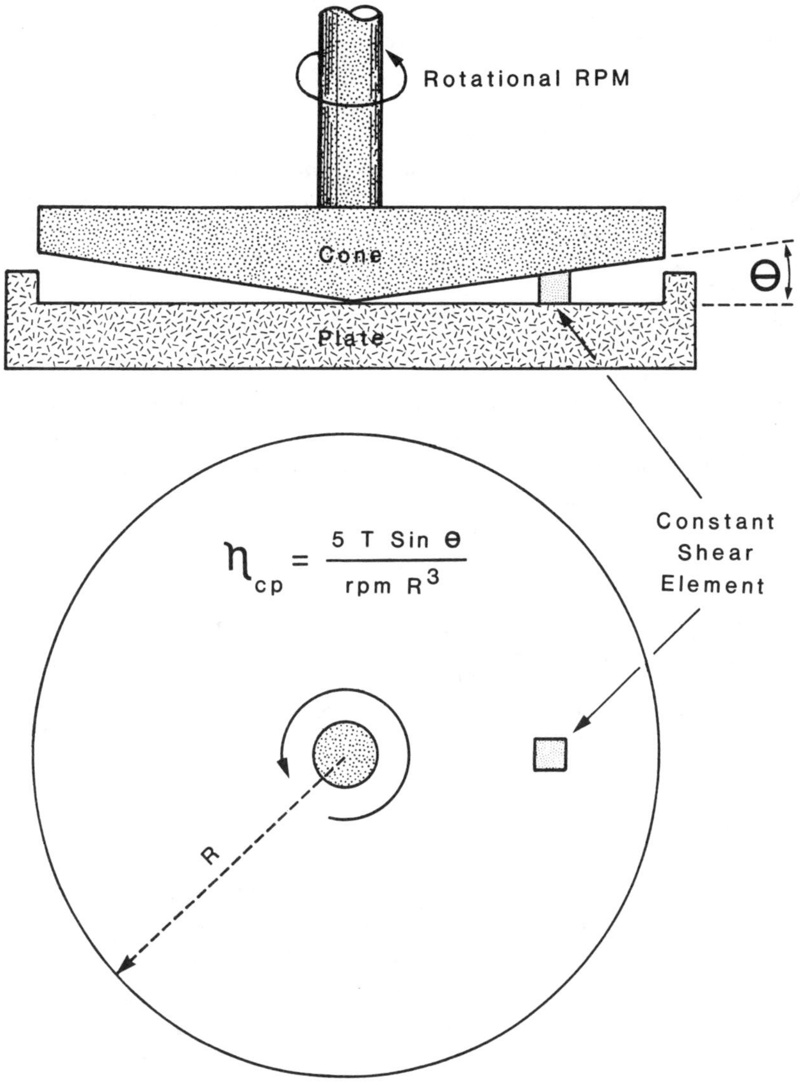

Figure 13.13 Plate and cone arrangement for low-shear viscometry.

type of instrumentation is moderately simple and is based on observable shear stress (torque over liquid volume) divided by shear rate[23] (rotational speed over the sine of the cone's angle):

$$\text{Viscosity}_{(\text{cp})} = \frac{5 \times \text{torque} \times \sin\theta}{\text{rpm} \times \text{radius}^3} \tag{13.8}$$

When torque is in dyne centimeters and the radius in centimeters, viscosity is in centipoise (grams per centimeter per second/100). Many cone and plate instruments are constructed with programmable speeds and electronic output to provide a broad range of recorded shear, hysteresis, and other viscodynamic properties.

Plate and cone design instrumentation is useful in assessing the influence on dispersive states during low-shear activities, thick slurry settling, viscous polymer sag (as with layup resins for molding and coating),[24] putties, and food products (jellies, syrups, slurried spices).

13.3.2 Dynamic (Online)-Type Units

Many viscometry investigations are made with the instruments described previously. However, for certain commercial applications, the stability or change of fluid flow viscosity must be known either on a frequent basis or in less time than sampling and manual process control permit. Either requirement negates the practice of extracting representative samples, carrying them to a laboratory bench, and conducting viscometry.

Online viscometers have several features not employed with bench-type units. They must be self-purging, unaffected by mass motion of slurry, and automatically temperature compensating. As a consequence of these needs and their time-of-flight data production, some sacrifice in accuracy is usually made, but the degree is not serious.[25,26] Although many methodologies exist for measuring viscosity in dynamic or kinetic fluids, the following four are the primary techniques:

1. Vibrational reed;
2. Capillary flow rate or pressure drop;
3. Plunger/piston displacement; and
4. Rotational disc.

The vibrational reed method sketched in Figure 13.14 employs a small-diameter rod or reed set into oscillation at a resonant frequency or one controlled by an electromagnetic driver with a frequency generator. Frequency is usually a few hundred hertz and may be designed around a harmonic of line frequency (60 hertz). Vibration imparts a shear wave into the slurry, shear being proportional to frequency. The viscosity of the fluid in the immediate vicinity of the reed damps reed movement. With higher viscosities the wave front penetrates further into the slurry, and the viscoelastic effects increasingly slow the reed movements. Two methods may be employed to sense this effect. Additional current may be supplied to maintain the frequency,[27] or the change in frequency may be sensed directly. Either technique induces an electrical signal into a sensing coil around or adjacent to the base of the reed. This signal is electrically processed and sent to a readout and to a control loop circuit (4–20 mA output).

Where the liquid flow displacement is greatly less than reed movement, readings are independent of flow changes. At 400 Hz and a 1 mm reed displacement, the movement is approximately 80 cm/sec. For slurry velocities exceeding this

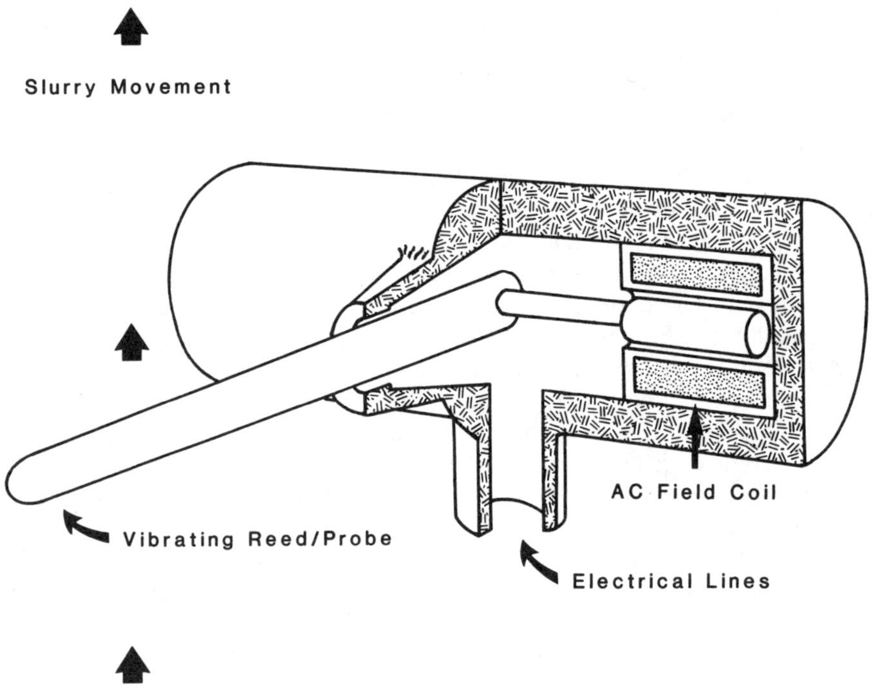

Figure 13.14 Vibrating reed instrumentation for in-line viscosity measurement (courtesy Dynatrol).

value, the reed must vibrate faster or perpendicular to the flow axis. An alternate procedure involves construction of a bypass loop with an enlargement section in which the viscometry measurement is conducted. An electronic thermometer in the probe corrects the output for thermal variations of the slurry.

Different reed sizes and probe frequencies can be employed to cover a broad viscosity range. However, any hysteresis within a fluid will produce the sheared value as a readout because reed oscillation time will be much shorter than slurry relaxation time. Also, frequencies are limited because of thermodynamic heating effects and cavitation phenomena.

Capillary in-line viscometers utilize a split stream accelerated through a capillary tube with a metering pump. At constant pump pressure the volumetric flow rate through the orifice is a direct measure of viscosity. The sensor output may be based on a measurement of flow (ultrasonic transducer) or by pressure drop (differential manometer) down the capillary. To change ranges, capillaries of different diameters are installed in the instrument. For continuous flow streams such systems rarely require cleaning unless undispersed particulate enters the capillary or its walls become slowly abraded from hard particles. Thereafter, periodic cleaning must be employed to keep the capillary surfaces particulate or flocculant free.

The falling plunger, or piston type, measures the time required for a given displacement of a fixed element of volume (piston). Slurry must bypass the piston in a loose-fitting, restricted cylindrical volume between the piston and its cylinder. Piston movement is usually perpendicular to fluid flow to eliminate flow pressure components. Either applied force or translation time is held constant, with the other being the dependent variable. The time for this slurry displacement increases directly with viscosity in the fashion of a powered Zahn cup. By changing the force applied, shear can be set and the instrument employed over a range of viscosities. However, displacement techniques are restricted to low-to-moderate viscosities. Displacement instrumentation requires moderately high maintenance because of periodic clogging and wear replacement.

A variant of this type of instrumentation is the orifice viscometer, in which pressure across a fixed orifice is applied, the pressure drop is measured, and the value mathematically related to viscosity. Such a monitor is a hybrid between the capillary and piston techniques. Different orifice diameters (usually several millimeters) allow a range of viscosities to be measured, from a few thousand to about 100,000 cp.

Rotational in-line viscometers employ methods similar to bench-type units. Most employ a free-moving cylinder concentric with a rotating drum armature. The drag force of the liquid, created by its viscosity, induces a set coupled cylinder velocity. Increasing the armature velocity and decreasing the separation distance between surfaces increases the shear. Velocity is a direct function of viscosity and can be plotted on a recorder. Many units operate solely at constant speed. However, some units employ a ramped armature velocity to obtain a measure of non-Newtonian character. The time to complete the shear range must be less than any viscosity fluctuation within the slurry. Because of constant liquid displacement between the surfaces, hysteresis cannot be obtained by a reverse reduction in rotational velocity. Also the instrumentation tends to find its major applications at low shear rates.

Most in-line instrumentation is designed to sense small variations within a stream having a limited viscosity variation range. Therefore, a full informational package on hysteresis, dilatancy, thixotropy, etc. can be sacrificed for operational quality control of a set product. In some applications the feedback loop output is employed to control dispersing agent addition, solids introduced into a mixer, residence time in a blender-mixer, or other process variable necessary to produce a consistent commercial product (paper coating machine, ink manufacture, food processing, cosmetic cream formulation, paint supply to a production line spray unit, etc.).

13.4 Interpreting Rheology Profiles

Interpreting rheograms is the single most important function of viscosity measurements. While a brief description has been made pertaining to basic dynamics in Chapter 2, a more detailed example will be given here. It is not possible to understand or work with rheological systems based solely on a single viscosity measurement. Thermodynamics and flow mechanics contain far too many variables and influential factors to be interpreted with such limited data assessment. For aqueous dispersions of solids, the multipointed rheogram may be considered to have three components, a descending element (underdispersed state), an ascending element (overdispersed state), and a trough with measurable breadth (region of workable viscosity). The steepness of the two elements or arms and the breadth of the trough provide considerable information on the interaction of dispersants with solid surfaces and their ability to overcome van der Waals forces between particles and to effect repulsion. The following discussion is illustrated by two representative rheograms in Figure 13.15.

When complex slurries (multiple solids, multiple dispersants, multiple morphologies) are measured, the ability of anisometric particles to align with system shear is a measure of shear response, while their inability to align at a critical level of applied shear appears as sudden dilatancy. Chapter 3 explained that a balance or equilibrium frequently exists between dispersant adsorbed (solid phase) and dispersant dissolved and solvated (liquid phase). As the dispersant requirement increases with solids, this equilibrium shifts to the solution phase. Eventually a condition arises where immobility of the liquid phase in conjunction with the inability of particles to orient along lines of shear stress dramatically reduces particle mobility under applied shear. The slurry suddenly (usually

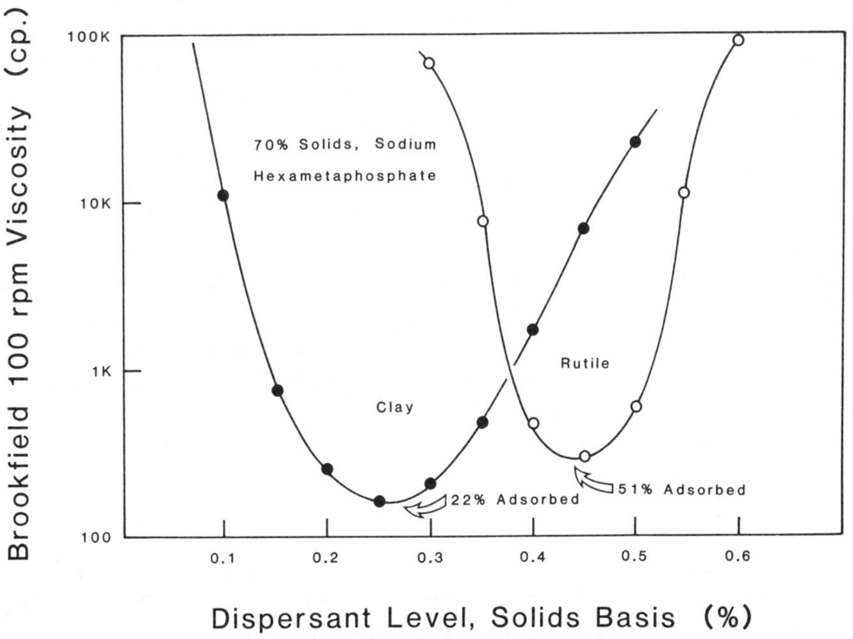

Figure 13.15 Typical pigment slurry rheograms.

$\pm 0.2\%$ solids differential) transforms from a creamy suspension to a solid. The ability of a particle system to flow is a volumetric function, the volume occupied by solid particles, and the free volume of liquid in which these are suspended. Conventional rheograms plot viscosity as a linear function against weight percent solids, also as a linear function. If plotted rather as a cubic function, or even on log–log coordinates, the change can be seen to be more of a continuous progression. The slope change at critical solids, however, is intensified by an increase in anisometry of the suspended particles.

With most ionic dispersants an equilibrium exists between surface and solution. At the commencement of the descending viscosity arms, the dispersant is nearly quantitatively adsorbed on both solids in the figure. The slopes, $d\eta/dC$ (where C is the dispersant concentration added), are nearly vertical. As additional aliquots of dispersant are added, progressively larger portions shift to the solution phase. In this range the slope diminishes drastically, eventually approaching zero (becoming horizontal). With progressively larger additions of dispersant, the agent shifts predominantly, then almost exclusively, to the solution phase, where it immobilizes water by solvation. The viscosity then begins to rise (ascending arm) for both materials. Because two separate phenomena are at work, the descending and ascending arms are rarely symmetrical.

Surface affinity for the dispersant can be measured by the curve steepness during descent and rise. Note in Figure 13.15 that titanium dioxide displays a polyphosphate (SHMP) viscosity curve that descends (underdosage) similarly but rises (overdosage) more steeply than does kaolinite. Further, the ascension is even steeper than the descension. The surface area for the titania is greater by nearly a factor of ten than for the kaolinite (36 vs. 3.9 M^2/g), yet the dispersant demand is only about twice. This in itself suggests the titania surface displays a higher affinity for SHMP than does kaolinite. Finally, the base of the rheogram is much sharper for titania. The particular titanium dioxide investigated was coated with alumina hydrate gel (and probably a small amount of zinc hydroxide). This coating apparently exerts a stronger affinity for phosphate than does the aluminum silicate.[28,29] Studies on the chemisorption of titanium dioxide at the viscosity minimum show that at equilibrium slightly under half the SHMP dispersant is shifted into solution. With most kaolinites the amount is over 75%.

Where the dispersant adsorbs essentially quantitatively onto the solid surface (as with alkanolamines on acidic substrates), the descending portion of the profile is steep, almost to the trough. Where dispersant equilibrium conditions exist, it is more shallow (curving toward higher concentration). For highly charged dispersant species the base, or trough, of the curve, is narrow. Such conditions make dispersion in critical viscosity operations (grinding, pumping, etc.) a more precise undertaking. If the dispersant has low solvation energy, the trough is much more broad.

The ascending arm is highly solvation dependent. Where the ionic dispersant continues to shift to the solution phase, the rise is steep, similar to the descending arm. Where an ionic dispersant is more tenaciously adsorbed, this rise is much less steep. Finally, where nonionics are employed and these are nearly quantitatively adsorbed, the ascendant curve is very shallow, not greatly removed from horizontal. A common corollary of dispersion is that shallow ascendant arms relate to broad viscosity trough minima. The two actions derive from similar solution mechanics.

The choice, where one must be made, between over- and underdispersion usually lies with the former. This rationale is based on the probability of dispersant being lost with time (surface decay, volatilization, bacterial breakdown) and excess concentration compensating by shifting the rheology downscale. If the choice must be made during a milling exercise, newly created surface will consume the excess. Underdispersed slurry systems have very few justifications for their existence because almost all chemical alterations adversely influence viscosity.

Where dispersing agent breaks down by any of the listed processes, the curve trough rises and narrows because nonfunctional chemicals are flushed into the aqueous phase to solvate water molecules further.

In finished formulations (paint, ceramics casting slurries, paper coating slurries, etc.) the rheological profile serves yet other purposes including drip potential, mold set speed, leveling capability, ease of spread, sag, etc.

Because viscosity is the ratio of shear stress to shear rate, a plot of viscosity (in poise) against shear rate (in sec^{-1}) should be a straight line, the slope of which is internal shear stress. For real world systems log–log paper is used, and the slope should approximate a 45 degree line (negative slope). With pure liquids this occurs. For thixotropic systems this curve breaks from linearity and approaches zero. With slurries having long recovery times (time to return to their original distribution and orientation following shearing), a determination of the breakpoint (which is gradational, not sharp) of a quiescent formulation immediately followed by rapid shearing, is a measure of drag, whether pumped, brushed, or blade coated. The employment of different dispersants or rheological additives will have a dramatic influence on these curves and will allow adjustment and tailoring of slurry systems for varied applications while retaining advantageous optical properties.

With dilatant systems the 45 degree line suddenly turns down vertically and approaches negative infinity. The determination of these curves with isometric–anisometric pigment mixtures, mixed or varied dispersants, and solids level will be a measure of the applicability of the formulation to a broad range of uses. With digital readout viscometers, data can be taken directly from the instrument and fed into a computer, then processed to reveal these phenomena clearly.

Viscometry can be employed for assessment of other application properties, as in the determination of leveling behavior at very low shear,[30] in the determination of dispersion stabilities,[31,32] and in conjunction with optodynamic measurements of particle orientation during flow[33] (near pipe walls, adjacent to impeller blades, velocity increase through constrictions, etc.).

Several ASTM standardization procedures have been adopted to measure and report rheological phenomena. These include ASTM D-562 (Stormer use) and D-2196 (Brookfield use).

13.5 Sedimentation Techniques

One of the simplest laboratory techniques for assessing dispersion quality lies in the use of sedimentation tubes. These can be as simple as a graduated cylinder or, where greater accuracy is demanded, the American Petroleum Institute's (API) sedimentation tube. The method embodies placing a well-agitated and dilute slurry of the solid in the liquid of choice with a given agent, pouring the slurry into a calibrated tube, and allowing it to sediment under gravity to a constant volume. This value is termed the *terminal sediment volume* (TSV). Dividing the TSV number by the solids true volume (weight divided by density) yields the terminal sediment volume ratio, or relative sediment volume, a true measure of packing fraction.

With 100 cm^3 of liquid at 1–2% solids, sediment volume is rarely over a few cubic centimeters. API tubes, as well as the common laboratory graduated cylinders, have 100 cm^3 capacity. While cylinders can be read with an accuracy of only about ± 0.5 cm^3, API tubes can be read to at least 0.05 cm^3. A standard API tube is shown in Figure 13.16. The method is equally applicable to aqueous and nonaqueous dispersions.

Typical dispersant level is high, based on surface area, usually about 10 weight percent of the solid. Overdispersion is not a problem because of the

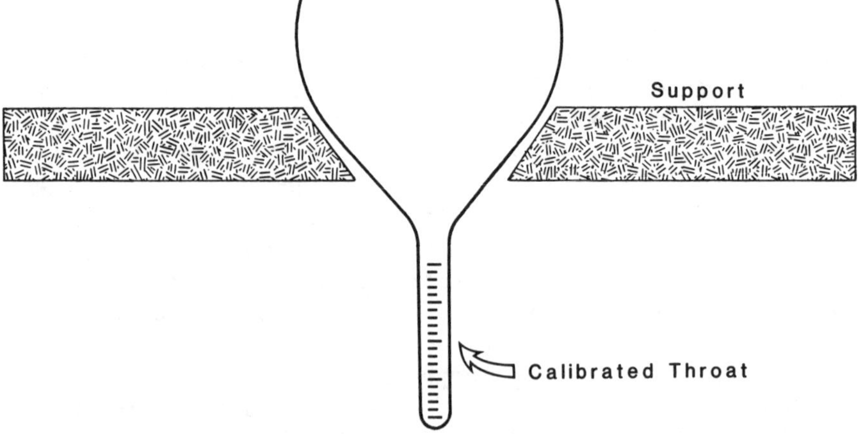

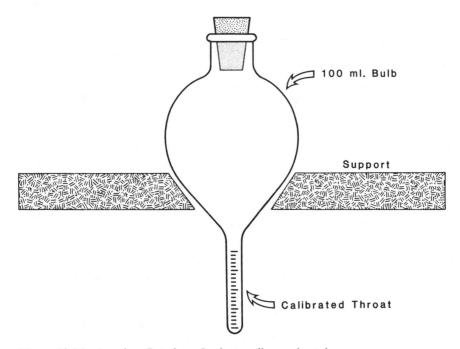

Figure 13.16 American Petroleum Institute sedimentation tube.

extremely low solids employed and low particle velocities. The particle separation and liquid envelope spacing with the technique is greater than 3 diameters. One exception to this practice lies with fatty acid and fatty amine treatments of powders for hydrocarbon dispersion. There the level in solution is limited to approximately 200% of the surface requirement, unless powder surface coverage is the specific parameter under evaluation.

An interpretation of the state of dispersion based upon sediment volume requires an understanding of the mechanics of sedimentation. For a log-normal size distribution of randomly packed spheres, final sediment will contain between 5% and 10% voids. Thus the terminal sediment volume ratio, equal to the measured volume divided by the true volume, will be about 1.1. For a monosize particle distribution, sediment will contain about 45–50% voids for a ratio of about 1.9. As particle anisometry increases (with needles or platelets), these values increase further. Because slurry formulated and introduced into the API tube is essentially homogeneous, initial sediment size composition is reflected by the true distribution of the particulate. However, as time increases, the sediment becomes progressively more monosize within any stratum of the sediment. A thoroughly dispersed, log-normal size distribution of platelets will sediment to an overall ratio of 2.0–2.1, needles slightly higher (2.5 and above depending upon anisometry), and spheres slightly less, generally about 1.7–1.8.

Fully flocculated systems will produce sediment volume ratios that are highly shape and distribution dependent, ranging from about 6 for spheres and equants to over 15 for platelets and needles.

The time required for sediment volume to assume a constant value is size dependent and is the chief drawback to the technique. A distribution ranging from 10 down to 0.1 μm requires a minimum of two weeks to sediment, time being dependent upon particle density. Time can be calculated from the Stokes equation[34]:

$$\text{Time (sec)} = 18nS/[\rho_s - \rho_l]\, gD^2 \tag{13.9}$$

where n is the viscosity (poise), S the distance (cm), ρ_s the density of solid (g/cm^3), ρ_l the density of liquid (g/cm^3), g the gravitational constant (cm/sec^2), and D the diameter (cm). With a 100 ml API tube (a maximum sedimentation distance of about 10 cm), a particle of 0.1 μm with a density of 2.5 g/cm^3 sedimenting in water requires a time of approximately 12 million seconds—over 4 months! In 30 days, however, 95–98% of the particles have settled. Unless extreme accuracy is required, this truncated period usually suffices for most evaluations, especially where simultaneous comparisons are being made with varying dispersants on a single solid.

Yet another approach to ascertaining terminal volume lies in plotting sedimentation rate (first derivative), dV/dT, against time or the reciprocal of volume against time, on log–log paper and extrapolating early data.

An alternate procedure, one of considerably shorter duration, employs a common laboratory centrifuge with a custom modified sedimentation tube (e.g., that of the MSA-Whitby instrument).[35,36] Such tubes hold about 10 ml and have

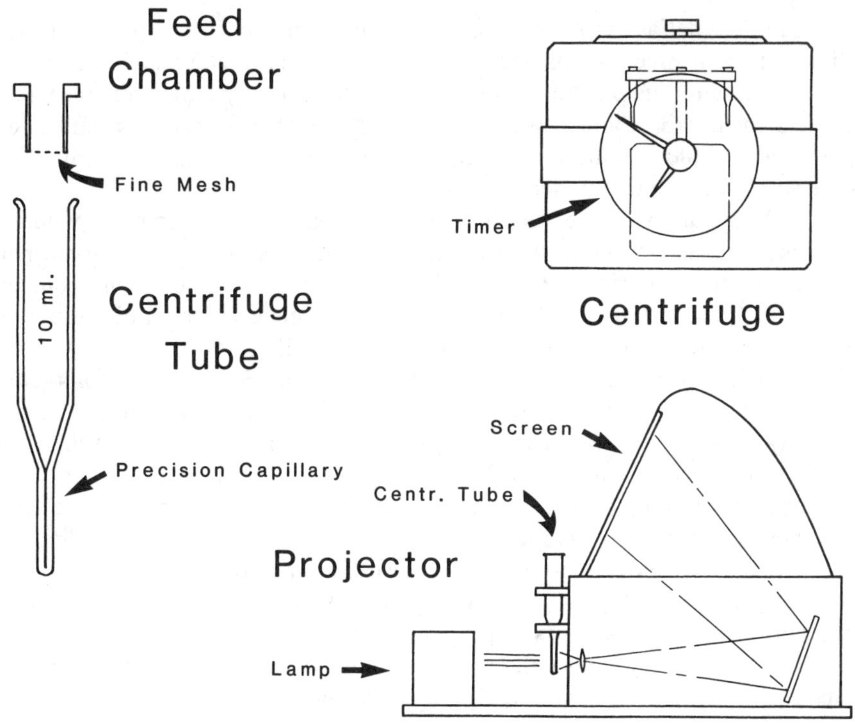

Figure 13.17 Sedimentation volume centrifuge (courtesy Mine Safety Appliances).

precision, calibrated-bore receptacles whose sediment volume can be measured with a magnifier or illuminated projector. One unit is shown schematically in Figure 13.17.

While time is shortened dramatically, typically to about 20–25 minutes, centrifugation above 10–15 *G*s artificially compacts flocculated structures (poorly dispersed) held together by relatively weak forces. Thus subtle differences which might exist within a broad survey of dispersants settling under gravity may be partially or wholly obscured. The MSA-Whitby laboratory unit functions best for this application if operated no faster than 300 rpm (12 *G*s).

Table 13.3 contains the results of a study with a 6:1 aspect ratio silicate platelet pigment dispersed with various agents in water and the sediment volume measured by the API tube technique. Potassium chloride is a common reference agent because it neither flocculates nor disperses oxide solids—neither anion or cation adsorbs to any significant extent. Any flocculation that may result derives from inherent particle charges or character.

Subtle variations between alkali phosphates show up with this technique, as do variations in chain lengths and active groups for the double dipole nonionic agents. Viscosity measurements at high solids (65–70%) may exhibit slightly

Table 13.3 Terminal Sediment Volume Estimates of Dispersiveness

Dispersion Agent Employed	Terminal Sediment Volume Ratio
Potassium chloride (ref. std.)	15.0
Hexamethylene diamine	13.0
Ammonium hydroxide	5.2
Trisodium phosphate	4.0
Sodium carbonate	3.9
Sodium orthosilicate	3.7
2-Amino 2-methyl 1-propanol (AMP)	3.6
Ethylene diamine	3.4
Hexanolamine	3.2
Sodium hexametaphosphate	2.3
Sodium polyacrylate (MW 2000)	2.2
Tetrasodium pyrophosphate	2.1

Solids concentration = 1.0% kaolinite; dispersant concentration = 100 μmoles.

different trends from the table due to interparticle interactions (hindered settling) but in general follow the pattern shown.

Particle wet out and milling effects also may be inferred from sedimentation volume measurements. In an investigation of milling methods versus liquid packing density, the hydrostatic technique was varied whereby slurries were mildly agitated above the sedimentation zone to maintain constant-distribution sedimentation.[37] This technique theoretically should yield sediment volume ratios approximating 1.0. Slurries were sedimented in water (with sodium polyacrylate) and in mineral spirits (with stearic acid). The results are given in Table 13.4. While the median sizes for the two powders are essentially equivalent, packing densities show variations derived either from morphology or distribution modifications in the milling operation.

While full interpretation of the several variables in this study would require additional TSV studies, the data illustrate the technique's value as a simple and economic tool for assessing variables relating to materials processing where dispersed states are goals.

Table 13.4 Sediment Packing Volume of Ground Calcium Carbonates

Grinding Mill Type	Liquid System Sedimented	Packing Density (g/cm^3)	TSV Ratio (rel.)
Ball mill	Water	0.759	3.5
Media mill	Water	0.990	2.7
Ball mill	Heptane	0.872	3.1
Media mill	Heptane	1.12	2.4

Feed median diameter = 7.3 μm; wt % + 44 μm = 0.9%.

13.6 Particle Size Analysis Techniques

Particle size analysis was described under specialty applications in Chapter 12 and to a lesser extent in Chapter 2. Those aspects of the methodology will not be repeated. What is important is dispersion assessment based on interpretations derived from particle size analytical data.

Methods employed in analysis may bias results because all reference size standards are common spheres, which few real-world particles resemble. Most methods employ the term "equivalent spherical diameter," with the various instrumentation techniques giving essentially equivalent results only with spheres. Sedimentation-based instruments (X-ray, density, and photon blockage) cause thin platelets and needles to appear as much smaller particles than their major diameters would suggest. Laser scattering instruments give an approximate average to the three ordinate dimensions. Thixotropy, which relies on fine particles (with their attendant surface area) and anisometry, cannot be predicted on the basis of these analyses.

Packing volume, which controls numerous optical, electrical, and mechanical properties of castings and coatings, cannot be predicted where morphology changes within a distribution (as with certain clays, zinc oxide, and carbon blacks). What particle size analysis does provide is an indication of uniformity between lots and a general concept of "average" size, a term in itself often nebulous.

To understand the intricate workings of particles in a fluid system fully, additional measurements must be made, including surface area, packing volume, and microscopy. Where a reasonable certainty exists pertaining to shape uniformity throughout the size distribution, the reduction of linear data to log size–weight probability plots provides a wealth of information, both on processing mechanics and in fluid flow dynamics. The method also becomes a highly sensitive tool for evaluating quality and uniformity of a given powder.

Most often the fine component in a size distribution will best reflect both dispersiveness and rheological influence. An effective approach involves assessing the median size (D_{50}) and ratioing this to an accurately defined fine component, typically a D_{25} or D_{10} value. Comparing this number with a theoretical log-normal distribution or a quality control standard provides a measure of dispersion assessment and/or rheological inference.

One grave danger in size analysis and interpretation lies in substituting new products for ones currently in use to achieve property equivalence, where size analyses have been conducted with different instrumentation or sample preparation procedures.

In a classical example of this a few decades past, a formulator replaced about half of a product component from Supplier A with a product having essentially identical color and particle size from Supplier B but at a considerable price reduction. As time passed B slowly displaced A competitively for total market share.

All went well until B was unexpectedly unable to supply product arising from plant labor problems. The formulator returned to A for the equivalent (original) component. B's product had been measured by instrumentation that gave a coarser size interpretation than that of A. Thus, to match A's theoretical size, B's product was actually 30% finer. While having equivalent spectral characteristics, the B product was much superior in light scattering (and hiding) because a far greater percentage of its particles lay in the optimum photon scattering range. After replacement with the coarser A component, the formulator's product quality and attendant sales dropped precipitously, resulting in a lawsuit being filed promptly against A for fraudulently stating equivalence in component matching.

Extensive litigation followed, best summarized as a technological draw, and particle size interpretation became the centerpiece of the courtroom drama. A technical witness in the proceedings contributed the wisdom of the moment with the opinion that particle size analysis bore a remarkable similarity to love making—it is not the elegance of the equipment that figures dominantly in the outcome but the skill and dexterity with which it is employed!

By establishing a relationship between size analyses by a set instrumentation method and microscopic measurements (shadowed transmission electron micrographs, or TEMs) for the same powders, particle size instrumentation can be given a more creditable role in predicting and controlling particle properties in final applications. Obviously, no size analysis is any better than its preparatory dispersion technique, and this consideration must be the first item on the quality control and quality assurance agenda.

13.7 Optical and Microscopic Methods

Numerous simple techniques that involve a direct observation of slurries are useful in evaluating dispersion and other flow properties. Microscopy employing polarized light to illuminate a liquid layer held between two glass slides (to minimize evaporation) enables an evaluation of aggregation and other interparticle association problems. The sample film thickness for this technique is critical. For dense slurries, individual particles within thick films are difficult to resolve, except with laser illumination, due to special diffractive effects. Very thin films have particle mobility more controlled by surface tension at the glass-slurry interfaces than by the more important hydrodynamic effects of wetting. This latter effect extends into the solution phase as far as a millimeter and can influence adversely many microscopic assessments.

In situations where dispersion of pigments and fillers are employed in film-forming applications, the degree of dispersion can be evaluated by optical density. Where dispersions are marginally stable, the optical density will change with time (nonsedimentation phenomenon). The quality of dispersion (Q) can be evaluated numerically by the equation:

$$Q(\%) = 100\,(D_0 - 0.3D_\infty)/D_0 \tag{13.10}$$

where D_0 is the initial dispersion optical density and D_∞ the optical density change at infinite time. The method, where D_∞ is approximated by a protracted period, has been found highly useful in the study of latex emulsion stability in paints, plastics, and food applications.[38]

A variation on the optical technique may be employed with a Wagner Sedimentometer, a simple cylinder having a light beam passing perpendicular to the sedimentation axis. A makeshift unit can be assembled readily with a standard graduated cylinder (or hydrometer cylinder), a fixed-position light beam (at about 30% of the slurry height up from the bottom), and a photodetector.

Stokes homogeneous sedimentation assumes the composition at this detection point remains homogeneous and constant until the largest particles have fallen past the point. Depending upon particle diameter and density, this requires several minutes, sufficient time for incipient floccules to initiate, if they are to form. Flocculation increases light transmission through the slurry, which the sensitive detector relays to a readout or recorder. Such techniques may be adapted to process streams (possibly with dilution) in the form of online dispersion sensors.

In another variation of this method, increments of slurry are diluted with potassium chloride (inert electrolyte), the system is passed through the aperture of a Coulter Counter, and particle population is counted with time. If flocculation commences, particle count falls quickly, and average size rises accordingly. This method, even though it involves expensive instrumentation, is unusually sensitive to flocculation, as it can detect variations of well under 0.1%. The reverse of the method has been employed to determine the state of dispersion as a function of agitator or ultrasonic bath time.

The state of dispersion with colored pigments is often assessed by direct film measurements of such factors as color strength with extenders or white diluent pigments, film gloss, transparency, reflectance at a specified wavelength, and internal texture.

The employment of or adaptation to laboratory instrumentation to estimate states of dispersion has been the subject of the International Standards Organization (ISO-9000) program to develop working standards based on relatively inexpensive and readily available equipment for networking laboratories and producer–client interactions. As this text is being prepared, only tentative (draft) standards have been promulgated.[39]

While photomicroscopy using high-speed film and high-intensity illumination yields excellent results in slurry dispersion assessment, both transmission and scanning electron micrographs (TEM and SEM) may contribute misleading results in the process of transferring slurry material to the dry state necessary for high vacuum in the electron microscope. The substrate onto which the slurry is placed must have a very low angle of wetting to ensure spreading, rather than constriction. Metal SEM posts and glass optical microscopy slides require extensive cleaning and prewetting with an appropriate surface tension reduction agent (anionics, alkanolamines) to ensure the dispersed state of the slurry being applied is not impaired.

Cellulose ester films cast from organic solvent onto water (e.g., collodion, cellulose acetate, etc.) are only slightly hydrophilic. Dispersed slurries, to which is added an additional quantity of a medium-chain alkanolamine, wet this film well, allowing particles to sediment or affix themselves at representative locales. Polyacrylate dispersants are somewhat less effective, and polyphosphates are prone to zonal coalescence.

Studies on sample surfaces have shown well-dispersed slurries, as assessed by sedimentation practice, to exhibit large floccule formation in optical and electron micrography due to poor film surface preparation.[40] Micrographs obtained from such specimens indicate a poorly dispersed state, formation of flocculi and coalescence zones, none of which is present in the bulk slurry. Preparation artifacts may suggest face-to-face flocculation of platelets, which in fact arise from film–sample surface tension factors. These do not form during free-environment flocculation, based on sediment volume studies.

Electron micrographs prepared by the technique of dusting specimens dry onto sample posts or slides (air dusting) are most misleading with regard to the association between mixed pigments and the form floccules take. Figure 13.18 is an example of tabular alumina prepared in this fashion. The suggestion the

Figure 13.18 Preparation influence on alumina hydrate aggregation.

micrograph gives that particles occur in tightly flocculated or otherwise aggregated bundles in a dispersed state is most likely false.

For accurate particle assessment, including diameter, aspect ratio, crystallinity, and other morphological measurements, particles must be fully separated on the plate, with an area coverage of no more than 15%. Shadowing at this population density provides at least 90% measurable images.

13.8 Calorimetric Evaluations

Several references have been made throughout this text to heats of wetting, predominantly by pure liquids, but secondarily by dispersant-bearing solutions, on solids of interest. Most calorimetry is performed either in academic circles or by pigment suppliers to tailor their products better to specific applications and industries. Commercial calorimeters of the type needed for solid powders have few manufacturers. Two are shown in Figures 13.19(a)[41] and (b).[42] Some are modifications of gas adsorption or cryostatic instruments for liquid–solid interaction assessment, and others have been constructed on a custom basis, especially in universities and materials research laboratories.

A comparison of heats of wetting by solvent with and without dispersing agents provides information on the dispersant stability, replacement, effects of mixed agents and mixed solvents, capability of dispersion, tendency for flocculation, wetting kinetics, polymer–solid film stability, and solvent selectivity.[43–45]

Most calorimetry, whether of the differential scanning or conventional adiabatic type, is slow, tedious, and requires extensive material preparation and instrumental time. The measurement technique is based on the thermodynamic property that ordering of liquid-phase molecules and solvated species to a fixed position (docking network) on a solid surface decreases their entropy, and a small compensatory temperature rise will occur within an adiabatic system.

While the principle of calorimetry is thermodynamically simple, instrumentally the technique is moderately complex. First, heats of adsorption are usually small on a molar basis. Second, the quantity of liquid involved, usually 2% or less of the solids, is quite small, and the heat capacities of the liquid and solid phase involved make actual temperature rise equally small and difficult to measure.

Heats of wetting most often fall in the range 10–1000 ergs/cm^2. For a powder having a surface area of 2 m^2/g, wetting energies are 0.005 to 0.5 calories per gram. At this low energy, and because ambient air fluctuations readily obscure such small changes, the measuring chamber for calorimetric instrumentation must be especially well insulated (adiabatic), the containment mass very small, and temperature transducers very sensitive.

A sealed chamber containing liquid is maintained separate from the powdered solid until thermal equilibrium has been obtained, usually within ±0.001°C. The solid is then introduced into the liquid, and the small temperature rise is mea-

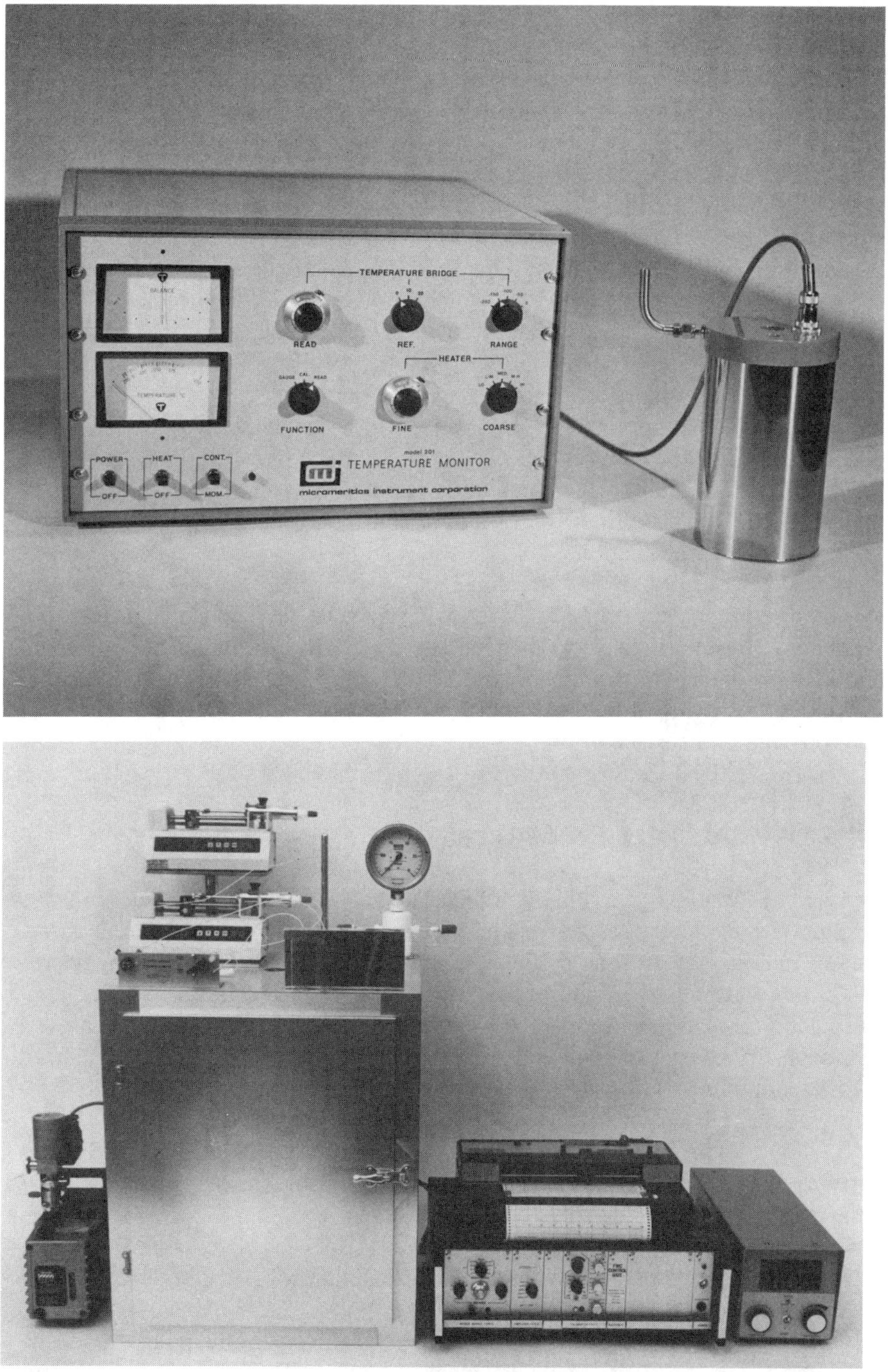

a)

b)

Figure 13.19 (a) Cryostat modified calorimeter (courtesy Micromeritics). (b) Commercial gas adsorption flow calorimeter (courtesy Microscal, Ltd.).

sured with a highly sensitive thermometer (platinum resistance or thermocouple bridge) placed in the liquid phase.

A difficulty shared with surface area equipment is that of cleaning the surface to be measured and voiding it of adsorbed molecules. This often can be accomplished by degassing in vacuum at elevated temperatures. Gas adsorption equipment may be employed to determine certain heats of wetting by liquids through volatilization of these liquids. Heat of adsorption appears as a volumetric expansion of the gas, a sensitive measurement technique. Correction for the heat of liquefication (vaporization) is then made to the liberated energy.

One application for calorimetry, which provides data for surface evaluation difficult by any other means, is that of surface hydrophobicity and hydrophilicity. Classic wet out rate from spreading a set quantity of powder on two liquid surfaces is neither quantitative nor definitive. Terminal sediment volume in disparate liquids is a fair procedure. However, comparison of the energies of adsorption of a dry surface with a nonpolar liquid (heptane or benzene) with a polar one (water or methanol) can yield a very quantitative numerical index[46,47] useful in determining the need for surface modification, solids level potential, and other factors in slurry formulation.

Because these energies are very small and their differences yet smaller, microcalorimetry is the singular method of providing the needed accuracy. Differential scanning calorimetry can also provide a population distribution of adsorption energies for both polar and nonpolar adsorbates, giving in addition the relative distribution of hydrophobic and hydrophilic sites, or patches, on anisotactic or heterotactic solid surfaces.

13.9 Special Laser Procedures

The use of lasers to assess particle size has been discussed previously in this chapter. It must be emphasized that laser sizing provides the most accurate diameter averaging value over all particle orientations of all commercial sizing instrumentation, with the exception of shadowed sample microscopy.

In addition to size analysis by diffraction, laser examination can be employed to assess surface charge, both in the dry state (electrostatic) and in the dispersed form (zeta potential). For dry evaluations, particles are passed perpendicularly through a laser beam in a stream scanning mode (one particle at a time). Laser diffraction provides an independent size estimate of each particle. Parallel plates carrying a variable potential are arranged mutually perpendicular to the laser axis and particle transit. Displacement of the particle from the axial path induces a phase lag in the monochromatic radiation, which is detected within the instrument and calculated back as charge. The mechanics are shown schematically in Figure 13.20.[48]

Charged particles in aqueous suspension migrate toward the pole of opposite charge in an applied electric field. By measuring migration velocity with a microscope (the Zetameter[49] principle), surface charge can be calculated and zeta

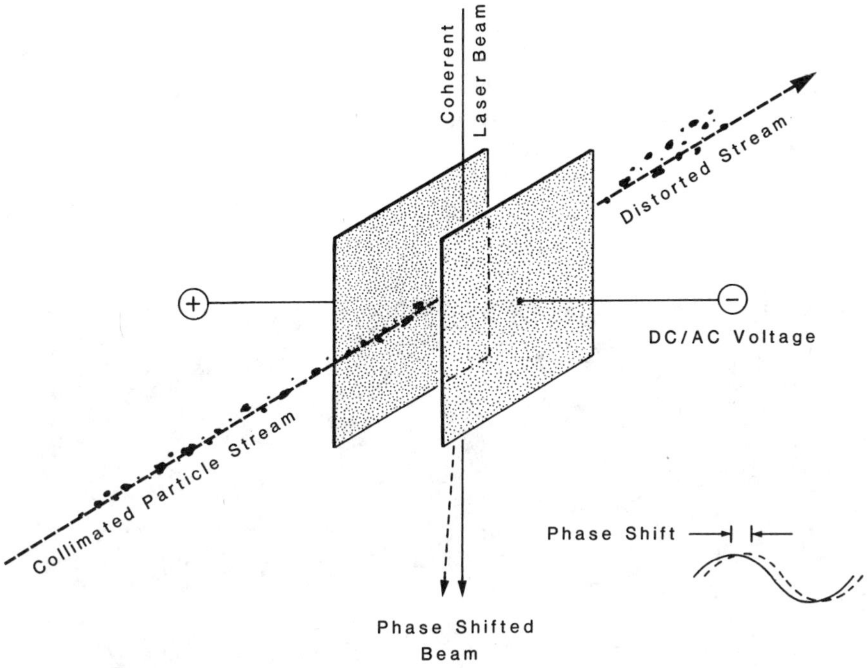

Figure 13.20 ESPART laser diffraction arrangement (courtesy Hosakawa Micron).

potential (actually streaming potential) determined. Microscopic methods utilize the zero flow point, commonly termed the *stationary layer*. This point is actually a ring around the interior of a tube in which electromigration is taking place. To understand this characteristic, imagine a particle moving under an electric field. Its counterions, whether inherent or contributed, move in the opposite direction. The ions, being much smaller, tend to travel along the wall, forcing a counter-directional flow. Liquid dynamics compensates for this motion by a core motion in the opposite direction, that is, parallel to the particles. By observing with a microscope the point of balanced liquid flow, the suspended particles' velocities are then solely the result of their surface charge. It is then necessary to count several dozen particles by measuring their velocity with a calibrated reticle and averaging, then computing the zeta potential from equations developed by Heuckel, Smoluchowski, and others.

$$\text{ZP(mV)} = 12.9 \times \text{electrophoretic mobility} \qquad (13.11)$$

This equation is based on an aqueous medium at 25°C with the electrophoretic mobility computed in μm sec^{-1}/cm^{-1}. The apparatus for conducting such measurements is rather easily assembled, using a good-quality microscope as the centerpiece. However, the actual measurement is tedious and eyestraining.

By employing a laser for particle illumination and by measuring the phase shift due to movement (Doppler principle), the charge can be evaluated more quickly and without the operator eyestrain associated with microscopy. Such an instrument is shown in Figure 13.21.

Zeta potential measurements, while conducted at dilute solids, but most often applied at concentrated solids, provide information on surface chemistry applicable in a broad variety of slurry processing.

In wastewater clarification, which may be conducted both at moderate and dilute solids, adjusting the pH or adding electrolyte to attain a zeta potential near

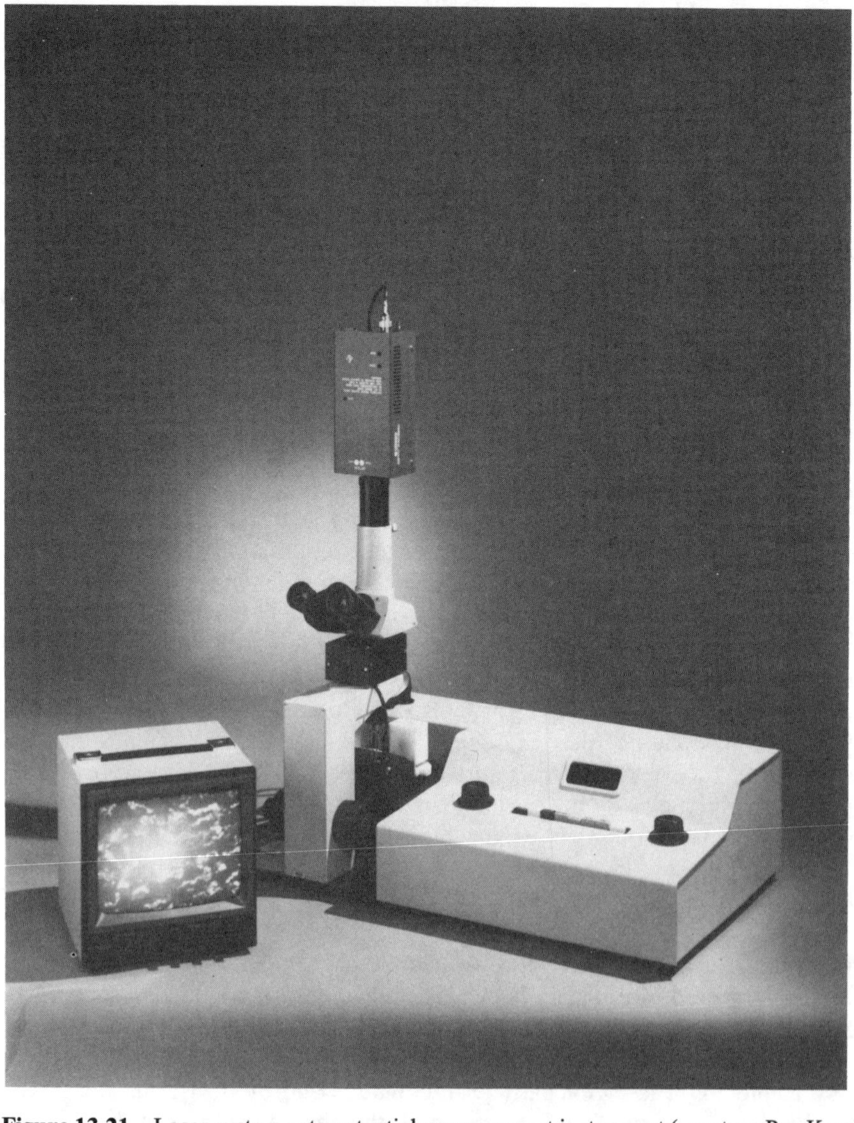

Figure 13.21 Laser system zeta potential measurement instrument (courtesy Pen Kem).

zero enables flocculation to proceed at its most rapid rate. This is illustrated by the curve relating to paper mill effluent (containing dispersed clays, titania, and cellulose fibers) in Figure 13.22. An addition of only 2.6 ppm of a cationic agent shifts the potential to zero, where large floccules form and filtration proceeds rapidly.

Abrasives employed in the wet state function best at removing material from work surfaces when all components, including the debris removed from the surface, possess a zeta potential at least 40 mV[50] (+ or −). As system fluidity generally improves on the alkaline side, as well as applicability to metal sanding and polishing, this potential is most often negative. The second example in Figure 13.22 shows the two components, alumina abrasive (corundum) and steel particles from metal sanding (having an oxide surface). The surface potential, both directly as a function of pH alone and with sodium polyacrylate, illustrates the improvement of the polishing compound at pH 7.2–9 in work effectiveness when dispersed. Without agent, neither component is thoroughly dispersed, even at pH 10 or above.

Food products, pharmaceutical, and cosmetic slurries, where ionic agents for dispersants are justified, can be stabilized and possess extended shelf lives when provided with a significant zeta potential.

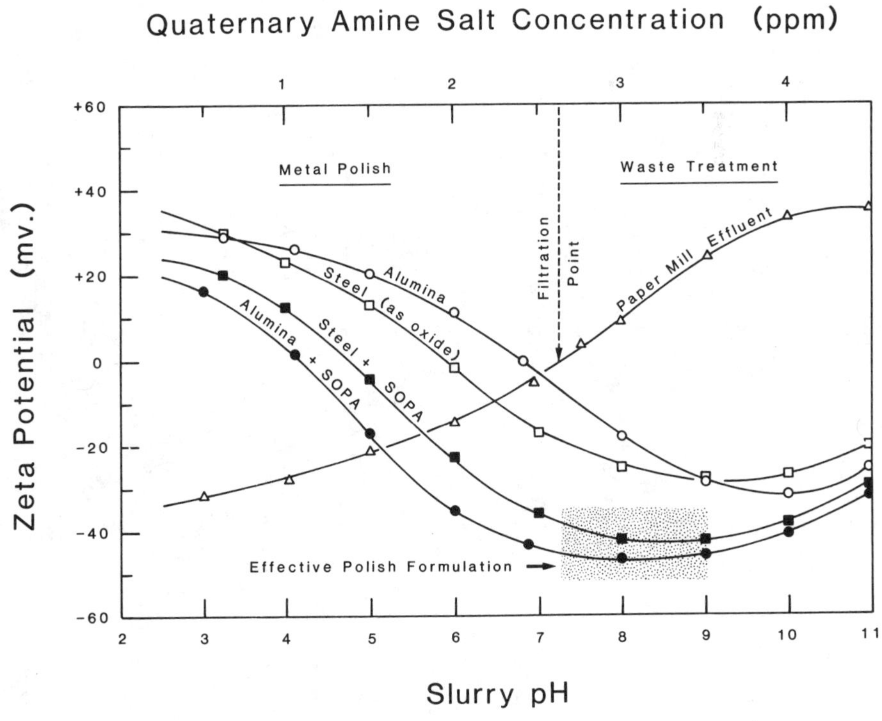

Figure 13.22 Zeta potential factors in two commercial operations.

Finally, two-component separations, especially those involved in mineral separation by way of foam and flotation, are improved dramatically where one species possesses high surface charge at a select pH and the second a zero value.[51]

A yet more sophisticated use of laser technology to study solids in suspension is that of LaserTweezers™ and LaserScissors.[52] Moving particles in suspension can be followed by laser diffraction. Their tendency to spin, their interactions, hydrodynamic, electrostatic, and magnetic, upon close approach of other particles, and their sedimentation effects can be measured and photographed in the dynamic state by specialized adaptation to a standard microscope of a low-light-flux video camera.[53]

A second, high-energy, laser beam is focused on the image in transit and can be employed to ablate, break apart, or shear off a specific section of the particle to observe its surface characteristics. The equipment during operation with a dilute particulate in suspension is shown in Figure 13.23. The technology has been used extensively to investigate surface and bulk properties of biological materials,[54,55] but has equal potential in evaluation of surface charge distribution,

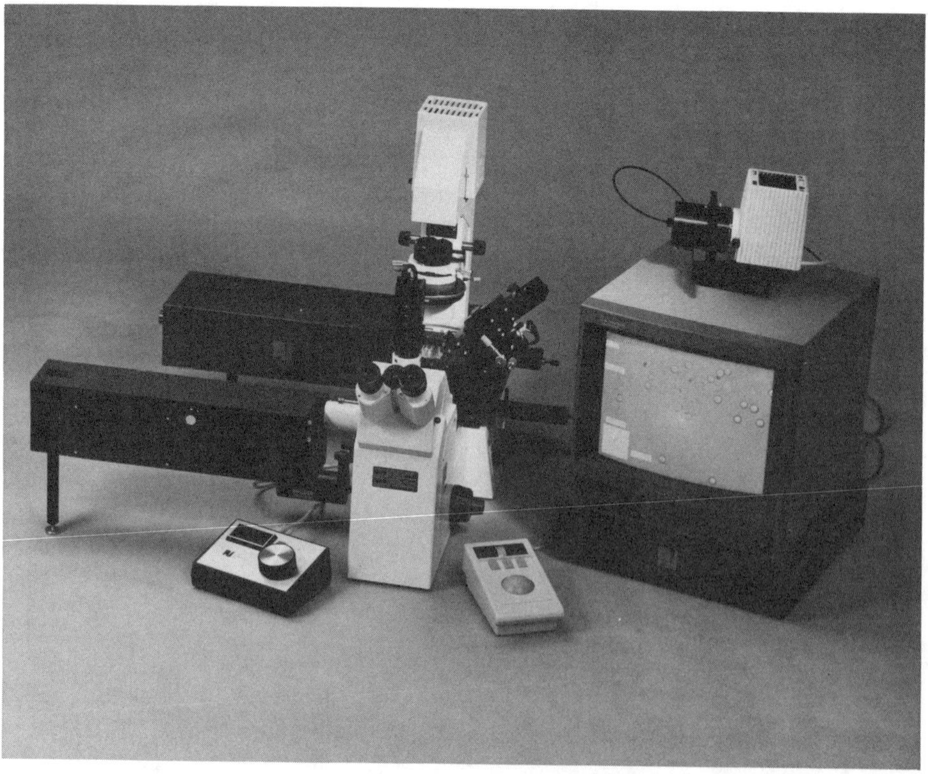

Figure 13.23 Laser particle manipulation instrument (courtesy Cell Robotics, Inc.).

adsorption of protective colloids, and interaction of mixed pigments during liquid-phase state alteration (drying, dispersant fugitivity, stratification, etc.).

13.10 Wetting and Contact Angle Measurements

The effectiveness of a dispersant in aiding liquid wetting of a solid can be estimated by determining the angle of wetting. However, because most solids of interest exist as fine powders, some at the boundary of resolution by optical microscopes, classical methods are not always practical. For liquids, torsiometers, precision capillary tubes, and surface balances may be employed. However, for slurry wet out, these techniques are less than attractive.

Where a single crystal, or at minimum a polished face, of a particle at least 500 μm in size is available, the contact angle can be estimated by examining a minute sessile drop of the liquid with a horizontal, optical axis microscope using a goniometer eyepiece. Although resolution decreases with finer particle sizes, the primary problem lies in obtaining a liquid drop sufficiently smaller than the particle for accurate measurement.

With powders, where contact angles cannot be measured readily, less accurate but considerably faster methods may be used to provide a bench chemistry estimate of relative wetting or dispersion of powders. One of these, viscosity, has been discussed in detail in several areas of this book. However, others are less common and considerably less complex.

These alternate methods of assessing wetting involve timed or quantitative mass processes. For hydrophobicity, a set quantity of powdered solid, nominally 10 g, is added slowly to 100 cm^3 of boiling 5% ammonia solution. The fraction (usually times 100) that remains floating after 20 min may be considered a measure of the integrity of the hydrophobic coating or surface. In some applications morpholine, AMP, or sodium carbonate is substituted for ammonium hydroxide to reduce alkali loss during boiling.

Where the polarity is reversed, the test is not applicable; that is, hydrophilic surfaces will not float on oleophilic liquids under prolonged boiling conditions.

The API tube referred to earlier may be used to assess hydrophobicity by comparing the terminal sediment volume ratio in water versus heptane or similar hydrocarbon. For a consistent size distribution, the terminal sediment volume ratio is approximately inversely related to wetting by the liquid phase.

In a modification of the D'Arcy flow method,[56] as refined for measuring particle size,[57] a glass tube of at least 8 mm inside diameter (to minimize capillary effects) is filled with a powder weight equal to one specific gravity and vibratory packed to a set multiple of 1.00 cm^3, usually 2.00 cm^3. This volume suffices as it produces a packing fraction of exactly 50%, which is fairly compact without stressed particle orientations (see terminal sediment volume). The tube is open at the top and has cemented or fused into its bottom a very fine stainless steel screen, usually 325 mesh.

The tube is placed vertically in a flat dish containing water (or other dispersion liquid) and dispersant at a level just sufficient to wet the screen. The time for the liquid to "wick" to a set point (typically halfway up) is measured. The wet portion is readily detectable through the glass with either dark or light powders. Following cleaning, the tube is refilled with the same dry powder at an equivalent packing density, another solution placed in the dish, and the procedure repeated. Time to reach a point at least halfway up the column is a function of the wetting energy. Times typically are of the order of a minute. As such the method represents a rapid technique for wetting assessment. A slight variant of this technique has been described earlier by Bartell[58] to assess liquid wettability of powders.

Dispersant effectiveness is obtained by a comparison of wetting times with various dispersion liquids to that with the pure liquid.

Time rates are theoretically related inversely to the energy of wetting. However, the time to reach the column midpoint (or higher) is a complex function of capillary forces, gravity, and heat of wetting and cannot be reduced mathematically to a simple, or at least linear, equation for surface applications. It is, however, sufficiently sensitive to discriminate between similar dispersants, for example TSPP versus SHMP on silicates.

13.11 Miscellaneous Tests

Each industry develops its own process and material testing procedures, both for production quality control and applications assessment. Some of these are incredibly crude (as observing a spit ball on the surface of a heated linseed oil–iron oxide makedown system), while others are of sufficient technical quality to be incorporated into A.S.T.M. procedures. A very large segment of such methods are industry specific. Because of the breadth and variation among these, it is not possible to include even a general survey in this chapter. However, a broad summary with some detail may be found in the survey chapter, "Functionality, Test Methods and Specifications for Inorganic and Mineral Fillers," in the publication cited in Ref. 59.

References

1. Boehm, H., *Disc. Faraday Soc.* **51**, 264 (1972).
2. Tamele, M., *Trans. Faraday Soc.* **8**, 270 (1950).
3. Mills, G., and Oblad, J., *JACS* **72**, 1559 (1950).
4. Schindler, P., *JACS* **72**, 268 (1950).
5. Other studies on anatase suggest a lower ionization value, pK_a between 5 and 6. This value may be an average of interactive patches of the two, or it may relate to variations between methods of titanium dioxide manufacture.
6. Conley, R., and Althoff, A., *J. Colloid Interface Sci.* **37**, 187 (1971).
7. Parfitt, G., *Trans. Faraday Soc.* **8**, 235 (1950).

8. Pham, T., and Brindley, G., *Clay and Clay Min.* **18**, 203 (1970).

9. Solomon, D., *Clay and Clay Min.* **16**, 31 (1968).

10. Brunauer, S., Emmett, P., and Teller, E., *JACS* **60**, 309 (1938).

11. Beebe, R., Beckwith, J., and Honig, J., *JACS* **67**, 1554 (1945).

12. Brunauer, S., Deming, R., and Teller, E., *JACS* **62**, 1723 (1940).

13. Lippens, B., and de Boer, J., *J. Catalysis* **4**, 319 (1965).

14. Conley, R., ''Laboratory Cryostats,'' *Amer. Laboratory* **3**(10), 33 (1971).

15. Perrygraf No. PG-940452-28, Hockmeyer Eqpt. Corp., Harrison, NJ (1994).

16. Brookfield Engineering Labs, ''Viscometer Comparison Chart,'' No. AR-15, Stoughton, MA (1967).

17. Paul N. Gardner, Cat. #58, p. 224 (1994).

18, Minard, R., 1st Intl. Congr. & Expos. of I.S.A., p. 3, Sept. (1954).

19. Minard, R., *Instr. & Contr. Systems* **22**, 6 (1959).

20. Millman, N., *TAPPI* **47**, 168 (1964).

21. Patton, T., *J. Paint Tech.* **38**, 656 (1966).

22. Gardner Laboratories, ''The Brushometer,'' Bethesda, MD (1982).

23. Wells, R., et al., *J. Lab. Clin. Med.* **57**, 645 (1961).

24. Pierce, P., *J. Paint Tech.* **43**, 35 (1971).

25. Tarrant, R., ''Instruments for Process Viscosity Measurement,'' AIChE Lecture Series, Nametre Co., Metuchen, NJ (1993).

26. Schoff, C., ''Rheological Measurements,'' Vol. 13, Encycl. Polymer Sci. & Eng., John Wiley & Sons (1988).

27. Fitzgerald, J., U.S. Patent #4,488,427, Oct. 7 (1984).

28. McCafferty, E., and Zettlemoyer, A., *Disc. Faraday Soc.* **51**, 239 (1972).

29. Hingston, F., Posner, A., and Quirk, J., *ibid.*, p. 334.

30. Smith, R., *J. Coatings Tech.* **54**, 21 (1982).

31. Bell, A., Jaramillo, J., and Markowski, T., *Cosmetics & Toiletries* **97**, 40 (1982).

32. Conley, R., *J. Paint Tech.* **46**, 58 (1974).

33. Whitmore, R., 1st Intl. Conf. on Rheology, Melbourne, Austr., p. 43, May (1979).

34. Stokes, G., *Cambridge Phil. Soc. Trans.* **8**, 287 (1849).

35. Conley, R., *Ceramic Ind.* **3**, 62 (1963).

36. Whitby, K., *ASHAE Journal* **27**, 231 (1955).

37. Mathur, K., Pfizer MPM, Easton, PA, private communication.

38. Menshikov, O., et al., *Lakokras Mater. Ikh. Primen.* **1**, 49 (1990).

39. Schmitz, O., and Spille, J., FATIPEC Congress (ISO Committee), Vol. 3, p. 321 (1988).

40. Suverov, V., *Zavod Lab.* **56**, 39 (1990).

41. Flow Microcalorimeter, courtesy Microscal Ltd., London, UK.

42. Microcalorimeter-cryostat, courtesy Micromeritics Instruments, Norcross, GA.
43. O'Rourke, D., *American Laboratory* **27**, 34 (1995).
44. Groszek, A., and Partyka, S., *Langmiur* **9**, 2721 (1993).
45. Goldner, H., *R&D Magazine* **12** (Dec.), 43 (1993).
46. Zettlemoyer, A., *J. Colloid & Interface Sci.* **28**, 345 (1968).
47. Bolis, V., et al., *J. Chem. Soc. Faraday Trans.* **87**, 498 (1991).
48. ESPART Instrument, courtesy Univ. of Wisconsin and Hosakawa Micron, Co., Summit, NJ.
49. Zetameter, Inc., New York, NY.
50. Rao, S., *Advances in Ceramics—Ceramic Powder Science*, Vol. 21 (1987).
51. Amankonal, J., and Somasundaran, P., *Colloids and Surfaces* **15**, 335 (1985).
52. Model 501 Lazer Zee meter, courtesy Pen Kem, Inc., Bedford Hills, NY (1993).
53. Cell Robotics, Inc., Albuquerque, NM (1992).
54. Neev, J., et al., *J. Assist. Reprod. Gen.* **9**, 315 (1992).
55. Conia, J., and Voekel, S., *Biotech.* **17**, 1162 (1984).
56. D'Arcy, H., ''Public Fountains in the City of Dijon,'' V. Dalamont, Paris (1856).
57. Blaine, R., A.S.T.M., Tech. Bull. No. 12-B (1943).
58. Bartell, F., et al., *J. Phys. Chem.* **38**, 503 (1934).
59. Conley, R., ''Functionality, Test Methods and Specifications for Inorganic and Mineral Fillers,'' Chap. 7, in *Inorganic and Mineral Fillers*, edited by Fallon, J., Fallon Research Assoc., Chatham, NJ (1981).

Appendix

A.1 Dispersant Manufacturers

Listed on the following pages is a short selection of major suppliers of dispersants cataloged by chemical functionality in the order discussed in the early chapters of this book. Many companies, some large, some small, manufacture broad categories of agents that serve as dispersants or surfactants. These include companies that produce highly specialized agents, for example, material designed exclusively for removing skid marks from commercial airport runways and materials designed for abating oil spills. Other companies modify waste products for limited dispersancy activity to reduce costly waste disposal requirements. And finally, some companies purchase chemical agents from major manufacturers, doctor the materials slightly with cosmetic alterations, or simply relabel them with their own tradenames.

Thus no attempt has been made to include all manufacturers in the following list because the number is exceptionally large and changes annually because of new entries, divisional selloffs, and discontinued product lines. Both the Thomas Register and McCutcheon's Emulsifiers and Detergents listings include many of the minor, reputable suppliers of such agents.

1. *Inorganic phosphates*

FMC Phosphorus Chem. Div.	440 North 9th St., Lawrence, KS 66044	913-749-8100
Monsanto Chemical	800 North Lindbergh Blvd., St. Louis, MO	800-325-4330

2. *Polyacrylates and polymaleates*

Rohm and Haas Co.	Independence Mall, W, Philadelphia, PA 19105 215-592-3000
Colloids, Inc.	112 Town Park Drive, Kennesaw, GA 30144 404-429-0822

3. *Simple and compound alkanolamines*

Angus Chemical Co.	1500 E. Lake Cook Road, Buffalo Grove, IL 60089 800-362-2580
DeForest, Inc.	6421 Congress Ave., Boca Raton FL 33487 718-991-7684

4. *Alkanolamides*

Chemron Corporation	Box 2299, Paso Robles, CA 93446 805-239-1550
DeForest, Inc.	(see above)
Henkel Corporation (Emery Group)	11501 Northlake Dr., Cincinnati, OH 45249 800-543-7370
McIntyre Ltd.	1000 Governors Hgwy, University Park, IL 60466 708-534-6200
Mona Industries, Inc.	76 East 24th St., Paterson, NJ 07544 201-345-8220
Norman-Fox & Company	5511 S. Boyle Ave., Vernon, CA 90058 800-632-1777
Reilly-Whiteman, Inc.	801 Washington St., Conshohocken, PA 19428 215-828-3800
Rhone-Poulenc, Inc.	CN 7500, Cranbury, NJ 08512-7500 609-860-4000
Sherex Chemical Company	5777 Frantz Rd., Dublin, OH 43017 614-764-6500
Witco Corporation	520 Madison Ave., New York, NY 10022 212-605-3800

5. *Polyethylene glycols and esters*

Calgene Chemical, Inc.	7247 N. Central Park Ave., Skokie IL 60076 708-675-3950
DeForest, Inc.	(see above)
Mona Industries, Inc.	(see above)
Stepan Chemical	(see above)
Reilly-Whiteman, Inc.	(see above)
Union Carbide Corp. Specialty Chem.	39 Old Ridgeway Rd., Danbury, CT 06817 203-794-3050

6. *Lecithins*

American Lecithin Co.	33 Turner Rd., Danbury, CT 06810 203-790-2705
Central Soya Co.	Box 2507, Ft. Wayne, IN 46801 219-425-5230

7. *Lignin derivatives*

Georgia-Pacific Chemical Company	1754 Thorne Rd., Tacoma, WA 98421 206-572-4721
Lignotech USA, Inc.	81 Holly Hill Lane, Greenwich CT 06830 203-625-0701
Westvaco Corp.	Box 70848, Charleston Hts, SC 29415-0848 803-740-2300

8. *Alkyl (aryl) sulfonates*

Ciby-Geigy Corp.	Box 18300, Greensboro, NC 27419 800-334-9481
E.I. DuPont de Nemours	29 Barley Mill Plaza, Wilmington, DE 19880-0029 800-441-7515
Emkay Chemical Company	319 Second St., Elizabeth, NJ 07206 908-352-7053
Henkel Corp. (Emery)	(see above)
Huls America, Inc.	80 Centennial Ave., Piscataway, NJ 08854 908-980-6800
Stepan Company	Edens & Winnetka Rd., Northfield IL 60093 708-446-7500
PPG Industries	3938 Porett Dr., Gurnee, IL 600341 708-244-3410
Witco Corporation	(see above)

9. *Quaternary anionics*

High Point Chemical Co.	243 Woodbine St., High Point, NC 27261 919-884-2214
ICI Surfactants	3411 Silverside Rd., Wilmington, DE 19850 800-822-8215
Scher Chemicals, Inc.	Box 4317, Clifton, NJ 07012 201-471-1300
Sherex Chemical Co.	(see above)
Witco Corporation	(see above)

10. *Organo-phosphates/phosphonates*

Harcros Chemicals, Inc.	5200 Speaker Rd., Kansas City MO 66106 913-321-3131

Intex Chemical	Box 5744, Greenville, SC 29606 803-877-5747
Rhone-Poulenc, Inc.	(see above)
Huls America, Inc.	(see above)
Reilly-Whiteman, Inc.	(see above)
Witco Corp.	(see above)
Texaco Chemical (Canada)	150 Research Lane, Ste. 150, Guelph, ONT N1E5R1 Canada 519-824-3280

A.2 Milling Equipment Manufacturers

The following list contains names, addresses, and phone numbers of the major manufacturers of dispersion equipment in the United States, and in some instances, Canada. Equipment designed for specific, rather than general, applications, for custom installations and unusual slurry properties may not be cited in this list, but may be found in other sources, including the Chemical Engineering Handbook and the Thomas Register.

1. *Impeller-type dispersors*

Admix, Inc.	23 Londonderry Rd., Londonderry, NH 03053 800-466-2369
Alsop Engineering Corp.	Box 3449, Kingston, NY 12401 800-366-0056
Barnant, Co.	28W092 Commercial Ave., Barrington, IL 60010 800-637-3739
Bematek Systems, Inc.	12 Tozer Rd., Beverly, MA 01915 509-927-3052
Black Bros. Co.	503 9th Avc., Mendota, IL 61342 815-539-7451
Brawn Mixers, Inc.	12764 Greenly St., Holland, MI 49424 616-399-5600
Buhler, Inc.	1102 Xenium Lane, Minneapolis, MN 55441 612-545-1401
Bukrans-Sharpe Co.	1541 South 92nd Place, Seattle, WA 98108 800-237-8815
Burnett Bros. Engineerng	950 E. Orangethorpe Ave., Anaheim CA 92801 714-526-2448
Chemineer, Inc.	Box 1123, Dayton, OH 45401 800-643-0641
Dorr-Oliver, Inc.	Box 3819, Milford, CT 06460 800-547-7809
Draiswerke, Inc.	40 Whitney Rd., Matawan, NJ 07430 908-847-0600

Eirich Machines, Inc.	4033 Ryan Rd., Gurnee, IL 60031 708-336-2444
EMI, Inc.	50 Heritage Park Rd., Clinton, CT 06413 800-243-1188
Entoleter, Inc.	251 Welton St., Hamden, CT 06517 203-787-3575
Epworth Manufacturing Co.	1402 Kalamazoo St., South Haven, MI 49090 616-637-2128
Glen Mills, Inc.	395 Allwood Road, Clifton, NJ 07012 201-777-0777
Greerco Corp.	2 Wentworth Drive, Hudson, NH 03051 603-883-5517
Grovehac, Inc.	4310 N. 126th St., Brookfield, WI 53003 800-369-2475
Henschel Mixers America	4500 S. Pinemont, Houston, TX 77401 713-690-3333
Hill Mixer, Inc.	295 Governor St., Paterson, NJ 07501 800-835-1509
Hochmeyer, Inc.	610 Worthington Ave., Harrison, NJ 07029 201-482-0225
Hosakawa Bepex Corp.	333 NE Tuft St., Minneapolis, MN 55413 800-607-2470
Jayco, Inc.	675 Rahway Ave., Union, NJ 07083 908-688-3600
Kady International	127 Pleasant Hill Rd., Scarborough, ME 04070 207-883-4141
Lightnin Mixer Co.	135 Mt. Read Blvd., Rochester, NY 14603 800-526-2209
Littleford Day, Inc.	7451 Empire Drive, Florence, KY 41022 606-525-7600
Marion Mixers, Inc.	Box 286, Marion, IA 52302 319-377-6371
Midwest Mixing, Inc.	5628 Pleasant Ave., Chicago Ridge, IL 60415 708-422-8100
Mixor, Inc.	3131 Casitas Ave., Los Angeles, CA 90039 800-335-4983
Morehouse Industries, Inc.	1600 Commonwealth Ave., Fullerton, CA 92633 714-738-5000
Myers Engineering	8376 Salt Lake Ave., Bell, Ca. 90201 213-560-4723
Philadelphia Mixers, Inc.	1223 East Main St., Palmyra, PA 17078 800-394-1341
Charles Ross & Sons, Inc.	712 Old Willets Path, Hauppauge, NY 11788 800-243-7677

Scott-Turbon Mixer, Inc.	Industrial Highway, Adelante, CA 92301-0160 800-285-8512
Shar, Inc.	3210 Freeman St., Ft. Wayne, IN 46889 800-863-7427
Silverson Machines, Inc.	355 Chestnut, East Longmeadows, MA 01028 800-204-6400
Sonic Corporation	1 Research Drive, Stratford, CT 06497 203-375-0063
Teledyne Readco, Inc.	901 South Richland Ave., York, PA 17405 800-395-4959
TK Products, Inc.	1565 Harmony Circle, Anaheim, CA 92807 714-693-3100
Welding Engineers, Inc. (Papenmeier)	Box 391, Norristown, PA 19404 215-272-6900

2. *Ball mills*

Paul O. Abbe, Inc.	139 Center Ave., Little Falls, NJ 07424 201-256-4242
Allis Mineral Systems, Inc.	240 Arch St., York, PA 17405 717-843-8671
Crossley-Economy Co.	57 East Taggart St., East Palestine, OH 44413 216-426-9486
Epworth Mfg. Co.	1402 Kalamazoo St., South Haven, MI 49090 616-637-2128
Patterson Industries, Ltd.	250 Danforth Rd., Scarborough, ONT M1L3X4 Canada 800-336-1110
U.S. Stoneware Corp.	700 East Clark St., East Palestine, OH 44413 800-426-8808

3. *Small-media mill dispersors*

Chicago Boilcr Co.	1225 Busch Pkwy, Buffalo Grove, IL 60089 800-522-7343
Draiswerke America	40 Whitney Rd., Matawan, NJ 07430 800-494-3151
Eiger Machinery, Inc.	1258 Allanson Rd., Mundelien, IL 60060 800-253-4437
Epworth Mfg. Co.	1402 Kalamazoo St., South Haven, MI 49090 616-637-2128
Netzsch, Inc.	119 Pickering Way, Exton, PA 19341 610-363-8010
Premier Mill Corp.	1 Birchmont Dr., Reading, PA 19606 610-779-9500
Charles Ross & Sons, Inc.	712 Old Willets Rd., Hauppauge, NY 11788 800-243-7677

Union Process, Inc.	1925 Akron-Peninsula Rd., Akron, OH 44313 216-929-3333

4. *Vibratory dispersing mills*

ABB Raymond (Palla) Co.	650 Warrenville Rd., Lyle, IL 60532 708-971-2500
Allis Mineral System, Inc.	240 Arch St., York, PA 17405 717-843-8671
Sweco, Inc.	7120 New Buffington Rd. Florence, KY 41042 606-727-5147

5. *Multiple-roll mills*

Burnett Bros. Engineering, Inc.	950 E. Orangethorpe Ave., Anaheim, CA 92801 714-526-2448
Fuller-Kovaco Co.	2158 Avenue C, Bethlehem, PA 18017 800-203-3417
Keith Machinery Corp.	34 Gear Ave. Lindenhurst, NY 11757 800-348-1769
Lehman Mills, Inc.	Box 1069, Salem, OH 44460 216-332-9951
Reliable Rubber and Plastic Machinery	2010 Union Tpk., North Bergen, NJ 07047 201-865-1073
Ross Engineering	32 Westgate Blvd., Savannah, GA 31405 800-524-7677
Sigma Engineering Corp.	39 Westmoreland Ave., White Plains, NY 10606 914-682-1820

6. *Ultrasonic-type dispersors*

Advanced Sonic Process Systems, Inc.	264 Bacon Pond Rd., Woodbury, CT 06798 203-266-4440
Blackstone Ultrasonics, Inc.	Box 220, Jamestown, NY 14702 716-665-2340
Crest Ultrasonics, Corp.	455 Montague Expy., Milpitas, Ca 95035 800-394-4057
Microsonic Engineering Devices Co.	5550 Merrick Rd., Massapequa. NY 11758 516-798-1873
Sonicor, inc.	100 Wartburg Ave., Copiaque, NY 11726 800-864-5022

7. *Viscometers and Rheometers*

Automation Products, Inc. (Dynatron)	3030 Max Roy St.,, Houston, TX 77008 800-231-2062
Bolin Instruments Div.	2540 Rt. 130, Cranbury, NJ 08512 609-655-4447

Brookfield Engineering Laboratories, Inc.	240 Cushing St., Stoughton, MA 02702 800-628-8139
Design Integrated Technology	100 Franklin St., Warrenton, VA 22186 703-349-9425
Erichsen Instruments	1340 Home Ave., Akron, OH 44310 216-633-3644
Haake Fisons, Inc.	53 West Century Rd., Paramus, NJ 07652 201-265-7865
Load Controls, Inc.	10 Picker Rd., Sturbridge, MA 01566 508-347-2606
Nametre Company	101 Forrest St., Metuchen, NJ 908-494-2422
Nordson Corp.	555 Jackson St. Amherst, OH 44001 800-241-8777
Paar Physica USA	400 Randal Way, Spring, TX 77388 713-350-3576
Schott Corp.	3 Odell Plaza, Yonkers, NY 10701 914-968-8900

A.3 Glossary and Abbreviations

Adiabatic: energy transformation with no heat added from or emitted to surroundings.

Agglomerate: weak association of particles usually held together by charge or polarity.

Aggregate: moderately strong association of particles often bound together by residual chemicals.

AMP: 2-amino, 2-methyl, 1-propanol.

Anisometric: having different dimensions along a particle's crystallographic or spatial axes.

Anisotactic: surface character (adsorption energy or chemical activity) that differs significantly between facets of the crystal.

Anisotropic: having structural (bulk) properties that vary in different directions.

Asperic: having sharp corners or projections.

Bimodal: composed of two distinct Gaussian distributions blended together.

Bingham yield point: transition point between set structure and fluid by shear in a gel.

Blooming: a film characteristic in which soluble salts in a damp film migrate to the surface, dry, crystallize, and discolor.

Breakdown potential: the voltage across a set thickness of material at which electrical conductivity will occur (usually in volts/mil).

Brightness: the percentage light reflected from a packed powder surface at a fixed wave length, usually at the blue spectral end, compared to a fixed standard (usually magnesium oxide or barium sulfate).

Centipoise: a unit of fluid friction equal to 1/100 of a gram per centimeter per second.

Centistoke: a unit of fluid friction equal to centistokes times the liquid or slurry density.

Chain scission: polymer chain breakup into fragments by chemical or bacterial action.

Chelation: a chemical complex formed by a metal atom and a molecule with two or more groups that enter into non-valence-electron orbitals of the metal forming a ring.

Chemical viscosity: viscosity derived from solution ions solvating water molecules reducing system mobility and particle movement.

Chemisorption: adsorption of a species onto a solid surface followed by a chemical bonding reaction between the two.

Cloud point: the temperature at which micelle formation of a dispersant reaches light scattering size, $\sim 0.2\ \mu m$.

Colloid: particles sufficiently small to overcome gravitational settling through Brownian movement and liquid random motion, usually between 0.001 and $0.1\,\mu m$ (1–100 nm).

Contact angle: the interior angle between a solid surface and a tangent drawn at the contact point with a bead of liquid.

Covalent solid: one in which atoms are interbonded by strong, covalent bonds.

cp: centipoise.

Crystal field: an electron-orbital distortion created by an orderly arrangement of atoms around the affected atom, usually in a well-defined crystal.

Debye unit: a measure of dipole moment having the units of 10^{-18} esu cm.

Defibrillation: the mechanical action of separating fibers from a fiber bundle without breaking up individual fibers.

Deflocculent: a chemical agent that by virtue of charge or surface character alteration prevents particles from aggregating (dispersant).

Delamination: the mechanical action of separating laminae or individual sheets from a sandwich-type structure without breaking individual laminae.

Dielectric loss: adsorption of alternating electrical energy by electrons or electronic structures.

Dilatancy: a viscosity characteristic where viscosity increases with applied shear.

Dipole moment: an electrical vector of an asymmetrical atom or molecule, the product of charge and distance.

Dispersant: an agent that stabilizes particulate separation in liquid suspension.

Enthalpy: the energy absorbed or evolved in a reaction that occurs at constant pressure (expressed as enthalpy change).

Entropic stabilization: dispersion arising from random movements and volume displacements by large or long molecular agents on solid surfaces.

Entropy: the quantity of disorder in a closed system (entropy times temperature is an energy term).

Eutrophication: a biological state in which algae reproduce with unusual speed due to overdoses of nutrients and form surface scum layers.

False body: development of apparent structure in a slurry that reduces particle mobility.

Fatty acid: an organic structure having the general formula R–COOH, in which R is a linear hydrocarbon with even members. R is usually greater than 4 atoms.

Fatty amine: compounds similar to fatty acids but with the general formula R–NH_2.

Faults: defects within solids, such as fractures, substituents, and disorder, which reduce strength.

Flocculent: an agent that reduces particulate dispersion through agglomeration.

Flocculation: low-energy agglomeration of particles in a liquid, often induced by attractive charges.

Floccule: an aggregation of particles held together by weak electrostatic or dipolar forces or their inability to be wet in a liquid suspension.

Flushing: the formulation technique of preparing a slurry in one liquid and introducing it into a second, immiscible liquid with high shear in which the particles selectively are wet out by the second liquid.

Fugitivity: a characteristic of a dispersant or treating agent whereby it evaporates during a conventional drying orperation.

Gaussian distribution: a grouping of sizes such that their population and size are related by the exponential or probability function.

Gel: a liquid–solid state in which particles are completely immobilized but which can be reduced to a liquid (usually thixotropic) upon sufficiently applied shear.

Gloss: the percentage reflectivity of a surface measured at a standard low angle.

Grinding aids: chemical agents that reduce the energy to grind by improving mill efficiency through wetting and viscosity reduction.

Hegman guage: a small precision metal block having a precision slope ground into its midsection whose depth from the block surface varies from $\sim 0.3 \mu m$ at one end to over 44 μm on the other. Slurry is poured onto the block and a separate level block rubbed across the grooved block. Particles will show as streaks when their size interferes with the clean slurry spread.

Heterocoagulation: flocculation caused by interaction between two differing surfaces, each dispersed, as between two separate pigments or oxides.

Heterodisperse systems: dispersion consisting of mixtures of particles having various sizes.

Heterotactic: having small areas of varying charge or polarity on the same surface or crystal facet (in contrast to anisotactic).

Hiding power: the ability to cover a substrate with a film or coating that prevents photons from passing through the film to that substrate and re-emerging (also opacity).

Homotactic: having a surface that is uniform throughout in charge or polarity (also isotactic).

Hydrophilic: having surfaces readily wet out by water.

Hysteresis: a rheological phenomenon in which structure created by or broken down by shear requires significant time to reform after shear is removed.

Intercalation: chemical propping apart of solid layers through agent coordination with interlayer surfaces.

Ion exchange: selective replacement between ions on a charged substrate because of ionic size or charge preference.

Ionic solid: one in which atoms are held together by electrostatic forces.

Irregular solid: one with no definable shape or order.

Isoelectric pH: that pH where a hydroxylated surface is equally ionized positive and negative.

Isometric solid: particles with equidimensionality.

Isotactic: having all facets of a crystal with essentially equally adsorption energy or chemical activity.

KTPP: potassium tripolyphosphate.

Laminar flow: a rheological slurry condition in which all particles and free liquid flow along parallel lines with no turbulence.

Langmuir adsorption: adsorption wherein the rate of adsorption is proportional to the free surface available to the adsorbate.

Lipophilic: having an oil affiliation (oil-like).

Log-normal distribution: a true Gaussian distribution of particle sizes whose breadth at the 25 and 75% by weight differs by a factor of 3.5.

Masterbatch: formation of a concentrated slurry to enhance shear, followed by dilution to an appropriate concentration in a formulation.

Median size: that size at which 50% of the particles by weight are finer and 50% are coarser.

Metallic solid: one in which atoms are held together by combined ionic and covalent bonding and which conducts electricity.

Micelle: a group of molecules having both oleophilic and hydrophilic ends that assemble into a sphere or closed surface with like ends on the exterior, presenting a lower energy surface to the liquid phase.

Micrometer: 10^{-6} meter (also called a micron).

Mineral flotation: a technique of separating one mineral from another through selective wetting of one surface type and allowing these particles to attach to entrained air bubbles (foam).

Molecular solid; one with large molecular clusters held together by van der Waals forces.

Monodisperse: a collection of particles all of the same size.

Monomodal: a collection of particles having a single Gaussian or near-Gaussian distribution.

Monosize: a collection of particles all of the same size (also monodisperse).

Nanometer: 10^{-9} m, 10^{-7} cm, 10^{-3} μm.

Newtonian flow: a rheological state in which viscosity is independent of applied shear.

nm: nanometer.

Oleophilic: having oil-like charactistics, not readily wet by water.

Opacity: a film characteristic in which photons are prevented by adsorption or refraction from passing through the film or coating.

Oxyirane: the chemical grouping $= \overbrace{\mathrm{C} - \mathrm{C}}^{\mathrm{O}} =$ (epoxide).

Parallaxis: an optical effect where light is reflected from two parallel surfaces causing uncertain depth perception (as in pearlescence).

Peptizing agent: another term for dispersant.

Pigmentation aid: an agent used to maintain separation of pigment particles to allow efficient light interaction (usually a dispersant).

pK_a, pK_b: the negative logarithmic value of the acid or base ionization constant.

ppb: parts per billion.

ppm: parts per million.

Pseudoplasticity: a rheological state in which viscosity decreases with shear but shows no time dependency in structure reformation.

Reynolds number: a flow constant relating to parallel liquid–solid movement under applied shear. Reynold numbers larger than 1 indicate progressive turbulence.

Rheopectic: an unusual rheological state in which a shear increase produces a viscosity decrease but the structure becomes yet more fluid as shear is slowly removed.

Rheogram: a plot of viscosity against shear or level of dispersant.

Rheometer: an instrument for measuring slurry flow under variable conditions, often automated.

Scatter function: a mathematical relationship between wavelength of illumination, refractive index of a slurry or solid phase, and the angle of illumination.

SEM: scanning electron micrograph, a technique that produces a synthesized three-dimensional representation of a surface or powder.

Shear: force between two differentially moving parallel planes divided by the distance of separation.

SHMP: sodium hexametaphosphate.

Sinter: to heat particles to a temperature and for a time just sufficient to cause particles to fuse together at contact points but not totally melt. Also the fused particles.

Sol: a dispersed fine particle too small to refract light but that may absorb color, generally in the size range from molecules to about 0.01 μm.

SOPA: sodium (salts) organopolyacid.

STP: sodium tripolyphosphate.

Surfactant: an agent that reduces the surface tension of a solid in a liquid or allows it to wet rapidly and with a low contact angle.

Syneresis: a state in which a gel or thixotrope settles sharply and permits clear liquid to form above its surface.

TEM: transmission electron micrograph that discloses the outlines only of objects.

Thixotropy: a rheological state in which viscosity decreases with shear but some time is required for the structure to reform as shear is slowly reduced.

Thixotropic index: the ratio of two viscosities at different rates of shear, often the 10 and 100 rpm Brookfield values.

TKPP: tetrapotassium pyrophosphate.

TSP: trisodium phosphate (ortho).

TSPP: tetrasodium pyrophosphate.

μm: micrometer.

van der Waals force: an induced dipole force in the electronic clouds around atoms and molecules that results in a net attractive force between atoms at close proximity.

Viscometer: an instrument that measures drag force at a set applied shear force.

Wetting agents: a specific group of surfactants that allow rapid contact and low contact angles between liquids and solid surfaces.

Whiteness: the percentage of light reflected by a powder averaged across all wavelengths — also measured as the difference between percentage reflectance at the blue end and red end of the visible spectrum.

Zeta potential: voltage at the shear plane of a surface in aqueous solution derived from selectively adsorbed ions.

Zwitterion: electrically neutral species having both positive and negative charges within the same molecule—usually an amino acid.

Index